D1246777

NUTRITIONAL GENOMICS

NUTRITIONAL GENOMICS

Discovering the Path to Personalized Nutrition

Edited by

Jim Kaput
Raymond L. Rodriguez

A JOHN WILEY & SONS, INC., PUBLICATION

Published by John Wiley & Sons, Inc., Hoboken, New Jersey

Published simultaneously in Canada

Library of Congress Cataloging-in-Publication Data:

Nutritional genomics : discovering the path to personalized nutrition / edited by James Kaput and Raymond L. Rodriguez.
p. cm.
Includes bibliographical references and index.
ISBN-13 978-0-471-68319-3 (cloth)
ISBN-10 0-471-68319-1 (cloth)
1. Nutrition. I. Kaput, Jim, 1952– . II. Rodriguez, Raymond L.
RA784.N855 2006
613.2—dc22 2005023391

Printed in the United States of America

10 9 8 7 6 5 4 3 2 1

CONTENTS

CONTRIBUTORS

David B. Allison Department of Biostatistics, Section on Statistical Genetics, Comprehensive Cancer Center, and Center for Nutrient–Gene Interaction, University of Alabama–Birmingham, Birmingham, Alabama

Bruce N. Ames Nutrition and Metabolism Center, Children's Hospital Oakland Research Unit, Oakland, California and University of California–Berkeley, Berkeley, California

Stephen Barnes Departments of Pharmacology and Toxicology, Comprehensive Cancer Center, and Center for Nutrient–Gene Interaction, University of Alabama–Birmingham, Birmingham, Alabama

Mark Carpenter Center for Nutrient–Gene Interaction, University of Alabama–Birmingham, Birmingham, Alabama and Department of Mathematics and Statistics, Auburn University, Auburn, Alabama

David Castle Department of Philosophy, University of Guelph, Guelph, Ontario, Canada

Sally Chiu Rowe Program in Genetics and Department of Pediatrics, University of California–Davis, Davis, California

Cheryl Cline Joint Centre for Bioethics, University of Toronto, Toronto, Ontario, Canada

Craig A. Cooney Department of Biochemistry and Molecular Biology, University of Arkansas for Medical Sciences, Little Rock, Arkansas

Dolores Corella Genetic and Molecular Epidemiology Unit, School of Medicine, University of Valencia, Valencia, Spain

Abdallah S. Daar Joint Centre for Bioethics and Program in Applied Ethics and Biotechnology, University of Toronto, Toronto, Ontario, Canada

Kevin Dawson Section of Molecular and Cellular Biology, Bioinformatics Shared Resources Core, and Center of Excellence in Nutritional Genomics, University of California–Davis, Davis, California

Adam L. Diament Rowe Program in Genetics and Department of Pediatrics, University of California–Davis, Davis, California

Cora J. Dillard University of California–Davis, Davis, California

James S. Felton Biosciences Program, Lawrence Livermore National Laboratory, Livermore, California

Erik Fernandez University of California–Davis, School of Medicine, Davis, California

Janis S. Fisler Department of Nutrition, University of California–Davis, Davis, California

Gary L. Gadbury Center for Nutrient–Gene Interaction, University of Alabama–Birmingham, Birmingham, Alabama and Department of Mathematics and Statistics, University of Missouri, Rolla, Missouri

Alfredo Galvez Section of Molecular and Cellular Biology, Bioinformatics Shared Resources Core, and Center of Excellence in Nutritional Genomics, University of California–Davis, Davis, California

J. Bruce German University of California–Davis, Davis, California and Nestlé Research Centre, Lausanne, Switzerland

Jeff Gregg Molecular Pathology Core, University of California–Davis Medical Center, and Center of Excellence in Nutritional Genomics, University of California–Davis, Davis, California

Shangqin Guo Department of Biochemistry and Women's Health Interdisciplinary Research Center, Boston University School of Medicine, Boston, Massachusetts

John L. Hartman IV Department of Genetics, University of Alabama–Birmingham, Birmingham, Alabama

Wayne Chris Hawkes Western Human Nutrition Research Center and Center of Excellence in Nutritional Genomics, University of California–Davis, Davis, California

S. Luke Hillyard University of California–Davis, Davis, California

Pamela Horn-Ross Center for Nutrient–Gene Interaction, University of Alabama–Birmingham, Birmingham, Alabama and Northern California Cancer Center, Fremont, California

Liping Huang Western Human Nutrition Research Center and Center of Excellence in Nutritional Genomics, University of California–Davis, Davis, California

Jim Kaput Center of Excellence in Nutritional Genomics, University of California–Davis, Davis, California, NutraGenomics, Chicago, Illinois, and Laboratory of Nutrigenomic Medicine, University of Illinois, Chicago, Illinois

Warren A. Kibbe Director of Bioinformatics, Robert H. Lurie Comprehensive Cancer Center, Center for Genetic Medicine, and NIH Center for Neurogenomics at Northwestern University, Chicago, Illinois

Helen Kim Department of Pharmacology and Toxicology, Comprehensive Cancer Center, and Center for Nutrient–Gene Interaction, University of Alabama–Birmingham, Birmingham, Alabama

Ronald M. Krauss Children's Hospital Oakland Research Institute, Oakland, California, Department of Genome Science, Lawrence Berkeley National Laboratory, Berkeley, California, and Department of Nutritional Sciences, University of California–Berkeley, Berkeley, California

Matthew C. Lange University of California–Davis, Davis, California

Coral A. Lamartinere Department of Pharmacology and Toxicology, Comprehensive Cancer Center, and Center for Nutrient–Gene Interaction, University of Alabama–Birmingham, Birmingham, Alabama

Su-Ju Lin Section of Microbiology, College of Biological Sciences, University of California–Davis, Davis, California

Jiankang Liu Nutrition and Metabolism Center, Children's Hospital Oakland Research Unit, Oakland, California and University of California–Berkeley, Berkeley, California

Mark Jesus M. Magbanua Section of Molecular and Cellular Biology, University of California–Davis, Davis, California

Michael A. Malfatti Biosciences Program, Lawrence Livermore National Laboratory, Livermore, California

Wasyl Malyj Section of Molecular and Cellular Biology, Bioinformatics Shared Resources Core, and Center of Excellence in Nutritional Genomics, University of California–Davis, Davis, California

Sreelatha Meleth Comprehensive Cancer Center and Center for Nutrient–Gene Interaction, University of Alabama–Birmingham, Birmingham, Alabama

Jose M. Ordovas Nutrition and Genomics Laboratory, Jean Mayer–U.S. Department of Agriculture Human Nutrition Research Center on Aging, Tufts University, Boston, Massachusetts

Grier P. Page Department of Biostatistics, Section on Statistical Genetics, Comprehensive Cancer Center, and Center for Nutrient–Gene Interaction, University of Alabama–Birmingham, Birmingham, Alabama

Raymond L. Rodriguez Section of Molecular and Cellular Biology, Bioinformatics Shared Resources Core, and Center of Excellence in Nutritional Genomics, University of California–Davis, Davis, California

Gertrud U. Schuster Nutrition Department, University of California–Davis, Davis, California

Peter A. Singer Joint Centre for Bioethics, University of Toronto, Toronto, Ontario, Canada

Patty W. Siri Children's Hospital Oakland Research Institute, Oakland, California

Jennifer T. Smilowitz University of California–Davis, Davis, California

Gail Sonenshein Department of Biochemistry and Women's Health Interdisciplinary Research Center, Boston University School of Medicine, Boston, Massachusetts

Jung H. Suh Nutrition and Metabolism Center, Children's Hospital Oakland Research Unit, Oakland, California and University of California–Berkeley, Berkeley, California

Melanie Tervalon Children's Hospital and Research Center at Oakland, Oakland, California

Charoula Tsamis Joint Centre for Bioethics, University of Toronto, Toronto, Ontario, Canada

Robert E. Ward University of California–Davis, Davis, California

Craig H. Warden Rowe Program in Genetics and Department of Pediatrics, and Section of Neurobiology, Physiology, and Behavior, University of California–Davis, Davis, California

Walter C. Willett Department of Nutrition, Harvard School of Public Health and Channing Laboratory, Department of Medicine, Brigham & Women's Hospital, Harvard Medical School, Boston, Massachusetts

Angela M. Zivkovic University of California–Davis Davis, California

FOREWORD

As director of the National Center for Minority Health and Health Disparities (NCMHD), I am pleased and honored to contribute this forward to *Nutrigenomics: Discovering the Path to Personalized Nutrition*. As one of the sponsors of the Bruce Ames International Symposium on Nutritional Genomics, from which many of the chapters of this volume were derived, it a reassuring to see innovative and multidisciplinary approaches being applied to address the problems of chronic disease and cancer. In 2002, NIH director Elias Zerhouni instituted the NIH Roadmap for the 21st Century which consists of the following three broad themes: new pathways to discover, research teams for the future, and re-engineering the clinical research enterprise. The Roadmap is designed to identify major opportunities and gaps in biomedical research by promoting high-risk, interdisciplinary research and public–private partnerships. I believe the editors and authors of this volume have captured the true spirit and highest aspirations of the NIH Roadmap. The volume's focus on diet–gene interactions is one that boldly crosses disciplinary, institutional, and organizational boundaries, to tackle complex biomedical problems and transform new scientific knowledge into tangible benefits for all people.

To a certain extent, this volume reflects the generous increase in the funding of biomedical and behavioral research at the National Institutes of Health over the past two decades. This era of "doubling" the NIH budget has resulted in a multitude of scientific advances and programs contributing to improved health and quality of life for many Americans. At the same time, this national focus on biomedical research has heightened our awareness that many individuals, both at home and abroad, still suffer disproportionately from a number of diseases such as cardiovascular disease, Type 2 diabetes, hypertension, asthma, and cancers of various kinds. These health disparity populations are typically characterized by higher incidence, earlier onset, and greater severity of a particular disease, as well as lower responsiveness to treatment and thus, lower survival rates than the general population. Moreover, health disparities are often most apparent among ethnic/racial groups, women, the poor, and the uninsured. As evidenced by the numerous articles in the popular press and scientific literature, it is clear that the American people, from patients to policy makers, are deeply concerned about these health inequities. Health disparities by definition are counter to our shared sense of fairness and our belief in equal access. But while health disparities may present a formidable challenge to the biomedical research community, they may also hold the key to the next scientific breakthrough or blockbuster drug. So whether its the Pima Indians of Arizona, coal miners in West

Virginia, or the Kosraeans of Micronesia, health disparities are both a challenge and opportunity that will require not only new technologies but new ways of thinking about how biological systems interface with lifestyle and culture.

Lastly, although the NIH is charged with the responsibility of addressing national health needs first and foremost, we cannot forget or ignore our responsibility to promote human health and wellness around the world. According to 2003 World Health Report, approximately 80% of all deaths from cardiovascular (CVD) disease occurred in low to middle-income countries and by 2010, CVD will be the leading cause of death in developing countries. In 1998, the World Health Organization (WHO) declared obesity a global epidemic, with more than one billion adults with BMIs greater than 25 and at least 300 million adults with BMIs greater than 30. At least 171 million people worldwide suffer from Type 2 diabetes and this figure is expected to more than double by 2030. Clearly health disparities are a global problem that must be viewed through the wider lens of "inclusion." I hope that the authors and readers alike will bear this in mind as they move forward in the development of new conceptual and methodological frameworks for reducing health disparities. By working locally, partnering nationally, and thinking globally, we can bring the benefits of cutting-edge biomedical, behavioral, and social science research and research training to the question of nutrition *and* genomics as risk factors for disease.

John Ruffin
Director, NCMHD

PREFACE

The link between food and health is a long and a well documented one. With over 24,000 people worldwide dying from hunger each day and obesity reaching epidemic proportions in developed counties, the consequences of too little or too much food are easily seen. While the tragedy of world hunger is beyond the scope of this volume, new scientific insights into how nutritional and genetic factors contribute to obesity, chronic disease, and cancer, will be addressed.

The focus and timing of this volume reflect a paradigm shift in the way people look to nutrition for its short- and long-term impacts on health and disease. People no longer view food as merely a source of calories but rather as a complex mixture of dietary chemicals, some of which are capable of preventing, mitigating, or treating disease. With the sequencing of the human genome, a new genetic dimension as been added to the equation linking the foods we eat to the good health we all hope to enjoy. This new genomic perspective on nutrition and health can be seen in recent marketing campaigns for drugs that address the "two sources of cholesterol—food and family history." Americans are beginning to understand that we bring two things to the dinner table—our appetite and our genotype. As we begin to understand the genetic diversity that makes each of us uniquely different, we are also beginning to understand why we respond to our nutritional environment differently and how these differences can, over time, lead to health or disease.

Genomic analysis reveals that humans are 99.9% identical at the DNA level. This implies that the remaining 0.1% of the human genome (or about three million single nucleotide polymorphisms (SNPs)) is responsible for all the morphological, physiological, biochemical and molecular differences between any two individuals. As will be discussed in this volume, common genetic variation in the form of SNPs in enzyme-encoding genes (or their promoters) can affect reaction rates in metabolic pathways that in turn, can create individual differences in the way we absorb, metabolize, store, and utilize nutrients. According to Bruce Ames to whom many of the chapters are dedicated; "single nucleotide polymorphisms provide a powerful tool for investigating the role of nutrition in human health and disease and . . . can contribute to the definition of optimal diets."

Some of our contributors discuss well-documented evidence that certain genotypes are more severely affected by specific types of dietary factors than other genotypes (although no genotype is completely immune to the deleterious effects of poor diet). However, it is unlikely that a single gene, SNP, mutation, biomarker, or risk factor will have the positive predictive value needed to show a predisposition for chronic disease or cancer. This is because diet–gene interactions are strongly influenced by epigenetic, environmental, socio-economic, and lifestyle filters that modify or potentiate genetic effects. For this reason, multidisciplinary approaches will be needed to develop accurate and reliable nutritional interventions using genome-based dietary recommendations.

The notion that interactions between dietary factors and genes (or their variants) can promote health or cause disease is perhaps best captured by the term "nutrigenomics" (a contraction of nutritional genomics). As one of the latest "omic" technologies to emerge from the post-genomic era, nutrigenomics adhere to the following precepts: (1) poor nutrition can be a risk factor for diseases; (2) common dietary chemicals can act on the human genome, either directly or indirectly, to alter gene expression and/or gene structure; (3) the degree to which diet influences the balance between health and disease depends on an individual's genetic makeup; (4) some diet-regulated genes (and their common variants) play a role in the onset, incidence, progression, and/or severity of chronic diseases, and (5) dietary intervention based on knowledge of nutritional requirement, nutritional status, and genotype can be used to prevent, mitigate, or cure chronic disease.

For nutrigenomics to grow and mature as a discipline, much research is needed to answer several important questions. For example, will the cost of omic technologies come down to a level that will make nutrigenomic testing affordable to everyone? How will researchers integrate dietary and medical histories with genotype, gene expression, and metabolomic datasets from large, diverse human populations? Can we assure human subjects and consumers of nutrigenomic services that these data will be secure, safe, and not exploited for legal/political reasons or financial gain? What are those genetic variants that keep us from deriving full benefit from our nutrition, versus those that will increase our risk of disease? What role will genetically modified foods play in dietary interventions and will the benefits of these genetically enhanced foods outweigh real or perceived risks? Can the health benefits of bioactive compounds in food be confirmed clinically and what are the safe upper limits for these bioactives? These are just a few of the challenges facing nutrigenomic researchers today. This volume should provide the conceptual and technical basis from which to tackle these difficult questions.

In closing, I would like to remind readers that good nutrition has been, and will continue to be, the cornerstone of good health and disease prevention—but good nutrition comes at a price. This is particularly true as new nutrigenomic tests come to market. Dietary interventions, including those using genetic tests, will play an important role in disease prevention and treatment, especially as populations around the world grow increasingly older and more obese. As we learn more about the health-promoting dietary chemicals we eat and how they interact with nutrient-regulated and disease-associated genes, we should be able to achieve optimal health and wellness earlier, maintain it longer, and at a lower cost. Just as pharmacogenomics has led to the development of "personalized drugs," so will nutrigenomics open the way for "personalized nutrition." This may be the single most important outcome to emerge from 100 years of nutrition research and the sequencing of the human genome.

Raymond L. Rodriguez, Ph.D.
Director, Center of Excellence in Nutritional Genomics
University of California, Davis

ACKNOWLEDGMENTS

This volume is largely an outgrowth of the Bruce Ames International Symposium on Nutritional Genomics held on the campus of the University of California, Davis on October 22–24, 2004. The purpose of the symposium, and, hence the focus of this volume, is to bring together the leading experts in the emerging discipline of nutritional genomics to discuss recent breakthroughs and future directions for the field. It became apparent to the symposium organizers and presenters after the first day that the symposium proceedings represented a unique and valuable educational resource and that the proceeding should to be documented for the future edification and benefit of a wider audience. Many individuals were involved in making this critical decision and the subsequent preparation of this volume. We would to thank first and foremost, Ms. Heather Bergman, Associate Editor at John Wiley and Sons for providing this opportunity to publish these proceeding as *Nutrtgenomics: Discovering the path to personalized nutrition.* Her vision and patience throughout this project were a tremendous encouragement to us. Of course, we owe a special debt of gratitude to the contributing authors who took time out of their busy schedules to share with the readers of this volume, their knowledge and insights on nutritional genomics. Lastly, we want to thank Ms. Liga Bivina who worked long and tirelessly on the preparation of this volume. Thanks to her exceptional skill and professionalism, this volume is more than just a collection of interesting scientific articles, but rather, it a carefully organized and highly readable compilation of chapters that will inform and enlighten both the student and practitioner of nutritional genomic research.

J.K.
R.L.R.

1

AN INTRODUCTION AND OVERVIEW OF NUTRITIONAL GENOMICS: APPLICATION TO TYPE 2 DIABETES AND INTERNATIONAL NUTRIGENOMICS

Jim Kaput

Center of Excellence in Nutritional Genomics, University of California-Davis, Davis, California and NutraGenomics, Chicago, Illinois and Laboratory of Nutrigenomic Medicine, University of Illinois, Chicago, Illinois

Nutritional genomics, or nutrigenomics, is the study of how foods affect the expression of genetic information in an individual and how an individual's genetic makeup metabolizes and responds to nutrients and bioactives. Not all individuals respond similarly to food, a concept crystallized by Galen about 1800 years ago: "No cause can be efficient without an aptitude of the body." In postgenome terminology, individuals inherit unique responses to food and, since food influences health, unique susceptibilities to chronic diseases. Differences in susceptibilities are caused by the same genetic variations that drive evolution. That food alters expression of genetic information and that genotypic differences result in different metabolic profiles are concepts central to nutritional genomics—and, indeed, provide the critical link between diet and health. The reciprocal interaction between genes and environments has been obvious for at least 2400 years. Hippocrates' maxim that ". . . food be your medicine and medicine be your food" all but foretold the increased incidence in obesity, meta-

Nutritional Genomics: Discovering the Path to Personalized Nutrition
Edited by Jim Kaput and Raymond L. Rodriguez

bolic syndrome, Type 2 diabetes mellitus (T2DM), cardiovascular diseases (CVDs), and, indeed, almost every chronic disease caused by overconsumption of calories and certain nutrients. Hence, the impact of nutritional genomics on society—from science to medicine to agricultural and dietary practices to social and public policies—is likely to exceed that of even the human genome project. Chronic diseases may be preventable, or at least delayed, by balanced, sensible diets, and knowledge gained from comparative nutrigenomics studies in different populations may provide information needed to address the larger problem of global malnutrition and disease.

1.1 INTRODUCTION

Hippocrates provided the directions that modern science has been pursuing on separate paths for much of the 20th century. Nutritionists have focused much attention on population studies that result in statistical associations between foods and disease incidences on a population level. Individual genetic variation within these populations could not be studied easily with pregenomic molecular tools. Similarly, geneticists and molecular biologists have tried to reduce biological complexity by developing model systems that assume the same environment for each subject. Both approaches have contributed significantly to our knowledge of disease risk factors but nevertheless fail to identify specific molecular mechanisms affected by diet and other environmental influences *in an individual* that contribute to health or produce chronic disease at a particular stage in life.

The biochemical details of how nutrients influenced individuals and, likewise, how individuality alters metabolism of nutrients began emerging in the early 1900s when Garrod suggested diet would influence disease differently in different individuals. As early as 1956, only three years after the elucidation of the structure of DNA, Williams [1] summarized these early studies in a book entitled *Biochemical Individuality*, which included a chapter on "Individuality in Nutrition." Medical researchers in various fields had discovered wide ranges in concentrations of insulin, cholesterol, ions, and other easily measured molecules in healthy individuals and in individuals with symptoms of disease (reviewed in [1]).

The concept of genetic uniqueness was solidified by sequencing the human genome [2, 3]. That achievement formed the foundation for one of science's most significant contributions to humankind—an evidence-based understanding that while humans are genetically similar, each individual retains a unique genetic identity that explains the wide array of biochemical, physiological, and morphological phenotypes observed in human populations. The diverse genetic variation in the human population produces a continuum for each human trait, challenging dichotomous social and metabolic groupings based solely on external phenotypes [4, 5]. The genome project provided not only the details and design for humans but, perhaps as importantly, the stimulus for interpreting and applying genomic information in studies of health and disease processes. Nutritionists and other life scientists proposed the study of nutrigenomics based on this new information. Since dietary recommendations (e.g., recommended daily allowances or RDAs) are developed through epidemiological analyses of populations, RDAs may not be optimal for any one individual. The need to understand nutritional influence on the genome and the genome's influence on metabolism led to the concept of nutritional genomics. We [6] published a comprehensive framework for the field with specific examples and suggested five tenets to explain nutrigenomics:

1. Common dietary chemicals act on the human genome, either directly or indirectly, to alter gene expression or structure.
2. Under certain circumstances and in some individuals, diet can be a serious risk factor for a number of diseases.
3. Some diet-regulated genes (and their normal, common variants) are likely to play a role in the onset, incidence, progression, and/or severity of chronic diseases.
4. The degree to which diet influences the balance between healthy and disease states may depend on an individual's genetic makeup.
5. Dietary intervention based on knowledge of nutritional requirement, nutritional status, and genotype (i.e., "personalized nutrition") can be used to prevent, mitigate, or cure chronic disease.

These tenets summarize concepts based on data and results from genetic, epidemiologic, nutritional, molecular, physiological, and other disciplines involved in health and disease research. The challenge in discovering the path to nutritional genomics is, and will be, integrating knowledge and approaches to address the molecular mechanisms of genotype–diet interactions [7] because biological processes are complex and influenced by environmental factors.

About 30 concept and review articles on nutrigenomics, published from 1994 through 2004, summarized various aspects of the field or the new tools available for nutritional genomics research. This chapter discusses some of the principles and concepts of nutrigenomics as they apply to Type 2 diabetes mellitus (T2DM). The focus of this chapter, however, is not on the many individual studies of mechanism, etiology, epidemiology, and genetics of T2DM since these studies have been extensively reviewed elsewhere [8–17]. Weaving information from the diverse disciplines of evolutionary history, genetics, molecular biology, epidemiology, nutrition, biochemistry, medicine, social sciences, and ethical implications requires intermittent excursions from the main topic of T2DM for explaining background information for interpreting or conducting nutrigenomics research. As a polygenic, multifactorial disease, T2DM serves as a model for cancer, obesity, cardiovascular disease, and other chronic diseases influenced by diet and environment.

1.2 UNDERSTANDING T2DM: THE CURRENT VIEW OF T2DM AND TREATMENT OPTIONS

Fasting glucose levels above 140 mg/dL on at least two occasions is one of the diagnostic indicators of T2DM since normal fasting levels are between 70 and 110 mg/dL (see EndocrineWeb.com—http://www.endocrineweb.com/diabetes/

diagnosis.html or [18, 19]). Individuals with high fasting glucose levels are often given an oral glucose tolerance test, which is administered in the fasted state (no food or drink except water for 10–16 hours). Blood is drawn before and 30 minutes, 1 hour, 2 hours, and 3 hours after consuming a high glucose (e.g., 75 grams of glucose) drink. Blood glucose levels rise quickly and fall quickly in individuals without T2DM because insulin is produced in response to the glucose. Glucose levels rise to higher concentrations for a longer time compared to the normal response in individuals with diabetes. The standard for T2DM diagnosis is a blood sugar level of >200 mg/dL at 2 hours. These two diagnostic measures—high fasting glucose and delayed metabolic clearing—are used as either/or classifications (e.g., normal versus diabetic). Individuals with fasting glucose levels between 110 and 126 mg/dL are diagnosed as having impaired fasting glucose and those who have 2 hour glucose levels between 140 and 200 mg/dL in the oral glucose test are diagnosed as having impaired glucose tolerance.

The important lesson from a description of diagnostic procedures for glucose levels and responses is that individuals are typically sorted into three groups: normal, intermediate, and diabetic. However, T2DM is often caused by, or associated with, obesity, a proinflammatory condition that will introduce additional variation in physiology and responsiveness to diet and drugs. Other symptoms associated with the metabolic syndrome—-dyslipidemia, hypertension, insulin resistance, hyperinsulinemia [8, 13, 20, 21]—further complicate simple classification schemes for diabetes. These physiological abnormalities may have overlapping molecular and genetic causes to further muddle diagnosis. If left untreated or managed poorly, many patients develop retinopathy leading to loss of vision, nephropathy resulting in renal failure, neuropathies, and cardiovascular disease [18].

Although the varying complications are well known and often acknowledged, the majority, if not all individuals, with diabetic symptoms are treated similarly [18], implying a single molecular cause: the first line of treatment recommendations are to alter diet and increase energy expenditure through exercise. Approximately 15% of patients control symptoms through these interventions. The patients not helped by diet and exercise are treated with sulfonylureas or biguanide, which are effective in 50% or 75% of patients, respectively. Other drugs are then used as second line treatment for individuals refractory to sulfonylurea or biguanide (Table 1.1). Four of the five major drug classes for controlling T2DM symptoms target different pathways and organs: insulin secretion by the pancreas (sulfonylurea and megtinilides), glucose absorption by the intestines (α-glucosidase inhibitors), glucose production in the liver (biguanide = glucophage = metformin), and insulin sensitivity in adipose and peripheral tissues (e.g., rosiglitazone). Approximately 50% of patients take oral medications only, and some take combinations of drugs plus (12%) or minus insulin. The remainder take no medications (15%) or insulin alone (19%) (http://www.diabetes.org/diabetes-statistics/national-diabetes-fact-sheet.jsp). Since (1) these drugs affect different metabolic pathways in different organs, (2) the treatments are found by trial and error for each patient, and (3) different patients control symptoms by different combinations of treatments, it follows that T2DM is not a single disease but rather multiple diseases caused by multiple genes affected by multiple environ-

TABLE 1.1. "Subtypes" of Type 2 Diabetes

Treatment	Target Tissue	Indications[a]	Effectiveness[b]
Life style	All	All	15%
Sulfonylurea	Pancreas	T2DM < 5 years	~50%
Meglitinides	Pancreas	TDM < 5 years, ↑PPG	?
Biguanide (glucophage)	Liver	Obese, insulin resistant	~75%
α-Glucosidase	Intestine	↑ PPG	Second line
Rosiglitazone	Adipose, muscle	Obese, insulin resistant	Second line

[a] Based on clinical indications of Type 2 diabetes (T2DM), PPG is postprandial glucose response.
[b] Percent of patients responding to treatment; from http://www.aafp.org/PreBuilt/monograph_diabetestreatment.pdf.

mental factors. More specifically, chronic diseases like T2DM are sets of metabolic processes and physiologies that present differently in different patients and change depending on genetic makeup, environmental influences, age, and the cumulative effects of the disease process itself. Physicians and patients alike tend to create artificial bins (e.g., responsiveness to diet or drugs) to help reduce this complexity. However, this simplification can impede development of diagnostics and treatments for the continuous physiologies within individuals and groups of patients. Nutrigenomics concepts can explain the complexities of a chronic disease such as T2DM, and data from well-designed nutritional genomics studies may lead to better diagnostics and treatments.

1.3 UNDERSTANDING T2DM: BEGIN BEFORE CONCEPTION

Mom and dad are the immediate donors of the chromosomes each of us receives at conception and our disease susceptibilities result from the genes we inherit. Disease susceptibilities are the products of collections of variants—single nucleotide polymorphisms (SNPs), deletions, and insertions—in coding and control regions of and near genes that influence the expression or RNA and activity of proteins and enzymes. The probability of inheriting that unique genetic makeup is limited because gene variants in human chromosomes are not uniformly distributed throughout the world. The specific distribution of gene variants results from the rich evolutionary history of human chromosomes.

Background: Out of Africa and Genetic Uniqueness

Individual genetic variation results from the inherent error rate of replicating DNA. No process is 100% accurate, so each time the 6 billion bases in our chromosomes (two each of each chromosome and therefore 2 × 3 billion base pairs) recombine, errors occur [22]. Since such processes occur in every living organism, it is not surprising that each human is genetically different.

Genetic variation among geographically separate populations results from human migration from east Africa in waves that ultimately led to the peopling of six continents [23]. These migrations did not happen uniformly—small groups established population centers, which then became the base for further migrations. The successive splitting off of a portion of the gene pool decreased genetic diversity in the migrating group. Food availability and other factors contributed selective pressures for specific gene variants during migration and dispersal into new environments. One example of such selection, and a model of gene–environment interactions (see below), is lactose tolerance. Humans, like other mammals, do not drink milk after weaning. However, mutations in the control sequences of lactase-phlorizin hydrolase (LPH), the enzyme that metabolizes lactose, allowed the gene to be expressed in adults. The mutations arose in Northern Europe about 9000 years ago [24]. Individuals inheriting these polymorphisms, or haplotype, have the selective advantage of an additional nutrient source: mammalian milk from other species. The lactose tolerance haplotype spread to individuals migrating to new environments, but also to previously established populations to the south and east of Northern Europe. The result is a geographic gradient from Europe to Southeast Asia, where only ~5% of the population is lactose tolerant.

Rich nutrient resources in other environments and human ingenuity led to subsequent population expansions to all parts of the world [25] that are continuing to this day. The combination of small founding groups and recent population growth produced geographically distinct populations who share 99.9% of genomic sequences. SNP and simple tandem repeat (STR) analyses have yielded more detailed information about human relatedness: on average, there is a 12–14% difference between geographically distinct populations—for example, between Asia and Europe (reviewed in [26]). Most genetic variation (estimated range of 86–88%) occurs within a geographic population (e.g., Asia) (reviewed in [26]).

Analyses of geographically isolated countries such as Iceland [27] have revealed significant population substructures. Natural geographical barriers, cultural practices of forming exclusive groups, and mating with individuals within a nearby area (e.g., [28] and reviewed in [29]) produce genetically distinct subgroups. Similar population substructures are predicted for virtually all rural cultures. The increasing tendency of humans to live in urban societies may result, ultimately, in homogenization of genetic substructures, a process that may require millennia and further changes in group and cultural practices. For the foreseeable future, nutrigenomicists will analyze the genetic architecture of individuals since many individuals maintain the preference of mating within like ethnic groups.

Where our parents came from then, and the evolutionary history that is implicit in that statement, produces each individual's chances of inheriting a specific gene variant. Table 1.2 provides an example. Genes 1, 2, and 3 have three variants each with 27 total possible combinations. Like other genes, variants of each of the genes are present in each population at different frequencies. The chance that a European would inherit the G variant of Gene 1, and the G variant of Gene 2, and the G variant of Gene 3 is smaller than an Asian inheriting that combination of gene variants. If these genes are involved in T2DM or differently regulated by diet, then the suscep-

TABLE 1.2. Hypothetical Allele Frequencies

Gene	Allele	Allele Frequencies[a]		
		European	Asian	African
1	A	50	25	33
	C	40	0	34
	G	10	75	33
2	T	20	1	70
	G	5	35	10
	A	75	64	20
3	G	4	55	2
	C	25	35	45
	T	71	10	33

[a] Hypothetical examples of allele distribution in different populations. See text.

tibility for individuals from different geographic areas may differ. An example may be obesity: Europeans and their descendents with body mass indices (BMIs) of 25 or greater have similar metabolic and diabetic profiles compared to Asians with a BMI of ~23 [30]. Although the genes underlying these differences have not yet been identified, it is likely that alleles of genes involved in producing BMI and symptoms of T2DM differ between Europeans and Asians because T2DM, and other chronic diseases, are not limited to individuals of a single geographic ancestry. Epidemiological data indicates that Asian and Hispanic populations seem to have insulin resistance as the predominant mechanism leading to diabetes rather than β-cell dysfunction. In African-Americans the opposite appears to hold (reviewed in [31]). Although these are broad generalizations, the results suggest genetic ancestry contributes to the differences in T2DM prevalence and by different mechanisms [31]. Identifying the genes that contribute to T2DM will provide a molecular and genetic understanding of differences among populations.

1.4 UNDERSTANDING T2DM: GENETIC COMPLEXITY

Understanding diseases caused by mutations in single genes—from inborn errors of metabolism to sickle cell anemia to the initiation of cancer—has guided much biomedical and disease research for the last 50 years. Chronic diseases such as T2DM were understood to present complex physiologies such as hyperglycemia, hyperinsulinemia, and dyslipidemias, but the early genetic research often focused on finding mutations in genes involved in the pathways. Genetic and molecular studies over the past 30 years have established that naturally occurring gene variants involved in many different processes contribute to chronic disease. Identifying the causative genes involved in polygenic diseases usually relies on genetic association studies in humans, but confirmatory data of the mechanisms often relies on cell culture and experimental animal (including transgenic) models. Genetic association studies are statistical analyses that link chromosomal regions with disease subphenotypes or incidence. A variation of association studies is quantitative trait locus analysis.

Background: Quantitative Trait Locus Analysis

Quantitative trait locus (QTL) analysis has been a key method in identifying chromosomal regions involved in complex diseases. A QTL is a polymorphic locus containing one or more genes that contribute to the phenotypic expression of a continuously variable trait. QTLs are typically found by statistical analyses of how frequently a region of chromosome is associated with a certain measurable phenotype such as insulin levels or glucose response. Since multiple genes are involved in those responses, a QTL and the gene(s) variant(s) encoded within contribute less than 100% to the phenotype. QTLs mapping is done in humans but the process is best explained with laboratory animals. Briefly, two parental inbred mouse strains, selected for differences in some observable or measurable phenotype—such as susceptibility to T2DM—are bred to produce an F1 generation. F1 mice are backcrossed to each of the parental strains producing an F2 generation differing in disease susceptibility because of independent assortment of chromosomes and chromosomal rearrangements that occur during meiosis. The incidence or severity of subphenotypes of the disease is measured in F2 mice and statistically associated with chromosomal regions from each of the original parental strains. A given pair of inbred mice may have 10–15 regions that contribute to the complex phenotype in those strains. Different pairs of strains may reveal new QTLs (e.g., [32]) because inbreeding selects for a subset of genetic variation.

If one or more of the genes within the QTLs regions were mutated, each parental strain would develop the particular disease. This does not occur. Rather, the sum of contributions from alleles of causative genes within different QTLs produces the specific trait or disease (reviewed in [33]). Two hypotheses were proposed to explain gene variants that contribute to quantitative traits. The first, called the common disease/common variant hypothesis (i.e., CDCV hypothesis [34, 35]), posits that combinations of normally occurring gene variants produce disease. These gene variants occur in greater than 1% of the population. Others dismiss this theory, suggesting that combinations of rare (<1% in the population) allelic variants cause common diseases. This theory is called the multilocus/multiallele hypothesis [36]. Regardless of the outcome of the controversy, the consequences of the polygenic nature of chronic disease complicate the search for genes involved in disease processes.

Jackson Laboratory (http://www.jax.org) currently lists 58 QTLs for insulin with subclasses for Type 1 diabetes (28 QTLs), insulin levels (13 QTLs), and T2DM (17 QTLs). Many QTLs found in different studies and with different models of diabetes overlap, providing a measure of confidence in their identification. Since QTL analysis is inherently more straightforward with inbred animals and different inbred strains are known to express different subphenotypes of disease (e.g., [37]), it is possible to identify more QTLs with smaller influences on each quantitative trait compared to analyses in outbred animals such as humans (see Sections 1.3 and 1.5).

As importantly, a subset of these murine QTLs map to syntenic regions of T2DM QTLs found in humans (e.g., [38]). Analyses of inbred strains with differing disease susceptibilities are therefore important components of nutrigenomics research because the genotypes are known and the diet and other environmental factors can be controlled.

1.5 UNDERSTANDING T2DM: QTLS IN HUMANS

The approximate locations of seven human T2DM QTLs with LOD (log of the odds, a measure of significance) of greater than 3.6 [16] are shown in Figure 1.1. A total of 17 QTLs distributed on chromosomes 1, 2, 4, 5, 7, 8, 9, 10, 11, 12, 20, and X (not shown) classified as near-suggestive linkage (LOD > 2.0) have also been identified [16]. Some of these T2DM QTLs may be population specific, which can be explained by evolutionary history (see above), by differences in defining phenotypes or subphenotypes in each study, and by methodological differences. Developing standard data elements is a critical need for the development of nutrigenomic research (see Chapters 3 and 16 in this volume). Table 1.3 shows examples of the

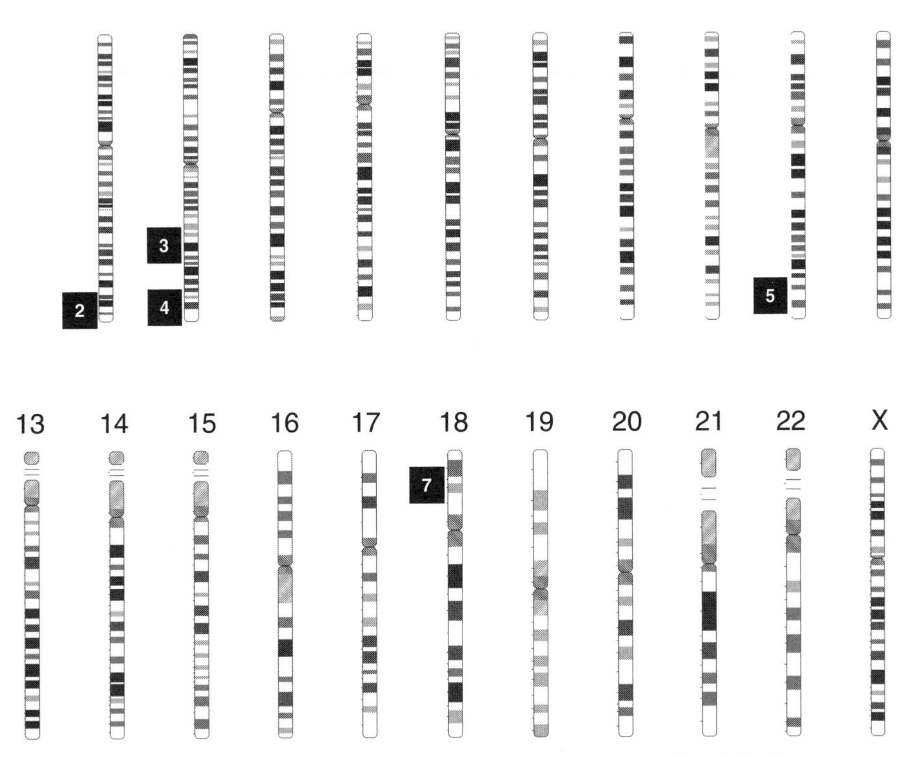

Figure 1.1. Type 2 diabetes mellitus (T2DM) quantitative trait loci (QTLs) for humans. See text.

TABLE 1.3. Hypothetical Genotypes of Six Individuals at Seven Disease Loci[a]

QTL	A	B	C	D	E	F
1	+	+	+	−	−	−
2	+	o	o	−	+	−
3	+	+	−	o	−	−
4	+	+	+	+	o	−
5	+	+	−	−	+	−
6	+	−	+	−	−	−
7	+	+	−	+	−	−

[a] Each individual inherits one of three alleles of each of the seven genes at QTLs 1 through 7. +, −, and o indicate protective allele, allele that contributes to disease, or allele that is neutral, respectively. Genetic susceptibility increases with increasing number of alleles that contribute to disease.

consequences and complexity for the seven QTLs shown in Figure 1.1. If there are only three alleles at each locus with one contributing to T2DM (designated −), providing protection (+), or neutral (O), the number of possible combinations for seven loci is 2187, but the actual number is constrained because of allele frequencies (e.g., Table 1.2). For purposes of illustration, individual A inherited all "protective" alleles and would thus have a low probability of developing T2DM (this speaks nothing of susceptibility to other diseases). Individual F is the other extreme and is genetically prone to develop symptoms. Individuals B, C, D, and E have intermediate risks and are likely to benefit from life-style changes (see below) and/or drug treatments. Perhaps the most interesting in this example are individuals D and E, who each have the same number of protective, neutral, and disease contributing alleles. They may not have the same genetic susceptibility, however, because one cannot predict a priori whether each QTL contributes the same amount to a given phenotype. Added to this complexity are epistatic interactions and epigenetic functions of DNA methylation and chromatin remodeling (see below). At least some of these genes are likely to be regulated directly or indirectly by diet and other environmental factors [7, 38, 39], further complicating analyses of their contribution to diseases or subphenotypes of disease.

Genetic analyses and studies in experimental systems have led to the identification of candidate genes (Table 1.4) that contribute to subphenotypes of T2DM [11, 16, 40] in at least some populations. Genes that map in or near T2DM QTLs are *IRS*, *CAPN10*, *PPARG*, *APM1*, and *HNF4*. Polymorphisms in these genes are associated with subphenotypes of T2DM. Given the number of QTLs identified in humans (with significant or suggestive linkage) and in rodents, it is likely that many other T2DM genes have yet to be identified. The reason that QTLs are not consistently found in different populations can be explained by epistasis and epigenetics.

Background: Epistasis Alters Statistical Associations

Epistasis, or gene–gene interactions, provides the explanation for how genetic ancestry is so important in understanding gene–nutrient (or gene–drug or gene–

TABLE 1.4. Subset of Likely T2DM Candidate Genes[c]

Gene	Function	Chromosome	Reference
SLC2A1	GLUT1 glucose transporter	1p35-p31.3	[16]
IRS	Insulin receptor substrates (IRS-1 has best evidence)	**2q36**[a]	[11]
CAPN10	Calpain 10	**2q37.3**[a]	[11]
PPARG	Peroxisome proliferators activated receptor γ	**3p25**[a]	[16]
APM1	Adiponectin	**3q27**[a]	[11]
PGC-1	Peroxisome proliferators activated receptor γ	4p15.1	[11]
ADRBR2	β2-Adrenergic receptor	5q32-q34	[11]
ENPP1	Glycoprotein PC-1	6q22-q23	[11]
GCK	Glucokinase	7p15-p13	[16]
ADRBR3	β3-Adrenergic receptor	8p12-p11.2	[11]
KCNJ11	Potassium inward rectifier channel $K_{ir}6.2$	11p15.1	[16]
ABCC8	Sulfonylurea receptor	11p15.1	[16]
VNTR-INS	Variable number tandem repeat in the insulin gene	11p15.5	[11]
FOXc2	Transcription factor	16q24.3	[11]
GCGR	Glucagon receptor	17q25	[16]
LIPE	Hormone sensitive lipase	19q13.1-q13.2	[11]
GYS1	Glycogen synthase	19q13.3	[11]
HNF4A	Hepatic nuclear factor 4α	**20q12-q13.1**[b]	[39, 157]

[a] Map position overlaps QTL shown in Figure 1.1.
[b] Map position overlaps QTL with near-suggestive LOD score [16].
[c] As of April 2005.

disease) interactions and, ultimately, health and disease (see Chapter 6 in this volume). Gene–gene interactions can occur through protein–protein, protein–gene, RNA–protein, or RNA silencing [41–44]. The molecular explanation for these genetic results is that proteins or enzymes produced by a gene or its variant do not act alone, but are usually part of a pathway, and many pathways are interconnected. As one example, a G to A polymorphism (IVS6 + G82A) in the tyrosine phosphatase 1B gene interacts statistically with a polymorphism (Gln223Arg) in the leptin receptor (*LEPR*) gene in a Finnish study of 257 individuals with T2DM and 285 controls [45]. *PTB1B* and *LEPR* may not interact directly but may be in the same signal transduction pathway and variants in one affect the activity of the other. A decrease in activity of one member of a pathway may be compensated for by another member of the same pathway, or by variations in a connected pathway. Compensation in the activity of parts to maintain the overall balance within the system is called "buffering" (discussed in [46, 47] and see Chapter 5 in this volume). Understanding and studying the impact of epistasis at the genetic and biochemical levels will be a key component of nutrigenomics research of health and chronic diseases.

Background: Epigenesis and Chromosome Structure—Another Level of Nutrient Control

Another layer of gene regulation is epigenesis. Epigenetics is the study of heritable changes in gene function that occur without a change in the sequence of nuclear DNA. X chromosome inactivation and gene silencing (imprinting) are examples of epigenesis [48]. Epigenetic mechanisms of altering gene regulation are DNA methylation and chromatin remodeling. Both mechanisms change the accessibility of DNA to regulatory proteins and complexes altering transcriptional regulation.

Methylated DNA is considered transcriptionally inactive although there are exceptions to this rule [49]. DNA can be methylated at specific sites, usually at CpG dinucleotides with islands rich in cytosine and guanosine. Methylated CpG within these islands [49] binds methyl-CpG binding proteins (MBDs) [50]. Other chromosomal proteins, some of which belong to the Polycomb group [48], bind to the MBDs forming a multicomponent complex. MBDs subsequently "recruit" histone deacetylases whose activities alter histone proteins, inducing further changes to chromatin structure [51, 52].

Nutrient intake affects DNA methylation status because DNA methyltransferases catalyze the transfer of a methyl group from *S*-adenosylmethionine (*S*-AM) to specific sites in DNA [53]. The products of the reaction are DNA methylated at (usually) CpG residues and *S*-adenosylhomocysteine (*S*-hcy). *S*-AM is generated by the one-carbon metabolic pathway, a network of interconnected biochemical reactions that transfer one-carbon groups from one metabolite to another [54]. Dietary deficiencies of choline, methionine, folate, vitamin B_{12}, vitamin B_6, and riboflavin affect one-carbon metabolism and DNA methylation and increase the risk of neural tube defects, cancer, and cardiovascular diseases [55]. Chapter 10 in this volume reviews how nutrients affecting one-carbon methyl pools alter epigenetic mechanisms in adults and in utero, which have long-term health implications.

Chromatin remodeling is regulated in part by the energy balance in a cell. Specifically, changing calorie intake has been shown to alter chromatin remodeling, changing the NADH/NAD^+ ratio (reviewed in [56]) and the activity of SIR2, an NAD^+-dependent histone deacetylase. Evidence from Guarente's laboratory indicates that the mammalian homologue of Sir2, SIRT1 (sirtuin 1), binds to the nuclear receptor corepressor (NCoR) and silencing mediator of retinoid and thyroid hormone receptors (SMRT) of the transcription factor PPAR-γ (peroxisome proliferators activated receptor gamma), a key regulator of fat mobilization in white adipocytes [57]. Since obesity is linked to T2DM, CVDs, cancer, and other chronic diseases [58], a reduction in fat via SIRT1 and PPAR-γ is likely to slow the aging process. Chapter 9 in this volume presents a review of caloric restriction and chromatin remodeling. Chromatin remodeling may also be affected by changing the expression of genes involved in chromosome structure or its regulation. Chapter 11 in this volume describes microarray data, showing that expression of genes involved in chromatin remodeling is altered by a lunasin, a peptide isolated from soybean. Altering the level of the proteins, enzymes, and RNAi (interfering RNA) [59] involved in chromatin remodeling by diet or other environmental factors will be another control point for regulating gene expression.

Long-term exposure to diets that remodel chromatin structure and DNA methylation could induce permanent epigenetic changes. Such changes might explain why certain individuals can more easily control symptoms of chronic diseases by changing life style but many seem to pass an irreversible threshold. Epigenetic changes may also explain "developmental windows"—key times during development, such as in utero, where short-term environmental influences may produce long-lasting changes in gene expression and metabolic potential (reviewed in [21]).

Developing experimental approaches for dissecting the environmental influences and the critical genes and pathways will be challenging.

1.6 UNDERSTANDING T2DM: FROM BIRTH ONWARD

Inheriting different combinations of genes and specific epigenetic modifications that may occur in utero means that each individual is born with a unique susceptibility to T2DM and other chronic diseases. We don't start life evenly. Once born, the external environment greatly affects the risk of developing chronic diseases.

Background: Genotype–Environment Interactions

Genes that cause chronic diseases must be regulated directly or indirectly by calorie intake and/or by specific chemicals in the diet because diet alters disease incidence and severity [7, 39]. The progressive and sometimes slow change in phenotype from health to disease must occur, at least in part, through changes in gene expression. These are gene–environment interactions and were defined in 1979 (e.g., [60]). The precise, statistical definition of gene–environment interaction is "a different effect of an environmental exposure on disease risk in persons with different genotypes" or, alternatively, "a different effect of a genotype on disease risk in persons with different environmental exposures" [61]. In other words, nutrients affect expression of genetic information and genetic makeup affects how nutrients are metabolized.

A dietary chemical or its metabolite may alter the expression of a susceptibility gene or its variant that in turn affects other gene–gene interactions. The consequence of such gene–environment interactions may not only be apparent on the gene of interest, but may also affect the action of that gene on other interacting genes. Different functional classes of genes may have greater importance than others: affecting the expression of a nuclear receptor or signal transduction pathways may affect more processes than altering the expression of a gene encoding an enzyme in a metabolic pathway. Examples of these concepts are more fully detailed for epigallocatechin gallate (EGCG) that modulates signal transduction pathways (Chapter 8 in this volume), for the multiple effects of other phytoestrogens (Chapter 14 in this volume), and for the many dietary chemicals that are ligands for various transcription factors (Chapter 7 in this volume). Unraveling the interactions among food, different QTLs, epistatic interactions, epigenetic regulation, and their various interactions is a challenging but tractable problem in the age of high-throughput technologies.

Gene–diet interactions have been found in experimental animals (e.g., [7, 38, 39, 62]) and in humans (reviewed in [63–67]). As one specific example, Krauss and co-workers (reviewed in [68]) relied on phenotype to show genotypic differences. Individuals with small, dense LDL particles (phenotype B) have an increased risk of coronary artery disease relative to those individuals exhibiting large, less dense LDL particles (phenotype A). The expression of phenotype A depended on diet: 12 out of 38 men who switched from a 32% fat diet to a diet containing 10% fat developed the phenotype B pattern [69]. At least three distinct genotypes were present in this group, one genotype each for the A or B phenotype and a third genotype that is responsive to low fat/high carbohydrate diets. This genotype produces the A phenotype when these individuals eat a diet containing 32% fat, but a B phenotype when fed 10% fat—a result that can be explained by genotype–environment interactions. Chapter 13 in this volume provides a more detailed description of recent advances in this field.

Although much attention in the nutrigenomics community is focused on gene regulation by dietary factors (Chapters 7, 11, and 14 in this volume), dietary chemicals also alter the activity of proteins and enzymes directly. Ames and colleagues noted that mutations or polymorphisms in genes often result in the corresponding enzyme having an increased K_m (Michaelis constant) for a coenzyme ([70–72], http://www.kmmutants.org/, and this volume). The K_m is a measure of affinity of ligand for its protein. Increases in K_m result in decreased affinity of coenzyme and therefore enzymes with increased K_m have a decreased activity. Increasing the concentration of the coenzyme, which may come from diet, can ameliorate the effect of the decreased K_m. This concept is called the K_m constant and is an example of how alterations in diet may influence individuals differently depending on their genetic makeup.

1.7 UNDERSTANDING T2DM: METABOLOMICS

The concentrations of other transcriptional ligands and coenzymes are controlled by their in vivo metabolism from a dietary chemical precursor (reviewed in [73]). Steroids, for example, are produced through a core of ten linked reactions from cholesterol. The concentration of any given steroid ligand [74] will be greatly influenced by specific combinations of alleles for the enzymatic steps in biosynthetic, branch, and degradative pathways. Alleles of these interacting and linked genes may be different among individuals from the same or different ancestral groups, creating differences in the levels of steroids and therefore expression of genes that these steroids regulate [75–78]. Hence, analyzing genotypes and haplotypes within genes of a pathway is unlikely to provide a reliable association with the amount of the end product. High-throughput analysis of metabolites and the effect of changes in their concentrations is a new field called metabolomics. Chapter 4 in this volume presents an overview of the concepts and procedures of this research area.

Background: The Need for Defined and Controlled Experimental Systems

Identifying genes, proteins, or metabolites regulated by diet and involved in or marked by chronic disease processes in humans is challenging because of the genetic variation among individuals, their long lifespan, and difficulty controlling and monitoring dietary intakes. Cell cultures, on the other hand, don't have livers, microflora in the alimentary tract, or the full metabolic repertoire of their complementary in vivo counterparts. That is, nutrient and bioactive metabolism and regulatory pathways are often affected by metabolism and its products in other organs. Animal studies are therefore necessary to verify the results from human studies and cell culture experiments.

A distinct advantage of using animal models is the array of genetically defined mouse strains, the result of a 100 year effort to produce and characterize inbred strains for biomedical research (see http://www.jax.org). Comparative genomic analyses (reviewed in [79]) have demonstrated that mice and rats share genes and diseases that are similar in other mammals. For example, 99% of mouse genes have human homologues [80] and obesity-induced diabetes (T2DM) occurs in mice (e.g., [81, 82]) and dogs [83]. Molecular responses to dietary chemicals can be analyzed or compared in strains of known genotypes with differing susceptibility to diet-induced disease, making previously unsuspected contributors to the disease process identifiable (e.g., [38, 62]).

Experimental laboratory animals are well suited for the study of nutrient–gene interactions because the genotype and environmental factors can be controlled and systematically altered. The use of diets with known compositions is of critical importance in nutritional genomics research. Dietary chemicals in chow diets have been shown to vary depending on the lot or manufacturer [84] and to affect gene expression [85]. The ability to compare between experiments conducted at different times and in different laboratories is an essential asset of using defined genotypes and reproducible diets.

Our laboratory developed, tested, and uses a strategy that varies diet and genotype systematically to identify diet versus genotype versus diet–genotype regulated genes [39, 62, 86–88]. This strategy relies on inbred strains of mice that differ in genetic susceptibility to diet-induced disease. Our design uses at least four groups: susceptible and nonsusceptible strains fed either control diets or diets that induce symptoms of chronic disease. Differences between these four groups of animals are then analyzed. Those differences could be measured as metabolites (metabolomics), proteins (proteomics), enzymes, or RNA (transcriptomics) or, ideally, some combination of those biological components (i.e, systems biology).

For example, gene expression was compared in livers of *A/a* (agouti) vs A^{vy}/A (obese yellow) segregants (i.e., littermates) from *BALB/cStCrlfC3H/Nctr* × *VYWffC3Hf/Nctr* – A^{vy}/A matings in response to 70% (caloric restriction) and 100% (AL) of ad libitum caloric intakes [38]. $A^{vy}/-$ mice are mottled yellow in coat color, more metabolically efficient, obese, and display subphenotypes of diabetes including mild hyperglycemia and hyperinsulinemia, with increased risk of spontaneous and chemically induced cancers ([89, 90] and reviewed in [91–94]). In contrast, *A/a* mice remain disease- and symptom-free for most of their lives. We also reported analyses of blood glucose with body and brain weights in mice from this experiment [95]. The results indicated that caloric restriction decreased metabolic efficiency associated with continuous ectopic expression of agouti.

Genes differentially expressed, based on diet, genotype, or their interactions, participate in producing the difference in phenotype because phenotype ultimately is an expression of genetic information. Our approach identified diet regulated genes within each of the strains (e.g., 70% vs 100% calories), genotype regulated genes (regulated in the same direction in *A/a* vs A^{vy}/A fed 70% calories or *A/a* vs A^{vy}/A fed 100% calories), and genotype–diet regulated genes whose regulation varies depending on genotype and diet [38]. To our knowledge, this is the only method that identifies genes regulated by genotype, diet, and their interactions, which in this case yields molecular information about physiological differences between agouti and obese yellow mice.

Changes in gene expression cannot discriminate cause from effect—that is, those genes that cause or contribute to the differences in phenotype from those that are affected by changes in expression of other genes. Targeting the regulation of genes or their proteins that cause disease by diet or drugs may ameliorate symptoms of the disease. Subsets of genes regulated by diet or diet–genotype that play a role in or cause the disease can be identified if they map to chromosomal regions associated with that complex phenotype, that is, within independently derived QTLs (see above) or within loci identified by other genetic methods [38, 62]. Our approach can be considered a variation of the common disease/common variant (CD/CV) hypothesis [34, 35], with the added proviso that diet influences expression or activity of variants of common genes.

Twenty-eight genes identified by differential expression in our system mapped to diabesity loci and eight of these loci were syntenic to T2DM QTLs in humans [38]. Some are expected to contribute to characteristics expressed in obese yellow mice. An additional 59 genes mapped to obesity and weight gain QTLs. A subset of the identified genes has been associated with diabetic subphenotypes and others can be linked to abnormal physiological conditions observed in obesity and diabetes. This novel approach can be applied to any chronic disease given careful selection of mice with differing susceptibilities to diet-induced disease.

1.8 UNDERSTANDING T2DM: ENVIRONMENTAL INFLUENCES

Genetic makeup and dietary influences are not the only determinant of health and disease susceptibility. Epidemiological studies have identified dietary and other environmental influences that contribute to the incidence and severity of chronic diseases. Epidemiology utilizes statistical methods that associate frequency and distribution of disease in human populations. Since the results are associations, they do not prove cause and effect. Nevertheless, the results from such studies have been invaluable to experimental scientists who analyze the genetic and molecular mechanisms of nutrient or environment influences in humans or model systems.

Background: Epidemiology

The incidence and severity of chronic diseases varies among countries (reviewed in [96]): stomach cancer is higher while breast and prostate cancer are lower in Japan compared with the United States (e.g., [97]); the incidence of the metabolic syndrome [98] and cardiovascular diseases [99] vary by country, for three specific examples. These disparities cannot be explained by either genes or diet alone because the average genetic makeup and environment varies among countries. However, disease incidences change when individuals migrate from one country and begin adopting the host country's diet and culture ([31, 97, 99, 100] and reviewed in [96]). Second generation descendents typically have disease incidences similar to the members of the principal culture, a time too short for Mendelian genetic changes to occur. By identifying food intake and cultural practices among individuals of different genetic ancestries and countries, epidemiological research aids in the identification of nutritional and environmental contributors to disease and provides essential information for guiding the design of molecular and genetic research. Chapter 2 in this volume describes results of the latest epidemiological research and their application to develop dietary recommendations for a population.

The next step in analyzing nutrient–gene interactions in humans became apparent even before the completion of the human genome project (see review in [101]): analyze naturally occurring variations in genes that affect aspects of disease processes (e.g., subphenotypes such as insulin or glucose levels) or nutrient utilization in different individuals in response to different diets. For example, the G-75A (a G to A change, 75 base pairs upstream of the transcription start site) polymorphism in *APOA1* is associated with increased HDL-cholesterol (HDL-C) concentrations in the serum in certain individual who consume higher amounts of polyunsaturated fatty acids (PUFAs) [102]. Individuals homozygous for the more common G SNP have lower levels of HDL-C with increased intake of PUFAs. This is a classic example of genotype–diet interactions: HDL-C levels are dependent on the genotype (G or A at −75) and the intake of a specific nutrient (PUFA). Chapter 3 in this volume critically reviews the methodology and progress of molecular nutrigenomic epidemiological studies [101, 103].

The advent of SNP–nutrient–phenotype analyses has led to the realization of the need for genomic controls—that is, testing whether individuals in a study have similar genetic makeup. Genomic controls test for population stratification and ensure better matches between cases and controls. Prior to the advent of high-throughput technologies and genome knowledge, genetic substructures existing in study populations could not be analyzed. Genetic variation in populations confounds molecular epidemiology studies that seek to analyze gene–disease or nutrient–gene associations. For example, many (597 out of 603) gene–disease associations are not replicated in more than three independent studies [104] although meta-analyses of

25 different reported associations (data from 301 published studies) showed statistically significant replication for eight gene associations [105]. Similarly, nonreplicable results associating diet with candidate gene variants are the norm (reviewed in [101–103, 106, 107]). In addition to stratification caused by population substructures, other confounders include sample sizes that lack appropriate statistical power, control groups not appropriately matched to cases, and overinterpretation of data [108–110]. Chapter 3 also provides a discussion of the best practices for analyzing nutrient–gene interactions [101].

1.9 UNDERSTANDING T2DM: ENVIRONMENT IS MORE THAN DIET

Capturing and assessing accurate food intakes continues to be a challenge for nutritional and nutrigenomics researchers. Food surveys and dietary histories are often inaccurate because of differences in ability to recall specifics (type and amounts) of food intakes and differences in dietary assessment methods (e.g., self-administered vs interviewed, food frequency questionnaires vs diet diaries), and variations in their definitions and analyses. In addition to accurate food intake, databases are needed for more detailed macro- and micronutrient content of local foods, a challenge for the diverse cultures and diets throughout the world. Surveys must also capture self-ascribed affiliations to religions, cultures, customs, or ethnic groups because of food restrictions and preferences. Added to this complexity are food preparation techniques, which alter chemical composition and nutrient availability and, in some cases, produce chemicals that affect health. Chapter 15 in this volume describes examples of food chemicals produced by different cooking methods which contribute to disease processes.

Although the primary focus of nutritional genomics is the understanding of nutrient–gene interactions, expression of genetic information also is influenced by numerous environmental factors. For example, cytokine levels are unusually sensitive to environmental changes and serve as good markers of environmental influences that may alter protein and RNA expression. Some examples of nonnutrient environmental factors affecting cytokine concentrations are:

- Overall sleep time and sleep continuity (e.g., [111, 112]).
- Oxygen tension [113], which is related to altitude.
- Over-the-counter drugs (e.g., nonsteroidal anti-inflammatory drugs) [114].
- Water intake relative to tea [115] and other beverages.
- Physical activity, including genetic fitness to activity [116–120].
- Psychological factors like stress [121].
- Exposure to allergens and pollutants (e.g., [122]).
- Circadian rhythm and seasonal changes [123].
- Balance between energy intake and expenditure (reviewed in [124]).

Many genetic and environmental influences change during life and aging (reviewed in [125]) with the net result that health and chronic diseases are not discrete, dichotomous states but are rather processes. Figure 1.2 schematically shows the theoretical paths of six individuals (A through F in Table 1.3) during aging. The different heights of the initial condition (left axis) reflect the differences in genetic susceptibility (including epigenetic factors) and the width of the paths were designed to suggest the influence of different environmental factors. Certain individuals (e.g., C and D) may be able to greatly influence onset or severity of disease by altering life style whereas others are destined for disease (e.g., F) or health (e.g., A) regardless of life style. Clinical measurements are taken at discrete time points along this curve and therefore are only a a single frame of a movie and may not accurately predict past physiological processes or future outcomes. While important, these snapshot diagnostics need to be supplemented with genetic analyses of susceptibility genes (at birth) and a greater understanding of their interactions with diet and the environment. The majority of individuals may therefore influence health processes since individuals with extreme susceptibilities are unlikely to be found at high frequencies as evidenced by the current epidemic of obesity and T2DM in many populations throughout the world [126, 127].

The combinations of inherited susceptibility genes and interactions (epistasis), in utero nutritional influences (epigenetics), aging, diet and other environmental

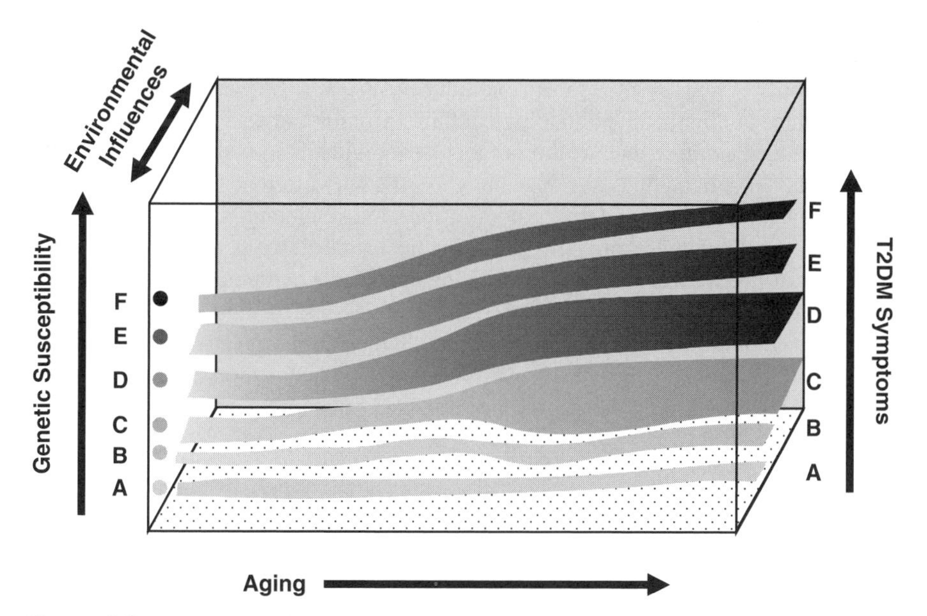

Figure 1.2. Complexity of T2DM. Individuals are born with differing susceptibilities to disease (*y* axis, location of starting ribbon). Life and aging (*x* axis) exposes each individual to environmental influences (*z* axis) that have differing effects (width) depending on genetic makeup. Modeled after Figure 2 in [124]. See text for details.

factors, and developmental windows, plasticity, and programming [21] are significant but tractable challenges to the field of nutrigenomics and, indeed, all health and disease research efforts.

The consequences of integrating differing genetic susceptibilities and environmental influences and in viewing health and disease as processes should have dramatic impact on developing personalized diets for prevention and treatment of disease. The path to personalized nutrition and disease management then requires the analysis of genetic susceptibility, with the concomitant identification of genes regulated by diet and other environmental influences.

1.10 UNDERSTANDING T2DM: DATA ACQUISITION AND ANALYSES

Large-scale studies that examine all or many physiological, genetic, and life-style parameters will be expensive and complex. Nevertheless, such studies, coupled with experiments in model systems, will be essential for obtaining a complete understanding of nutrient–gene interactions underlying T2DM or other chronic diseases. Current funding strategies may not be sufficient for large-scale population studies designed to capture large and disparate data sets. A possible solution to this challenge would be to combine data from many smaller and independent studies. Combining and analyzing data across studies, or meta-analyses, is routinely used to synthesize evidence and produce meaningful associations. However, methodological flaws have led to criticism of this statistical procedure (see [128]). To overcome such limitations, standards may be developed for independent, smaller scale projects, which might reduce or eliminate methodological flaws. Combining data from many studies is a two-edged sword, since as the number of individuals in a study increases, the greater the likelihood that variance may be due to differences in environment, population stratification, and differences in food intake among populations in many countries. Each of these problems is solvable by developing collaborations and sharing best practices across studies, funding agencies, governments, and countries. An international effort to coordinate such activities was begun in late 2004 as a consequence of the Bruce Ames International Symposium on Nutritional Genomics (see [129]). Regardless of the exact path forward, data acquisition and analyses will be critical for developing the path to personalized nutrition.

1.11 BIOINFORMATICS AND BIOCOMPUTATION

Highly developed databases exist for genomic and molecular data, which serve as models for the development of clinical databases linked to genetic and nutritional information. Researchers in a variety of fields are reconsidering the nature of experimental designs and tools for capturing diverse datasets of genetic (e.g., SNP analyses), molecular (e.g., gene expression), proteomic including enzymatic, metabolite (e.g., mass spectra or NMR), dietary history, nutritional status, and the many life-style factors described above. The integration and analyses of these seemingly disparate datasets are required for the research phase of nutrigenomics and for the implementation of the data for clinical use.

Bioinformatics is defined here as the capture, manipulation, and access of high-dimensional datasets (see Chapter 16 in this volume). Warehousing this diverse set of data may seem challenging for biologists, but other disciplines have developed scalable databases on an even larger scale (see Chapters 16 and 17 in this volume).

The analyses of these datasets are likely to pose more significant challenges for nutrigenomics researchers. Biocomputation is used here to highlight the differences between data capture, storage, and access and the algorithms for analyzing those data. Biocomputation then converts data into knowledge. Progress in developing new software tools for such analyses is growing rapidly as more and more high-dimensional datasets are acquired (see Chapter 17 in this volume).

1.12 CONVERTING SCIENCE INTO PRACTICE

Until our understanding of diet–gene interactions is transformed into useful applications for societal benefit, nutrigenomics will remain more promise than practice. Some of the products and applications envisioned by nutritional genomics researchers include diagnostics, preventative life-style guidelines, more efficacious dietary recommendations, health-promoting food supplements, and drugs. In all cases, involving the commercial sector will be necessary since scientific advances made in the laboratory require a substantial investment of financial resources before they are ready for the marketplace.

Diagnostics

Since the goal of nutritional genomics is to provide safe, accurate, and informative diet and life-style recommendations specific for an individual, diagnostics will be critical for developing genome-specific foods and for determining treatment options for disease. Chemicals in food often influence the same pathways involved in disease development (e.g., [7, 39]), which means that nutrigenomics testing of individuals will overlap with and be complementary to pharmacogenomics testing (genetic testing for drug efficacy in each individual). As one example, the natural ligands that activate peroxisome proliferators activated receptor gamma 2 (PPAR-γ2; see Chapter 7 in this volume for discussion of transcriptional regulators) are eicosapentaenoic acid [130] and its derivatives (e.g., 15-deoxy-$\Delta^{12,14}$-prostaglandin J_2 (PGJ_2)) [131, 132]. Rosiglitazone, a member of the thiazolitazone (TZD) class of drugs for T2DM, also binds and activates PPAR-γ. To date, no polymorphisms (>1% allele frequency in population) in this *PPAR* gene that would alter ligand binding have been identified (although several dominant negative mutations have been discovered; reviewed in [133]). Although binding of ligand to nuclear receptor is a key initiating step in transcription, other nuclear receptors (e.g., RXR) and transcription factors are involved in forming an active transcriptional complex. One such protein, PGC-1α, is a coactivator of PPAR-γ2 and polymorphisms in this gene have been linked to insulin resistance and susceptibility to T2DM in a Japanese population [134]. Activation of PPAR-γ by lipids or drugs is followed by interaction with PGC-1α, which varies activity depending on the sequence an individual inherits. Hence, the diet/drug–gene interaction may be indirect through the secondary binding of a coregulator as well as direct, by differential binding of ligand to receptor.

For consumers, the initial introduction to the practical applications of nutrigenomics will be through clinical diagnostics of disease susceptibility and severity. More precisely, testing may not be for a complex phenotype such as obesity or diabetes, but for "subphenotypes" such as insulin levels, glucose tolerance test, or some defined and well-accepted "intermediate" biomarker of a disease. Genetic testing in the clinic will be no different than the analyses of other diagnostic biomarker such as cholesterol levels, which are familiar to consumers. Once enough data and knowledge are extracted from experimental animal and human association studies, genetic testing may be used to identify individuals with specific needs for vitamins (see Chapter 12 in this volume) or types of diets (see Chapter 13 in this volume). Since the "average" dietary chemical interacts with membrane, cell, and serum transporters, may be metabolized to produce other bioactive compounds, or may alter regulation via specific or nonspecific interactions (e.g., selenium [135]), accurate predictive tests for specific nutrients may not be available soon. Nevertheless, the use of genotype–phenotype–biomarker diagnostics is beginning to appear and will be fundamental in the translation of nutrigenomics beyond the laboratory to the consumer market.

Health and Wellness

Many (75–90%) consumers state that they make foods choices with the intent of benefiting their or their families' health. The complexity of analyzing multiple genes in multiple metabolic pathways will be difficult to interpret for the average consumer. Hence, a key need for translating nutrigenomics tests from academic knowledge to real-world utility will be genetic and nutritional counselors capable of interpreting genetic tests and linking that knowledge to specific nutrients and diets.

In the not too distant future, sufficient data will be available for food companies to design medical or consumer-oriented foods targeted to specific genotypes. It is not difficult to imagine a medical food targeted to patients with one of the T2DM subphenotypes such as hyperinsulinemia, and a different medical food designed for another T2DM patient whose metabolic profile suggests the primary cause of disease is insulin resistance. This is an example of market segmentation for health and wellness. These products may be sold direct to the consumer once genetic data can provide better certainty for genetic susceptibility for different subtypes of disease. Some have predicted that a shift in cultural attitudes will be required because of the probabilistic nature of genetic testing (see Chapter 17 in this volume).

Blurring the Line: Food and Drugs

Our growing knowledge of nutrient–gene interactions and the bioactive components of foods that prevent disease is causing researchers, government regulators, and public policy makers to rethink the distinction between medicine and foods, as defined by the Food and Drug Administration. Naturally occurring chemicals in foods have dramatic effects on the cell processes responsible for health and disease. Hence, purified or complex mixtures of bioactive nutrients are likely to be used for disease prevention and management.

Modern pharmaceuticals evolved from thousands of years of traditional lore concerning the uses of plants and herbs as medicines. Research efforts over the past 100 years have led to the widespread adoption of Paul Elrlich's principles of (chemo)therapy [136]:

- Drugs need not be of natural origin and could be developed by planned chemical synthesis.
- Systematic exploration of structure–function relationship distinguishing therapeutic activity from toxicity.
- Maximization of the ratio of dose required to cure disease to dose producing toxicity (broad therapeutic index).
- The importance of developing animal models of diseases for quantitative measurements of both therapeutic potency and toxicity.

Nutrients from foods fit these criteria and large segments of the population are consuming supplements in the desire to improve health, prevent disease, and delay the onset of age-related disease. However, the growing interest, acceptance, and use of dietary supplements, not to mention herbal medicines [137], has outpaced the scientific, medical [138], and food industries' ability to carefully analyze their combined and independent activities, effectiveness, and safety [136, 139]. Although government agencies like the FDA are charged with ensuring the purity, truth in labeling, safety, and effectiveness of foods and drugs, the 1994 Dietary Supplement Health and Education Act (DSHEA) essentially eliminated all government controls of safety and efficacy for dietary supplements [136]. The DSHEA changed the regulatory climate, shifting to the FDA the responsibility for certifying that a product is unsafe. The ramifications of the DSHEA are significant for the present market and for the future of marketing food products based on nutritional genomics research. The most important need will be for safety and efficacy data for isolated nutrients or nutrient mixtures since dosages cannot be ascertained easily from genotype alone (at least as currently analyzed), and certain genotypes may be at risk for those doses or combinations of nutrients. Since traditional nutrients and some phytonutrients will be considered as generally regarded as safe (GRAS), the distinction between medicine and food will be further blurred. Nutrigenomics, by definition, will require clinical validation of effects, including safety, in the target market segment. There is an opportunity and a growing need for a new regulatory framework that will accommodate emerging science as well as deliver consumer benefits and afford consumer protection [140].

SEARCH ETHICS AND GENETIC PRIVACY

... and eventually applying nutritional genomics data and information poses special ethical and public policy concerns for individuals and for society. To discover gene–nutrient–phenotype associations, it is necessary to develop a complete biological and cultural description of an individual. This may include highly personal and confidential genetic information, a medical history, cultural and religious practices, and eating habits. Since any one individual is a unique combination of genes and environment (and their interactions), the probability of a given genotype, diet, and life style for health or disease prevention must be calculated from data obtained from a large number of individuals. Different cultures, groups, and individuals may have different beliefs and laws concerning the acquisition and use of such data. Chapter 18 in this volume describes a life-long learning goal called cultural humility, which may be applied not only for its intended goals for reducing health disparities (see below), but also for research efforts in nutritional genomics.

The application of nutrigenomics in clinics or the marketplace presents moral and ethical issues similar to those faced by pharmacogenomics or other forms of genetic testing. The fears are that the confidentiality of DNA testing information will be violated and this information may be used by employers, insurance companies, and the criminal justice system to discriminate against certain individuals, particularly members of certain racial/ethinic groups, the disabled, the poor, and the uninsured. In spite of the fact that every individual is susceptible to one or more chronic diseases, these fears and the potential misuse of a person's genetic data are real and must be addressed individually and by society. The Joint Centre for Bioethics at the University of Toronto has taken the lead in these early discussions and a review of their well-researched thoughts are presented in Chapter 19 in this volume.

1.14 HEALTH DISPARITIES

The landmark 1985 report, the U.S. Secretary's Task Force Report on Black and Minority Health [141], revealed that certain minority populations exhibit higher incidence and severity of many chronic diseases including diabetes [142], obesity [143, 144], asthma [144], cardiovascular diseases (CVDs) [145, 146], and certain cancers [147]. Similar health disparities exist in New Zealand, Canada, Australia [148], and England [149]. For example, men with African admixture in the United States have 60% greater risk of prostate cancer diagnosis and two to three times greater mortality than European admixture men [150]. Genetic differences alone, however, cannot explain these health disparities since the incidence of prostate cancer in Africa has increased from 1960 to 1997 [151] and is now approximately as prevalent as in the United States [152, 153]. Social and cultural attitudes among health-care workers, researchers who design health studies to identify causative genes, access to medical care, insurance, environmental factors, and attitudes in both the minority and general population also contribute significantly to minority health disparities (e.g., [149, 154]).

The Center of Excellence in Nutritional Genomics at the University of California at Davis and its collaborators at the Children's Hospital of Oakland Research Institute (CHORI), the Ethnic Health Institute of Oakland, and the USDA's Western Human Nutrition Research Center (WHNRC) at UC Davis are investigating how individual genetic variation may exacerbate diet as a risk factor for disease and how dietary intervention based on knowledge of nutritional status, nutritional requirements, and genotype can remedy or ameliorate disease symptoms.

Nutrigenomics researchers seek to identify and characterize genes regulated by naturally occurring chemicals in foods and the subset of those genes that influence balance between healthy and disease states. Such knowledge is necessary but not sufficient to address health disparities observed in ethnic populations and the poor. Achieving and maintaining optimum health care can be assisted by (1) developing better approaches to human association studies that recognize the importance of population stratification in ethnically mixed populations, (2) incorporating concepts and data from dietary histories in association studies, (3) educating health-care professionals about the nonbiological factors contributing to health disparities such as stereotyping, bias, racism, and other bad practices in biomedical research and health-care delivery (see Chapter 18 in this volume), and (4) establishing community outreach and education programs intended to inform ethnic groups and the poor about the importance of good nutritional habits as they relate to their particular genetic makeup, and to disseminate existing information about health matters, including health disparities. Unraveling the causes of health-care disparities will require multidisciplinary teams communicating and collaborating on all aspects of the societal and biological problem.

1.15 PUBLIC AND INTERNATIONAL POLICIES

Nutritional genomics research, in its early stage, has focused on chronic diseases of economically developed nations such as obesity, CVDs, and T2DM. These are afflictions that could be avoided by ~90% of the population by eating less and exercising more. Chronic diseases affect several billion individuals in developed countries, and recently the incidence of these diseases has been increasing dramatically in less developed countries. T2DM is a case in point: by 2025, 300 million people age 20 and older will have diabetes. The incidence is expected to increase 42% in developed countries, but 170% in developing countries [126]. The increases are concentrated in urban areas, but rural areas are facing or will face similar pressures in the near future. Most of the increase in these diseases is caused by individuals in developing countries adopting Western foods and life styles. Although these increases may seem to have little to do with nutritional genomics on the surface, certain thrifty genotypes are thought to be more susceptible to high-fat and high-calorie foods and inactivity levels found in developed countries [8]. The genetic and molecular differences that produce the thrifty genotype [21, 155, 156], which are likely to be a collection of variants in genes involved in regulatory and metabolic functions, have not been identified.

At the same time that developing countries are struggling with surges in chronic diseases, over two billion of the world's citizens face undernourishment—a significant health problem especially for the young, old, and women. Malnutrition early in life establishes a pattern where children cannot reach full mental and metabolic potential (e.g., [21])—a condition passed on from generation to generation. Breaking this cycle has proved to be a challenge for governments and the international nutrition and public health communities. Although most knowledgeable observers contend that political action and public policy decisions could alleviate malnutrition worldwide (e.g., [157]), nutritional genomics nevertheless contributes significantly to alleviation of world health problems by a two-step procedure: analyzing data of chronic diseases in socioeconomically privileged societies, and comparing responses to nutrients in individuals from different geographical ancestral groups.

The rationale for these two steps is simple although complex in execution and costs: analyzing nutrient–gene interactions is best done through comparative analyses—that is, analyzing biological responses between individuals of one ancestral group (i.e., similar genetic makeup [158]) against the response of individuals from another ancestral group. Valuable insights into biology are often discovered by comparing between states: a whole generation of scientists learned to compare events and processes between disease and normal tissue, for example. Similarly, understanding biological processes is best accomplished by comparing responses to diet (or physical activity, or other life-style choices) between different individuals who differ significantly in genotype but who are otherwise healthy. That is, the association between diet and nutrients reveals itself when it differs among genetically unique individuals.

Many nutrigenomics researchers therefore expect that addressing the health issues in developed countries must be done by comparisons among different genetic makeups and environments throughout the world [129]. As a part of that analysis, the differences observed among different ancestral populations is likely to identify nutrients and supplements best suited for individuals sharing similar nutrigenomic profiles regardless of their geographic origin. Such evidence-based nutrient recommendations can then be developed for subgroups in a population, allowing those individuals to bypass the deleterious diets developed and eaten in more privileged countries. Such studies may identify the gene variants of the thrifty genotypes [155].

Neither nutrigenomics research nor its applications can by themselves solve the complex problems faced by individuals in either developed or developing countries. Nutrigenomics can, however, provide data, explanations, and advice for public policy experts throughout the world for treating and preventing chronic diseases, and for targeting specific policies for reducing or eliminating malnutrition. The latter is of particular importance with the growing awareness that maternal and early childhood nutrition may program health or disease (reviewed in [21]). Such knowledge will allow policy planners and implementers to spend limited resources in ways to effect the most change in both developing and developed countries.

1.16 CONCLUSION

It is a truism that progress in science and technologies will generate useful knowledge from genomic, nutritional, disease, social, and ethical endeavors. The integration of these efforts is a systems approach unlike any in human history. The newly expanded breadth of nutritional genomics research that encompasses environmental, social and cultural, and dietary influences on expression of genetic information can direct the development of comprehensive answers to questions of individual susceptibility to health and wellness. These efforts can, with foresight and persistence, be implemented for the benefit of all the world's peoples.

ACKNOWLEDGMENT

Every biochemistry student learns some of the principles of genetics and gene regulation from the Ames test, which Bruce developed early in his career to study mutagenesis. When I had the honor of meeting Bruce the first time in the early 1990s, I was struck by how young he was—a tribute to the many contributions he has made to science and society: he started making significant contributions at a young age and continues to this day. It is a pleasure to be associated with a true scientist and gentleman.

REFERENCES

1. R. P. Williams (1956). *Biochemical Individuality: The Basis for the Genetotrophic Concept.* McGraw Hill, New York.
2. J. C. Venter, M. D. Adams, E. W. Myers, et al. (2001). The sequence of the human genome, *Science* **291**:1304–1351.
3. E. S. Lander, L. M. Linton, B. Birren, et al. (2001). Initial sequencing and analysis of the human genome, *Nature* **409**:860–921.
4. S. O. Keita, R. A. Kittles, C. D. Royal, et al. (2004). Conceptualizing human variation, *Nat. Genet.* **36**(Suppl 1):S17–S20.
5. E. J. Parra, R. A. Kittles, and M. D. Shriver (2004). Implications of correlations between skin color and genetic ancestry for biomedical research, *Nat. Genet.* **36**(Suppl 1): S54–S60.
6. J. Kaput and R. L. Rodriguez (2004). Nutritional genomics: The next frontier in the postgenomic era, *Physiol. Genomics* **16**:166–177.
7. J. Kaput (2004). Diet–disease gene interactions, *Nutrition* **20**:26–31.
8. J. Curtis and C. Wilson (2005). Preventing type 2 diabetes mellitus, *J. Am. Board Fam. Pract.* **18**:37–43.
9. L. Pirola, A. M. Johnston, and E. Van Obberghen (2004). Modulation of insulin action, *Diabetologia* **47**:170–184.
10. M. E. Patti (2004). Gene expression in humans with diabetes and prediabetes: What have we learned about diabetes pathophysiology? *Curr. Opin. Clin. Nutr. Metab. Care* **7**:383–390.

11. H. Parikh and L. Groop (2004). Candidate genes for type 2 diabetes, *Rev. Endocr. Metab. Disord.* **5**:151–176.
12. M. Laakso (2004). Gene variants, insulin resistance, and dyslipidaemia, *Curr. Opin. Lipidol.* **15**:115–120.
13. A. Steinmetz (2003). Treatment of diabetic dyslipoproteinemia, *Exp. Clin. Endocrinol. Diabetes* **111**:239–245.
14. R. L. Hanson and W. C. Knowler (2003). Quantitative trait linkage studies of diabetes-related traits, *Curr. Diabetes Rep.* **3**:176–183.
15. L. Hansen (2003). Candidate genes and late-onset type 2 diabetes mellitus. Susceptibility genes or common polymorphisms? *Dan. Med. Bull.* **50**:320–346.
16. J. C. Florez, J. Hirschhorn, and D. Altshuler (2003). The inherited basis of diabetes mellitus: Implications for the genetic analysis of complex traits, *Annu. Rev. Genomics Hum. Genet.* **4**:257–291.
17. M. I. Mccarthy and P. Froguel (2002). Genetic approaches to the molecular understanding of type 2 diabetes, *Am. J. Physiol. Endocrinol. Metab.* **283**:E217–E225.
18. D. M. Nathan (2002). Clinical practice. Initial management of glycemia in type 2 diabetes mellitus, *N. Engl. J. Med.* **347**:1342–1349.
19. A. J. Ahmann and M. C. Riddle (2002). Current oral agents for type 2 diabetes. Many options, but which to choose when? *Postgrad. Med.* **111**:32–34, 37–40, 43–46.
20. L. Cordain, M. R. Eades, and M. D. Eades (2003). Hyperinsulinemic diseases of civilization: More than just Syndrome X, *Comp. Biochem. Physiol. A Mol. Integr. Physiol.* **136**:95–112.
21. I. C. Mcmillen and J. S. Robinson (2005). Developmental origins of the metabolic syndrome: Prediction, plasticity, and programming, *Physiol. Rev.* **85**:571–633.
22. T. A. Kunkel and K. Bebenek (2000). DNA replication fidelity, *Annu. Rev. Biochem.* **69**:497–529.
23. S. A. Tishkoff and K. K. Kidd (2004). Implications of biogeography of human populations for "race" and medicine, *Nat. Genet.* **36**(Suppl 1):S21–S27.
24. N. S. Enattah, T. Sahi, E. Savilahti, et al. (2002). Identification of a variant associated with adult-type hypolactasia, *Nat. Genet.* **30**:233–237.
25. L. A. Zhivotovsky, N. A. Rosenberg, and M. W. Feldman (2003). Features of evolution and expansion of modern humans, inferred from genomewide microsatellite markers, *Am. J. Hum. Genet.* **72**:1171–1186.
26. L. B. Jorde and S. P. Wooding (2004). Genetic variation, classification and "race," *Nat. Genet.* **36**(Suppl 1):S28–S33.
27. A. Helgason, B. Yngvadottir, B. Hrafnkelsson, J. Gulcher, and K. Stefansson (2005). An Icelandic example of the impact of population structure on association studies, *Nat. Genet.* **37**:90–95.
28. R. Ploski, M. Wozniak, R. Pawlowski, et al. (2002). Homogeneity and distinctiveness of Polish paternal lineages revealed by Y chromosome microsatellite haplotype analysis, *Hum. Genet.* **110**:592–600.
29. M. Pagel and R. Mace (2004). The cultural wealth of nations, *Nature* **428**:275–278.
30. Y. Kim, Y. K. Suh, and H. Choi (2004). BMI and metabolic disorders in South Korean adults: 1998 Korea National Health and Nutrition Survey, *Obes. Res.* **12**:445–453.

31. N. Abate and M. Chandalia (2003). The impact of ethnicity on type 2 diabetes, *J. Diabetes Complications* **17**:39–58.
32. G. A. Brockmann and M. R. Bevova (2002). Using mouse models to dissect the genetics of obesity, *Trends Genet.* **18**:367–376.
33. J. Flint, W. Valdar, S. Shifman, and R. Mott (2005). Strategies for mapping and cloning quantitative trait genes in rodents, *Nat. Rev. Genet.* **6**:271–286.
34. E. S. Lander (1996). The new genomics: global views of biology, *Science* **274**:536–539.
35. F. S. Collins, M. S. Guyer, and A. Charkravarti (1997). Variations on a theme: Cataloging human DNA sequence variation, *Science* **278**:1580–1581.
36. A. F. Wright and N. D. Hastie (2001). Complex genetic diseases: Controversy over the Croesus code, *Genome Biol.* **2**:Comment2007.1–2007.8.
37. P. M. Nishina, J. Wang, W. Toyofuku, et al. (1993). Atherosclerosis and plasma and liver lipids in nine inbred strains of mice, *Lipids* **28**:599–605.
38. J. Kaput, K. G. Klein, E. J. Reyes, et al. (2004). Identification of genes contributing to the obese yellow Avy phenotype: Caloric restriction, genotype, diet × genotype interactions, *Physiol. Genomics* **18**:316–324.
39. J. Kaput, D. Swartz, E. Paisley, et al. (1994). Diet–disease interactions at the molecular level: An experimental paradigm, *J. Nutr.* **124**:1296S–1305S.
40. I. Barroso, J. Luan, R. P. Middelberg, et al. (2003). Candidate gene association study in type 2 diabetes indicates a role for genes involved in beta-cell function as well as insulin action, *PLoS Biol.* **1**:E20.
41. L. He and G. J. Hannon (2004). MicroRNAs: Small RNAs with a big role in gene regulation, *Nat. Rev. Genet.* **5**:522–531.
42. K. Nakahara and R. W. Carthew (2004). Expanding roles for miRNAs and siRNAs in cell regulation, *Curr. Opin. Cell Biol.* **16**:127–133.
43. M. Scherr and M. Eder (2004). RNAi in functional genomics, *Curr. Opin. Mol. Ther.* **6**:129–135.
44. A. Goto, S. Blandin, J. Royet, et al. (2003). Silencing of Toll pathway components by direct injection of double-stranded RNA into drosophila adult flies, *Nucleic Acids Res.* **31**:6619–6623.
45. M. Santaniemi, O. Ukkola, and Y. A. Kesaniemi (2004). Tyrosine phosphatase 1B and leptin receptor genes and their interaction in type 2 diabetes, *J. Intern. Med.* **256**:48–55.
46. N. E. Caporaso (2002). Why have we failed to find the low penetrance genetic constituents of common cancers? *Cancer Epidemiol. Biomarkers Prev.* **11**:1544–1549.
47. J. L. Hartman, B. Garvik, and L. Hartwell (2001). Principles for the buffering of genetic variation, *Science* **291**:1001–1004.
48. K. Delaval and R. Feil (2004). Epigenetic regulation of mammalian genomic imprinting, *Curr. Opin. Genet. Dev.* **14**:188–195.
49. M. J. Fazzari and J. M. Greally (2004). Epigenomics: Beyond CpG islands, *Nat. Rev. Genet.* **5**:446–455.
50. E. Prokhortchouk and B. Hendrich (2002). Methyl-CpG binding proteins and cancer: Are MeCpGs more important than MBDs? *Oncogene* **21**:5394–5399.

51. C. Dennis (2003). Epigenetics and disease: Altered states, *Nature* **421**:686–688.
52. G. Egger, G. Liang, A. Aparicio, and P. A. Jones (2004). Epigenetics in human disease and prospects for epigenetic therapy, *Nature* **429**:457–463.
53. T. W. Sneider, W. M. Teague, and L. M. Rogachevsky (1975). *S*-adenosylmethionine: DNA-cytosine 5-methyltransferase from a Novikoff rat hepatoma cell line, *Nucleic Acids Res.* **2**:1685–1700.
54. J. B. Mason (2003). Biomarkers of nutrient exposure and status in one-carbon (methyl) metabolism, *J. Nutr.* **133**(Suppl 3):941S–947S.
55. P. J. Stover and C. Garza (2002). Bringing individuality to public health recommendations, *J. Nutr.* **132**:2476S–2480S.
56. G. Blander and L. Guarente (2004). The Sir2 family of protein deacetylases, *Annu. Rev. Biochem.* **73**:417–435.
57. F. Picard, M. Kurtev, N. Chung, et al. (2004). Sirt1 promotes fat mobilization in white adipocytes by repressing PPAR-gamma, *Nature* **429**:771–776.
58. I. Darnton-Hill, C. Nishida, and W. P. James (2004). A life course approach to diet, nutrition and the prevention of chronic diseases, *Public Health Nutr.* **7**:101–121.
59. Y. H. Jiang, J. Bressler, and A. L. Beaudet (2004). Epigenetics and human disease, *Annu. Rev. Genomics Hum. Genet.* **5**:479–510.
60. V. R. Young and N. S. Scrimshaw (1979). Genetic and biological variability in human nutrient requirements, *Am. J. Clin. Nutr.* **32**:486–500.
61. R. Ottman (1996). Gene–environment interaction: Definitions and study designs, *Prev. Med.* **25**:764–770.
62. E. I. Park, E. A. Paisley, H. J. Mangian, et al. (1997). Lipid level and type alter stearoyl CoA desaturase mRNA abundance differently in mice with distinct susceptibilities to diet-influenced diseases, *J. Nutr.* **127**:566–573.
63. S. Q. Ye and P. O. Kwiterovich, Jr. (2000). Influence of genetic polymorphisms on responsiveness to dietary fat and cholesterol, *Am. J. Clin. Nutr.* **72**:1275S–1284S.
64. L. F. Masson, G. Mcneill, and A. Avenell (2003). Genetic variation and the lipid response to dietary intervention: a systematic review, *Am. J. Clin. Nutr.* **77**:1098–1111.
65. J. M. Ordovas, D. Corella, S. Demissie, et al. (2002). Dietary fat intake determines the effect of a common polymorphism in the hepatic lipase gene promoter on high-density lipoprotein metabolism: Evidence of a strong dose effect in this gene–nutrient interaction in the Framingham Study, *Circulation* **106**:2315–2321.
66. J. M. Ordovas and A. H. Shen (2002). Genetics, the environment, and lipid abnormalities, *Curr. Cardiol. Rep.* **4**:508–513.
67. S. Vincent, R. Planells, C. Defoort, et al. (2002). Genetic polymorphisms and lipoprotein responses to diets, *Proc. Nutr. Soc.* **61**:427–434.
68. R. M. Krauss (2001). Dietary and genetic effects on LDL heterogeneity, *World Rev. Nutr. Diet* **89**:12–22.
69. D. M. Dreon, H. A. Fernstrom, P. T. Williams, and R. M. Krauss (1999). A very-low-fat diet is not associated with improved lipoprotein profiles in men with a predominance of large, low-density lipoproteins, *Am. J. Clin. Nutr.* **69**:411–418.
70. B. N. Ames, I. Elson-Schwab, and E. A. Silver (2002). High-dose vitamin therapy stimulates variant enzymes with decreased coenzyme binding affinity (increased K_m): Relevance to genetic disease and polymorphisms, *Am. J. Clin. Nutr.* **75**:616–658.

71. B. N. Ames (2003). The metabolic tune-up: Metabolic harmony and disease prevention, *J. Nutr.* **133**:1544S–1548S.
72. C. Courtemanche, A. C. Huang, I. Elson-Schwab, et al. (2004). Folate deficiency and ionizing radiation cause DNA breaks in primary human lymphocytes: A comparison, *FASEB J.* **18**:209–211.
73. G. A. Francis, E. Fayard, F. Picard, and J. Auwerx (2003). Nuclear receptors and the control of metabolism, *Annu. Rev. Physiol.* **65**:261–311.
74. S. Nobel, L. Abrahmsen, and U. Oppermann (2001). Metabolic conversion as a pre-receptor control mechanism for lipophilic hormones, *Eur. J. Biochem.* **268**:4113–4125.
75. S. S. Tworoger, J. Chubak, E. J. Aiello, et al. (2004). Association of CYP17, CYP19, CYP1B1, and COMT polymorphisms with serum and urinary sex hormone concentrations in postmenopausal women, *Cancer Epidemiol. Biomarkers Prev.* **13**:94–101.
76. C. A. Haiman, S. E. Hankinson, I. De Vivo, et al. (2003). Polymorphisms in steroid hormone pathway genes and mammographic density, *Breast Cancer Res. Treat.* **77**:27–36.
77. Y. Miyoshi and S. Noguchi (2003). Polymorphisms of estrogen synthesizing and metabolizing genes and breast cancer risk in Japanese women, *Biomed. Pharmacother.* **57**:471–481.
78. D. Kang (2003). Genetic polymorphisms and cancer susceptibility of breast cancer in Korean women, *J. Biochem. Mol. Biol.* **36**:28–34.
79. C. C. Linder (2001). The influence of genetic background on spontaneous and genetically engineered mouse models of complex diseases, *Lab. Anim. (NY)* **30**:34–39.
80. R. H. Waterston, K. Lindblad-Toh, E. Birney, et al. (2002). Initial sequencing and comparative analysis of the mouse genome, *Nature* **420**:520–562.
81. M. Rossmeisl, J. S. Rim, R. A. Koza, and L. P. Kozak (2003). Variation in type 2 diabetes-related traits in mouse strains susceptible to diet-induced obesity, *Diabetes* **52**:1958–1966.
82. M. L. Hribal, F. Oriente, and D. Accili (2002). Mouse models of insulin resistance, *Am. J. Physiol. Endocrinol. Metab.* **282**:E977–E981.
83. L. M. Fleeman and J. S. Rand (2001). Management of canine diabetes, *Vet. Clin. North Am. Small Anim. Pract.* **31**:855–880, vi.
84. C. K. Lardinois, T. Caudill, and G. H. Starich (1989). Dissimilar fatty acid composition of standard rat chow, *Am. J. Med. Sci.* **298**:305–308.
85. J. M. Naciff, G. J. Overmann, S. M. Torontali, et al. (2004). Impact of the phytoestrogen content of laboratory animal feed on the gene expression profile of the reproductive system in the immature female rat, *Environ. Health Perspect.* **112**:1519–1526.
86. T. S. Elliott, D. A. Swartz, E. A. Paisley, et al. (1993). F1Fo-ATPase subunit e gene isolated in a screen for diet regulated genes, *Biochem. Biophys. Res. Commun.* **190**:167–174.
87. E. A. Paisley, E. I. Park, D. A. Swartz, et al. (1996). Temporal-regulation of serum lipids and stearoyl CoA desaturase and lipoprotein lipase mRNA in BALB/cHnn mice, *J. Nutr.* **126**:2730–2737.
88. D. A. Swartz, E. I. Park, W. J. Visek, and J. Kaput (1996). The e subunit gene of murine F1F0-ATP synthase. Genomic sequence, chromosomal mapping, and diet regulation, *J. Biol. Chem.* **271**:20942–20948.

89. G. L. Wolff, R. L. Kodell, A. M. Cameron, and D. Medina (1982). Accelerated appearance of chemically induced mammary carcinomas in obese yellow (Avy/A) (BALB/c X VY) F1 hybrid mice, *J. Toxicol. Environ. Health.* **10**:131–142.
90. G. L. Wolff (1987). Body weight and cancer, *Am. J. Clin. Nutr.* **45**:168–180.
91. J. Voisey and A. Van Daal (2002). Agouti: From mouse to man, from skin to fat, *Pigment Cell Res.* **15**:10–18.
92. G. L. Wolff, D. W. Roberts, and K. G. Mountjoy (1999). Physiological consequences of ectopic agouti gene expression: The yellow obese mouse syndrome, *Physiol. Genomics* **1**:151–163.
93. M. B. Zemel (2002). Regulation of adiposity and obesity risk by dietary calcium: Mechanisms and implications, *J. Am. Coll. Nutr.* **21**:146S–151S.
94. N. M. Moussa and K. J. Claycombe (1999). The yellow mouse obesity syndrome and mechanisms of agouti-induced obesity, *Obes. Res.* **7**:506–614.
95. G. L. Wolff, R. L. Kodell, J. A. Kaput, and W. J. Visek (1999). Caloric restriction abolishes enhanced metabolic efficiency induced by ectopic agouti protein in yellow mice, *Proc. Soc. Exp. Biol. Med.* **221**:99–104.
96. F. A. N. B. Committee on Diet and Health, Commission on Life Sciences, National Research Council, Diet and Health (1989). *Implications for Reducing Chronic Disease Risk.* National Academy Press, Washington, DC.
97. L. N. Kolonel, D. Altshuler, and B. E. Henderson (2004). The multiethnic cohort study: Exploring genes, lifestyle and cancer risk, *Nat. Rev. Cancer* **4**:519–527.
98. A. J. Cameron, J. E. Shaw, and P. Z. Zimmet (2004). The metabolic syndrome: Prevalence in worldwide populations, *Endocrinol. Metab. Clin. North Am.* **33**:351–375, table of contents.
99. S. Yusuf, S. Reddy, S. Ounpuu, and S. Anand (2001). Global burden of cardiovascular diseases: Part I: general considerations, the epidemiologic transition, risk factors, and impact of urbanization, *Circulation* **104**:2746–2753.
100. L. H. Kuller (2004). Ethnic differences in atherosclerosis, cardiovascular disease and lipid metabolism, *Curr. Opin. Lipidol.* **15**:109–113.
101. J. M. Ordovas and D. Corella (2004). Nutritional genomics, *Annu. Rev. Genomics Hum. Genet.* **5**:71–118.
102. J. M. Ordovas (2004). The quest for cardiovascular health in the genomic era: Nutrigenetics and plasma lipoproteins, *Proc. Nutr. Soc.* **63**:145–152.
103. D. Corella and J. M. Ordovas (2004). The metabolic syndrome: A crossroad for genotype–phenotype associations in atherosclerosis, *Curr. Atheroscler. Rep.* **6**:186–196.
104. J. N. Hirschhorn, K. Lohmueller, E. Byrne, and K. Hirschhorn (2002). A comprehensive review of genetic association studies, *Genet. Med.* **4**:45–61.
105. K. E. Lohmueller, C. L. Pearce, M. Pike, E. S. Lander, and J. N. Hirschhorn (2003). Meta-analysis of genetic association studies supports a contribution of common variants to susceptibility to common disease, *Nat. Genet.* **33**:177–182.
106. A. Loktionov, S. Scollen, N. Mckeown, and S. A. Bingham (2000). Gene–nutrient interactions: Dietary behaviour associated with high coronary heart disease risk particularly affects serum LDL cholesterol in apolipoprotein E epsilon4-carrying free-living individuals, *Br. J. Nutr.* **84**:885–890.
107. A. Loktionov (2003). Common gene polymorphisms and nutrition: Emerging links with pathogenesis of multifactorial chronic diseases (review), *J. Nutr. Biochem.* **14**:426–451.

108. E. Lander and L. Kruglyak (1995). Genetic dissection of complex traits: Guidelines for interpreting and reporting linkage results [see comments], *Nat. Genet.* **11**:241–247.

109. N. Risch (1997). Evolving methods in genetic epidemiology. II. Genetic linkage from an epidemiologic perspective, *Epidemiol. Rev.* **19**:24–32.

110. L. R. Cardon and J. I. Bell (2001). Association study designs for complex diseases, *Nat. Rev. Genet.* **2**:91–99.

111. M. Irwin (2002). Effects of sleep and sleep loss on immunity and cytokines, *Brain Behav. Immun.* **16**:503–512.

112. L. Redwine, R. L. Hauger, J. C. Gillin, and M. Irwin (2000). Effects of sleep and sleep deprivation on interleukin-6, growth hormone, cortisol, and melatonin levels in humans, *J. Clin. Endocrinol. Metab.* **85**:3597–3603.

113. N. R. Prabhakar and Y. J. Peng (2004). Peripheral chemoreceptors in health and disease, *J. Appl. Physiol.* **96**:359–366.

114. C. N. Serhan, C. B. Clish, J. Brannon, et al. (2000). Novel functional sets of lipid-derived mediators with antiinflammatory actions generated from omega-3 fatty acids via cyclo-oxygenase 2-nonsteroidal antiinflammatory drugs and transcellular processing, *J. Exp. Med.* **192**:1197–1204.

115. M. Tomita, K. I. Irwin, Z. J. Xie, and T. J. Santoro (2002). Tea pigments inhibit the production of type 1 (T(H1)) and type 2 (T(H2)) helper T cell cytokines in CD4(+) T cells, *Phytother. Res.* **16**:36–42.

116. D. C. Nieman, J. M. Davis, D. A. Henson, et al. (2003). Carbohydrate ingestion influences skeletal muscle cytokine mRNA and plasma cytokine levels after a 3-h run, *J. Appl. Physiol.* **94**:1917–1925.

117. D. C. Nieman, J. M. Davis, V. A. Brown, et al. (2004). Influence of carbohydrate ingestion on immune changes after 2 h of intensive resistance training, *J. Appl. Physiol.* **96**:1292–1298.

118. D. C. Nieman, C. I. Dumke, D. A. Henson, et al. (2003). Immune and oxidative changes during and following the Western States Endurance Run, *Int. J. Sports Med.* **24**:541–547.

119. M. V. Chakravarthy and F. W. Booth (2004). Eating, exercise, and "thrifty" genotypes: Connecting the dots toward an evolutionary understanding of modern chronic diseases, *J. Appl. Physiol.* **96**:3–10.

120. M. Gleeson, D. C. Nieman, and B. K. Pedersen (2004). Exercise, nutrition and immune function, *J. Sports Sci.* **22**:115–125.

121. M. Irwin, C. Clark, B. Kennedy, et al. (2003). Nocturnal catecholamines and immune function in insomniacs, depressed patients, and control subjects, *Brain Behav. Immun.* **17**:365–372.

122. R. J. Pandya, G. Solomon, A. Kinner, and J. R. Balmes (2002). Diesel exhaust and asthma: Hypotheses and molecular mechanisms of action, *Environ. Health Perspect.* **110**(Suppl 1):103–112.

123. U. Albrecht and G. Eichele (2003). The mammalian circadian clock, *Curr. Opin. Genet. Dev.* **13**:271–277.

124. R. J. Seeley, D. L. Drazen, and D. J. Clegg (2004). The critical role of the melanocortin system in the control of energy balance, *Annu. Rev. Nutr.* **24**:133–149.

125. C. F. Sing, J. H. Stengard, and S. L. Kardia (2003). Genes, environment, and cardiovascular disease, *Arterioscler. Thromb. Vasc. Biol.* **23**:1190–1196.

126. T. Costacou and E. J. Mayer-Davis (2003). Nutrition and prevention of type 2 diabetes, *Annu. Rev. Nutr.* **27**:147–170.

127. P. Zimmet, K. G. Alberti, and J. Shaw (2001). Global and societal implications of the diabetes epidemic, *Nature* **414**:782–787.

128. D. A. Bennett (2003). Review of analytical methods for prospective cohort studies using time to event data: Single studies and implications for meta-analysis, *Stat. Methods Med. Res.* **12**:297–319.

129. J. Kaput, L. Allen, B. Ames, et al. (2005). The Case for Strategic International Alliances to Harness Nutritional Genomics for Public and Personal Health, *submitted.*

130. C. Chambrier, J. P. Bastard, J. Rieusset, et al. (2002). Eicosapentaenoic acid induces mRNA expression of peroxisome proliferator-activated receptor gamma, *Obes. Res.* **10**:518–525.

131. B. M. Forman, P. Tontonoz, J. Chen, et al. (1995). 15-Deoxy-delta 12, 14-prostaglandin J2 is a ligand for the adipocyte determination factor PPAR gamma, *Cell* **83**:803–812.

132. O. Nosjean and J. A. Boutin (2002). Natural ligands of PPARgamma: Are prostaglandin J(2) derivatives really playing the part? *Cell Signal* **14**:573–583.

133. C. Knouff and J. Auwerx (2004). Peroxisome proliferator-activated receptor-{gamma} calls for activation in moderation: Lessons from genetics and pharmacology, *Endocr. Rev.* **25**:899–918.

134. K. Hara, K. Tobe, T. Okada, et al. (2002). A genetic variation in the PGC-1 gene could confer insulin resistance and susceptibility to type II diabetes, *Diabetologia* **45**:740–743.

135. K. T. Suzuki and Y. Ogra (2002). Metabolic pathway for selenium in the body: Speciation by HPLC-ICP MS with enriched Se, *Food Addit. Contam.* **19**:974–983.

136. P. Talalay (2001). The importance of using scientific principles in the development of medicinal agents from plants, *Acad. Med.* **76**:238–247.

137. R. J. Blendon, C. M. Desroches, J. M. Benson, M. Brodie, and D. E. Altman (2001). Americans' views on the use and regulation of dietary supplements, *Arch. Intern. Med.* **161**:805–810.

138. D. D. Silverstein and A. D. Spiegel (2001). Are physicians aware of the risks of alternative medicine? *J. Community Health* **26**:159–174.

139. H. B. Matthews, G. W. Lucier, and K. D. Fisher (1999). Medicinal herbs in the United States: Research needs, *Environ. Health Perspect.* **107**:773–778.

140. N. Fogg-Johnson and J. Kaput (2003). Nutrigenomics: An emerging scientific discipline, *Food Technol.* **57**:61–67.

141. M. Heckler (1985). Task Force on Black and Minority Health. Report of the Secretary's Task Force on Black and Minority Health. U.S. Department of Health and Human Services, Washington, DC.

142. S. A. Black (2002). Diabetes, diversity, and disparity: What do we do with the evidence? *Am. J. Public Health* **92**:543–548.

143. A. L. Rosenbloom, D. V. House, and W. E. Winter (1998). Non-insulin dependent diabetes mellitus (NIDDM) in minority youth: Research priorities and needs, *Clin. Pediatr. (Phila.)* **37**:143–152.

144. J. Gennuso, L. H. Epstein, R. A. Paluch, and F. Cerny (1998). The relationship between asthma and obesity in urban minority children and adolescents, *Arch. Pediatr. Adolesc. Med.* **152**:1197–1200.

145. J. Sundquist, M. A. Winkleby, and S. Pudaric (2001). Cardiovascular disease risk factors among older black, Mexican-American, and white women and men: an analysis of NHANES III, 1988–1994, *J. Am. Geriatr. Soc.* **49**:109–116.

146. M. A. Winkleby, H. C. Kraemer, D. K. Ahn, and A. N. Varady (1998). Ethnic and socioeconomic differences in cardiovascular disease risk factors: Findings for women from the Third National Health and Nutrition Examination Survey, 1988–1994, *JAMA* **280**:356–362.

147. C. J. Bradley, C. W. Given, and C. Roberts (2001). Disparities in cancer diagnosis and survival, *Cancer* **91**:178–188.

148. D. Bramley, P. Hebert, R. Jackson, and M. Chassin (2004). Indigenous disparities in disease-specific mortality, a cross-country comparison: New Zealand, Australia, Canada, and the United States, *N. Z. Med. J.* **117**:U1215.

149. C. Schoen and M. M. Doty (2004). Inequities in access to medical care in five countries: findings from the 2001 Commonwealth Fund International Health Policy Survey, *Health Policy* **67**:309–322.

150. B. K. Edwards, H. L. Howe, A. G. Ries, et al. (2002). Annual Report to the Nation on the status of cancer, 1973–1999, featuring implications of age and aging on U.S. cancer burden, *Cancer* **94**:2766–2792.

151. H. R. Wabinga, D. M. Parkin, F. Wabwire-Mangen, and S. Nambooze (2000). Trends in cancer incidence in Kyadondo County, Uganda, 1960–1997, *Br. J. Cancer* **82**:1585–1592.

152. A. K. Echimane, A. A. Ahnoux, I. Adoubi, et al. (2000). Cancer incidence in Abidjan, Ivory Coast: First results from the cancer registry, 1995–1997, *Cancer* **89**:653–663.

153. D. Dawam, A. H. Rafindadi, and G. D. Kalayi (2000). Benign prostatic hyperplasia and prostate carcinoma in native Africans, *BJU Int.* **85**:1074–1077.

154. D. Bramley, P. Hebert, L. Tuzzio, and M. Chassin (2005). Disparities in indigenous health: a cross-country comparison between New Zealand and the United States, *Am. J. Public Health* **95**:844–850.

155. J. V. Neel (1962). Diabetes mellitus: a "thrifty" genotype rendered detrimental by "progress"? *Am. J. Hum. Genet.* **14**:353–362.

156. L. S. Lieberman (2003). Dietary, evolutionary, and modernizing influences on the prevalence of type 2 diabetes, *Annu. Rev. Nutr.* **23**:345–377.

157. J. D. Sachs (2005). *The End of Poverty. Economic Possibilities for Our Time.* The Penguin Press, New York.

158. L. D. Love-Gregory, J. Wasson, J. Ma, et al. (2004). A common polymorphism in the upstream promoter region of the hepatocyte nuclear factor-4 alpha gene on chromosome 20q is associated with type 2 diabetes and appears to contribute to the evidence for linkage in an Ashkenazi Jewish population, *Diabetes* **53**:1134–1140.

2

THE PURSUIT OF OPTIMAL DIETS: A PROGRESS REPORT

Walter C. Willett

Department of Nutrition, Harvard School of Public Health and Channing Laboratory, Department of Medicine, Brigham & Women's Hospital, Harvard Medical School, Boston, Massachusetts

2.1 INTRODUCTION

In this brief presentation I will attempt to summarize from the perspective of an epidemiologist what we have learned in the last ten to fifteen years about the role of nutrition in the prevention of major chronic disease. I will point out ways in which the integration of genetic information into research on diet and health is already adding to our insights.

2.2 CONSIDERATIONS IN DEFINING AN OPTIMAL DIET

Traditionally, animal experiments and small human metabolic studies formed the basis of dietary recommendations. The primary objective was to prevent signs and symptoms of nutrient deficiencies. Evidence is now clear that intakes of micronutrients that are sufficient to prevent classical clinical manifestations of deficiency may still be suboptimal for long-term optimal function and health. The most striking

Nutritional Genomics: Discovering the Path to Personalized Nutrition
Edited by Jim Kaput and Raymond L. Rodriguez

example has been the fact that the majority of neural tube defects in newborns can be prevented by supplementation of folic acid, even in populations without clinical evidence of folic acid deficiency. Many other examples relating to cancers, cardiovascular disease, fractures, and other chronic diseases have been identified.

Inevitably, the study of chronic disease in humans has required epidemiologic approaches. Until recently, these largely consisted of international comparisons and case–control studies, which examined dietary factors retrospectively in relation to cancer and other diseases. Now, large prospective studies of many thousands of persons are providing data based on both biochemical indicators of diet and dietary questionnaires that have been rigorously validated [1]. Ideally, each potential relationship between diet and a health outcome would be evaluated in a randomized trial, but this is often not feasible due to practical constraints. The best available evidence will be based on a synthesis of epidemiologic, metabolic, animal, and mechanistic studies.

For many years nutritionists have recognized that individuals differ in their response to nutrient intakes. Well-documented examples include the response of serum cholesterol to dietary cholesterol [2] or the change in blood pressure with sodium intake [3]. The elucidation of the human genome and rapid identification of polymorphisms in almost all genes are creating new opportunities to individualize dietary guidance. For example, a homozygous polymorphism in the methylenetetrahydrofolate reductase (MTHFR) gene, present in about 10% of the population, increases the amount of dietary folic acid needed to minimize blood levels of homocysteine [4]. However, this does not necessarily mean that special dietary guidance needs to be given to those persons. Although we could now easily screen for MTHFR polymorphisms and give individualized dietary advice, this is probably not a logical strategy. First, other functionally important polymorphisms in genes related to folic acid requirements will probably be discovered. Second, having different dietary advice for folic acid for different persons would create considerable complexity within populations and even within families. Because these variations probably exist for almost every nutrient, the possible combinations are almost infinite and would mean that each person would have a unique dietary recommendation. An alternative is to define healthy diets that would be sufficiently high in folic acid to meet the needs of this subset of the population. This has been the general approach in setting RDAs, whereby a margin of error has been added above average requirements to include individual variations in nutrient needs. This is an appropriate approach when variation in requirements is known to exist and we have no practical way of identifying individuals with different requirements or the reason for these differences; it will often still be a reasonable strategy even though we have the potential to identify individual differences in requirements. For treatment, individualized approaches are more readily justified because a specific pathway is perturbed and needs to be targeted, whereas in prevention many potential pathways need to be protected. Thus, for example, different dietary approaches based on genetic information may be prescribed for the treatment of hypercholesterolemia.

Even if genetic characteristics are not used to individualize dietary advice, the ability to identify individuals with different requirements will allow more detailed studies to ensure that their needs are being met by overall recommendations. Also, the integration of information on genetic polymorphisms into studies of diet and disease can enhance the power of a study and greatly aid in the interpretation of findings. As described by Ames [5], for example, the observation that a functionally important polymorphism in the MTHFR gene is associated with colon cancer provides important evidence that a relation between low folate intake and risk of colon cancer is causal.

2.3 DIETARY FAT AND SPECIFIC FATTY ACIDS

Recommendations on diet and health have until recently emphasized reductions in total fat intake, usually to 30% of energy or less [6, 7], to decrease coronary heart disease (CHD) and cancer. The classical diet–heart hypothesis has rested heavily on observations that total serum cholesterol levels predict CHD risk; serum cholesterol has thus functioned as a surrogate marker of risk in hundreds of metabolic studies. These studies, summarized as equations by Keys [8] and Hegsted [9], indicated that, compared to carbohydrates, saturated fat increases and polyunsaturated fat decreases serum cholesterol, whereas monounsaturated fat has no influence. These widely used equations, while valid for total cholesterol, have become less relevant with the recognition that the high-density lipoprotein cholesterol fraction (HDL) is strongly and inversely related to CHD risk, and that the ratio of total cholesterol to HDL is a better predictor [10–13]. Substitution of carbohydrate for saturated fat (the basis of the American Heart Association diets) tends to reduce HDL as well as total and low-density lipoprotein (LDL) cholesterol; thus, the ratio does not change appreciably [14]. In contrast, substituting monounsaturated fat for saturated fat reduces LDL without affecting HDL, thus providing an improved ratio [14].

In Keys' pioneering ecologic study of diets and CHD in seven countries [15, 16], total fat intake had little association with population rates of CHD. Indeed, the lowest rate was in Crete, which had the highest fat intake due to the large consumption of olive oil, but saturated fat intake was positively related to CHD. In contrast to international comparisons, little relationship has been seen with saturated fat intake in prospective studies of individuals [1, 17, 18]. Some studies, however, tend to support a modest association between dietary cholesterol and CHD risk [19], and inverse associations have been seen with polyunsaturated fat [17, 18]. Similarly, dietary intervention trials have generally shown little effect on CHD incidence when carbohydrate replaces saturated fat, but replacing saturated fat with polyunsaturated fat has reduced incidence of CHD [20–23]. At intakes within the dietary range, the benefits of omega-3 fatty acids appear to be primarily in the prevention of fatal arrhythmias that can complicate CHD, rather than in prevention of infarction [24–26]. The amount of omega-3 fatty acids needed to prevent arrhythmia is remarkably small—on the order of 1 g/day or less [26]—and fish consumption twice a week appears to provide most of the potential reduction of sudden death [27].

Trans fatty acids are formed by the partial hydrogenation of liquid vegetable oils in the production of margarine and vegetable shortening and can account for as much as 40% of these products. Trans fatty acids increase LDL and decrease HDL [28–33], raise the proportion of small, dense, and atherogenic LDL particles [34], raise Lp(a) [32, 35], and increase inflammatory markers that have been related to CHD risk [36, 37]. In the most detailed prospective study, trans fatty acid intake was strongly associated with risk of CHD [18] and, as predicted by metabolic studies, this association was stronger than for saturated fat. The association between trans fatty acid intake [38] and risk of CHD has been confirmed in other prospective studies. The FDA has announced that food labeling will be required to include the trans fat content as of 2006.

The relation between dietary fat and risk of type 2 diabetes appears to be similar to that for CHD [39]. The overall percentage of fat does not appear to be related to risk. However, consistent with its effect on insulin resistance, polyunsaturated fat is inversely associated with risk; and trans fat has been positively associated with risk [39], which may be explained by its effects on inflammatory markers noted above. Consumption of red meat, particularly processed red meat, has been associated with greater risk [40].

The belief that decreases in dietary fat would reduce the incidence of cancers of the breast, colon and rectum, and prostate has been one justification for low-fat diets [6, 41]. The primary evidence has been that countries with low fat intake (also the less affluent areas) have had low rates of these cancers [41, 42]. These correlations have been primarily with animal fat and meat intake, rather than with vegetable fat consumption. The hypothesis that fat intake increases breast cancer risk has been supported by most animal models [43, 44], although no association was seen in a large study that did not use an inducing agent [45]. Moreover, much of the effect of dietary fat in the animal studies appears to be due to an increase in total energy intake, and energy restriction profoundly decreases incidence [43, 45, 46]. Data from many large prospective studies, including approximately 8000 cases in over 300,000 women, have been published [47]. In none of these studies was the risk of breast cancer significantly elevated among those with the highest fat intake, and the summary relative risk for the highest vs lowest category of dietary fat composition was 1.03 [47]. In the largest study [48], no reduction in risk was seen even below 20% of energy from fat. Thus, over the range of fat intake consumed by middle-aged women in these studies, which included the present dietary recommendations, dietary total fat does not appear to increase breast cancer risk. Recently, higher intake of animal fat, particularly from dairy products, during the premenopausal years was associated with a greater risk of breast cancer. Vegetable fat was not associated with risk of breast cancer in this study, suggesting that some components of animal foods rather than fat per se may increase risk [49].

Associations between animal fat consumption and colon cancer incidence have been seen in some [50–52], although not all, studies [53], whereas little relation has been seen with vegetable fat. However, the associations between red meat consumption, particularly processed meats, and colon cancer have been even stronger than the association of fat in some analyses [51, 54]. These data suggest that relationships with red meat are due to components other than fat, such as heat-induced carcinogens [55] or the high content of readily available iron [56]. Like breast and colon cancer, prostate cancer rates are much higher in affluent compared to poor and Eastern countries [42]. More detailed epidemiologic studies are few, but associations with animal fat or red meat consumption have been suggested in some prospective studies [57, 58]. A positive association has been seen between intake of alpha-linolenic acid, primarily attributable to consumption of fat from red meat [59].

Overweight is an important cause of morbidity and mortality, and short-term studies have suggested that reducing the fat content of the diet induces weight loss [60]. However, in randomized studies lasting a year or longer, reductions in fat to 20–25% of energy had minimal effects on overall long-term body weight [61].

In summary, there is little evidence that dietary fat per se is associated with risk of CHD. Metabolic and epidemiological data are consistent in suggesting that intake of partially hydrogenated vegetable fats should be minimized. Metabolic studies, epidemiologic observations, and randomized trials support a reduction in saturated fats, but these data suggest that the benefits will be small if carbohydrate rather than unsaturated fat replaces the saturated fat. Definitive data are not available on the optimal intake of polyunsaturated and monounsaturated fats, but the metabolic data as well as the experience of Southern European populations suggest that consuming a substantial proportion of energy as monounsaturated fat would be desirable. Available evidence also suggests that total fat reduction would have little effect on breast cancer risk, although reducing red meat intake may well decrease the incidence of colon cancer and possibly prostate cancer.

2.4 CARBOHYDRATES

As protein varies only modestly across a wide range of human diets, higher carbohydrate consumption is, in practice, the reciprocal of a low-fat diet. For reasons discussed under the topic of fat, a high-carbohydrate diet can have adverse metabolic consequences. In particular, such diets are associated with an increase in triglycerides and a reduction in HDL cholesterol [13], and these adverse responses are aggravated in the context of insulin resistance [62, 63].

The traditional distinction between simple and complex carbohydrates is not useful in dietary recommendations, because some forms of complex carbohydrates, like starch in potatoes, are rapidly metabolized to glucose. Instead, emphasis is better placed on whole grain and other less-refined complex carbohydrates as opposed to the highly refined products and sugar generally consumed in the United States. Adverse consequences of highly refined grains appear to result from the rapid digestion and absorption of these foods, as well as from the loss of fiber and micronutrients in the milling process. The glycemic response after carbohydrate intake, which has been characterized by the glycemic load, is greater with highly refined foods as compared to less-refined, whole grains [64]. The greater glycemic response due to highly refined carbohydrates is accompanied by increased plasma insulin levels and augments the other adverse metabolic changes due to carbohydrate consumption noted above to a greater degree than with less-refined foods [65]. Higher intakes of refined starches and sugar, particularly when associated with low fiber intake, appear to increase the risk of noninsulin-dependent diabetes [66, 67] and possibly risk of CHD [68].

The adverse metabolic response to high intake of refined starches and sugar, which we have characterized as a high glycemic load, appears to be modified by the underlying degree of insulin resistance. This was described by Jeppesen and colleagues [63, 69], who noted that the adverse effects of high carbohydrate intake on metabolic markers of the insulin resistance syndrome were strongly correlated with baseline insulin resistance. This relation has been confirmed in population studies showing a much stronger relation between the dietary glycemic load and blood triglyceride levels [65] and risk of CHD [68] among persons with a greater body mass index, a major determinant of insulin resistance. The implication is that a person who is lean and active can better tolerate a high-carbohydrate diet than someone who is less active and overweight. This also has important implications on a population basis because of strong evidence that most Asian groups have a higher prevalence of insulin resistance for genetic reasons compared to European populations [70]. Neel [71] had earlier described this as the "thrifty gene." Until recently, these populations were generally highly active and lean, and thus protected from the adverse effects of this genetic predisposition. However, with the reductions in activity and gains in body weight that typically accompany a modern life style, the ability to tolerate a diet high in refined carbohydrates diminishes.

In contrast to refined starches, higher intake of fiber from grain products has consistently been associated with lower risks of CHD and diabetes [39, 72]. Whether these benefits are mediated by only fiber per se or in part by the accompanying micronutrients is not clear, but for practical reasons this distinction is not essential. Anticipated reductions in colon cancer risk by diets high in grain fiber have not been supported in most prospective studies [73, 74]. However, reduced constipation and risk of colonic diverticular disease [75] are clear benefits of such diets.

The importance of micronutrients in the prevention of many chronic conditions has reemphasized the problem of "empty calories" associated with diets high in sugar and highly refined carbohydrates. In the standard milling of white flour, as much as 60–90% of vitamins B_6 and E, folate, and other nutrients are lost [76]; this may be nutritionally critical for persons with otherwise marginal intakes. Thiamin, riboflavin, folate, and niacin are presently replaced by fortification, but other nutrients remain substantially reduced.

2.5 PROTEIN

Average protein consumption in the United States substantially exceeds conventional requirements [6], and adequate intake can be maintained on most reasonable diets. Optimal protein intake has been widely debated and high intakes are advocated in many popular diets, but long-term data are limited. Substituting protein for carbohydrate improves blood lipids and has been associated with lower risk of CHD [77].

The specific sources of dietary protein do have important implications for long-term health, probably more related to the other constituents of these foods than to the protein per se. As noted above, fish consumption is related to lower risk of sudden cardiac death, probably due to its content of N-3 fatty acids. Also, regular consumption of nuts has been inversely related to risk of CHD in multiple studies [78] and type 2 diabetes [79], likely due to their high content of unsaturated fatty acids and possibly also to their high content of micronutrients and other phytochemicals. Soy products are high in polyunsaturated fatty acids and would presumably be beneficial with regard to CHD risk, but little direct evidence is available, and the same applies to other legumes. Poultry fat is relatively unsaturated compared to that of red meat, which is the primary contributor to saturated fat intake in the U.S. diet. Not surprisingly, the dietary ratio of red meat to chicken plus fish has been positively related to risk of CHD [80]. As noted above, consumption of red meat, particularly processed meats, has also been related to risks of several cancers and type 2 diabetes. This extensive body of evidence supports the replacement of red meat with a combination of nuts, fish, poultry, and legumes as protein sources for overall long-term health.

2.6 VEGETABLES AND FRUITS

A generous intake of vegetables and fruits [6] has been largely justified by anticipated reductions in cancer and cardiovascular disease. However, more recent cohort studies have tended to show a much weaker—or no—relation between overall fruit and vegetable consumption and risks of common cancers [81, 82]. The possibility remains for a small overall benefit, or benefits only of specific fruits or vegetables, against specific cancers. For example, considerable evidence suggests that lycopene, mainly from tomato products, reduces risk of prostate cancer, but overall consumption of fruits is unrelated to risk [83].

In contrast to the data for cancer, the epidemiologic evidence quite consistently supports a benefit of higher intake of fruit and vegetable consumption for the prevention of cardiovascular disease [72, 82, 84]. Evidence that elevated blood homocysteine is an independent risk factor for coronary heart and cerebrovascular disease [85, 86], and that homocysteine levels can be reduced by increasing folic acid intake [87], suggest one mechanism. Evidence that the low-activity genotype of MTHFR is associated with increased risk of coronary heart disease adds indirect but important support for the benefit of higher folic acid intake [88]. High intake of vegetables reduces blood pressure [89]; potassium is likely the most important contributing factor [90].

Other benefits of higher fruit and vegetable intake are likely to include lower risk of neural tube defects, the most common severe birth defect [91], due to higher folic acid intake. Intake of the carotenoids lutein and zeaxanthin, which are high in green leafy vegetables, has been inversely related to risk of cataracts [92, 93].

2.7 CALCIUM AND DAIRY PRODUCTS

Recommendations to maintain adequate calcium intake [6] and to consume large amounts of dairy products on a daily basis [94] derive primarily from the importance of calcium in maintaining bone strength. Calcium supplements when combined with vitamin D have reduced fracture incidence in some studies [95], but benefits of calcium cannot be distinguished from those of vitamin D. Uncertainty remains regarding the optimal calcium intake. Intakes of 1200 mg/day or higher are recommended for those over 50 years of age in the United States [96], whereas a more recent review in the United Kingdom concluded that 700 mg/day was adequate [97]. Many populations have low fracture rates, despite minimal dairy product consumption and low overall calcium intake by adults [98]. Several large prospective studies have directly addressed the relation of dairy product consumption to fracture incidence; higher consumption of calcium or dairy products as an adult has consistently not been associated with lower fracture incidence [99, 100]. At best, the benefits of high calcium intake are minor compared with those from regular physical activity [101, 102] or additional vitamin D [99].

Inverse associations have been reported between calcium intake and blood pressure in some studies [103], but in a review of supplementation trials little overall effect was seen [104]. Low calcium and low dairy product consumption is associated with a modestly elevated risk of colon cancer [105], but most benefits appear to be achieved with calcium intake of about 800 mg/day. Evidence from a randomized trial in which calcium supplementation modestly reduced colon adenoma recurrence adds important evidence of causality to the epidemiologic studies [106].

Although recommended calcium intakes can be achieved by a high consumption of greens and certain other vegetables, greatly increased intakes would be required for most women to achieve currently recommended levels by diet without regular use of milk and other dairy products. However, calcium supplements are an inexpensive form of calcium without accompanying calories or saturated fat. Thus, dairy product consumption can be considered an optional rather than a necessary dietary component. Enthusiasm regarding high dairy consumption should also be tempered by the suggestion in many studies that this is associated with increased risks of advanced or fatal prostate cancer [107, 108]. Whether an increased risk is due to the calcium, endogenous hormones, or other factors in milk remains uncertain.

2.8 SALT AND PROCESSED MEATS

Reduction of salt (sodium chloride) intake from an average of approximately 8–10 g/day to less than 6 g/day will, on average, modestly decrease blood pressure. In a comprehensive review, Law et al. [109] estimated that a 3 g/day reduction would translate to a reduction in the incidence of stroke by 22% and of CHD by 16%. Although the decrease in risk of cardiovascular disease achieved by reducing salt consumption is modest for most individuals, the overall number of deaths potentially avoided is large, supporting policies to reduce salt consumption, particularly in processed foods and by institutions.

In many case–control studies, the consumption of salty and pickled foods has been associated with stomach cancer [58]. However, as this cancer is relatively rare in the United States, further benefit from reducing salt intake would be small.

2.9 ALCOHOL

Many adverse influences of heavy alcohol consumption are well recognized, but moderate consumption has both beneficial and harmful effects, greatly complicating decisions for individuals. Overwhelming epidemiologic data indicate that moderate consumption (1–2 drinks/day) reduces risk of myocardial infarction [110, 111] by approximately 30–40%. Although this effect has been hypothesized to be the result of antioxidants in red wine, similar protective effects for equivalent amounts of alcohol have been seen for all types of alcoholic beverages [112, 113]. The relation between alcohol and cardiovascular disease provides another example where information on genotype can contribute to the interpretation of epidemiologic evidence. Specifically, the inverse relation between alcohol consumption and risk of myocardial infarction was seen primarily among persons with the slow metabolizing genotype of alcohol dehydrogenase-3 [114], which provides further evidence that alcohol per se is the primary protective factor in alcoholic beverages.

On the other hand, modest positive associations with risk of breast cancer incidence have been observed in more than 30 studies [115] for similar levels of alcohol intake, possibly in part because alcohol appears to increase endogenous estrogen levels [116, 117]. The overall effect of alcohol, as represented by total mortality, appears beneficial for up to about 2 drinks/day in men [118]. Overall, a similar relation with total mortality is seen among women, but no net benefit was observed among those at low risk of CHD because of younger age or lack of coronary risk factors [119]. Several studies suggest that the adverse effects of alcohol on cancer risk may be mitigated by adequate intake of folate [120].

2.10 VITAMIN AND MINERAL SUPPLEMENTS

The role of vitamin supplements and food fortification has been debated on both a philosophical and scientific level. Some nutritionists have believed as a matter of principle that nutritional needs should be met by diet alone, and the U.S. Dietary Guidelines are set based on this philosophy. However, often this is not possible, for example, when iodine levels are low in the soil, and iodine fortification has been a great public health advance. Also, a large percentage of the U.S. population appears to have suboptimal blood levels of vitamin D, largely due to limited solar exposure at northern latitudes during the winter. In addition, low incomes and limited access can be serious barriers to optimal food intakes; to achieve the recommended 400 μg/day of folic acid by diet alone can be difficult for many low-income groups. Many of these shortcomings can be remedied efficiently and effectively by some combination of fortification and supplementation.

Only recently have data become available that address the effects of vitamin supplements against the background of actual diets in the United States, which appear far from ideal [121]. The most firmly established benefit is that folic acid supplements in the amounts contained in multiple vitamins can reduce the risks of neural tube defects by approximately 70% [122]. This is probably an indicator of more widespread consequences of suboptimal folate intakes; inverse associations have also been seen with colonic neoplasias [123] and breast cancer [120]; and low folate intake, along with suboptimal vitamin B_6, is likely to contribute to elevated blood homocysteine levels and risk of cardiovascular disease [86, 124]. Since 1998, grain products in the United States have been fortified with folic acid; the extent to which this has reduced the value in taking supplemental folic acid is unclear, but there are still some Americans who have intakes below the RDA of 400 μg/day [125]. Many elderly persons have suboptimal vitamin B_{12} status, due mainly to loss of stomach acid, which is needed to liberate vitamin B_{12} from food sources. In contrast, vitamin B_{12} in supplements or from fortified food sources is readily absorbed without stomach acid.

The majority of the U.S. population appears to have suboptimal blood levels of vitamin D, as assessed by their relation to bone density [126]. That is, vitamin D levels in blood are directly and nearly linearly related to bone density, suggesting that the majority of Americans do not have optimal vitamin D levels. As natural sources of vitamin D are few (mainly fish), and recommendations to increase sun exposure would lead to elevations in skin cancer, increases in supplementation and fortification are the only viable options. The current RDA of 400 μg/day is far too low to achieve optimal levels, and intakes of at least several times this amount are probably desirable [127].

In a randomized trial conducted in a region of China with low consumption of fruits and vegetables, a supplement containing beta-carotene, vitamin E, and selenium reduced incidence of stomach cancer [128]. In a recent study conducted among Tanzanian women infected with HIV, a multiple vitamin containing B vitamins, vitamin E, and vitamin C reduced progression of the disease and HIV-related mortality [129]. Whether these benefits would be seen in the background of dietary intakes in the United States is not clear.

Any recommendation for use of nutritional supplements should carefully consider possible adverse effects. One of the few adverse effects of vitamin supplement use at the RDA level appears to be an increase in risk of hip fractures due to vitamin A, when consumed as 5000 IU/day in the form of retinol. Higher intake of preformed vitamin A (retinol) has been associated with excess risk of hip fracture in prospective studies [130, 131], and elevated risks were seen for both use of multiple vitamins and specific supplements of vitamin A. Also, serum retinol levels have been associated with future risk of fractures [132]. These effects may be due to competition at the vitamin D receptor [133] and might not have occurred if vitamin D levels were adequate. The amount of retinol in most multiple vitamins has been reduced.

Current evidence, although far from complete, suggests that supplements of folate and possibly other vitamins, at the RDA level contained in most nonprescription multivitamin preparations, may have substantial benefits for at least an important subgroup of the U.S. population, perhaps characterized in part by increased requirements or by suboptimal diets. As intakes of folate as well as other micronutrients appear marginal for many Americans [121, 134], the risks of using multivitamins appear minimal; and as the cost of supplements is low (especially compared to that of fresh fruits and vegetables), the use of a daily multiple vitamin appears rational for the majority of Americans. This case has been made earlier by Ames [135], and recent data have added further support for this strategy. Continued research to optimize the contents of multiple vitamin preparations is important.

2.11 POTENTIAL IMPACT OF OPTIMAL DIET AND LIFE-STYLE CHANGES

The various lines of evidence described above have identified many aspects of diet that can have important health benefits. From comparisons of disease rates in different countries and the experience of populations who migrate from low- to high-risk countries, we know that the potential for disease prevention by nongenetic factors is large. For example, in Keys' Seven Countries study the rates of heart disease in Crete and rural Japan were only about one-tenth those of the United States, and rates of colon cancer in the past also differed to a similar degree [42]. The specific aspects of life style, and whether such changes are practically attainable in a modern society, cannot be determined from such data. We have therefore conducted a series of analyses in our large cohort studies to address these questions.

To identify the potential for prevention of myocardial infarction or death due to coronary heart disease, we defined a low-risk group within the Nurses' Health Study using variables for which the evidence of causality is strong [136]. Specifically, the low-risk group was not to currently smoke, have a BMI below 25 kg/m^2, have 30 min/day of moderate to vigorous physical activity, be in the upper half of a healthy diet score (based on low trans fat, high P:S fatty acid ratio, high whole grain intake, fish intake twice a week or more, and on attaining the RDA for folic acid), and drink 5 grams or more per day of alcohol. Although these all involved quite modest changes, we found that only 3.1% of the Nurses' Health Study was in this low-risk group. However, based on 14 years of follow-up, we calculated that over 80% of coronary heart disease in this population could have been prevented had the whole group adopted these behaviors. Most of these same variables (with the exception of folic acid and fish intake) are also related to risk of type 2 diabetes, and we found that over 90% of type 2 diabetes could potentially have been avoided by adoption of these behaviors [137]. In a similar analysis that also included consumption of red meat, we found that over 70% of colon cancer cases were potentially avoidable [138]. Thus, the potential for disease prevention by modest dietary and life-style changes that are readily compatible with life in the 21st century is enormous. Further refinements, for example, by including adequate vitamin D intake, should further improve our ability to prevent major diseases.

2.12 CONCLUSION

Any description of a healthy diet must be made with the recognition that information is currently incomplete and conclusions are subject to change with new data. Most of the major diseases contributing to morbidity and mortality in the United States develop over many decades, and large-scale nutritional epidemiologic studies have only begun in the last 25 years; thus, a full picture of the relation between diet and disease will require additional decades of careful investigation. Nevertheless, in combining available metabolic, clinical, and epidemiologic evidence, several general conclusions that are unlikely to change substantially can be drawn.

1. Staying lean and active throughout life will have major health benefits. Because most members of developed countries work at sedentary jobs, weight control will usually require conscious daily physical activity, as well as some effort to avoid overconsumption of calories.
2. Dietary fats should be primarily in the form of nonhydrogenated plant oils. Butter and lard, and fat from red meat should be used sparingly, if at all, and trans fatty acids from partially hydrogenated vegetable oils should be minimized.
3. Grains should be consumed primarily in a minimally refined, high-fiber form, and intakes of refined starches and simple sugars should be low.
4. Vegetables and fruits should be consumed in abundance (five servings/day is minimal) and should include green leafy and orange vegetables daily.
5. Red meat should be consumed only occasionally and in low amounts if at all; nuts, legumes, poultry, and fish in moderation are healthy alternatives.
6. The optimal consumption of dairy products and calcium intake is not known, but dairy products should be considered as optional. High consumption of milk (e.g., more than two servings/day) is not likely to be beneficial for middle-aged and older adults and may increase risk of prostate cancer. Adequate calcium intake may be particularly important for growing children, adolescents, and lactating women; supplements should be considered if dietary sources are low.
7. For most people, taking a daily RDA-level (DV) multiple vitamin containing folic acid provides a sensible nutritional safety net. Because menstrual losses of iron may not be adequately replaced by iron intake on the low-energy diets of women in a sedentary society, it is sensible for most premenopausal women to use a multiple vitamin that also contains iron.
8. Salt intake should be kept low.

Although the exact contribution of each aspect of diet will become clearer as evidence emerges, we can be confident that the overall benefits of adopting these behaviors will be large. The integration of genomic information into research on diet and health has only begun and will surely provide additional insight and more precise nutritional guidance.

ACKNOWLEDGMENT

This chapter is presented in honor of Bruce Ames. Dr. Ames' vision, imagination, generosity, and humor remain unsurpassed in all of American science. Like many others here, I feel extremely fortunate to know Bruce Ames and to benefit continuously from his thoughts.

REFERENCES

1. W. C. Willett (1998). *Nutritional Epidemiology*, 2nd edition. Oxford University Press, New York.
2. M. B. Katan, A. C. Beynen, J. H. de Vries, et al. (1986). Existence of consistent hypo- and hyperresponders to dietary cholesterol in man, *Am. J. Epidemiol.* **123**:221–234.
3. E. Beeks, A. G. Kessels, A. A. Kroon, et al. (2004). Genetic disposition to salt-sensitivity: A systematic review, *J. Hypertens.* **22**:1243–1249.
4. L. B. Bailey and J. F. Gregory, 3rd. (1999). Polymorphisms of methylenetetrahydrofolate reductase and other enzymes: Metabolic significance, risks and impact on folate requirement, *J. Nutr.* **129**:919–922.
5. B. N. Ames (1999). Cancer prevention and diet: Help from single nucleotide polymorphisms. *Proc. Natl. Acad. Sci.* **96**:12216–12218.
6. National Research Council—Committee on Diet and Health (1989). *Diet and Health: Implications for Reducing Chronic Disease Risk*, National Academy Press, Washington, DC.
7. Department of Health and Human Services (1988). *The Surgeon General's Report on Nutrition and Health*, Government Printing Office, Washington, DC. (DHHS publication [PHS] 50210).
8. A. Keys (1984). Serum-cholesterol response to dietary cholesterol, *Am. J. Clin. Nutr.* **40**:351–359.
9. D. M. Hegsted (1986). Serum-cholesterol response to dietary cholesterol: A re-evaluation, *Am. J. Clin. Nutr.* **44**:299–305.
10. W. P. Castelli, R. D. Abbott, and P. M. McNamara (1983). Summary estimates of cholesterol used to predict coronary heart disease, *Circulation* **67**:730–734.
11. H. N. Ginsberg, S. L. Barr, A. Gilbert, et al. (1990). Reduction of plasma cholesterol levels in normal men on an American Heart Association Step 1 diet or a Step 1 diet with added monounsaturated fat, *N. Engl. J. Med.* **322**:574–579.
12. R. P. Mensink and M. B. Katan (1987). Effect of monounsaturated fatty acids versus complex carbohydrates on high-density lipoprotein in healthy men and women, *Lancet* **1**:122–125.

13. R. P. Mensink and M. B. Katan (1992). Effect of dietary fatty acids on serum lipids and lipoproteins: A meta-analysis of 27 trials, *Arterioscler. Thromb.* **12**:911–919.
14. R. P. Mensink, P. L. Zock, A. D. Kester, et al. (2003). Effect of dietary fatty acids and carbohydrates on the ratio of serum total to HDL cholesterol and on serum lipids and apolipoproteins: A meta-analysis of 60 controlled trials, *Am. J. Clin. Nutr.* **77**: 1146–1155.
15. A. Keys (1980). *Seven Countries: A Multivariate Analysis of Death and Coronary Heart Disease.* Harvard University Press, Cambridge, MA.
16. W. M. Verschuren, D. R. Jacobs, B. P. Bloemberg, et al. (1995). Serum total cholesterol and long-term coronary heart disease mortality in different cultures. Twenty-five year follow-up of the Seven Countries Study. *JAMA* **274**:131–136.
17. R. B. Shekelle, A. M. Shryock, O. Paul, et al. (1981). Diet, serum cholesterol, and death from coronary disease: The Western Electric Study, *N. Engl. J. Med.* **304**:65–70.
18. F. Hu, M. J. Stampfer, J. E. Manson, et al. (1997). Dietary fat intake and the risk of coronary heart disease in women, *N. Engl. J. Med.* **337**:1491–1499.
19. R. B. Shekelle and J. Stamler (1989). Dietary cholesterol and ischemic heart disease, *Lancet* **1**:1177–1179.
20. Multiple Risk Factor Intervention Trial Research Group (1982). Multiple Risk Factor Intervention Trial: Risk factor changes and mortality results, *JAMA* **248**:1465–1477.
21. J. Stamler, D. Wentworth, and J. D. Neaton (1986). Is the relationship between serum cholesterol and risk of premature death from coronary heart disease continuous and graded? Findings in 356,222 primary screenees of the Multiple Risk Factor Intervention Trial (MRFIT), *JAMA* **256**:2823–2828.
22. I. D. J. Frantz, E. A. Dawson, P. L. Ashman, et al. (1989). Test of effect of lipid lowering by diet on cardiovascular risk: The Minnesota Coronary Survey, *Arteriosclerosis* **9**:129–135.
23. F. Sacks (1994). Dietary fats and coronary heart disease. Overview, *J. Cardiovasc. Risk* **1**:3–8.
24. M. de Lorgeril, S. Renaud, N. Mamelle, et al. (1994). Mediterranean alpha-linolenic acid-rich diet in secondary prevention of coronary heart disease [Erratum in: *Lancet* (1995), **345**:738], *Lancet* **343**:1454–1459.
25. A. Leaf (1995). Omega-3 fatty acids and prevention of ventricular fibrillation, *Prostaglandins Leukot. Essent. Fatty Acids* **52**:197–198.
26. GISSI–Prevention Investigators (1999). Dietary supplementation with n-3 polyunsaturated fatty acids and vitamin E after myocardial infarction: Results of the GISSI-Prevenzione trial, *Lancet* **354**:447–455.
27. C. M. Albert, C. H. Hennekens, C. J. O'Donnell, et al. (1998). Fish consumption and risk of sudden cardiac death, *JAMA* **279**:23–28.
28. J. Booyens and C. C. Louwrens (1986). The Eskimo diet: Prophylactic effects ascribed to the balanced presence of natural *cis* unsaturated fatty acids and to the absence of unnatural *trans* and *cis* isomers of unsaturated fatty acids, *Med. Hypoth.* **21**:387–408.
29. R. P. M. Mensink and M. B. Katan (1990). Effect of dietary *trans* fatty acids on high-density and low-density lipoprotein cholesterol levels in healthy subjects, *N. Engl. J. Med.* **323**:439–445.
30. P. L. Zock and M. B. Katan (1992). Hydrogenation alternatives: Effects of *trans* fatty acids and stearic acid versus linoleic acid on serum lipids and lipoproteins in humans, *J. Lipid Res.* **33**:399–410.

31. J. T. Judd, B. A. Clevidence, R. A. Muesing, et al. (1994). Dietary trans fatty acids: Effects of plasma lipids and lipoproteins on healthy men and women, *Am. J. Clin. Nutr.* **59**:861–868.
32. P. Nestel, M. Noakes, and B. Belling (1992). Plasma lipoprotein and Lp[a] changes with substitution of elaidic acid for oleic acid in the diet, *J. Lipid. Res.* **33**:1029–1036.
33. K. Sundram, A. Ismail, K. C. Hayes, et al. (1997). *Trans* (elaidic) fatty acids adversely affect the lipoprotein profile relative to specific saturated fatty acids in humans, *J. Nutr.* **127**:514S–520S.
34. A. H. Lichtenstein, L. M. Ausman, S. M. Jalbert, et al. (1999). Effects of different forms of dietary hydrogenated fats on serum lipoprotein cholesterol levels in moderately hypercholesterolemic females and male subjects, *N. Engl. J. Med.* **340**:1933–1940.
35. R. P. Mensink, P. L. Zock, M. B. Katan, et al. (1992). Effect of dietary cis and trans fatty acids on serum lipoprotein [a] levels in humans, *J. Lipid Res.* **33**:1493–1501.
36. D. Mozaffarian, T. Pischon, S. E. Hankinson, et al. (2004). Dietary intake of trans fatty acids and systemic inflammation in women, *Am. J. Clin. Nutr.* **79**:606–612.
37. D. J. Baer, J. T. Judd, B. A. Clevidence, et al. (2004). Dietary fatty acids affect plasma markers of inflammation in healthy men fed controlled diets: A randomized crossover study, *Am. J. Clin. Nutr.* **79**:969–973.
38. A. Ascherio, M. B. Katan, P. L. Zock, et al. (1999). Trans fatty acids and coronary heart disease, *N. Engl. J. Med.* **340**:1994–1998.
39. F. B. Hu, R. M. van Dam, and S. Liu (2001). Diet and risk of Type II diabetes: The role of types of fat and carbohydrate, *Diabetologia* **44**:805–817.
40. R. M. van Dam, W. C. Willett, E. B. Rimm, et al. (2002). Dietary fat and meat intake in relation to risk of type 2 diabetes in men, *Diabetes Care* **25**:417–424.
41. R. L. Prentice and L. Sheppard (1990). Dietary fat and caner. Consistency of the epidemiologic data, and disease prevention that may follow from a practical reduction in fat consumption, *Cancer Causes Control* **1**:81–97.
42. B. Armstrong and R. Doll (1975). Environmental factors and cancer incidence and mortality in different countries, with special reference to dietary practices, *Int. J. Cancer* **15**:617–631.
43. C. Ip (1990). Quantitative assessment of fat and calorie as risk factors in mammary carcinogenesis in an experimental model. In: C. J. Mettlin and K. Aoki eds. *Recent Progress in Research on Nutrition and Cancer*, proceedings of a workshop sponsored by the International Union Against Cancer, held in Nagoya, Japan, November 1–3, 1989. Wiley-Liss, Hoboken, NJ, pp. 107–117.
44. L. S. Freedman, C. Clifford, and M. Messina (1990). Analysis of dietary fat, calories, body weight, and the development of mammary tumors in rats and mice: A review, *Cancer Res.* **50**:5710–5719.
45. B. S. Appleton and R. E. Landers (1986). Oil gavage effects on tumor incidence in the National Toxicology Program's 2-year carcinogeneis bioassay, *Adv. Exp. Med. Biol.* **206**:99–104.
46. C. W. Welsch (1992). Relationship between dietary fat and experimental mammary tumorigenesis: A review and critique, *Cancer Res.* **52**(Suppl 7):2040S–2048S.
47. S. A. Smith-Warner, D. Spiegelman, H. O. Adami, et al. (2001). Types of dietary fat and breast cancer: A pooled analysis of cohort studies, *Int. J. Cancer* **92**:767–774.
48. M. D. Holmes, D. J. Hunter, G. A. Colditz, et al. (1999). Association of dietary intake of fat and fatty acids with risk of breast cancer, *JAMA* **281**:914–920.

49. E. Cho, D. Spiegelman, D. J. Hunter, et al. (2003). Premenopausal fat intake and risk of breast cancer, *J. Natl. Cancer Inst.* **95**:1079–1085.
50. A. S. Whittemore, A. H. Wu-Williams, M. Lee, et al. (1990). Diet, physical activity and colorectal cancer among Chinese in North America and China, *J. Natl. Cancer Inst.* **82**:915–926.
51. W. C. Willett, M. J. Stampfer, G. A. Colditz, et al. (1990). Relation of meat, fat, and fiber intake to the risk of colon cancer in a prospective study among women, *N. Engl. J. Med.* **323**:1664–1672.
52. E. Giovannucci, E. B. Rimm, A. Ascherio, et al. (1995). Alcohol, low-methionine–low-folate diets, and risk of colon cancer in men, *J. Natl. Cancer Inst.* **87**:265–273.
53. R. L. Phillips and D. A. Snowdon (1983). Association of meat and coffee use with cancers of the large bowel, breast, and prostate among Seventh-Day Adventists: Preliminary results, *Cancer Res.* **43**(Suppl):2403S–2408S.
54. T. Norat, A. Lukanova, P. Ferrari, et al. (2002). Meat consumption and colorectal cancer risk: Dose-response meta-analysis of epidemiological studies, *Int. J. Cancer* **98**:241–256.
55. M. Gerhardsson de Verdier, U. Hagman, R. K. Peters, et al. (1991). Meat, cooking methods and colorectal cancer: A case-referent study in Stockholm, *Int. J. Cancer* **49**:520–525.
56. C. F. Babbs (1990). Free radicals and the etiology of colon cancer, *Free Radic. Biol. Med.* **8**:191–200.
57. L. N. Kolonel (1996). Nutrition and prostate cancer, *Cancer Causes Control* **7**:83–94.
58. World Cancer Research Fund, American Institute for Cancer Research (1997). *Food, Nutrition and the Prevention of Cancer: A Global Perspective.* American Institutue for Cancer Research, Washington, DC.
59. E. Giovannucci, E. B. Rimm, M. J. Stampfer, et al. (1994). Intake of fat, meat, and fiber in relation to risk of colon cancer in men, *Cancer Res.* **54**:2390–2397.
60. G. A. Bray and B. M. Popkin (1998). Dietary fat intake does affect obesity!, *Am. J. Clin. Nutr.* **68**:1157–1173.
61. W. C. Willett and R. L. Leibel (2002). Dietary fat is not a major determinant of body fat, *Am. J. Med.* **113**(Suppl 9B):47S–59S.
62. J. Jeppesen, C. B. Hollenbeck, M. Y. Zhou, et al. (1995). Relation between insulin resistance, hyperinsulemia, postheparin plasma lipoprotein lipase activity, and postprandial lipemia, *Arterioscler. Thromb. Vasc. Biol.* **15**:320–324.
63. J. Jeppesen, Y. D. I. Chen, M. Y. Zhou, et al. (1995). Effect of variations in oral fat and carbohydrate load on postprandial lipemia, *Am. J. Clin. Nutr.* **61**:787–791.
64. D. J. Jenkins, T. M. Wolever, R. H. Taylor, et al. (1981). Glycemic index of foods: A physiological basis for carbohydrate exchange, *Am. J. Clin. Nutr.* **34**:362–366.
65. S. Liu, J. E. Manson, M. J. Stampfer, et al. (2001). Dietary glycemic load assessed by food frequency questionnaire in relation to plasma high-density lipoprotein cholesterol and fasting triglycerides among postmenopausal women, *Am. J. Clin. Nutr.* **73**:560–566.
66. J. Salmeron, J. E. Manson, M. J. Stampfer, et al. (1997). Dietary fiber, glycemic load, and risk of non-insulin-dependent diabetes mellitus in women, *JAMA* **277**:472–477.
67. J. Salmeron, A. Ascherio, E. B. Rimm, et al. (1997). Dietary fiber, glycemic load, and risk of NIDDM in men, *Diabetes Care* **20**:545–550.

68. S. Liu, W. C. Willett, M. J. Stampfer, et al. (2000). A prospective study of dietary glycemic load, carbohydrate intake, and risk of coronary heart disease in U.S. women, *Am. J. Clin. Nutr.* **71**:1455–1461.
69. J. Jeppesen, Y. D. Chen, M. Y. Zhou, et al. (1995). Postprandial triglyceride and retinyl ester responses to oral fats effects of fructose, *Am. J. Clin. Nutr.* **62**:1201–1205.
70. S. Dickinson, S. Colagiuri, E. Faramus, et al. (2002). Postprandial hyperglycemia and insulin sensitivity differ among lean young adults of different ethnicities, *J. Nutr.* **132**: 2574–2579.
71. J. Neel (1962). Diabetes mellitus: A "thrifty" geneotype rendered detrimental by "progress,"? *Am. J. Hum. Genet.* **14**:353–362.
72. F. B. Hu and W. C. Willett (2002). Optimal diets for prevention of coronary heart disease, *JAMA* **288**:2569–2578.
73. C. S. Fuchs, G. A. Colditz, M. J. Stampfer, et al. (1999). Dietary fiber and the risk of colorectal cancer and adenoma in women, *N. Engl. J. Med.* **340**:169–176.
74. P. Terry, E. Giovannucci, K. B. Michels, et al. (2001). Fruit, vegetables, dietary fiber, and the risk of colorectal cancer, *J. Natl. Cancer. Inst.* **93**:525–533.
75. W. H. Aldoori, E. L. Giovannucci, H. R. Rockett, et al. (1998). A prospective study of dietary fiber types and symptomatic diverticular disease in men, *J. Nutr.* **128**:714–719.
76. H. A. Schroeder (1971). Losses of vitamins and trace minerals resulting from processing and preservation of foods, *Am. J. Clin. Nutr.* **24**:562–573.
77. F. B. Hu, M. J. Stampfer, J. E. Manson, et al. (1999). Dietary protein and risk of ischemic heart disease in women, *Am. J. Clin. Nutr.* **70**:221–227.
78. F. B. Hu and M. J. Stampfer (1999). Nut consumption and risk of coronary heart disease: A review of epidemiologic evidence, *Curr. Atheroscler. Rep.* **1**:204–209.
79. R. Jiang, J. E. Manson, M. J. Stampfer, et al. (2002). Nut and peanut butter consumption and risk of type 2 diabetes in women, *JAMA* **288**:2554–2560.
80. F. B. Hu, M. J. Stampfer, J. E. Manson, et al. (1999). Dietary saturated fats and their food sources in relation to the risk of coronary heart disease in women, *Am. J. Clin. Nutr.* **70**:1001–1008.
81. K. B. Michels, E. Giovannucci, K. J. Joshipura, et al. (2002). Fruit and vegetable consumption and colorectal cancer incidence, *IARC Sci. Publ.* **156**:139–140.
82. H.-C. Hung, K. Joshipura, R. Jiang, et al. (2004). Fruit and vegetable intake and risk of major chronic disease, *J. Natl. Cancer Inst.* **96**(21):1577–1584.
83. E. Giovannucci (1999). Tomatoes, tomato-based products, lycopene, and cancer: Review of the epidemiologic literature, *J. Natl. Cancer Inst.* **91**:317–331.
84. K. J. Joshipura, F. B. Hu, J. E. Manson, et al. (2001). The effect of fruit and vegetable intake on risk of coronary heart disease, *Ann. Intern. Med.* **134**:1106–1114.
85. M. J. Stampfer, M. R. Malinow, W. C. Willett, et al. (1992). A prospective study of plasma homocyste(e)ine and risk of myocardial infarction in US physicians, *JAMA* **268**:877–881.
86. J. Selhub, P. F. Jacques, A. G. Bostom, et al. (1995). Association between plasma homocysteine concentrations and extracranial carotid-artery stenosis, *N. Engl. J. Med.* **332**:286–291.
87. K. L. Tucker, B. Olson, P. Bakun, et al. (2004). Breakfast cereal fortified with folic acid, vitamin B-6, and vitamin B-12 increases vitamin concentrations and reduces homocysteine concentrations: A randomized trial, *Am. J. Clin. Nutr.* **79**:805–811.

88. M. Klerk, P. Verhoef, R. Clarke, et al. (2002). MTHFR 677C → T polymorphism and risk of coronary heart disease. A meta-analysis [Review], MTHFR Studies Collaboration Grp **288**:2023–2031.
89. F. M. Sacks, L. P. Svetkey, W. M. Vollmer, et al. (2001). Effects on blood pressure of reduced dietary sodium and the dietary approaches to stop hypertension (DASH) diet, *N. Engl. J. Med.* **344**:3–10.
90. F. M. Sacks, W. C. Willett, A. Smith, et al. (1998). Effect on blood pressure of potassium, calcium, and magnesium in women with low habitual intake, *Hypertension* **31**:131–138.
91. M. M. Werler, S. Shapiro, A. A. Mitchell, et al. (1993). Periconceptional folic acid exposure and risk of ocurrent neural tube defects, *JAMA* **269**:1257–1261.
92. L. Chasan-Taber, W. C. Willett, J. M. Seddon, et al. (1999). A prospective study of carotenoid and vitamin A intakes and risk of cataract extraction in US women, *Am. J. Clin. Nutr.* **70**:509–516.
93. L. Brown, E. B. Rimm, J. M. Seddon, et al. (1999). A prospective study of carotenoid intake and risk of cataract extraction in US men, *Am. J. Clin. Nutr.* **70**:517–524.
94. S. Welsh, C. Davis, and A. Shaw (1992). Development of the food guide pyramid, *Nutr. Today* **27**:12–23.
95. M. C. Chapuy, M. E. Arlof, F. Duboeuf, et al. (1992). Vitamin D3 and calcium to prevent hip fractures in elderly women, *N. Engl. J. Med.* **327**:1637–1642.
96. Standing Committee on the Scientific Evaluation of Dietary Reference Intakes (1997). *Dietary Reference Intakes for Calcium, Phosphorus, Magnesium, Vitamin D, and Fluoride*. Institute of Medicine, Washington, DC.
97. Scientific Advisory Committee on Nutrition (2002). Key dietary recommendation. SACN, Annual Report, p. 15.
98. B. E. C. Nordin (1966). International patterns of osteoporosis, *Clin. Orthop.* **45**: 17–20.
99. D. Feskanich, W. C. Willett, and G. A. Colditz (2003). Calcium, vitamin D, milk consumption, and hip fractures: A prospective study among postmenopausal women, *Am. J. Clin. Nutr.* **77**:504–511.
100. K. Michaelsson, H. Melhus, R. Bellocco, et al. (2003). Dietary calcium and vitamin D intake in relation to osteoporotic fracture risk, *Bone* **32**:694–703.
101. D. Feskanich, W. Willett, and G. Colditz (2002). Walking and leisure-time activity and risk of hip fracture in postmenopausal women, *JAMA* **288**:2300–2306.
102. C. A. C. Wickham, K. Walsh, C. Cooper, et al. (1989). Dietary calcium, physical activity, and risk of hip fracture: A prospective study, *Br. Med. J.* **299**:889–892.
103. D. A. McCarron, C. D. Morris, H. J. Henry, et al. (1984). Blood pressure and nutrient intake in the United States, *Science* **224**:1392–1398.
104. J. A. Cutler and E. Brittain (1990). Calcium and blood pressure: An epidemiologic perspective, *Am. J. Hypertens.* **3**:137S–146S.
105. E. Cho, S. Smith-Warner, D. Spiegelman, et al. (2004). Dairy foods and calcium and colorectal cancer: A pooled analysis of 10 cohort studies, *Nat. Cancer Inst.* In press.
106. J. A. Baron, M. Beach, J. S. Mandel, et al. (1999). Calcium supplements for the prevention of colorectal adenomas. The Calcium Polyp Prevention Study Group, *N. Engl. J. Med.* **340**:101–107.
107. E. Giovannucci, E. B. Rimm, A. Wolk, et al. (1998). Calcium and fructose intake in relation to risk of prostate cancer, *Cancer Res.* **58**:442–447.

108. E. Giovannucci (2002). Nutritional and environmental epidemiology of prostate cancer. In: P. W. Kantoff, P. R. Carroll, and A. V. D'Amico, eds. *Prostate Cancer: Principles and Practice.* Lippincott/Williams & Wilkins, Philadelphia, pp. 117–139.
109. M. R. Law, C. D. Frost, and N. J. Wald (1991). By how much does dietary salt reduction lower blood pressure? III-Analysis of data from trials of salt reduction, *Br. Med. J.* **302**:819–824.
110. A. L. Klatsky, M. A. Armstrong, and G. D. Friedman (1990). Risk of cardiovascular mortality in alcohol drinkers, ex-drinkers, and nondrinkers, *Am. J. Cardiol.* **66**: 1237–1242.
111. E. B. Rimm, E. L. Giovannucci, W. C. Willett, et al. (1991). Prospective study of alcohol consumption and risk of coronary disease in men, *Lancet* **338**:464–468.
112. L. M. Hines and E. B. Rimm (2001). Moderate alcohol consumption and coronary heart disease: A review, *Postgrad. Med. J.* **77**:747–752.
113. K. J. Mukamal, K. M. Conigrave, M. A. Mittleman, et al. (2003). Roles of drinking pattern and type of alcohol consumed in coronary heart disease in men, *N. Engl. J. Med.* **348**:109–118.
114. L. M. Hines, M. Stampfer, J. Ma, et al. (2001). Genetic variation in alcohol dehydrogenase and the beneficial effect of moderate alcohol consumption on myocardial infarction, *N. Engl. J. Med.* **344**:549–555.
115. S. A. Smith-Warner, D. Spiegelman, S.-S. Yaun, et al. (1998). Alcohol and breat cancer in women: A pooled analysis of cohort studies, *JAMA* **279**:535–540.
116. M. E. Reichman, J. T. Judd, C. Longcope, et al. (1993). Effects of alcohol consumption on plasma and urinary hormone concentrations in premenopausal women, *J. Natl. Cancer Inst.* **85**:722–727.
117. S. E. Hankinson, W. C. Willett, J. E. Manson, et al. (1995). Alcohol, height, and adiposity in relation to estrogen and prolactin levels in postmenopausal women, *J. Natl. Cancer Inst.* **87**:1297–1302.
118. P. Boffetta and L. Garfinkel (1990). Alcohol drinking and mortality among men enrolled in a American Cancer Society prospective study, *Epidemiology* **1**:342–348.
119. C. S. Fuchs, M. J. Stampfer, G. A. Colditz, et al. (1995). Alcohol consumption and mortality among women, *N. Engl. J. Med.* **332**:1245–1250.
120. S. M. Zhang, W. C. Willett, J. Selhub, et al. (2003). Plasma folate, vitamin B6, vitamin B12, and homocysteine and risk of breast cancer, *J. Natl. Cancer Inst.* **95**:373–380.
121. G. Block and B. Abrams (1993). Vitamin and mineral status of women of childbearing potential, *Ann. N. Y. Acad. Sci.* **678**:244–254.
122. MRC Vitamin Study Research Group (1991). Prevention of neural tube defects: Results of the Medical Research Council Vitamin Study, *Lancet* **338**:131–137.
123. E. Giovannucci (2002). Epidemiologic studies of folate and colorectal neoplasia: A review, *J. Nutr.* **132**:2350S–2355S.
124. E. B. Rimm, W. C. Willett, F. B. Hu, et al. (1998). Folate and vitamin B6 from diet and supplements in relation to risk of coronary heart disease among women, *JAMA* **279**:359–364.
125. S. F. Choumenkovitch, J. Selhub, P. W. F. Wilson, et al. (2002). Folic acid intake from fortification in United States exceeds predictions, *J. Nutr.* **132**:2792–2798.
126. H. A. Bischoff-Ferrari, T. Dietrich, E. J. Orav, et al. (2004). Positive association between 25-hydroxy vitamin D levels and bone mineral density: A population-based study of younger and older adults, *Am. J. Med.* **116**:634–639.

127. M. F. Holick (2004). Vitamin D: Importance in the prevention of cancers, type 1 diabetes, heart disease, and osteoporosis, *Am. J. Clin. Nutr.* **79**:362–371.

128. W. J. Blot, J. Y. Li, P. R. Taylor, et al. (1993). Nutrition intervention trials in Linxian, China: Supplementation with specific vitamin/mineral combinations, cancer incidence, and disease-specific mortality in the general population, *J. Natl. Cancer Inst.* **85**:1483–1492.

129. W. W. Fawzi, G. I. Msamanga, D. Spiegelman, et al. (2004). A randomized trial of multivitamin supplements and HIV disease progression and mortality, *N. Engl. J. Med.* **351**:23–32.

130. D. Feskanich, V. Singh, W. C. Willett, et al. (2002). Vitamin A intake and hip fractures among postmenopausal women, *JAMA* **287**:47–54.

131. H. Melhus, K. Michaelsson, A. Kindmark, et al. (1998). Excessive dietary intake of vitamin A is associated with reduced bone mineral density and increased risk for hip fracture, *Ann. Intern. Med.* **129**:770–778.

132. K. Michaelsson, H. Lithell, B. Vessby, et al. (2003). Serum retinol levels and the risk of fracture, *N. Eng. J. Med.* **348**:287–294.

133. S. Johansson and H. Melhus (2001). Vitamin A antagonizes calcium response to vitamin D in man, *J. Bone Miner. Res.* **16**:1899–1905.

134. G. Block, B. Patterson, and A. Subar (1992). Fruits, vegetables, and cancer prevention: A review of the epidemiological evidence, *Nutr. Cancer* **18**:1–29.

135. B. N. Ames (2003). The metabolic tune-up: Metabolic harmony and disease prevention, *J. Nutr.* **133**:1544S–1548S.

136. M. J. Stampfer, F. B. Hu, J. E. Manson, et al. (2000). Primary prevention of coronary heart disease in women through diet and lifestyle, *N. Engl. J. Med.* **343**:16–22.

137. F. B. Hu, J. E. Manson, M. J. Stampfer, et al. (2001). Diet, lifestyle, and the risk of type 2 diabetes mellitus in women, *N. Engl. J. Med.* **345**:790–797.

138. E. A. Platz, W. C. Willett, G. A. Colditz, et al. (2000). Proportion of colon cancer risk that might be preventable in a cohort of middle-aged US men, *Cancer Causes Control* **11**:579–588.

3

GENE–ENVIRONMENT INTERACTIONS: DEFINING THE PLAYFIELD

Jose M. Ordovas[1] and Dolores Corella[1,2]

[1] *Nutrition and Genomics Laboratory, Jean Mayer–U.S. Department of Agriculture Human Nutrition Research Center on Aging, Tufts University, Boston, Massachusetts*

[2] *Genetic and Molecular Epidemiology Unit, School of Medicine, University of Valencia, Valencia, Spain*

3.1 INTRODUCTION

All species encounter multiple biological challenges for the duration of their existence. Some adapt successfully, while many are not as effective, leading to their extinction. Although still young as a species (~10,000 generations), humans appear to have been able to deal with scores of primarily environmental challenges that include climate, food supply, and societal changes. Moreover, despite new infectious agents and increased incidence of age-related disorders such as cancer, osteoporosis, and cardiovascular and neurological diseases, humans today "enjoy" their longest life expectancy. The major age-related disorders that affect contemporary humans involve interactions of multiple genetic and environmental factors and nonhuman genomes such as bacteria and viruses [1]. The notion of health, however, should not be equated with the lack of disease. The World Health Organization defined health in 1947 as the state of complete physical, mental, and social well-being, and not

Nutritional Genomics: Discovering the Path to Personalized Nutrition
Edited by Jim Kaput and Raymond L. Rodriguez

merely the absence of disease. This concept highlights that health is an expression of the total conditions of human life: genetic, chemical, physical, environmental, cultural, economic, psychosocial, and mental. Accordingly, from a holistic point of view, health has also been defined as the product of a perfect interaction between the human being and his/her environment. Aside from formal definitions, it is becoming evident that the way we feel about our health is also an important factor when examining the interplay between the individual's genome and his/her environment, adding a layer of complexity to the interactions. From a pragmatic point of view, Hubert Laframboise [2] proposed that, for public health policies, the many interactive factors that influence health and disease could be grouped in more manageable clusters, identified with the term "determinants of health." Laframboise defined four major groups determining the "health field". These groups are: *human biology, environment, life style,* and *health care organization.* Figure 3.1 shows these four quadrants of the health field as well as the adaptation of the Laframboise model of health determinants to include the current concept of gene X–environment interactions. The Laframboise model was applied in the highly influential Lalonde Report of 1974 [3]. According to these authors, "*human biology* was the determinant involving all aspects of health, physical and mental, developed within the human body as result of organic make-up." *Environment* referred to "all matters related to health external to the human body and over which the individual has little or no control," which includes physical (air pollution, ionizing radiations, water contamination, electromagnetic fields, temperature, microorganisms, chemicals, etc.) and the social environment. *Life style* was the most important determinant of health and was

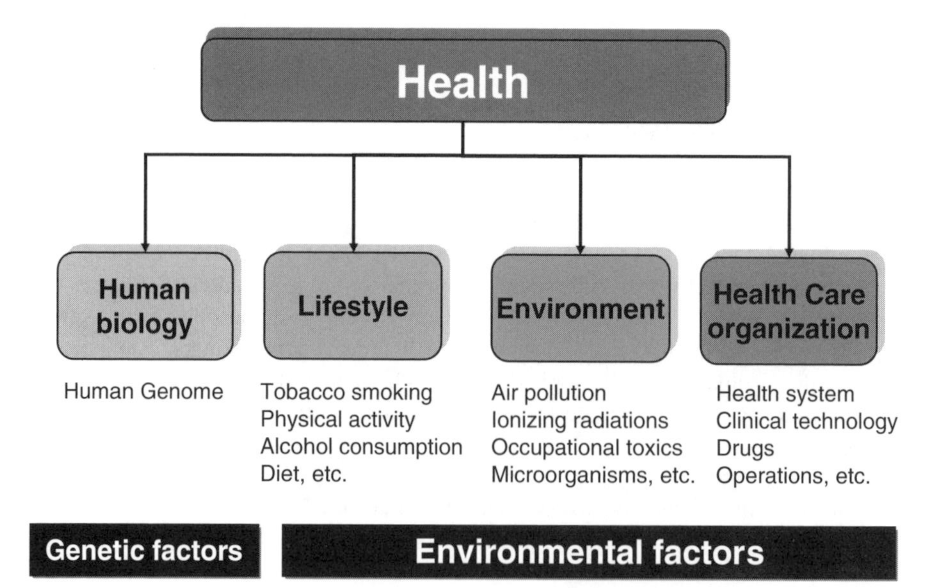

Figure 3.1. The four major determinants of health according to Laframboise [2] and the adjustment to the most simplified model of gene–environment interaction.

defined as "the aggregation of personal decisions, over which the individual has control" (tobacco smoking, physical activity, alcohol consumption, dietary intake, etc.). Unhealthy life-style choices were considered as self-imposed risks. The *health care organization* included "the quantity, quality, arrangement, nature and relationships of people and resources in the provision of health care" (hospital beds, health technology, specialized units, diagnostics, treatments, etc.). Human biology was the determinant less defined by Laframboise because of the lack of information about the human genome in the 1970s. The name of this determinant changed several times from "basic human biology" to "endogenous" and, finally, to "human biology." Laframboise included the genetic inheritance and all the complex internal body systems in this category in order to create a conceptual home for the many health conditions, good or bad, which are not a consequence of life style or environmental factors [2]. The scientific achievements of the past 30 years in genetics and molecular biology have produced an unprecedented volume of genetic information to be integrated in the so-called human biology determinant of health. The Human Genome Project has provided the consensus sequence that comprises the human genome [4]. Using this information as the springboard, the following steps include the structuring and deciphering of the 3 billion base sequence in order to gain insights into its functional role. Moreover, great emphasis is being placed on the search for DNA sequence variations underlying common/complex diseases. As expected, the genetic architecture of these traits shows tremendous complexity, and the discovery and characterization of susceptibility alleles constitute a real challenge for the geneticist. Nonetheless, following the model of determinants proposed by Laframboise [2], it is important to realize that life style and environment as well as health-care organization can mediate the effect of genes and influence health and disease. Accordingly, in the postgenomic era, a simplified model of determinants of health is proposed. In this simplified model, only genetic and environmental factors are considered (Figure 3.1). Genetic factors refer to the genome (human biology), whereas environmental factors refer to nongenetic factors and include the other three determinants of the Laframboise model (life style, environment, and health-care organization). Thus, the different outcomes or intermediate phenotypes of health and disease are the results of complex interactions between genetic and environmental factors. Currently, the research is still at a formative stage with only a few of these genetic risk factors or gene–environment interactions positively identified, but the paradigm is established and the case for further investment is compelling.

As indicated above, a major goal of human genetic research is to understand the genetic factors predisposing or causing disease and to develop effective predictive markers and appropriate therapies. Under the umbrella of this major goal, there are two emerging and related disciplines known as pharmacogenetics and nutrigenetics. Their scope and goals have been reviewed [5]. The focus of this chapter is to highlight the challenges that lie ahead of us before we can reach the proposed goal of improving global health by focusing on the individual. Four areas will be discussed and illustrated with some examples from the cardiovascular disease field. First, we will concentrate on the variation in the human genome. Second, we will focus on the environmental aspects. The third part will be the reunion of the first and second parts in the gene–environment interaction. Finally, we will touch upon the underdeveloped area of the interaction with the gut flora or microbiome.

3.2 GENETIC VARIABILITY

The language and nomenclature for describing gene variants (mutations or polymorphisms) is inconsistent among published reports and therefore confusing. A mutation involves any change in the hereditary material—from a point mutation to a chromosomal loss. In some disciplines, the term "mutation" is used to indicate this change, while in other disciplines mutation indicates a disease-causing change. Similarly, the term "polymorphism" is used both to indicate a non-disease-causing change or a change found at a higher frequency in the population. To prevent this confusion some authors [6] recommend using the terms "sequence variant" or "allelic variant" for any genomic change regardless of the frequency of phenotypic effects. However, in large epidemiological studies, the term polymorphism is still used as it was originally defined (a locus in which the least common allele occurs with a frequency equal to or higher than 1%). Thus, polymorphisms can be caused by mutations ranging from a single nucleotide base change to variations in several hundred bases, such as large deletions and/or insertions. These ancestrally "old" mutations that occurred in one or a few individuals were subsequently spread throughout the population. The simpler and most common type of polymorphism, known as single nucleotide polymorphism (SNP), results from a single base mutation replacing one base for another. Other types of genetic polymorphisms result from the insertion or deletion of a section of DNA, which include microsatellite repeat sequences and gross genetic losses and rearrangements. SNPs are estimated to represent about 90% of all human DNA polymorphisms. SNPs can result from either the transition or transversion of nucleotide bases. Transitions are substitutions between A and G (purines) or between C and T (pyrimidines). Transversions are substitutions between a purine and a pyrimidine. SNPs are scattered throughout the genome and are found in both coding and noncoding regions. Nucleotide substitutions occurring in protein-coding regions can also be classified as synonymous and nonsynonymous according to their effect on the resulting protein. A substitution is synonymous if it causes no amino acid change, while a nonsynonymous substitution results in alteration in the encoded amino acid. The latter type can be further classified into missense and nonsense mutations. A missense mutation results in amino acid changes due to the change of codon used, while a nonsense mutation results in a termination codon. SNPs occur with a very high frequency, with estimates ranging from about 1 in 1000 bases to 1 in 100–300 bases. Overall, it has been estimated that the human genome contains about ten million SNPs [7], many of them already available in public databases such as dbSNP, provided by the National Center for Biotechnology Information (NCBI); HGVbase (Human Genome Variation database), which is produced by a European consortium; or the Human SNP Database, provided by the Center for Genome Research at the Whitehead Institute for Biomedical Research in Cambridge, Massachusetts.

SNP discovery/detection differs conceptually and experimentally from SNP analyses (also called SNP genotyping or genotyping). One strives to identify new SNPs in the genome, while the other involves methods to determine the genotypes of one to many individuals for particular SNPs that have already been discovered. Once a SNP is discovered, validated in many chromosomes from ancestrally diverse individuals, and it is deemed to have some potential interest as a genetics marker, association studies are conducted to assess statistical associations between the SNP and the specific phenotypes known or suspected to be affected by the gene variant. In the examples used herein, the goal is to identify relevant genetic loci that influence lipid metabolism. Until recently, the number of identified polymorphisms in candidate genes for lipid metabolism was scarce and considerable effort went into SNP discovery. Dozens of studies focused on a very limited number of specific genetic variants in different populations and subsets of patients. This provided the opportunity to check the replication of the observed associations and potentially increase the level of evidence. A case in point is the association between *APOE* polymorphisms and plasma LDL-C concentrations in different populations worldwide in which, in most studies, carriers of the E4 allele have higher LDL-C concentrations than subjects homozygotic for the E3 allele, whereas carriers of the E2 allele have much lower LDL-C concentrations [8].

Currently, the vast number of SNPs available for candidate genes can be seen as an advantage, but a level of complexity is added to the data interpretation as more and more investigators focus association studies on different SNP sets even within a single gene locus. Humphries et al. [9] have strongly recommended distinguishing between functional (i.e., altering an amino sequence or a transcription factor binding element) and nonfunctional SNPs, avoiding the use of nonfunctional genetic variants in association studies. They suggested that before inclusion in an epidemiological study, every SNP used should have good in vitro data supporting its functionality. However, as these in vitro data may be difficult to obtain, some researchers have developed and/or used bioinformatic tools to infer functionality [10]. Although several simulation programs have been designed to predict whether a nonsynonymous SNP is likely to have a functional effect on the phenotype, Tchernitchko et al. [11] evaluated some of these predicted functional SNPs and suggested caution from these in silico predictions. This warning obviously applies to those SNPs present in putative transcription sites and other regulatory elements. One of the problems with the results obtained from association studies using apparently nonfunctional SNPs, which in some cases may be markers of an unknown functional variant, is (potentially) differences in the degree of linkage disequilibrium with the true functional SNP among different ancestral populations [12]. Thus, a nonfunctional variant may have different impact in association studies depending on the linkage disequilibrium with the functional polymorphism, contributing to the lack of replication of these studies in different populations, especially those encompassing different ethnic groups. One such example is the *HindIII* intronic polymorphism of the LPL gene. This SNP was associated with lower HDL-C and higher TAG concentrations in healthy men and women from Iceland [13]. However, Larson et al. [14] did not find an association in men and women from North America. Lack of replication is the most excruciating problem in the hundreds of published reports of gene–lipid associations involving lipid metabolism [15]. Moreover, publication bias tilts the balance toward results showing significant associations, prolonging the survival of spurious initial results.

In addition to linkage disequilibrium, a very important cause of variability between studies is the potential influence of environmental factors. Since the classical definition of phenotype as observable physical or biochemical characteristics of an organism determined by both genetic makeup and environmental influences, including descriptions of environmental factors that include diet, it is of critical importance for genetic association sudies. Hence, controversial and nonreplicable findings from studies involving genotype–phenotype association are likely due to the lack of inclusion of gene–environment interactions, which still occurs even in the most recent studies and publications [16].

3.3 HOW TO DETECT GENETIC VARIABILITY

Most of the past and many of the current SNP detection techniques involve the analyses of one genetic marker, or at most a few genetic markers, per DNA sample. However, at this time about 3 million SNPs are deposited in the public databases. These resources now allow a much more comprehensive linkage analysis and candidate gene association and interaction studies in complex diseases. Therefore, accurate, efficient, flexible, and low-cost SNP genotyping techniques are becoming vital—first for the identification of informative markers and second for their use for further screening or for clinical applications. Although important at every stage of scientific experimentation and assay development, accuracy is a must for clinical implementation of gene–nutrient interactions.

Unlike DNA sequencing technology, SNP genotyping techniques come in many flavors [17], including differences in the hybridization dynamics [7, 18], enzymatic discrimination using ligation [19, 20], cleavase assay [21], and polymerase incorporation of allele-specific nucleotides to discriminate allele differences [22]. Some of these genotyping systems can be multiplexed, such as microarray and flow cytometry methods. The experimental approaches are drastically different for array and microtiter plate platforms. The arrays allow for all SNPs of interest to be analyzed in one experiment for each sample [23], and they provide the capacity needed for high-throughput projects, albeit at very high cost, especially if the arrays are custom made for the SNPs of interest. Most commercial arrays available in 2004 are designed for mapping studies (i.e., candidate locus identification) using mainly noncoding SNPs. Conversely, the current microtiter plate methods typically determine one genotype at a time and usually focus on coding or noncoding SNPs within the gene locus. The microtiter plate format provides more flexibility in designing the project, requires less upfront cost, and is more appropriate for small to medium size projects. However, they cannot provide the same number of data points per sample as those based on arrays. Several recent papers have compared different platforms [24, 25]; however, it is not clear that any one platform can be globally recommended, as the decision is dependent on laboratory facilities, budget, and scope of the project. These methodological advances have eliminated genotyping as the once major technical challenge and bottleneck of genetic epidemiology. Indeed, even academic laboratories can process thousands of DNA samples per day.

As discussed above, misclassification of genotyping results is of critical importance and can occur due to several factors. Among the most obvious sources of error are laboratory based: the tremendous flow of data coming from high-throughput techniques requires effective quality control procedures in the laboratory. A misclassification of the genotype (i.e., a data set containing less than 95% reproducibility) can bias the measure of the association and largely affect the risk assessment and, even more markedly, the study of gene–environment interactions. Quality control procedures should be reported in the Methods section and should include information about internal validation, blinding, duplicates, test failure rate, assessment of Hardy–Weinberg equilibrium, and blind of data entry [26].

3.4 WHAT TO ANALYZE

Data from the Human Genome and HapMap projects present major challenges for present and future studies in selecting which SNPs to assay after balancing cost, time, and information generated: in other words, finding the "allelic architecture" of human disease [27]. The choice of methodology and SNPs depends on choosing one of several hypotheses that predict the genetic contribution to complex disease. Two of these hypotheses are the focus of most of the attention [28]. The first is known as the "common disease/common variant" (CD/CV) hypothesis; the second is known as the "multiple rare variants" (MRVs) hypothesis (see discussion in [29]).

The CD/CV hypothesis implies that much of the genetic variation contributing to disease is old and shared by most human populations. Under this premise, differences in the health status of population groups (health disparities) should primarily be the result of differences in environmental exposures. Conversely, according to the MRVs hypothesis, a substantial proportion of genetic polymorphisms will be rare and will probably be specific to groups who experienced similar evolutionary forces of selection or genetic drift. MRVs would require a much more comprehensive sampling of multiple human populations to show that a differential distribution of genes underlies the disease risk. Most probably, these hypotheses are not exclusive and the true answer will be a combination of both with more weight toward one or the other depending on the specific disease being considered. For those disorders in which the MRVs model is more appropriate, the current efforts of the HapMap project, which involves 270 DNA samples from Utah, Nigeria, Tokyo, and Beijing, will probably not be comprehensive enough to explain the complex interplay between multiple genetic variants and multiple environmental factors in disease etiology across all global populations.

These uncertainties about the "allelic architecture" of human disease also affect haplotype tagging SNPs (htSNPs), haplotypes, and haplotype block (a discrete chromosome region of restricted haplotype diversity) structures [30, 31] since they are derived and interpreted using algorithms [32–39] rather than from empirical data [30, 40, 41].

3.5 ENVIRONMENTAL FACTORS

Before achieving our current level of understanding of the human genome, most of the emphasis in terms of evaluating disease risk was placed on the environmental factors. Gender and age are the main risk factors for cardiovascular disease (CVD), while the best known environmental risk factors include diet, alcohol, smoking, and physical activity. These variables are the most commonly examined in gene–environment interaction studies. However, they may just be the tip of the iceberg of environmental exposures that modulate CVD risk by themselves or, most probably, by interacting with alleles of specific causative and modifier genes. A relatively old comprehensive examination of the factors involved in CVD risk revealed 177 CVD risk factors classified in ten categories [42]. The categories can be sorted as follows:

- 33 nutrition-related CVD risk factors.
- 35 internal CVD risk factors identifiable by laboratory tests (abnormal blood and tissue chemistry tests).
- 34 related to drug, chemical, hormonal, and nutritional supplement intake (including drug–drug interaction and drug–food interaction).
- 33 were classified as signs and symptoms associated with high CVD incidence.
- 13 noninvasively detectable abnormal laboratory findings.
- 5 hereditary factors.
- 14 environmental factors (other than nutritional), including air pollution, electromagnetic fields, materials that contact the body surface, and poisonous venoms.
- 7 socioeconomic and demographic factors.
- 2 related to medical care.
- 1 related to coexistence of multiple CVD risk factors.

In addition, an eleventh category included seven factors previously regarded as CVD risk factors that are now considered questionable. Considering the progress made in the decade since this compilation was made, one can predict that an update of this information will include a considerable number of additional factors, especially for those with a hereditary basis or environmental exposures in utero or during growth and development. This enumeration of CVD risk factors illustrates the potential quagmire of trying to use the current knowledge about gene–environment interactions and disease outcomes to provide personalized risk assessment, disease prevention, and therapy.

To date, our laboratory has taken the piecemeal approach of examining one variant (i.e., −75 G/A) at a single candidate gene (*APOA1*) interacting with a single dietary component (i.e., PUFA) to modulate a single biochemical risk factor (i.e., HDL-C) in a single ethnic group (i.e., Framingham whites) [43]. These first steps have uncovered major limitations, with the lack of replication as one of the most important and worrisome. The current approach and results are far from the intended goal of providing personalized holistic disease risk assessment and behavioral advice to achieve optimal health. Our path to achieve the goal of personalized nutrition has to include large genetic epidemiological studies in which data collection is done more carefully and comprehensively than has been done in most published studies [44]. This is not trivial. Data gathering problems exist even for traditional epidemiological studies as highlighted by a recent study that analyzed the shortcomings about data collection, presentation of results, and interpretation of the analyses [45]. We have identified the following limitations of particular concern:

1. The participant selection process (i.e., information on exclusions and refusals) often lacks details.
2. The quality of data collected and any problems therein are often insufficiently described.
3. Some studies are too small and may be prone to exaggerated claims, while few provide power calculations to justify sample size.
4. Quantitative exposure variables are commonly grouped into ordered categories, but few state the rationale for choice of grouping and analyses.
5. The terminology for estimates of association (i.e., the term "relative risk") is used inconsistently.
6. Confidence intervals are in widespread use but are presented excessively in some articles.
7. P values are used more sparingly, but there is a tendency to overinterpret arbitrary cutoffs such as $P < 0.05$.
8. The selection of and adjustment for potential confounders need greater clarity, consistency, and explanation.
9. Subgroup analyses to identify effect modifiers often lack appropriate methods (i.e., interaction tests) and are frequently overinterpreted.
10. Studies exploring many associations tend not to consider the increased risk of false positive findings.
11. The epidemiological literature seems prone to publication bias.
12. There are insufficient epidemiological studies and therefore publications on diseases other than cancer and cardiovascular diseases.
13. There are not enough studies of chronic diseases in developing countries where subphenotype (e.g., Asians are considered overweight at a different BMI than Europeans) and genetic differences may find different pathways and gene associations.

14. Overall, there is a serious risk that some epidemiological publications reach misleading conclusions.

It should be noted that these caveats were identified in 73 epidemiological studies, and that only 5 included genetic data. The specific needs surrounding genetic studies, especially those examining gene–environment interactions, should add a number of items to the list above, such as some of those suggested by Hall [46].

15. A full, three-generation family history with particular emphasis on the sign and symptoms of the disorder and features of pathogenic relevance.
16. Natural history of the participant including other illnesses, pregnancy and birth history, and system-by-system review, including behavior, with changes over time in the areas of interest, noting the ages at which signs or symptoms develop.

Hall [46] suggests another important issue: it's "now or never" for including as many physiological and life-style variables in the design of future studies. She points out that if full clinical descriptions have not been collected and recorded for the study participant at the time of DNA collection, it probably never will be collected due to the difficulty of recontacting people. Although the full set of clinical information may not find its way into the printed version of the manuscript due to space constraints, it should be a standard feature of the journals to have it accessible as supplementary web-based information. The lack of this information may seriously hinder the ability to advance our understanding of biology and disease.

An aspect that was not specifically considered by most researchers is the very early environmental exposure that some have suggested may determine future health of the individual. According to the current knowledge and thought, details about environmental influences during gestation and following birth could contribute significantly to a more complete picture of the complex puzzle of chronic disease development (see below). Capturing such data in the near term may not be possible, but an awareness of these potentially important contributions to long-term health risks necessitates efforts to include such variables in study designs.

3.6 GENE–ENVIRONMENT INTERACTIONS: FOCUS ON DIET

Measurement of Gene–Diet Interaction

The nature vs nurture controversy in disease risk is being replaced by systematic evaluation of gene–environment interactions that are the basis of the discussion. Although the methods used in genetics and epidemiology to study gene–environment interactions continue to evolve, the most used approaches are the unmeasured genotype approach (genetic epidemiology) and measured genotype approach (molecular epidemiology).

The unmeasured genotype approach is based on statistical analysis of the distribution of phenotypes in individuals and families and does not rely on any direct measure of DNA variation [47]. The purpose of such studies is to provide clues to the importance of genetic factors embodied in the concept of heritability and measured by analysis of variance or correlation. No specific genes are measured at this stage. Heritability is defined as the proportion of phenotypic variance attributable to genetic variance. A value of 1.0 indicates that all of the population variation is attributable to genetic variation. A value of 0.0 indicates that genes do not contribute at all to phenotypic individual differences. The quantity (1.0 – heritability) gives the environmental contribution to the trait, and it is the proportion of phenotypic variance attributable to environmental variance or the extent to which individual differences in the environment contribute to individual differences in behavior. There are several twin and family studies analyzing the genetic and environmental influence on lipid traits. As an example, Heller et al. [48] examined both genetic and environmental influences on serum lipid levels in twins reared either together (158 pairs) or apart (146 pairs) who participated in the Swedish Adoption/Twin Study of Aging. Statistical analyses revealed substantial heritability for the serum levels of each lipid measured (total cholesterol, HDL-C, TAG, apo A-I and apo B), ranging from 0.28 to 0.78. In addition, they concluded that the environment of rearing had a substantial impact on the level of total cholesterol (accounting for 0.15–0.36 of the total variance).

With advances in molecular genetics, the study of gene–environment interactions is now performed at the gene level. With this approach, genetic variations in one or in various candidate genes are investigated for their role in determining gene–diet interactions. Direct measurement of DNA variants (i.e., SNPs) is needed to give precision and meaning to the genetic effects in gene–diet interactions. The concept of gene–diet interactions can be used in several different ways [49, 50]. For example, statistical interaction exists if the degree or direction of the effect of one dietary factor (e.g., dietary fat) differs according to values of a second factor (gene variants). In recent years, an epidemiological framework for evaluating gene–diet interaction has been proposed [51]. In a simple gene–diet interaction model, in which both the susceptibility genotype at a single locus and the dietary exposure are considered dichotomous, one can construct an extended 2-by-2 table incorporating genetic and environment factors in studying disease etiology (e.g., hypercholesterolemia). The interaction of these two factors can be measured as a departure from a multiplicative model of disease risk [52, 53]. Using such an approach, it is relatively easy to find parsimonious models by keeping statistical interactions to a minimum. However, in the study of lipid metabolism, continuous variables such as plasma cholesterol concentrations and continuous measures of dietary intake are analyzed rather than just dichotomous traits. Then, in addition to a logistic regression model, covariance regression models with interaction terms are employed [52]. Hierarchical multilevel interaction models can then be fitted, and the presence of a gene–diet interaction is identified by the detection of a statistically significant interaction term between the dietary exposure and the corresponding genetic variable. However, from a strict point of view, it is not clear what ultimately constitutes interaction or synergy on the biological level and any statistical approach is inherently arbitrary and model dependent [49].

Methodological Issues in Assessing Gene–Diet Interaction

Gene–diet interactions in epidemiological studies are currently detected based on the statistical significance of the interaction terms. Taking into account that a statistically significant association does not imply causality, special attention to bias is required. This issue is crucial and determines the study design as well as adequate control of misspecification. Interventional studies and particularly randomized controlled trials offer the highest level of confidence in data and results [54]. In interventional study designs, the environmental variables are controlled by the investigators, including the amount of food or nutrient administered and the time period. This control minimizes the possibility of bias and increases the level of causality. However, in observational studies (i.e., cohort, case–control, cross–sectional), the researcher does not provide food to study participants and obtains the information about the type and amount of food consumed by dietary questionnaires. Such measurement tools are widely used [5] but they increase the likelihood of confounding and decrease the evidence level of data and results. The large populations needed for such studies have led some authors to suggest the importance of Mendelian randomization (the random assortment of genes from parents to offspring that occurs during gamete formation and conception) in increasing the level of evidence when genetic polymorphisms and dietary intake are analyzed [55]. Mendelian randomization can increase the level of evidence of observational studies because this results in population distributions of genetic variants that are generally independent of environmental factors that typically confound epidemiological associations between putative risk factors and disease. In some circumstances, this can provide a study design akin to randomized comparisons. In fact, Tobin et al. [56] have suggested that the approach should be termed "Mendelian deconfounding."

Misclassification of Dietary Exposures

Precise measurement of an individual's exposure to dietary factors is often difficult because of the individual's poor recall of previous exposures, the complex pattern of nutrient intake in today's cultures, and the lack of good biological indicators of exposure levels for many nutrients. Therefore, in the study of gene–diet interaction, the consequences of a dietary mismeasurement can lead to bias in the estimation associated with interaction effects and possible loss of precision and power in the estimation of interactions. Nondifferential misclassification usually leads to biases of interaction toward the null value, and differential misclassification may produce biased interactions in either direction (increasing or decreasing risk). Therefore, high-quality dietary information in epidemiological studies is a key factor for improving causality. More information about validity of the specific dietary instruments (i.e., diet records, 24 hour recalls, food frequency questionnaires) can be found in [5]. Another important question regards the type of dietary information that is most relevant in gene–diet interaction studies. Should we be using nutrients, foods, or dietary patterns? There is no unique response to this question, and the choice depends on the aims of the study and on the specific hypothesis being investigated.

Each one of these approaches has advantages and methodological limitations [57]. The acquisition of the many other environmental variables (Figure 3.1) that may affect physiological processes is at least as problematic as the capture of dietary information.

Sample Size Considerations

One of the most frequent errors in even well-designed and well-conducted epidemiological studies aimed at investigating gene–diet interactions in lipid metabolism is the lack of statistical power (Type II error). Power can be defined as the ability of a test to detect an effect, given that the effect actually exists. In an epidemiological study of a given sample size, the power to detect statistical interactions is less than the power to detect main effects, and the variance of the interaction estimate will also be greater than the variance of the main effects estimate under a no-interaction model. More information about sample size and power calculation needed to detect gene–environment interaction for categorical and continuous variables can be found in [58].

3.7 COMMON GENETIC VARIANTS AND GENE–DIET INTERACTIONS MODULATING PLASMA LIPOPROTEIN CONCENTRATIONS

We summarize the current evidence for gene–diet interactions in the field of lipid metabolism, which include the association between dietary factors (total fat, specific fatty acids, carbohydrates, total energy intake, and alcohol intake) with specific gene variants. The evidence from observational, intervention studies and differences between fasting and postprandial states are presented.

Results from Interventional Studies

Interventional studies in which subjects receive a controlled dietary intake provide the best available approach to ascertain true dietary intake under controlled conditions. However, these well-controlled feeding studies have several limitations including the small number of participants and the brief duration of the interventions. Results from scores of interventional studies analyzing the impact of gene–diet interactions on different parameters of lipid metabolism have been reported. However, little consensus exists among studies analyzing the same genetic variation. Several reasons have been proposed to justify the lack of replication, including the different characteristics of study subjects, length of intervention, sample size limitation, and heterogeneity in the design. Most of the past and recent evidence is collected in a comprehensive systematic review (from 1966 to 2002) by Masson et al. [59]. These authors selected 74 relevant articles, including dietary intervention studies that had measured the lipid and lipoprotein response to dietary intervention in different genotype groups and 17 reviews on gene–diet interactions. They analyzed the published articles according to seven groups of genetic loci: apolipoprotein (apo) A-I, C-III, and A-IV cluster; apo B; apo E; enzymes (lipoprotein lipase, hepatic lipase, and cholesterol-7α-hydroxylase); low density lipoprotein (LDL) receptor gene; and other genes, including CETP, fatty acid binding protein (FABP), and neuropeptide Y (NPY). At each locus, one or more polymorphisms were included and the response to a defined dietary intervention (high-fat diet versus a low-fat diet; high SFA versus high polyunsaturated fatty acids (PUFAs); high monounsaturated fatty acids (MUFAs) versus low MUFAs, etc.) was examined. Based on the selected papers, these authors concluded that there is evidence to suggest that variations in genes for apo A-I (*APOA1*), apo A-IV (*APOA4*), apo B (*APOB*), and apo E (*APOE*) contribute to the heterogeneity in the lipid response to dietary intervention. However, the effects of the reported gene–diet interactions are not consistently seen and are sometimes conflicting. For example, for the *APOE* locus, 46 studies were analyzed. Of those, significantly different LDL-C responses to changes in the fat content of the diet by *APOE* genotype were reported in 11 studies. Since 2002, several other papers have been published [60–74] examining gene–diet interactions in the fasting state and they are presented in more detail in recent reviews [5, 75, 76].

Results from Observational Studies

Observational studies have the advantage of the large number of subjects who can be studied and whose long-term dietary habits can be estimated. However, their level of evidence has been traditionally considered to be lower than in experimental studies. We have recently reviewed this area regarding nutrigenetics [75, 76]. In general and as pointed out for the intervention studies, consensus between studies is low [43, 77–95]. A classical example of gene–diet statistical interaction is between the –75G/A polymorphism of the apolipoprotein A-I gene (*APOA1*) and polyunsaturated fatty acid (PUFA) intake in the Framingham Study [43]. We conducted a population-based study in 755 men and 822 women from the Framingham Offspring Study. The frequency of the A allele in this population was 0.165 and dietary PUFA intake was measured by a validated food frequency questionnaire. In women, a statistically significant gene–diet interaction ($P < 0.05$) was found between PUFA intake as a three category variable (<4% of energy from fat, 4–8%, and >8%) and the APOA1 SNP in determining plasma HDL-C concentrations. This interaction shows that PUFA intake clearly modulates the effect of the –75A/G polymorphism on HDL-C concentrations. Thus, in carriers of the A allele, higher PUFA intakes were associated with higher HDL cholesterol concentrations, whereas the opposite effect was observed in G/G homozygotes. One of the most consistent examples of gene–diet interaction in lipid metabolism is that between the –514C/T polymorphism at the hepatic lipase gene (LIPC) promoter and dietary fat on HDL metabolism. Ordovas et al. [88] found a strong interaction between this polymorphism and total fat intake in determining HDL-related measures in 1020 men and 1110 women participating in the Framingham Study. When total fat intake was ≥30% of energy, mean HDL-C concentrations and HDL particle size were lowest among those with the TT genotype, and no differences were observed between CC and CT individuals. This gene–diet interaction was further examined in a multiethnic Asian population from Singapore consisting of 1324 Chinese, 471 Malays, and 375 Asian Indians [93]. Tai et al. [93] found a highly significant interaction ($P = 0.001$) between the –514C/T polymorphism and total fat intake in determining TG concentration and the HDL-C/TG ratio ($P = 0.001$) in the three ethnic groups. Thus, TT subjects showed higher TG concentrations only when fat intake was higher than 30% of total energy. For HDL-C concentrations, the gene–diet interaction was significant ($P = 0.015$) only in subjects of Indian origin, suggesting additional mechanisms to be explored. However, intervention studies are needed to evaluate the potential significance of these results before nutritional recommendations are implemented at the individual level based on this information. In addition, these findings need increased interfacing with other types of nutrigenomic studies (e.g., model systems such as laboratory animals) to acquire mechanistic knowledge for the reported statistical interactions.

Gene–Diet Interactions in the Postprandial State

Humans are typically in a nonfasting state. Postprandial lipemia, characterized by a rise in TRL (triglyceride-rich lipoprotein) after eating, is a dynamic, nonsteady-state condition in which humans spend the majority of time [96]. During prolonged and exaggerated postprandial lipemia, there is accelerated neutral lipid exchange between triglyceride-rich lipoproteins and LDL and HDL, leading to a preponderance of small, dense LDL particles and a reduced concentration of HDL-C, which are all CVD risk factors. Several studies have investigated the potential interaction between some polymorphisms in candidate genes (*APOA1, APOA5, APOC3, APOE, LPL, NYP*, and *SCARB1*) and diet on postprandial lipids [97–106]. In postprandial studies, subjects usually receive a fat-loading test meal that has not been widely standardized in terms of size, physical shape, and nutrient composition. After the test meal, blood samples are again taken for the measures of postprandial lipids. As meal absorption is a complex phenomenon, and postprandial hyperlipidemia and hyperglycemia are simultaneously present in the postabsorptive phase, particularly in diabetics and in subjects with impaired glucose tolerance, the distinct role of these two factors is a matter of debate. For this reason, an oral glucose tolerance test was also carried out in some of these studies, as it was the case of Couillard et al. [97] in obese men. Although some of these postprandial studies clearly show that genetic polymorphisms can modulate the nonfasting plasma response to diet, consistency is still very low and, similar to the conclusions reached above for the other experimental approaches, replication is needed for postprandial studies where the number of subjects and the complexity of the designs may add even more bias than for other experimental approaches.

3.8 GENE–MICROORGANISMS INTERACTIONS

It has been shown repeatedly that infection could be another factor involved in the risk and pathogenesis of atherosclerosis and CVD [107]. The century-old "infectious" hypothesis of atherosclerosis has implicated a number of microorganisms that may act by triggering inflammation. Inflammatory cells such as T lymphocytes, macrophages, and monocytes play a key role not only in the initiation of atherosclerosis but also during plaque rupture. Although *Cytomegalovirus*, *Helicobacter pylori*, and *Chlamydia pneumoniae* are the three microorganisms most extensively studied, serologic studies involving polymerase chain reaction, immunocytochemistry, and electron microscopy largely support the hypothesis of an association between *Chlamydia pneumoniae* and atherosclerosis [108]; however, much work remains to be done, especially when considering the individual susceptibility to infections. Currently, infectious diseases are considered as another situation of gene–environment interactions, where the environmental factor is the microorganism (virus, bacteria, fungus, etc.). Moreover, all clinical phenotypes in the course of an infectious disease result from the complex interaction between environmental (microbial and nonmicrobial) and host (genetic) factors. However, for most microorganisms, the host factors that modulate their susceptibility to infect are poorly understood [109]. Historical accounts of the plague tell of individuals who survived unscathed in households where almost everyone else died. In addition, over a million African children die each year of malaria, but many more remain in relatively good health despite being continually infected with the parasite. Genetic factors may explain, at least in part, why some people resist infection more successfully than others. However, it is difficult to answer to what extent our genetic makeup may determine the different ways that we respond to the same infectious agent because of the many other contributory factors involved, such as nutritional status, acquired immunity, and even variability in the genome of the microorganism. Nevertheless, there is compelling evidence for a genetic component, including twin studies of tuberculosis, leprosy, malaria, and primary immunodeficiency diseases [110].

Primary immunodeficiency diseases consist of a group of more than 100 inherited conditions, mostly monogenic, predisposing individuals to different sets of infections, allergy, autoimmunity, and cancer. These diseases have led to the identification of more than 100 genes that are crucial in the immune system. Thus, a plethora of information about the development, function, and regulation of both innate and adaptive immunity has been generated [110]. Multiple examples exist in which primary immunodeficiency arises when B cell development is arrested at the pre-B stage. Mutations that prevent signaling through the pre-B cell receptor (pre-BCR) result in a lack of mature B cells and, thus, an inability to produce serum immunoglobulins, which can lead to defects in any aspect of the immune response to different microorganisms. The study of these so-called Mendelian "holes" in immunity to infection has revealed the existence of "pathogen-specific" genes in natural conditions of infection. Examples include the group of genetic defects of the terminal components of complement, associated with *Neisseria meningitidis* infections, and epidermodysplasia verruciformis, associated with a selective susceptibility to human papillomaviruses [111]. The classic example of genetic variation in the immune system is the major histocompatibility complex (MHC) on chromosome 6.

This includes the highly polymorphic human leukocyte antigen (HLA) genes, best known in the context of organ transplantation and autoimmune disease but increasingly recognized as a correlate of susceptibility to various infections including malaria, tuberculosis, HIV, and hepatitis B.

With the advances in the human genome project and in high-throughput genotyping technology, tremendous progress has been achieved in the study of immunity to infection in the past ten years. A beautiful example is the relation between the Duffy blood group antigen and susceptibility to *Plasmodium vivax,* a species of *Plasmodium*, a unicellular protozoan that causes malaria [112]. Malaria is an internationally devastating disease, producing nearly 600 million new infections and 3 million deaths each year. The protozoan *Plasmodium* is transmitted through the bite of the female *Anopheles* mosquito. Then, the malaria sporozoites are released into the bloodstream from the mosquito's salivary glands. From the bloodstream, the sporozoites enter liver parenchymal cells. In the hepatocytes, the sporozoites undergo asexual amplification. During this pre-erythrocytic stage, no illness is induced by malaria. The liver schizont bursts, releasing the merozoites into the bloodstream, where the beginning of the erythrocytic phase begins. In the erythrocyte the merozoite goes through ring, trophozoite, and schizont stages. When the erythrocytic schizont ruptures, the merozoites spill into the blood once again; it is during this phase that malaria-associated morbidity and mortality occurs. The merozoites continue in a repeated cycle of infecting erythrocytes, multiplying, and bursting the erythrocytes. The malaria parasite *Plasmodium vivax* invades human erythrocytes by binding to Duffy antigen/chemokine receptor (DARC) expressed on the erythrocyte surface. Many West Africans have a single nucleotide polymorphism in the DARC promoter region that prevents binding of the erythroid transcription factor GATA-1, thus suppressing DARC expression in erythrocytes but not other cell types. This confers complete protection against infection with *Plasmodium vivax*. However, there are other species of malaria parasites such as *Plasmodium palciparum*, which invade erythrocytes through different receptors, and subjects with the polymorphism in the Duffy antigen are not genetically protected. Interestingly, it has been reported that subjects who have the sickle cell trait (heterozygotes for the abnormal hemoglobin gene HbS) are relatively protected against *Plasmodium falciparum* malaria and thus enjoy a genetic advantage [113]. This explains the higher allele frequency of sickle cell related mutations in black populations.

There are other examples involving common polymorphism conferring protection against infectious diseases highly prevalent in developed countries such as pneumonia. Currently, *Streptococcus pneumoniae* is a major cause of infectious morbidity and mortality. Recently, Roy et al. [114] have reported a genetic locus associated with susceptibility to invasive pneumococcal disease. This locus is a dinucleotide repeat polymorphism located in an intron of the C reactive protein gene. C reactive protein is an acute phase protein that is important in the early stages of this infection because it binds the C polysaccharide of the cell wall of *Streptococcus pneumoniae* and activates the classical complement pathway. They studied 205 cases (patients in whom *S. pneumoniae* had been isolated from a normally sterile site: blood, cerebrospinal fluid, or joint fluid) and 345 controls (selected randomly from local blood donors) and found that the most common allele at the C reactive protein locus was present more often in cases than controls (odds ratio 1.52, 95% confidence interval 1.18–1.96; $p < 0.05$).

Additional evidence comes from the *APOE* locus. The virus, herpes simplex virus type 1 (HSV1), when present in the brain, acts together with the *APOE4* allele to confer a strong risk of Alzheimer disease (AD); however, in carriers of the *APOE2* and *APOE3* alleles, the virus does not confer risk [115]. In addition to HSV1 infection, the *APOE* locus has been shown to influence infection for several other diseases known to be caused by viruses. Thus, Wozniak et al. [116] found that APOE2 homozygotes became infected by malaria at an earlier age than those carrying the other genotypes, suggesting that APOE2 may be a risk factor for early infection.

Likewise, it is well known that the outcome of infection with hepatitis C virus (HCV) varies greatly. Results from the same group [117] suggest that carriers of the *APOE4* allele may be protected against liver damage caused by HCV. These different susceptibilities to infection have probably played a major role in defining the differences in allele frequencies for the *APOE* locus observed around the world [118]. However, these are only statistical associations that do not necessarily indicate a causal mechanism.

Despite the abundant evidence supporting a different susceptibility to classical infectious diseases depending on the genotype, the number of studies analyzing the gene–pathogen interactions predisposing to CVD is scarce. In 1999, two independent groups in Europe [119, 120] first reported that genetic variations in the receptor for lipopolysaccharides (LPSs; endotoxins) produced by gram-negative bacteria (such as *Chlamydia pneumoniae*), CD14, was a risk factor for myocardial infarction, suggesting that the sensitivity of individuals to infection as an eventual risk factor for atherosclerosis is at least partially genetically determined. These groups from Germany [120] and the Czech Republic found that the common polymorphism in the upstream, untranslated region of the CD14 gene (–260C/T) within the Spl transcription factor binding site, was related to the CD14 expression. Carriers of the T allele had higher density of monocyte CD14 and a risk factor for myocardial infarction as compared with CC homozygotes. The possible mechanistic connection between the receptor for bacterial endotoxins and clinical coronary disease is clearly through inflammation. Recent work carried out by Georges et al. [121] has analyzed the number of infectious pathogens to which an individual has been exposed (pathogen burden) in relation to the development and the prognosis of coronary artery disease (CAD) depending on the genetic host susceptibility. They found that the pathogen burden was strongly associated with the risk of coronary artery disease. They also found that the association between CAD and pathogen burden was modulated by the interleukin (IL)-6/G-174C polymorphism. This interaction appeared to be mediated by variations in serum IL-6 levels. These data call for confirmation in other populations and can explain previous inconsistent results when analyzing the influence of infections on cardiovascular risk.

3.9 THE MICROBIOME (MICROBIOTA)

Microbiome refers to the specific microbial community in an individual mammalian host. In human normal adults, the microbiome may be composed of about 100 trillion cells and weighs up to 1 kg. Hence, over 1000 known species of symbionts probably contain more than 100 times as many genes as exist in the human host [1]. These interacting genomes can be considered as a superorganism, with coordinated interactions, especially in terms of the gut–liver and the gut–immune systems. Based on the potential influence of the microbiome on human health, some investigators have suggested that "sequencing the components of the microbiome can be viewed as a logical albeit ambitious expansion of the human genome project" [122].

Diet modulates the complex internal community of gut microorganisms and this may be another factor related to the interindividual response to diet [123]. Thus, perturbations in the gastrointestinal microbiota composition that occur as a result of antibiotics and diet in "westernized" countries are strongly associated with allergies and asthma. The microbiome–host relationship has recently been discussed in terms of drug metabolism and toxicity and it has been proposed that mammalian genome–microflora interactions may account for some of the individual variation in drug responses as well as toxicity of certain compounds [1]. The gut microbiome influences cytochrome P450 levels in the host, has drug and nutrient metabolizing capabilities, and can influence the immune status [1, 124] as well as such factors as peroxisome proliferator activated receptor gamma location and activity [125]. Most interestingly, the gut microbiota have recently been proposed as another target for weight control [126]. The microbiota are essential for processing dietary polysaccharides. Backhed et al. [126] found that colonization of adult germ-free (GF)C57BL/6 mice with normal microbiota harvested from the distal intestine (cecum) of conventionally raised animals produces a 60% increase in body fat content and insulin resistance within 14 days despite reduced food intake, probably due the fact that microbiota promote absorption of monosaccharides from the gut lumen, with resulting induction of de novo hepatic lipogenesis. These findings suggest that the gut microbiota are an important environmental factor that affects energy harvest from the diet and energy storage in the host. Therefore, the intestinal environment must be visualized as an entire ecosystem where chemical interactions occur at multiple organizational levels with crosstalk between the mammalian, parasitic, and microbial systems as well as with nutrients and drugs.

3.10 CONCLUSION

Systems biology is not a new concept and it was already used in the 19th century by Hermann von Helmholtz whose studies of metabolism led to the first law of thermodynamics [127]. Helmholtz explored whole human physiology and contributed to the understanding of whole body energy balance. Today, the scientific offspring of Helmholtz have the clear goal of defining all of the elements present in a given system and to create an interaction network between these components so that the behavior of the system, as a whole and in parts, can be explained under specified conditions [128]. But, unlike the 19th century German scientist, contemporary researchers have access to an immense wealth of technological resources and accumulated knowledge that will have to be combined into mathematical models that should lead to a deeper understanding of the biological networks.

ACKNOWLEDGMENTS

This study is supported by grant NIH/NHLBI#HL54776, contracts 53-K06-5-10 and 58-1950-9-001 from the U.S. Department of Agriculture Research Service, and grants G03/140 and PI02-1096 (ISCIII) and Grupos 04/43 from the Direcció General d'Universitat, Spain.

REFERENCES

1. J. K. Nicholson, E. Holmes, J. C. Lindon, and I. D. Wilson (2004). The challenges of modeling mammalian biocomplexity, *Nat. Biotechnol.* **22**:1268–1274.
2. H. L. Laframboise (1973). Health policy: breaking the problem down into more manageable segments, *Can. Med. Assoc. J.* **108**:388–391.
3. M. Lalonde (1974). *A New Perspective on the Health of Canadians.* Minister of Supply and Services, Ottawa.
4. F. S. Collins, E. D. Green, A. E. Guttmacher, and M. S. Guyer (2003). U.S. National Human Genome Research Institute: a vision for the future of genomics research, *Nature* **422**:835–847.
5. J. M. Ordovas and D. Corella (2004). Nutritional genomics, *Annu. Rev. Genomics Hum. Genet.* **5**:71–118.
6. S. E. Antonarakis and the Nomenclature Working Group (1998). Recommendations for a nomenclature system for human gene mutations, *Hum. Mutat.* **11**:1–3.
7. M. Cargill, D. Altshuler, J. Ireland, P. Sklar, K. Ardlie, N. Patil, N. Shaw, C. R. Lane, E. P. Lim, N. Kalyanaraman, J. Nemesh, L. Ziaugra, L. Friedland, A. Rolfe, J. Warrington, R. Lipshutz, G. Q. Daley, and E. S. Lander (1999). Characterization of single-nucleotide polymorphisms in coding regions of human genes, *Nat. Genet.* **22**:231–238.
8. J. M. Ordovas and V. Mooser (2002). The APOE locus and the pharmacogenetics of lipid response, *Curr. Opin. Lipidol.* **13**:113–117.
9. S. E. Humphries, P. M. Ridker, and P. J. Talmud (2004). Genetic testing for cardiovascular disease susceptibility: a useful clinical management tool or possible misinformation? *Arterioscler. Thromb. Vasc. Biol.* **24**:628–636.
10. T. R. Rebbeck, M. Spitz, and X. Wu (2004). Assessing the function of genetic variants in candidate gene association studies, *Nat. Rev. Genet.* **5**:589–597.
11. D. Tchernitchko, M. Goossens, and H. Wajcman (2004). In silico prediction of the deleterious effect of a mutation: proceed with caution in clinical genetics, *Clin. Chem.* **50**:1974–1978.

12. B. M. Neale and P. C. Sham (2004). The future of association studies: gene-based analysis and replication, *Am. J. Hum. Genet.* **75**:353–362.
13. R. E. Peacock, A. Temple, V. Gudnason, M. Rosseneu, and S. E. Humphries (1997). Variation at the lipoprotein lipase and apolipoprotein AI-CIII gene loci are associated with fasting lipid and lipoprotein traits in a population sample from Iceland: interaction between genotype, gender, and smoking status, *Genet. Epidemiol.* **14**: 265–282.
14. I. Larson, M. M. Hoffmann, J. M. Ordovas, E. J. Schaefer, W. Marz, and J. Kreuzer (1999). The lipoprotein lipase HindIII polymorphism: association with total cholesterol and LDL-cholesterol, but not with HDL and triglycerides in 342 females, *Clin. Chem.* **45**:963–968.
15. J. P. Ioannidis, E. E. Ntzani, T. A. Trikalinos, and D. G. Contopoulos-Ioannidis (2001). Replication validity of genetic association studies, *Nat. Genet.* **29**:306–309.
16. L. Bastone, M. Reilly, D. J. Rader, and A. S. Foulkes (2005). MDR and PRP: a comparison of methods for high-order genotype–phenotype associations, *Hum. Hered.* In press.
17. A. C. Syvanen (2001). Accessing genetic variation: genotyping single nucleotide polymorphisms, *Nat. Rev. Genet.* **2**:930–942.
18. D. J. Cutler, M. E. Zwick, M. M. Carrasquillo, C. T. Yohn, K. P. Tobin, C. Kashuk, D. J. Mathews, N. A. Shah, E. E. Eichler, J. A. Warrington, and A. Chakravarti (2001). High-throughput variation detection and genotyping using microarrays, *Genome Res.* **11**:1913–1925.
19. X. Chen, K. J. Livak, and P. Y. Kwok (1998). A homogeneous, ligase-mediated DNA diagnostic test, *Genome Res.* **8**:549–556.
20. J. Jarvius, M. Nilsson, and U. Landegren (2003). Oligonucleotide ligation assay, *Methods Mol. Biol.* **212**:215–228.
21. D. Ryan, B. Nuccie, and D. Arvan (1999). Non-PCR-dependent detection of the factor V Leiden mutation from genomic DNA using a homogeneous invader microtiter plate assay, *Mol. Diagn.* **4**:135–144.
22. J. Ihalainen, H. Siitari, S. Laine, A. C. Syvanen, and A. Palotie (1994). Towards automatic detection of point mutations: use of scintillating microplates in solid-phase minisequencing, *Biotechniques* **16**:938–943.
23. H. Matsuzaki, H. Loi, S. Dong, Y. Y. Tsai, J. Fang, et al. (2004). Parallel genotyping of over 10,000 SNPs using a one-primer assay on a high-density oligonucleotide array, *Genome Res.* **14**:414–425.
24. D. C. Chen, J. Saarela, I. Nuotio, A. Jokiaho, L. Peltonen, and A. Palotie (2003). Comparison of GenFlex tag array and pyrosequencing in SNP genotyping, *J. Mol. Diagn.* **5**:243–249.
25. S. J. Sawcer, M. Maranian, S. Singlehurst, T. Yeo, A. Compston, et al. (2004). Enhancing linkage analysis of complex disorders: an evaluation of high-density genotyping, *Hum. Mol. Genet.* **13**:1943–1949.
26. J. Little, L. Bradley, M. S. Bray, M. Clyne, J. Dorman, et al. (2002). Reporting, appraising, and integrating data on genotype prevalence and gene-disease associations, *Am. J. Epidemiol.* **156**:300–310.
27. J. K. Pritchard and N. J. Cox (2002). The allelic architecture of human disease genes: common disease–common variant . . . or not? *Hum. Mol. Genet.* **11**:2417–2423.

28. S. O. Keita, R. A. Kittles, C. D. Royal, G. E. Bonney, P. Furbert-Harris, G. M. Dunston, and C. N. Rotimi (2004). Conceptualizing human variation, *Nat. Genet.* **36**(11 Suppl): S17–S20.
29. A. F. Wright and N. D. Hastie (2001). Complex genetic diseases. Controversy over the croeus code. *Genome Biol.* **2**:comment 2007. (http://genomebiology.com/2001/2/8/comment/2007).
30. L. R. Cardon and G. R. Abecasis (2003). Using haplotype blocks to map human complex trait loci, *Trends Genet.* **19**:135–140.
31. S. B. Gabriel, S. F. Schaffner, H. Nguyen, J. M. Moore, J. Roy, et al. (2002). The structure of haplotype blocks in the human genome, *Science* **296**:2225–2229.
32. T. Niu, Z. S. Qin, X. Xu, and J. S. Liu (2002). Bayesian haplotype inference for multiple linked single-nucleotide polymorphisms, *Am. J. Hum. Genet.* **70**:157–169.
33. Z. S. Qin, T. Niu, and J. S. Liu (2002). Partition-ligation EM algorithm for haplotype inference with single nucleotide polymorphisms, *Am. J. Hum. Genet.* **71**:1242–1247.
34. H. Kang, Z. S. Qin, T. Niu, and J. S. Liu (2004). Incorporating genotyping uncertainty in haplotype inference for single-nucleotide polymorphisms, *Am. J. Hum. Genet.* **74**:495–510.
35. G. R. Abecasis, S. S. Cherny, W. O. Cookson, and L. R. Cardon (2002). Merlin—rapid analysis of dense genetic maps using sparse gene flow trees, *Nat. Genet.* **30**:97–101.
36. E. C. Anderson and J. Novembre (2003). Finding haplotype block boundaries by using the minimum-description-length principle, *Am. J. Hum. Genet.* **73**:336–354.
37. D. O. Stram (2004). Tag SNP selection for association studies, *Genet. Epidemiol.* **27**:365–374.
38. D. O. Stram, C. A. Haiman, J. N. Hirschhorn, D. Altshuler, L. N. Kolonel, et al. (2003). Choosing haplotype-tagging SNPs based on unphased genotype data using a preliminary sample of unrelated subjects with an example from the Multiethnic Cohort Study, *Hum. Hered.* **55**:27–36.
39. Z. Lin and R. B. Altman (2004). Finding haplotype tagging SNPs by use of principal components analysis, *Am. J. Hum. Genet.* **75**:850–861.
40. I. Furman, M. J. Rieder, S. Da Ponte, D. P. Carrington, D. A. Nickerson, L. Kruglyak, and K. Markianos (2004). Sequence-based linkage analysis, *Am. J. Hum. Genet.* **75**:647–653.
41. M. S. Phillips, R. Lawrence, R. Sachidanandam, A. P. Morris, D. J. Balding, et al. (2003). Chromosome-wide distribution of haplotype blocks and the role of recombination hot spots, *Nat. Genet.* **33**:382–387.
42. Y. Omura, A. Y. Lee, S. L. Beckman, R. Simon, M. Lorberboym, H. Duvvi, S. I. Heller, and C. Urich (1996). 177 cardiovascular risk factors, classified in 10 categories, to be considered in the prevention of cardiovascular diseases: an update of the original 1982 article containing 96 risk factors, *Acupunct. Electrother. Res.* **21**:21–76.
43. J. M. Ordovas, D. Corella, L. A. Cupples, et al. (2002). Polyunsaturated fatty acids modulate the effects of the APOA1 G-A polymorphism on HDL-cholesterol concentrations in a sex-specific manner: the Framingham Study, *Am. J. Clin. Nutr.* **75**:38–46.
44. F. S. Collins (2004). The case for a U.S. prospective cohort study of genes and environment, *Nature* **429**:475–477.

45. S. J. Pocock, T. J. Collier, K. J. Dandreo, B. L. de Stavola, M. B. Goldman, et al. (2004). Issues in the reporting of epidemiological studies: a survey of recent practice, *BMJ* **329**:883.
46. J. G. Hall (2003). A clinician's plea, *Nat. Genet.* **33**(4):440–442.
47. M. de Andrade, C. I. Amos, and T. J. Thiel (1999). Methods to estimate genetic components of variance for quantitative traits in family studies, *Genet. Epidemiol.* **17**: 64–76.
48. D. A. Heller, U. de Faire, N. L. Pedersen, G. Dahlen, and G. E. McClearn (1993). Genetic and environmental influences on serum lipid levels in twins, *N. Engl. J. Med.* **328**:1150–1156.
49. S. Greenland (1993). Basic problems in interaction assessment, *Environ. Health Perspect.* **101**(Suppl 4):59–66.
50. K. J. Rothman, S. Greenland, and A. M. Walker (1980). Concepts of interaction, *Am. J. Epidemiol.* **112**:467–470.
51. T. H. Beaty and M. J. Khoury (2000). Interface of genetics and epidemiology, *Epidemiol. Rev.* **22**:120–125.
52. L. D. Botto and M. J. Khoury (2001). Commentary: facing the challenge of gene–environment interaction: the two-by-four table and beyond, *Am. J. Epidemiol.* **153**: 1016–1020.
53. S. Greenland (1993). Basic problems in interaction assessment, *Environ. Health Perspect.* **101**(Suppl 4):59–66.
54. J. LeLorier, G. Gregoire, A. Benhaddad, J. Lapierre, and F. Derderian (1997). Discrepancies between meta-analyses and subsequent large, randomized, controlled trials, *N. Engl. J. Med.* **337**:536–542.
55. G. Davey Smith and S. Ebrahim (2003). Mendelian randomization: Can genetic epidemiology contribute to understanding environmental determinants of disease? *Int. J. Epidemiol.* **32**:1–22.
56. M. D. Tobin, C. Minelli, P. R. Burton, and J. R. Thompson (2004). Development of Mendelian randomization: from hypothesis test to "Mendelian deconfounding," *Int. J. Epidemiol.* **33**:26–29.
57. W. C. Willett (2000). Nutritional epidemiology issues in chronic disease at the turn of the century, *Epidemiol. Rev.* **22**:82–86.
58. M. Garcia-Closas and J. H. Lubin (1999). Power and sample size calculations in case–control studies of gene–environment interactions: comments on different approaches, *Am. J. Epidemiol.* **149**:689–692.
59. L. F. Masson, G. McNeill, and A. Avenell (2003). Genetic variation and the lipid response to dietary intervention: a systematic review, *Am. J. Clin. Nutr.* **77**:1098–1111.
60. C. Atkinson, W. Oosthuizen, S. Scollen, A. Loktionov, N. E. Day, and S. A. Bingham (2004). Modest protective effects of isoflavones from a red clover-derived dietary supplement on cardiovascular disease risk factors in perimenopausal women, and evidence of an interaction with ApoE genotype in 49–65 year-old women, *J. Nutr.* **134**:1759–1764.
61. P. Couture, W. R. Archer, B. Lamarche, et al. (2003). Influences of apolipoprotein E polymorphism on the response of plasma lipids to the ad libitum consumption of a high-carbohydrate diet compared with a high-monounsaturated fatty acid diet, *Metabolism* **52**:1454–1459.

62. A. Halverstadt, D. A. Phares, R. E. Ferrell, K. R. Wilund, A. P. Goldberg, and J. M. Hagberg (2003). High-density lipoprotein-cholesterol, its subfractions, and responses to exercise training are dependent on endothelial lipase genotype, *Metabolism* **52**:1505–1511.

63. M. K. Hofman, R. M. Weggemans, P. L. Zock, E. G. Schouten, M. B. Katan, and H. M. Princen (2004). CYP7A1 A-278C polymorphism affects the response of plasma lipids after dietary cholesterol or cafestol interventions in humans, *J. Nutr.* **134**:2200–2204.

64. R. Jemaa, A. Mebazaa, and F. Fumeron (2004). Apolipoprotein B signal peptide polymorphism and plasma LDL-cholesterol response to low-calorie diet, *Int. J. Obes. Relat. Metab. Disord.* **28**:902–905.

65. V. Lindi, U. Schwab, A. Louheranta, M. Laakso, B. Vessby, et al. (2003). Impact of the Pro12Ala polymorphism of the PPAR-gamma2 gene on serum triacylglycerol response to n-3 fatty acid supplementation, *Mol. Genet. Metab.* **79**:52–60.

66. A. M. Lottenberg, V. S. Nunes, E. R. Nakandakare, M. Neves, M. Bernik, et al. (2003). The human cholesteryl ester transfer protein I405V polymorphism is associated with plasma cholesterol concentration and its reduction by dietary phytosterol esters, *J. Nutr.* **133**:1800–1805.

67. J. A. Moreno, F. Perez-Jimenez, C. Marin, P. Gomez, P. Perez-Martinez, et al. (2004). The effect of dietary fat on LDL size is influenced by apolipoprotein E genotype in healthy subjects, *J. Nutr.* **134**:2517–2522.

68. B. J. Nicklas, R. E. Ferrell, L. B. Bunyard, D. M. Berman, K. E. Dennis, and A. P. Goldberg (2002). Effects of apolipoprotein E genotype on dietary-induced changes in high-density lipoprotein cholesterol in obese postmenopausal women, *Metabolism* **51**:853–858.

69. P. Perez-Martinez, P. Gomez, E. Paz, C. Marin, E. Gavilan Moral, et al. (2001). Interaction between smoking and the SstI polymorphism of the apo C-III gene determines plasma lipid response to diet, *Nutr. Metab. Cardiovasc. Dis.* **11**:237–243.

70. P. Perez-Martinez, J. M. Ordovas, J. Lopez-Miranda, P. Gomez, C. Marin, et al. (2003). Polymorphism exon 1 variant at the locus of the scavenger receptor class B type I gene: influence on plasma LDL cholesterol in healthy subjects during the consumption of diets with different fat contents, *Am. J. Clin. Nutr.* **77**:809–813.

71. J. Plat and R. P. Mensink (2002). Relationship of genetic variation in genes encoding apolipoprotein A-IV, scavenger receptor BI, HMG-CoA reductase, CETP and apolipoprotein E with cholesterol metabolism and the response to plant stanol ester consumption, *Eur. J. Clin. Invest.* **32**:242–250.

72. M. Rantala, M. L. Silaste, A. Tuominen, J. Kaikkonen, J. T. Salonen, et al. (2002). Dietary modifications and gene polymorphisms alter serum paraoxonase activity in healthy women, *J. Nutr.* **132**:3012–3017.

73. N. Tamasawa, H. Murakami, K. Yamato, J. Matsui, J. Tanabe, and T. Suda (2003). Influence of apolipoprotein E genotype on the response to caloric restriction in type 2 diabetic patients with hyperlipidaemia, *Diabetes Obes. Metab.* **5**:345–348.

74. S. Vincent, R. Planells, C. Defoort, M. C. Bernard, M. Gerber, et al. (2002). Genetic polymorphisms and lipoprotein responses to diets, *Proc. Nutr. Soc.* **61**:427–434.

75. D. Corella and J. M. Ordovas (2005). Single nucleotide polymorphisms that influence lipid metabolism: interaction with dietary factors, *Annu. Rev. Nutr.* In press.

76. J. M. Ordovas and D. Corella (2004). Genes, diet and plasma lipids: the evidence from observational studies, *World Rev. Nutr. Diet.* **93**:41–76.

77. S. Brown, J. M. Ordovas, and H. Campos (2003). Interaction between the APOC3 gene promoter polymorphisms, saturated fat intake and plasma lipoproteins, *Atherosclerosis* **170**:307–313.

78. H. Campos, M. D'Agostino, and J. M. Ordovas (2001). Gene–diet interactions and plasma lipoproteins: role of apolipoprotein E and habitual saturated fat intake, *Genet. Epidemiol.* **20**:117–128.

79. H. Campos, J. Lopez-Miranda, C. Rodriguez, M. Albajar, E. J. Schaefer, and J. M. Ordovas (1997). Urbanization elicits a more atherogenic lipoprotein profile in carriers of the apolipoprotein A-IV-2 allele than in A-IV-1 homozygotes, *Arterioscler. Thromb. Vasc. Biol.* **17**:1074–1081.

80. D. Corella, M. Guillen, C. Saiz, O. Portoles, A. Sabater, et al. (2001). Environmental factors modulate the effect of the APOE genetic polymorphism on plasma lipid concentrations: ecogenetic studies in a Mediterranean Spanish population, *Metabolism* **50**:936–944.

81. D. Corella, M. Guillen, C. Saiz, O. Portoles, A. Sabater, et al. (2002). Associations of LPL and APOC3 gene polymorphisms on plasma lipids in a Mediterranean population: interaction with tobacco smoking and the APOE locus, *J. Lipid Res.* **43**: 416–427.

82. D. Corella, K. Tucker, C. Lahoz, O. Coltell, L. A. Cupples, et al. (2001). Alcohol drinking determines the effect of the APOE locus on LDL-cholesterol concentrations in men: the Framingham Offspring Study, *Am. J. Clin. Nutr.* **73**:736–745.

83. A. T. Erkkila, E. S. Sarkkinen, V. Lindi, S. Lehto, M. Laakso, and M. I. Uusitupa (2001). APOE polymorphism and the hypertriglyceridemic effect of dietary sucrose, *Am. J. Clin. Nutr.* **73**:746–752.

84. A. Ganan, D. Corella, M. Guillen, J. M. Ordovas, and M. Pocovi (2004). Frequencies of apolipoprotein A4 gene polymorphisms and association with serum lipid concentrations in two healthy Spanish populations, *Hum. Biol.* **76**:253–266.

85. J. A. Hubacek, J. Pitha, Z. Skodova, R. Poledne, V. Lanska, et al. (2003). Polymorphisms in CYP-7A1, not APOE, influence the change in plasma lipids in response to population dietary change in an 8 year follow-up; results from the Czech MONICA study, *Clin. Biochem.* **36**:263–267.

86. A. Loktionov, S. Scollen, N. McKeown, and S. A. Bingham (2000). Gene–nutrient interactions: dietary behaviour associated with high coronary heart disease risk particularly affects serum LDL cholesterol in apolipoprotein E epsilon4-carrying free-living individuals, *Br. J. Nutr.* **84**:885–890.

87. A. Memisoglu, F. B. Hu, S. E. Hankinson, S. Liu, J. B. Meigs, et al. (2003). Prospective study of the association between the proline to alanine codon 12 polymorphism in the PPARgamma gene and type 2 diabetes, *Diabetes Care* **26**:2915–2917.

88. J. M. Ordovas, D. Corella, S. Demissie, L. A. Cupples, P. Couture, O. Coltell, P. W. Wilson, E. J. Schaefer, and K. L. Tucker (2002). Dietary fat intake determines the effect of a common polymorphism in the hepatic lipase gene promoter on high-density lipoprotein metabolism: evidence of a strong dose effect in this gene–nutrient interaction in the Framingham Study, *Circulation* **106**:2315–2321.

89. M. E. Paradis, P. Couture, Y. Bosse, J. P. Despres, L. Perusse, et al. (2003). The T111I mutation in the EL gene modulates the impact of dietary fat on the HDL profile in women, *J. Lipid Res.* **44**:1902–1908.

90. G. J. Petot, F. Traore, S. M. Debanne, A. J. Lerner, K. A. Smyth, and R. P. Friedland (2003). Interactions of apolipoprotein E genotype and dietary fat intake of healthy older persons during mid-adult life, *Metabolism* **52:**279–281.
91. J. Robitaille, C. Brouillette, A. Houde, S. Lemieux, L. Perusse, et al. (2004). Association between the PPARalpha-L162V polymorphism and components of the metabolic syndrome, *J. Hum. Genet.* **49**:482–489.
92. J. Robitaille, C. Brouillette, S. Lemieux, L. Perusse, D. Gaudet, and M. C. Vohl (2004). Plasma concentrations of apolipoprotein B are modulated by a gene–diet interaction effect between the LFABP T94A polymorphism and dietary fat intake in French-Canadian men, *Mol. Genet. Metab.* **82**:296–303.
93. E. S. Tai, D. Corella, M. Deurenberg-Yap, J. Cutter, S. K. Chew, et al. (2004). Dietary fat interacts with the –514C>T polymorphism in the hepatic lipase gene promoter on plasma lipid profiles in a multiethnic Asian population: the 1998 Singapore National Health Survey, *J. Nutr.* **133**:3399–3408.
94. E. S. Tai, J. M. Ordovas, D. Corella, M. Deurenberg-Yap, E. Chan, et al. (2003). The TaqIB and –629C>A polymorphisms at the cholesteryl ester transfer protein locus: associations with lipid levels in a multiethnic population. The 1998 Singapore National Health Survey, *Clin. Genet.* **63**:19–30.
95. M. Tomas, M. Senti, R. Elosua, J. Vila, J. Sala, et al. (2001). Interaction between the Gln-Arg 192 variants of the paraoxonase gene and oleic acid intake as a determinant of high-density lipoprotein cholesterol and paraoxonase activity, *Eur. J. Pharmacol.* **432**:121–128.
96. D. Hyson, J. C. Rutledge, and L. Berglund (2003). Postprandial lipemia and cardiovascular disease, *Curr. Atheroscler. Rep.* **5**:437–444.
97. C. Couillard, M. C. Vohl, J. C. Engert, I. Lemieux, A. Houde, et al. (2003). Effect of apoC-III gene polymorphisms on the lipoprotein–lipid profile of viscerally obese men, *J. Lipid Res.* **44**:986–993.
98. Y. Jang, J. Y. Kim, O. Y. Kim, J. E. Lee, H. Cho, et al. (2004). The –1131T→C polymorphism in the apolipoprotein A5 gene is associated with postprandial hypertriacylglycerolemia; elevated small, dense LDL concentrations; and oxidative stress in nonobese Korean men, *Am. J. Clin. Nutr.* **80**:832–840.
99. J. Lopez-Miranda, G. Cruz, P. Gomez, C. Marin, E. Paz, et al. (2004). The influence of lipoprotein lipase gene variation on postprandial lipoprotein metabolism, *J. Clin. Endocrinol. Metab.* **89**:4721–4728.
100. C. Marin, J. Lopez-Miranda, P. Gomez, E. Paz, P. Perez-Martinez, et al. (2002). Effects of the human apolipoprotein A-I promoter G-A mutation on postprandial lipoprotein metabolism, *Am. J. Clin. Nutr.* **76**:319–325.
101. S. Martin, V. Nicaud, S. E. Humphries, P. J. Talmud, and EARS Group (2003). Contribution of APOA5 gene variants to plasma triglyceride determination and to the response to both fat and glucose tolerance challenges, *Biochim. Biophys. Acta.* **1637**: 217–225.
102. J. A. Moreno, F. Perez-Jimenez, C. Marin, P. Gomez, P. Perez-Martinez, et al. (2004). The effect of dietary fat on LDL size is influenced by apolipoprotein E genotype in healthy subjects, *J. Nutr.* **134**:2517–2522.
103. P. Perez-Martinez, J. Lopez-Miranda, J. M. Ordovas, C. Bellido, C. Marin, et al. (2004). Postprandial lipemia is modified by the presence of the polymorphism present in the exon 1 variant at the SR-BI gene locus, *J. Mol. Endocrinol.* **32**:237–245.

104. I. Reiber, I. Mezo, A. Kalina, G. Palos, L. Romics, and A. Csaszar (2003). Postprandial triglyceride levels in familial combined hyperlipidemia. The role of apolipoprotein E and lipoprotein lipase polymorphisms, *J. Nutr. Biochem.* **14**:394–400.

105. U. S. Schwab, J. J. Agren, R. Valve, M. A. Hallikainen, E. S. Sarkkinen, et al. (2002). The impact of the leucine 7 to proline 7 polymorphism of the neuropeptide Y gene on postprandial lipemia and on the response of serum total and lipoprotein lipids to a reduced fat diet, *Eur. J. Clin. Nutr.* **56**:149–156.

106. S. K. Woo and H. S. Kang (2003). The apolipoprotein CIII T2854G variants are associated with postprandial triacylglycerol concentrations in normolipidemic Korean men, *J. Hum. Genet.* **48**:551–555.

107. E. Gurfinkel (1998). Link between intracellular pathogens and cardiovascular diseases, *Clin Microbiol Infect.* **4**(Suppl 4):S33–S36.

108. J. Ngeh, V. Anand, and S. Gupta (2002). *Chlamydia pneumoniae* and atherosclerosis—what we know and what we don't, *Clin. Microbiol. Infect.* **8**:2–13.

109. D. Kwiatkowski (2000). Science, medicine, and the future: susceptibility to infection, *BMJ* **321**:1061–1065.

110. A. Fischer (2004). Human primary immunodeficiency diseases: a perspective, *Nat. Immunol.* **5**:23–30.

111. J. L. Casanova and L. Abel (2004). The human model: a genetic dissection of immunity to infection in natural conditions, *Nat. Rev. Immunol.* **4**:55–66.

112. C. Tournamille, Y. Colin, J. P. Cartron, and C. Le Van Kim (1995). Disruption of a GATA motif in the Duffy gene promoter abolishes erythroid gene expression in Duffy-negative individuals, *Nat. Genet.* **10**:224–228.

113. B. Lell, J. May, R. J. Schmidt-Ott, L. G. Lehman, D. Luckner, B. Greve, P. Matousek, D. Schmid, K. Herbich, F. P. Mockenhaupt, C. G. Meyer, U. Bienzle, and P. G. Kremsner (1999). The role of red blood cell polymorphisms in resistance and susceptibility to malaria, *Clin. Infect. Dis.* **4**:794–799.

114. S. Roy, A. V. Hill, K. Knox, D. Griffiths, and D. Crook (2002). Research pointers: association of common genetic variant with susceptibility to invasive pneumococcal disease, *BMJ* **324**:1369.

115. R. F. Itzhaki, C. B. Dobson, S. J. Shipley, and M. A. Wozniak (2004). The role of viruses and of APOE in dementia, *Ann. N. Y. Acad. Sci.* **1019**:15–18.

116. M. A. Wozniak, E. B. Faragher, J. A. Todd, K. A. Koram, E. M. Riley, and R. F. Itzhaki (2003). Does apolipoprotein E polymorphism influence susceptibility to malaria? *J. Med. Genet.* **40**:348–351.

117. M. A. Wozniak, R. F. Itzhaki, E. B. Faragher, M. W. James, S. D. Ryder, and W. L. Irving (2002). Trent HCV Study Group. Apolipoprotein E-epsilon 4 protects against severe liver disease caused by hepatitis C virus, *Hepatology* **36**:456–463.

118. R. M. Corbo and R. Scacchi (1999). Apolipoprotein E (APOE) allele distribution in the world. Is APOE*4 a "thrifty" allele? *Ann. Hum. Genet.* **63**(Pt 4):301–310.

119. J. A. Hubacek, G. Rothe, J. Pit'ha, Z. Skodova, V. Stanek, R. Poledne, and G. Schmitz (1999). C(–260)→T polymorphism in the promoter of the CD14 monocyte receptor gene as a risk factor for myocardial infarction, *Circulation* **99**:3218–3220.

120. K. Unkelbach, A. Gardemann, M. Kostrzewa, M. Philipp, H. Tillmanns, and W. Haberbosch (1999). A new promoter polymorphism in the gene of lipopolysaccharide receptor CD14 is associated with expired myocardial infarction in patients with low atherosclerotic risk profile, *Arterioscler. Thromb. Vasc. Biol.* **19**:932–938.

121. J. L. Georges, H. J. Rupprecht, S. Blankenberg, O. Poirier, C. Bickel, G. Hafner, V. Nicaud, J. Meyer, F. Cambien, L. Tiret, and AtheroGene Group (2003). Impact of pathogen burden in patients with coronary artery disease in relation to systemic inflammation and variation in genes encoding cytokines, *Am. J. Cardiol.* **92**:515–521.

122. J. Xu, H. C. Chaing, M. K. Bjursell, and J. I. Gordon (2004). Message from a human gut symbiont: sensitivity is a prerequisite for sharing, *Trends Microbiol.* **12**:21–28.

123. F. Guarner and J. R. Malagelada (2003). Gut flora in health and disease, *Lancet* **361**:512–519.

124. M. C. Noverr and G. B. Huffnagle (2004). Does the microbiota regulate immune responses outside the gut? *Trends Microbiol.* **12**:562–568.

125. D. Kelly, J. I. Campbell, T. P. King, G. Grant, E. A. Jansson, A. G. P. Coutts, S. Pettersson, and S. Conway (2003). Commensal anaerobic gut bacteria attenuate inflammation by regulating nuclear-cytoplasmic shuttling of PPArg and RelA, *Nat. Immunol.* **5**:104–112.

126. F. Backhed, H. Ding, T. Wang, L. V. Hooper, G. Y. Koh, A. Nagy, C. F. Semenkovich, and J. I. Gordon (2004). The gut microbiota as an environmental factor that regulates fat storage, *Proc. Natl. Acad. Sci. U.S.A.* **101**:15718–15723.

127. C. Delisi (2004). Systems biology, the second time around, *Environ. Health Perspect.* **112**, A926–A927.

128. A. Ghazalpour, S. Doss, X. Yang, J. Aten, E. M. Toomey, A. Van Nas, S. Wang, T. A. Drake, and A. J. Lusis (2004). Thematic review series: the pathogenesis of atherosclerosis. Toward a biological network for atherosclerosis, *J. Lipid Res.* **45**:1793–805 (2004).

4

METABOLOMICS: BRINGING NUTRIGENOMICS TO PRACTICE IN INDIVIDUALIZED HEALTH ASSESSMENT

J. Bruce German,[1,2] Cora J. Dillard,[1] S. Luke Hillyard,[1] Matthew C. Lange,[1] Jennifer T. Smilowitz,[1] Robert E. Ward,[1] and Angela M. Zivkovic[1]

[1] *University of California–Davis, Davis, California*
[2] *Nestlé Research Centre, Lausanne, Switzerland*

4.1 INTRODUCTION

Genomics is the study of all the genes (and gene products—RNA and proteins) as a dynamic system, over time, determining how they interact and influence biological pathways, networks, and physiology, in a global sense. The field of genomics is propelling all of life science research and with it bringing unprecedented opportunities to improve foods for individual health. Research on biological processes is moving forward in such disparate areas as individual dietary needs and properties of biomaterials that comprise foods. The knowledge that this research is building is revealing that needs for dietary components differ not only among individuals but also for a single individual under different conditions. In parallel with this growing knowledge of varying dietary needs of individuals, there is a growing understanding of the biomaterial properties of agricultural commodities and foods. Bringing this knowledge together into practice in the food and health-care communities by design-

Nutritional Genomics: Discovering the Path to Personalized Nutrition
Edited by Jim Kaput and Raymond L. Rodriguez

ing foods and diets according to individual health needs will require a new approach to human assessment using both genetic and phenotypic analyses.

If it is expected to predict future health outcomes, an individual's immediate health status will need to be defined in highly quantitative and comprehensive terms. Research into human genetic polymorphisms is revealing that individuals do indeed vary in their responses to dietary inputs. These genetic differences in many cases underlie varying susceptibilities to diet-responsive diseases. However, it will be some time before our understanding of genetic polymorphisms alone can accurately predict overall metabolism in an individual and, in particular, predict his/her overall metabolic responses to complex diets. Actually measuring metabolism using metabolic profiling—the global, quantitative measurement of metabolites, a technology known as metabolomics [1]—will be the logical partner to take the principles of human variation revealed by genomics research into routine health monitoring. Furthermore, knowledge of varying predispositions to essential nutrient needs and deficiency states alone is not sufficient to assure overall and long-term health of individuals. In addition to inadequate intakes of essential nutrients, diet is known to affect health through variations in the intake of nonessential components, variations in the matrix in which essential and nonessential components are consumed, and even variations in the time over which they are consumed. Metabolomics methodologies for monitoring health and the influence of such dietary variables must have sufficient accuracy to anticipate an individual's health trajectory for all aspects of health in order to effect dietary and life-style changes prior to onset of disease. This chapter discusses how partnering genomics and metabolomics technologies together is able to further accelerate our basic understanding of metabolic health, the mechanisms of action of dietary components, and the interactions between diet and human health.

4.2 OPPORTUNITIES FOR FOODS AND HEALTH

Nutrition research and its applications are expanding beyond the diagnosis of diseases caused by exogenous microbial pathogens, exposures to toxins, and deficiencies of essential nutrients. New research includes the means to recognize and prevent diseases arising from consuming unbalanced diets and the resulting dysregulated metabolism. A major objective of research on diet-related metabolic dysfunction is to gain an understanding of the optimal dietary intakes of all food components so that individuals will exhibit a lower incidence of metabolic diseases.

Realizing such an obviously valuable goal will not be easy, and research to date has identified some key hurdles. These hurdles can be categorized as due to (1) *extensiveness*, the multiplicity of effects of diet on overall health; (2) *individuality*, the specificity of responses to diet among different individuals; (3) *reference*, the lack of detailed quantitative data on normal human metabolites and on the quantitative responses of these metabolites to diets mapped to their respective phenotypes; (4) *mechanisms*, an understanding of the relationships between molecular and metabolic changes and trajectories of health and disease; and (5) *archetype*, a comprehensive exemplar of diets that genuinely improve health rather than reverse existing disease.

Initial applications of the comprehensive tools of genomics to analyses of gene expression have demonstrated that even the most innocuous of dietary components—fats, proteins, and carbohydrates—have wide ranging effects on the overall regulation of genomic expression [2, 3]. Unavoidably, designing foods that contain functional ingredients for one targeted benefit cannot be guaranteed to have no potential deleterious side effects on other pathways of metabolism. If overall risk of disease is to be lessened, an intervention that improves one aspect of metabolism must not have deleterious effects on any other. This means that, in both research on and implementation of dietary modification through food choices, the effects of dietary change on metabolism must be examined comprehensively. The principle of comprehensive analysis of biological samples is a possibility in terms of gene expression as a result of the development of genome-wide gene expression arrays—gene chips (see Chapter 14 in this volume). Similar strategies are being developed for measuring proteins—proteomics—and metabolites—metabolomics—although a single analytical platform, comparable to gene arrays, has not been developed for either [4–8].

4.3 NUTRIGENOMICS

The initial successes of nutrigenomics have revealed that indeed the natural variation in the human genome is responsible for significant variations in response to diets [9]. Therefore, the optimal diet for one individual in a population will not be the same for every individual in that population. The age of personalized diets has arrived, if not in practice, at least in research. The arrival of personalized diets for health could not have come at a better time. The devastating effects of getting diets "wrong," not by deficiency of essential nutrients but by imbalance of macronutrients and nonessential food components, has led the public health organizations of the world to recognize that diseases caused by dietary imbalances, such as atherosclerosis, obesity, diabetes type 2, hypertension, and osteoporosis, are as much a threat to human health and economic stability as epidemics of infectious diseases. Discouragingly, the practical success of building a food supply that largely achieves the goal of eliminating nutritional deficiencies has been followed within a generation by global epidemic diseases of nutritional excess—obesity and diabetes. Although diet plays an indisputable role in the problems of metabolic dysregulation, the varying expression of dietary imbalances in different people with different life styles makes it impossible to impose either population-wide solutions or genetically based preventions. Health will have to be monitored and diets implemented one individual at a time. Nutrigenomics will be the cornerstone of the research engines that build this new paradigm of health and nutrition. Nonetheless, comprehensive analysis of phenotype, such as by metabolomic analyses, will be necessary to bring this new scientific knowledge to practice. Therefore, newer methods of human assessment must be developed that are capable of recognizing subtle, quantitative differences among humans in their metabolism and of looking comprehensively at metabolism to distinguish health as a continuum. Metabolism itself must be viewed comprehensively within each individual. This approach to health will require development of metabolic profiling by global, quantitative measurement of metabolites—a technology now known as metabolomics—as well as databases of normal metabolite levels in humans and their variations as a function of varying health status.

4.4 METABOLOMICS

Nomenclature

The emerging field of metabolomics has the potential to provide the knowledge and tools that health-care professionals can use to assess and guide individuals toward their personal health goals and to prevent the development of diet-related metabolic dysfunction. The definition of terms used in this emerging field is still undergoing change. Metabolomics has been variously defined, and it is necessary to build a nomenclature of terms that capture the breadth and complexity of concepts within this new field. The newly organized Metabolomics Society (metabolomicssociety.org) will hopefully guide the expansion of nomenclature and consolidate terms and their definitions. To maintain consistency with the conventions of the other "omic" sciences—gene–genome–genomics and protein–proteome–proteomics—the terms metabolite, metabolome, and metabolomics would be most appropriate. At the present time, metabolomics is defined as the global analysis of metabolites—small molecules generated in the process of metabolism—that represent the sum total of all the metabolic pathways in an organism, with a focus on the identification of each pathway and its role in an organism's function.

Goals of Metabolomics

There are several goals of this new field of metabolomics. The technologies that are developing are providing a comprehensive snapshot of dynamic metabolism, whereas an individual's genotype is fixed throughout life. Comprehensive metabolic measurements have numerous advantages:

- To estimate nutritional status, including both essential and nonessential nutrients and their effects on endogenous metabolism.
- To follow the compliance, progress, and success of dietary intervention.
- To identify side effects, unexpected metabolic responses, or lack of responses to specific dietary intervention.
- To recognize metabolic shifts in individuals due to environmental changes, life-style modifications, and normal progression of aging and maturation.
- To predict metabolic trajectories of individuals and assign likely metabolic consequences of interventions or failing to intervene.
- To assess metabolic stress.

- To explore the range of metabolic states accessible to individuals as measures of their ultimate health potential.

Tools of Metabolomics

The major objective of metabolomics is to measure and quantify essentially all metabolites within a biological sample. There is presently no single technology that can simultaneously identify and quantify all metabolites in a sample. There are two platforms in use that rely on spectroscopic detection—nuclear magnetic resonance (NMR) spectroscopy [10], which is not yet qualitative, and mass spectrometry (MS), which at present requires compound-targeted sample cleanup and prefractionation [5, 11]. Both NMR spectroscopy and mass spectrometry are extremely powerful analytical platforms well suited to measuring small molecule analytes. They nonetheless bring very different perspectives to metabolomic analyses and each has strengths and weaknesses with respect to delivering on the ultimate objective of identifying and quantifying all metabolites in a single sample in high throughput. Thus, the field of metabolomics is moving toward these goals in stages.

Among the strengths of NMR as an analytical system are that it requires little sample preparation and no explicit metabolite separation and produces a relatively accurate and finite signal from every NMR active atom in a sample. The power of NMR even extends to nonliquid samples and tissues with the potential of NMR spectroscopy and imaging to add structural information to metabolite analyses. Unfortunately, the proton or carbon NMR spectra of complex mixtures of biological samples remains practically undecipherable and all metabolites cannot be assigned to specific resonances. Therefore, at present, NMR is able to generate complex NMR spectra but as the majority of resonances are unassigned they serve as metabolite fingerprints. That is, resonance intensities are typically binned into intensity × ppm datafiles and these data are treated as inputs to statistical analyses. Considerable successes have been achieved in resolving groups of samples from diseased human patients versus normals, animals exposed to toxins, developmental stages of various animals and plants, and genetically modified animals and plants (reviewed in [4, 8, 10]). The field of metabolomics will move forward dramatically when NMR resonances can be annotated in terms of actual metabolites.

Mass spectrometry brings remarkable strengths to the task of measuring metabolites as well. With the astonishing accuracy of modern mass spectrometers, metabolite identification is now ostensibly an automated process. Combining various separation technologies in line with mass spectroscopy (liquid, electrophoretic, gas, and mass spectrometry itself) makes it possible for MS to identify all metabolites in even highly complex biological samples (reviewed in [7, 11]). However, because MS techniques invariably separate the molecules to analyze them, any higher structural information in the sample is lost and unavailable. The most important weakness of mass spectrometry as currently applied to metabolomic analyses is that mass detectors require precise standardization and calibration, and, at present, the absolute concentration of metabolites cannot be successfully defined due to variations in ionization parameters of the instruments.

Mass spectrometry, nuclear magnetic resonance spectroscopy, and chromatographic technologies are all high-throughput platforms that are capable of simultaneously producing highly quantitative data on important metabolites. The field of metabolomics still faces its major challenge in the dynamic range of natural biological metabolites, both the variation in chemical properties and their absolute abundance in biofluids. This dynamic range requires that new tools be developed that can accurately measure the entire range of components, particularly minor, and frequently important, bioactive products. These new tools (likely to be elaborations on current techniques) will produce quantitative data to add to databases that are being assembled today. Pending the development of a single analytical platform, it is likely that combinations of analytical platforms will be necessary to acquire the first-generation metabolomic databases.

While it is beyond the scope of this chapter's discussion, bioinformatics tools that are capable of building biological information out of the massive metabolomic data sets are also being developed. The great potential for these tools is that metabolomics informatics can join with knowledge management tools to link an individual's metabolite analysis to his/her present and future health status and to the metabolic knowledge of entire populations of samples and health outcomes. The growing science of managing (see Chapter 16 in this volume) and analyzing (see Chapter 17 in this volume) biological data using advanced computing techniques will allow scientists to have a "field day" analyzing the high-density data that are obtained through metabolomic data. In fact, it can be stated that complex biological data will resuscitate the entire field of applied mathematics as much as mathematics will bring understanding to biological systems.

The Future of Metabolomics

The field of metabolomics has great potential for metabolic research, and this potential has been recognized by the NIH by the inclusion of metabolomics in the NIH roadmap [12]. All "omics" technologies—genomics, proteomics, and metabolomics—have as a goal to understand the effects of exogenous compounds on human metabolic regulation. In spite of a number of challenges faced by metabolomics, there is great potential in the food and health fields [13, 14]. Foods contain myriad nonnutrient molecules that are absorbed, metabolized, and released into body fluids. Intestinal microflora also produce metabolites that can alter the metabolome of human biofluids. Metabolic profiling by nuclear magnetic resonance spectroscopy will detect, but not necessarily identify, these components. Nuclear magnetic resonance spectroscopy pattern recognition, identification of metabolites by mass spectrometry, and conventional high-throughput chromatographic techniques can identify metabolites resulting from different dietary treatments. Genetics, diet, and nutritional status all play a significant role in the development of chronic disease [15, 16]. The emerging field of metabolomics is well suited to the assessment of dysfunction and metabolic balance [17] and contaminating [18] dietary components. The unique ability to study the complex relationship between nutrition and metabolism can aid investigation of the roles that dietary components, as well as drugs used in therapy, play in health and disease [19]. When all the tools of metabolomics are in place—analytical platforms, databases, and bioinformatics applications—investigators will be able to explore homeostatic control and to determine how metabolic balance may be disturbed by genetic variations, disease processes, drug and therapeutic interventions, and deficiencies or excesses of dietary components [20]. Metabolomic studies have demonstrated important effects of nutrients on endogenous lipid metabolism [6, 21, 22] and the effects on, and the effects of, dietary isoflavones [17] that may be beneficial although not essential. The new technologies for analysis of metabolites together with the developing databases, computational tools, and data processing will in the near future provide health-care professionals with uniquely valuable resources to detect predispositions to disease, will guide therapeutic approaches to disease cure, and ultimately will manage the entire spectrum of endogenous and exogenous influences to improve individual health.

4.5 GENOMICS

Using Evolution to Guide Diet and Health

The impact of the science of genomics to all of life sciences, and most certainly the field of nutrition, is truly revolutionary. Within the first three years of the post-human-genome era, some of the most perplexing questions about metabolic health and nutritional science have been answered with astonishing ease. A compelling example was described by Goldstein and Brown (for review see [23]). For much of the 20th century, researchers were unaware that there existed four monogenetic causes of premature atherosclerosis in humans. Decades of intensive research amounting to tens of thousands of scientific publications were necessary to identify two of the four causes—mutations in the LDL receptor [24] and its ligand, apo B100 (for review see [25]). The third genetic cause explained the rare sensitivity of certain individuals to dietary phytosterols (sitosterolimia), and the fourth was tied to intracellular cholesterol trafficking (ARH adaptor protein). These latter two genetic causes were identified using the techniques of genomics, and investigations by a single research group led to two publications within a period of four months [26, 27]! As another example of the power of genomics to reveal the underlying mechanisms of diet, the effect of saturated and trans fats to dysregulate hepatic cholesterol and lipoprotein metabolism has been the basis of a worldwide agricultural agenda to reduce saturated fat abundance in the diet. Yet the basic mechanism underlying the effect of saturated fat on metabolism was not known. The target or basic mechanism by which the effects of saturated and trans fats act was revealed by diverse genomics technologies assembled and applied as the basis of a single publication [28]. Genomics is thus providing unprecedented opportunities for developing both detailed and comprehensive understanding to biology. The success of these strategies in food applications, however, is extending beyond genomics as the measurement of gene expression.

Using Genomic Tools to Discover Nutritional "Gold" in Milk

The new field of nutrigenomics is bringing a new set of tools to gain an understanding of the effects of genetic diversity on the digestion, absorption, transport, and utilization of food chemicals, and how these processes affect health and disease at the level of the individual. An implicit promise of such an effort is that once this understanding is acquired, it will enable sound scientific approaches to change discrete aspects of health via dietary intervention. At this point, however, the dietary ingredients with which to rationally manipulate human metabolic systems and deliver on the promise of improved health are still undeveloped. The lack of understanding highlights the need to identify additional bioactive nutrients and in particular their molecular mechanisms of action, and to determine for which individuals and under what metabolic conditions these bioactive nutrients are appropriate. Traditionally, bioactive nutrients have been discovered empirically, through association of foods with particular biological effects or through epidemiological correlation of food consumption with health outcomes. Fortunately, genomics brings a new opportunity to discover unrecognized benefits by which diet influences health but to take a rational, genetics and evolutionary approach to the discovery of new food ingredients and diets as well. Evolution did in fact produce specific animal foods, and the genomes of various animals, including humans, contains this exceptionally valuable knowledge resource.

Nonessential bioactive dietary chemicals can be classified as phytonutrients—derived originally from the plant kingdom—or zoonutrients—derived solely from animal tissues—according to their biological tissue of origin [29]. All of these molecules arose through evolution via selective pressure on their actions and functions. The plant kingdom provides vitamins, minerals, and other essential nutrients that humans cannot synthesize. Plants also contain toxic secondary metabolites that do not participate in their own basic metabolism but that are the result of a strong evolutionary pressure to avoid predation [30]. The physiological mechanisms of action by which these toxins dissuade animal and microbial predation range from enzyme inhibition to physiological irritation to endocrine disruption. The defining activity of these secondary plant metabolites is to disrupt a physiological process in animals or microorganisms consuming them. This underlying theme of producing chemicals that on ingestion disrupt the normal physiological processes of animals thus makes plant secondary metabolites, in many cases, intriguing drug candidates. Some of these actions of plant metabolites as drugs have been discovered empirically, and plant materials have been used by humans for centuries for their medicinal actions on individuals who are ill or diseased. The tools of genomics are allowing researchers to understand how these molecules emerged and how to use their targets of action to modify metabolism in curing diseases. However, although plant secondary metabolites have been successfully used as therapeutic drugs targeting a variety of diseases, as a principle, they cannot be considered the logical conceptual basis for improving health. The actions of plant metabolites, even when dramatically successful in curing disease, cannot be considered useful to optimize health. Optimizing health through dietary ingredients will need a different genetic model than plants evolving to avoid being consumed. Now that metabolomics provides the promise of measuring metabolites that can be measured precisely and comprehensively, it is possible to build an unprecedented understanding of the mechanisms by which diet can affect metabolic processes. What is necessary is a genetic model of diet for improving health. Fortunately, there is an ideal model—milk—and beyond this, there is a wealth of examples that are broadly termed zoonutrients [29].

Zoonutrients are those molecules that evolved under Darwinian selective pressure to support particular physiological processes in animals and thus tend to promote and not to disrupt metabolic pathways. The mechanisms can broadly be divided into three categories: (1) promotion of optimal growth, (2) protection from pathogens, toxins, and stresses, and (3) promotion of correct response to environmental stimuli and stress [29]. Of the sources of zoonutrients, the richest in bioactive components is mammalian milks. The basic role of milk in biology is to provide a complete source of nutrition to infant mammals, and to do so at a low metabolic cost to the mother. Mammalian milks have traditionally been viewed as a good source of nutrition because they deliver essential nutrients at the correct time in infant development, and the beneficial activities attributed to milk are becoming increasingly more recognized [31]. As mammals are underdeveloped at birth, they are vulnerable to many stresses, and they rely on an exogenous source of protective molecules. Milk is a source of passive and active immune factors and other bioactive components that promote health and survival of infants. Milk has not only been a participant in evolutionary selection, it has been a driver, facilitating the emergence of mammals as successful animals.

The composition of mammalian milks, which includes nonessential but bioactive components, has resulted from the myriad processes of evolution and it follows that the history of evolutionary pressure is contained within the mammalian genomes. Thus, the emergence of genomics as a field of scientific discovery should provide an increasingly transparent window into the processes and benefits of milk. In parallel, the functionality of milk constituents should yield valuable information for the field of nutrigenomics. Milks are complete foods for the infants they support. Deconstruction of milk components (i.e., separating and studying the functions of all components) across all important nutrients will provide strategies by which to deliver specific nutrients when incorporated into a new generation of foods.

Studies into the functionality of milk constituents will broaden the understanding of nutritional targets. The bioactivities of milk components have traditionally been studied one molecule at a time, yet a more integrated approach is now possible because of the new separation and measurement technologies available [32]. A first priority for the application of genome research to nutrition is the opportunity to discover all the genes responsible for lactation and to annotate the functions of all the constituents of mammalian milks in an unbiased way. The sequencing of many mammalian genomes will allow comparative genomics to identify the genes involved in lactation. By taking a genomic approach, nutrition investigations will not only further characterize the previously identified components in milk, but will be equipped with a new toolset to discover milk's nutritional functions.

The composition of milk varies significantly across species and during the stages of lactation. Genomics-based approaches to milk component analysis using the genes that produce the components and their regulation will provide insight into which components change over the course of lactation. These changes and their genetic regulation can then be compared with known aspects of a particular mammal's developmental biology, and this knowledge can lead to new hypotheses about the function of the nutrient component. As an example of such an approach, recent research in this laboratory has focused on human milk, which is particularly rich in soluble oligosaccharides. After lactose and fat, oligosaccharides are the third most prevalent component of human milk. These molecules range in degree of polymerization from three monomers to over 32 and are comprised of glucose, galactose, *N*-acetylglucosamine, fucose, and sialic acid. To date, over 130 different human milk oligosaccharides (HMOs) have been identified in a pooled sample using mass spectrometry [33], yet the reasons for such a high concentration and diversity are not known. HMOs vary among human maternal genotypes, with oligosaccharide species present in each mother's milk varying with Lewis and secretor genes [34]. Despite the impressive concentration and diversity of these molecules in breast milk, current understanding of their nutritional function does not explain such a metabolic energy investment by the human mother. Unraveling the bioactivities of oligosaccharides in the infant will provide a broader understanding of their nutritional function and suggest ways to utilize these molecules as nutritional components in other foods.

Although oligosaccharides are a dominant constituent in human milk, their role does not seem to be the simple provision of calories. They resist digestion by host hydrolases in the small intestine [35, 36], and those that are absorbed to a small extent are largely excreted in an infant's urine [37]. Not surprisingly, due to their lack of hydrolysis, the majority of HMOs arrive in an infant's colon relatively intact. Work dating back to the 1950s with the mutant strain *Bifidobacterium bifidum* var. *pennsylvanicum* indicated that milk oligosaccharides may contain, or in themselves be, a necessary factor for growth of this organism in vitro [36]. The fractionation of milk and subsequent assay of the individual fractions for growth promotion indicated that the necessary component was *N*-acetlyglucosamine. More recently, the results of these studies have been interpreted to indicate that the HMOs are in part responsible for the higher concentration of bifidobacteria in breast-fed infants compared with formula-fed infants due to their selective fermentability (for review see [38]). Nonetheless, interpretations of this early work will need to be substantially revisited since the property being measured was not fermentability of human milk oligosaccharides, as the media used to support the bacterial growth also contained lactose.

In the last decade, there was increasing interest in the composition of the gut microflora, and in potentially modulating the composition through components of the diet [39]. Prebiotics are indigestible carbohydrates that are fermented by bacteria in the colon and that selectively stimulate the growth of bacterial species considered beneficial to health [40]. Although it has often been suggested that HMOs are fermented by species of *Bifidobacterium*, there has been little characterization of this activity reported in the literature. From a comparison of maximum growth of six species of *Bifidobacterium* grown on lactose, inulin—a well-characterized prebiotic—and human milk oligosaccharides, it can be concluded that the oligosaccharides were somewhat fermentable by all six species of *Bifidobacterium*, yet to a lesser degree than were lactose and inulin (Figure 4.1). Milk oligosaccharides contain at least 13 different glycosidic bonds, whereas inulin contains two, and lactose one. Thus, it stands to reason that these molecules would resist digestion by one species of bacteria. Ongoing efforts are underway to determine which oligosaccharides are fermented by which bacteria (strain, species, genera) and to determine the effect of incubation of these molecules with mixtures of bacteria of fecal origin.

4.6 METABOLOME ASSEMBLY AND ANNOTATION

The early phases of building metabolomics databases and metabolic relationships are using techniques that sample defined subsets of metabolites. When the identities of metabolites are unknown, the techniques are termed metabolic fingerprinting. When the determination of subsets of metabolites or when the concentrations of all individual metabolites is not complete, the process is termed metabolic profiling. Metabolic fingerprinting is a rapid classification of samples according to their origin or their biological relevance by high-throughput analysis [41]. These techniques can be used to develop and assemble larger metabolomes, to annotate specific pathways for their biochemical and physiological relationships, and to follow developmental processes, toxicological responses, and nutritional processes. Use of these principles in nutrition research and metabolite assessment as metabolic profiling is particularly

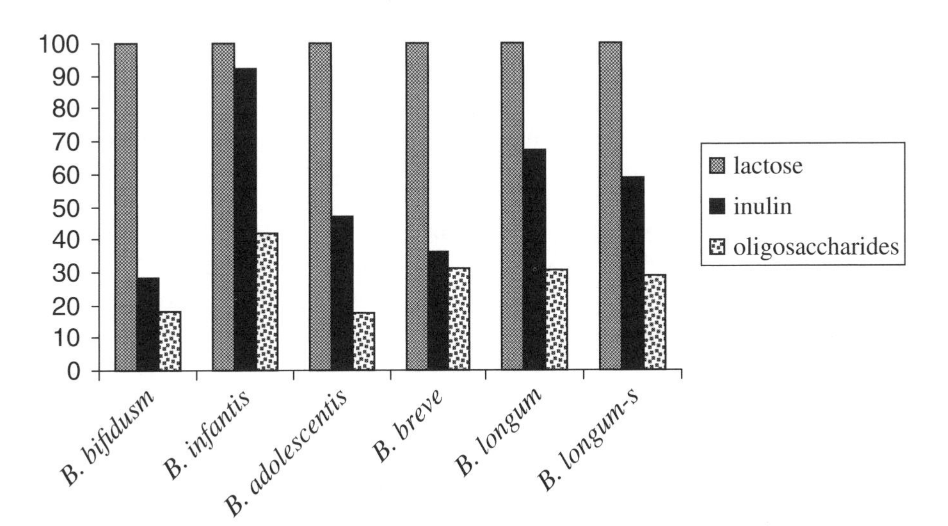

Figure 4.1. Fermentability of lactose, inulin, and human milk oligosaccharides by selected *Bifidobacteria* species as a percentage relative to lactose. *B. longum-s* has been sequenced. Genbank ID AABM02000000.

appropriate to highly dynamic metabolic conditions, in which expensive and comprehensive metabolomic analyses are impractical.

An obvious example of a dynamic condition is the postprandial state: it is recognized as a metabolic window in which variations in diet have significant effects on multiple aspects of physiologic regulation. Metabolic profiling is now being applied to the description of the myriad, time-dependent processes of metabolism during the postprandial state with the intent of understanding individual variations in responses to diet that are not discernible in the fasted condition and yet are nonetheless important to long-term health. Such profiling will be linked with studies that have linked specific genetic variations among humans to incidence and susceptibility of disease risk.

Metabolic Profiling of Postprandial Lipoprotein Metabolism in Humans

The majority of information on the composition of plasma lipids and lipoproteins has been obtained from fasted individuals. New approaches will be needed to describe and ultimately understand the metabolic events occurring after consumption of both fat-rich and carbohydrate-rich meals. A pilot study to precisely quantify and compare the lipid composition of lipoprotein fractions in plasma as a function of the fasted or fed state illustrates the opportunities and challenges afforded by metabolomic-style analysis.

The pilot study was designed to address the following questions: (1) Do individuals differ in their response to the same meal in terms of postprandial lipid metabolism? (2) Does the postprandial state represent a discrete metabolic window that elucidates features of metabolism that cannot be observed in the fasting state, and that can be discerned by compositional analysis of blood lipids? (3) Could unique features of lipid metabolism in subjects in the fed state be used as a metabolic diagnostic tool as in the fasted state currently used in most clinical diagnostics.

Blood was obtained from volunteers after an overnight fast and 3.5 h after consuming a high-fat (65%) breakfast consisting of 1 egg, 2 tbsp margarine, 1/2 English muffin, and 8 oz of whole milk. The lipoprotein classes in plasma were separated by density ultracentrifugation using a rapid isolation procedure [42]. The VLDL, IDL, LDL, sub-1 LDL, sub-2 LDL, and HDL layers were isolated and their respective lipid classes separated by HPLC, quantified by GC-FID (gas chromatography with flame ionization detection) analysis of fatty acid methyl esters, and identified by LC/MS. The sheer volume of compositional lipid data produced by these analyses requires software tools for display and interpretation. All lipoproteins responded to the postprandial state. Fatty acid composition (nanomoles/gram of lipid) data for each class of lipid in the VLDL for Subjects 1 and 2 in the fasting state and in the postprandial state reflected in part the consumption of the high-fat meal (Figure 4.2). In spite of consuming the same meal, there were significant variations in the blood lipid responses that are discernible in just two subjects. Hence, some key aspects of the nature of intraindividual and interindividual variation in lipoprotein lipid composition were revealed. The metabolomics data from studies of this type are capable

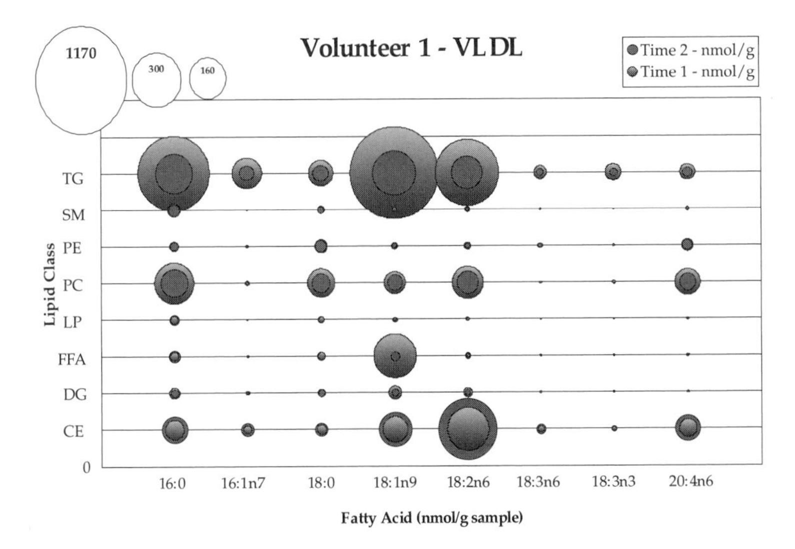

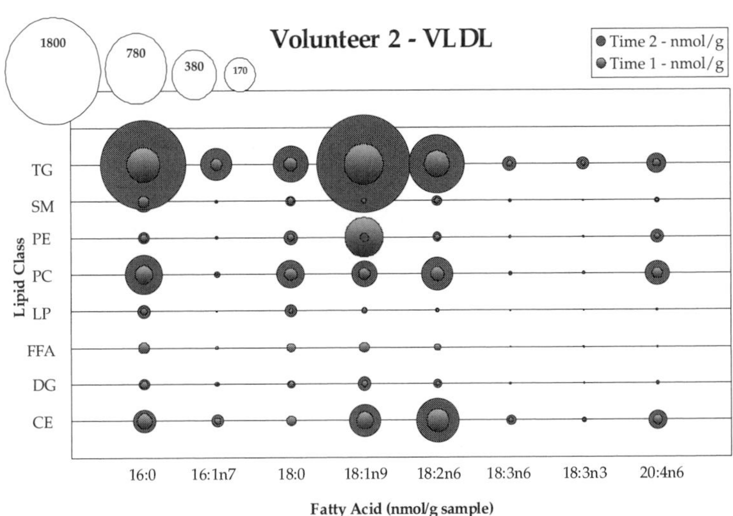

Figure 4.2. The figures display the results for the VLDL fraction of volunteers 1 and 2 and show the response of each volunteer to the high-fat meal in terms of the change in nanomoles (nmol) of selected fatty acids (FAs) per gram of lipid within each lipid class. The bubbles represent the size of the FA pool within each lipid class at fasting (in dark gray), and at 3.5 hours after the meal (in black).

of being assembled into mineable databases, suitable for building and testing dynamic models of metabolism, for establishing statistical estimates of metabolic data in humans in the postprandial state, and for exploring the consequences of specific dietary components on metabolic fluxes in the fed state as opposed to the fasted condition.

Thus, in evaluating differences in health risk and diet responsiveness, features of lipid metabolism not discernible in the fasting state can be revealed in the postprandial state. These results imply that future experiments designed to establish data sets of metabolic phenotypes of populations will need to include considerations of standardized times and meals to include the postprandial state. Such standardization will be necessary for data from future investigations to be comparable but also to build models of metabolism and include estimates of the dynamic variations in individual responses to diet as additional, quantitative predictors of health status.

Metabolic Profiling of Human Lipid Responses to Calcium Sources

Although considerable research has documented the effects of diet on the composition of lipoproteins during the fasted condition, other aspects of plasma lipids have not been as intensively investigated. Importantly, the abundance and composition of free fatty acids has only recently been recognized as an indication of health risk within the overall context of fasted blood lipids. Thus, it is not known to what extent diet per se affects the concentrations of free fatty acids in the fasted state. A study of the effects of the form of dietary calcium on weight loss and lipid metabolism revealed that the composition of unesterified fatty acids was in fact the lipid class that was most quantitatively responsive to dietary variation in calcium from different sources. Overweight subjects randomized into a control (cornstarch) group, a high-supplemented (calcium carbonate) group, or a high-dairy (3–4 servings/day) group were maintained on a 500 kcal/day deficit diet for 12 weeks, with weight and fat loss as endpoints. Comprehensive lipid analyses in this study revealed distinctive variations in unesterified fatty acids (Figure 4.3). Currently, there is a focus on the measurements of single endpoints associated with biomarkers of disease producing an incomplete and frequently misleading snapshot of the metabolic effects of various interventions—dietary or pharmacologic. With the use of currently available metabolomic tools, it is as straightforward to measure and quantify each and every fatty acid in a lipid class as to measure one (Figure 4.3). With such extensive data, the individual's metabolic profile provides a window to understand a much greater cross section of the biochemical processes involved in the individual's response to dietary inputs. Figure 4.3 illustrates the fatty acid composition of one lipid class—FFA—in three subjects—one from each intervention: control, high calcium, and high dairy. The goal of future studies is to bring a mechanistic understanding (i.e., individual annotation) to each of the lipid species in human biofluids.

Metabolic Profiling of Dietary Response to Long-Chain Fatty Acids in *Caenorhabditis Elegans*

Genomics knowledge and tools are making it possible to combine in vivo biological models with genetically defined molecular targets to address nutrition-related questions not possible in humans. One area of emphasis in this laboratory has been to understand how and why different fatty acids accumulate in specific subcellular

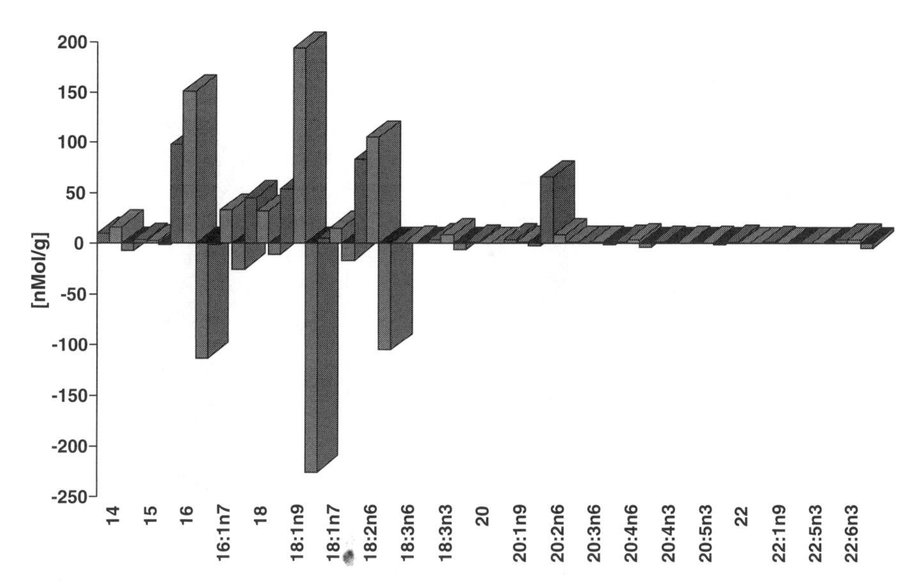

Figure 4.3. Change in FA composition of FFAs in Individuals 1 (control), 2 (high calcium), 3 (high dairy).

compartments in cells. This pursuit is illustrated by investigations of the conspicuous tendency of docosahexaenoic acid to accumulate in the cardiolipin class of lipids synthesized and residing within the inner mitochondrial membrane [43]. Studies on isolated human cells identified the accumulation of docosahexaenoic acid in cardiolipin and documented that this resulted in increased free radical leakage from mitochondria [44]. These studies pointed to a seemingly paradoxical behavior of these cells—they selectively accumulated docosahexaenoic acid in mitochondrial lipids, but this accumulation appeared to be deleterious to the integrity of cells.

These studies were extended to *C. elegans*, a nematode model that is relatively genetically controlled and that provides easy access to an in vivo study of the physiological properties of the organism. Knocking out the *fat-3* gene, which encodes a delta-6 desaturase, blocked de novo biosynthesis of polyunsaturated fatty acid past linoleic acid and alpha-linolenic acid [45]. When *fat-3 C. elegans* mutants were grown in the absence of long-chain polyunsaturated fatty acid, the inability to accumulate these polyunsaturated fatty acids into mitochondria depressed mitochondria energetics to the extent that whole-body locomotion was dramatically reduced, as measured by counting the body bends per minute of individual worms in a buffer solution [45]. Providing polyunsaturated fatty acid to the nematodes in the form of a strain of *Escherichia coli* (JMS-1) that contained a plasmid for eicosapentaenoic acid production led to complete recovery of normal locomotion (Figure 4.4). Additionally, a small amount of an omega-3 fatty acid supplement that contained both eicosapentaenoic acid and docosahexaenoic acid in triglyceride form, when spread on the surface of the culture plates on which the mutants were grown, also led to complete recovery of the mutants to wild-type levels of locomotion (Figure 4.5).

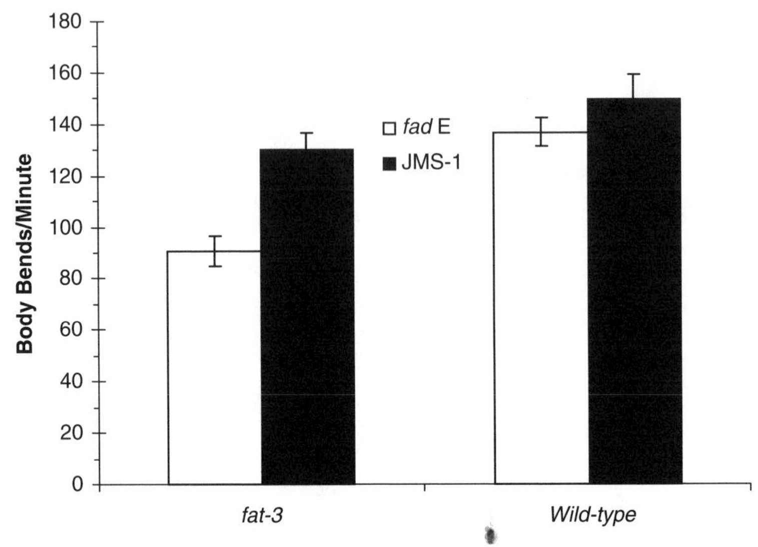

Figure 4.4. Body bends per minute of both wild-type and *fat-3 C. elegans* grown on an *E. coli* strain containing a plasmid for EPA production, JMS-1, and a control strain, *fad* E. Error bars represent standard deviation.

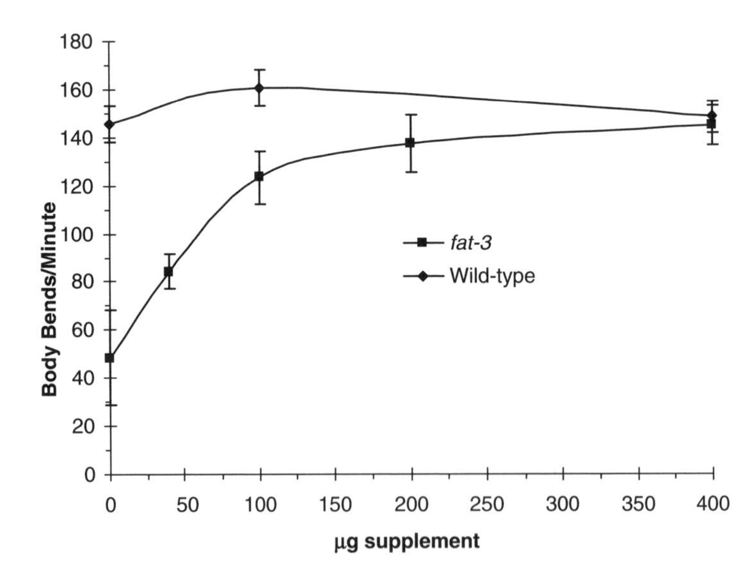

Figure 4.5. Body bends per minute of *fat-3* and wild-type *C. elegans* grown on increasing amounts of an omega-3 human dietary supplement and *fad* E *E. coli*. Error bars represent standard deviation.

Further research into the additional mechanisms underlying the pleiotropic phenotypes of the *fat-3* mutant will address the functions of polyunsaturated fatty acid in higher organisms, including humans.

4.7 BIOINFORMATICS: KNOWLEDGE MANAGEMENT FROM GENOMICS AND METABOLOMICS TO HEALTH ASSESSMENT

Metabolic information has been a cornerstone of health assessments and clinical recommendations for decades. Measurements of blood cholesterol, lipoproteins, triglycerides, and glucose have been the basis of population recommendations and for individual diet and drug interventions. Metabolomics will provide far greater insights into an individual's existing health and metabolic status. Nonetheless, metabolomics first delivers dramatically greater densities of data that require computerized data collection, analyses, and interpretation. Thus, automated informatic tools will be needed to provide an interface between metabolic data and the individual for whom they are diagnostic. There are various examples of high data density platforms in which software tools integrate large quantities of information into manageable units (see Chapter 16 in this volume). The data content of satellite-based global positioning systems is, by any criterion, massive. Yet users access precisely the information they desire via increasingly user-friendly interfaces on a "need-to-know" basis (Where am I now and how do I get where I want to go?). It is now time to consider how the knowledge that will form the basis for the future of diet and health will be managed.

At present, databases of metabolites are narrow, crude, and only rarely absolute; thus, it is not possible to compare data accurately within studies, much less between different experiments and populations. The promise of the metabolic information obtained from broader metabolic profiling has the potential to revolutionize human health. Nevertheless, there are important considerations necessary to move quickly to success. Data must be accurate, absolute, and comprehensive. For individuals to remain in good health, all of their metabolic pathways must function appropriately and metabolic needs must be balanced by appropriate nutritional inputs.

An understanding of the complex and multifactorial nature of chronic metabolic diseases requires coupling of metabolic pathways with new analytical tools, all within an integrative knowledge management system. New technologies allow measurement of most known endogenous metabolites by existing platforms that are being integrated, mined, and annotated by computational technologies [46–48]. The technologies constituting metabolomics are providing a first look at integrated metabolism and how it affects human health. The pattern of various metabolites within biochemical pathways must be understood or at least be comparable to existing databases to predict how the patterns can be altered by diet, drugs, or life style [6, 49]. The basics of biochemical pathways in humans have been assembled, and metabolomics knowledge tools must coalesce this information into comprehensive knowledge.

Metabolomic Databases

A major goal of life science research is to understand the mechanistic basis for biological differences—for example, healthy versus diseased, pathogenic versus nonpathogenic microorganisms. The presence or absence of a gene, a gene message, or a protein is typical as the basis of differences between samples with differing phenotype; thus, semiquantitative analyses are useful in genomics and proteomics. However, in metabolomics, quantitative analysis is essential [19]. Virtually all endogenous metabolites are present in discrete concentrations in biological samples. Annotation of genomes means assigning functions to the identities of genes. Annotation of metabolomics will mean assigning functions and consequences to the varying quantities of each metabolite in particular samples. It is the difference in the absolute concentration of metabolites that distinguishes biologically important differences in phenotype/outcome. Therefore, metabolite data must be quantitative to produce databases that provide unbiased biological information about a sample. Such quantitative databases can be used to compare different samples, phenotypes, or outcomes to serve as input variables for metabolic modeling and to assess the integrated flux through all metabolic pathways [6, 50]. Such data can continue to be compared, mined, and fitted to mathematical models to pursue and ultimately to resolve various hypotheses. Databases constructed with quantitative metabolite data are therefore permanent and are not obsolesced by new analytical platforms as they are developed.

4.8 CONCLUSION

Genetic variation in humans dictates in part their individual predispositions to disease and health potential. The field of nutrigenomics is seeking to understand the interaction between diet and human health using genomics. The expression of individual genetics, according to environment, gives rise to the phenotypic differences among people, and these differences are in part due to variations in metabolism. The field of metabolomics seeks to measure and understand the variation in metabolites in organisms, including humans. Metabolomics as an assessment strategy is designed to estimate the real-time realization of an individual's health potential. When the variation in metabolism and how such variation relates to health are understood, diet can be employed to control metabolism. Faulty enzyme activity (genetics, toxicology), improper substrate balance (nutrition), and faulty metabolic regulation (genetics, nutrition, life style, etc.) are the influences that relate to development of diseases. These effects are observable through quantitative metabolic assessment. The means to interpret changes in metabolites according to specific metabolic pathways is largely at hand because the knowledge of biochemical pathways has been a research achievement of the life sciences over the last century. The value of this knowledge is that observable, metabolic phenomena are potentially understandable and ultimately controllable from an informed perspective. If substrate imbalances are primarily responsible for deranged metabolism, then nutritional and drug intervention are logical strategies for therapy. By comprehensive measurement of metabolites to assess health, dietary advice can be tailored to the molecular basis for potential disease processes rather than assessing only the endpoint symptoms of a disease's consequences. This informed perspective will eventually provide actionable information for managing health and disease.

REFERENCES

1. O. Fiehn, J. Kopka, P. Dormann, T. Altmann, R. N. Trethewey, and L. Willmitzer (2000). *Nat. Biotechnol.* **18**:1157–1161.
2. A. Berger, D. M. Mutch, J. B. German, and M. A. Roberts (2002). *Genome Biol.* **3**(7): preprint 0004.1–0004.53.
3. D. M. Mutch, M. Grigorov, A. Berger, L. B. Fay, M. A. Roberts, et al. (2005). *FASEB J.* **19**:599–601.
4. J. K. Nicholson, J. C. Lindon, and E. Holmes (1999). *Xenobiotica* **29**:1181–1189.
5. J. W. Newman, T. Watanabe, and B. D. Hammock (2002). *J. Lipid Res.* **43**:1563–1578.
6. S. M. Watkins, P. R. Reifsnyder, H.-J. Pan, J. B. German, and E. H. Leiter (2002). *J. Lipid Res.* **43**:1809–1817.
7. R. J. Bino, R. D. Hall, O. Fiehn, J. Kopka, K. Saito, et al. (2004). *Trends Plant Sci.* **9**:418–425.
8. H. Jenkins, N. Hardy, M. Beckmann, J. Draper, A. R. Smith, et al. (2004). *Nat. Biotechnol.* **22**:1601–1606.
9. J. M. Ordovas and D. Corella (2004). *Annu. Rev. Genomics Hum. Genet.* **5**:71–118.
10. J. K. Nicholson and I. D. Wilson (2003). *Nat. Rev. Drug Discov.* **2**:668–676.
11. J. van der Greef, P. Stroobant, and R. van der Heijden (2004). *Curr. Opin. Chem. Biol.* **8**:559–565.
12. National Institutes of Health (2004). NIH Roadmap Initiatives 2004: Accelerating medical discovery to improve health: http://nihroadmap.nih.gov/ (accessed 14 February 2005).
13. J. B. German, D. E. Bauman, D. Burrin, M. Failla, H. C. Freake, et al. (2004). *J. Nutr.* **134**:2729–2732.
14. P. D. Whitfield, A. J. German, and P. J. M. Noble (2004). *Br. J. Nutr.* **92**:549–555.
15. V. L. Go, R. R. Butrum, and D. A. Wong (2003). *J. Nutr.* **133**:3830S–3836S.
16. J. A. Milner (2003). *J. Nutr.* **133**:3820S–3826S.
17. K. S. Solanky, N. J. Bailey, B. M. Beckwith-Hall, A. Davis, S. Bingham, et al. (2003). *Anal. Biochem.* **323**:197–204.
18. C. Teague, E. Holmes, E. Maibaum, J. Nicholson, H. Tang, et al. (2004). *Analyst* **129**:259–264.
19. S. M. Watkins, B. D. Hammock, J. W. Newman, and J. B. German (2001). *Am. J. Clin. Nutr.* **74**:283–286.
20. J. B. German, M. A. Roberts, and S. M. Watkins (2003). *J. Nutr.* **133**:2078S–2083S.
21. S. M. Watkins, X. Zhu, and S. H. Zeisel (2003). *J. Nutr.* **133**:3386–3391.
22. J. B. German, M. A. Roberts, and S. M. Watkins (2003). *J. Nutr.* **133**:4260–4266.
23. J. L. Goldstein and M. S. Brown (2001). *Science* **292**(5520):1310–1312.

24. M. S. Brown and J. L. Goldstein (1987). *Nature* **330**(6144):113–114.
25. J. P. Kane and R. J. Havel (2001). In C. R. Scriver, A. L. Beaudet, W. S. Sly, and D. Valle, eds., *The Metabolic and Molecular Bases of Inherited Disease*. McGraw Hill, New York, Chap. 115.
26. K. E. Berge, H. Tian, G. A. Graf, L. Yu, N. V. Grishin, et al. (2000). *Science* **290**(5497):1771–1775.
27. C. K. Garcia, K. Wilund, M. Arca, G. Zuliani, R. Fellin, et al. (2001). *Science* **292**(5520):1394–1398.
28. J. Lin, R. Yang, P. T. Tarr, P. H. Wu, C. Handschin, S. Li, et al. (2005). *Cell* **120**:261–273.
29. R. E. Ward and J. B. German (2003). *Food Technol.* **57**:30–36.
30. A. C. Leopold and R. Ardrey (1972). *Science* **176**:512–514.
31. J. B. German, C. J. Dillard, and R. E. Ward (2002). *Curr. Opin. Clin. Nutr. Metab. Care* **5**:653–658.
32. R. E. Ward and J. B. German (2004). *J. Nutr.* **134**:962S–967S.
33. B. Stahl, S. Thurl, J. Zeng, M. Karas, F. Hillenkamp, M. Steup, and G. Sawatzki (1994). *Anal. Biochem.* **223**:218–226.
34. Z. Jarkovsky, D. M. Marcus, and A. P. Grollman (1970). *Biochemistry* **9**:1123–1128.
35. M. B. Engfer, B. Stahl, B. Finke, G. Sawatzki, and H. Daniel (2000). *Am. J. Clin. Nutr.* **71**:1589–1596.
36. C. Kunz, S. Rudloff, W. Baier, N. Klein, and S. Strobel (2000). *J. Nutr.* **130**:3014–3020.
37. S. Obermeier, S. Rudloff, G. Pohlentz, M. J. Lentze, and C. Kunz (1999). *Isotopes Environ. Health Stud.* **35**:119–125.
38. C. Kunz, S. Rudloff, W. Baier, N. Klein, and S. Strobel (2000). *Annu. Rev. Nutr.* **20**:699–722.
39. T. S. Manning and G. R. Gibson (2004). *Best Pract. Res. Clin. Gastroenterol.* **18**:287–298.
40. G. R. Gibson and M. B. Roberfroid (1995). *J. Nutr.* **125**:1401–1412.
41. O. Fiehn (2001). *Comp. Funct. Genomics* **2**:155–168.
42. J. M. Ordovas, E. J. Schaefer, D. Salem, R. H. Ward, C. J. Glueck, et al. (1986). *N. Engl. J. Med.* **314**:671–677.
43. A. Berger, J. B. German, B. L. Chiang, A. A. Ansari, C. L. Keen, et al. (1993). *J. Nutr.* **123**:225–233.
44. S. M. Watkins, T. Y. Lin, R. M. Davis, J. R. Ching, E. J. DePeters, et al. (2001). *Lipids* **36**:247–254.
45. J. L. Watts, E. Phillips, K. R. Griffing, and J. Browse (2003). *Genetics* **63**:581–589.
46. R. J. Lamers, J. DeGroot, E. J. Spies-Faber, R. H. Jellema, V. B. Kraus, et al. (2003). *J. Nutr.* **13**:1776–1780.
47. W. Weckwerth (2003). *Annu. Rev. Plant Biol.* **54**:669–689.
48. L. Hood (2003). *Mech. Ageing Dev.* **124**:9–16.
49. S. M. Watkins, X. Zhu, and S. H. Zeisel (2003). *J. Nutr.* **133**:3386–3391.
50. O. Fiehn (2003). *Phytochemistry* **62**:875–886.

5

GENETIC AND MOLECULAR BUFFERING OF PHENOTYPES

John L. Hartman IV

Department of Genetics, University of Alabama-Birmingham, Birmingham, Alabama

5.1 INTRODUCTION

Robustness in Biological Systems

As a result of natural selection, cells and ultimately organisms are robust, meaning that they retain the ability to execute their genetic programs optimally in the face of stochastic variations, environmental perturbations, and mutations of the genetic code itself [1–6]. The properties of biological systems that confer such stability are of interest, since human disease can be viewed as simply the partial loss of robustness against various genetic or environmental alterations, which the system may encounter. Genetic buffering is the compensatory process whereby particular gene activities confer phenotypic stability against genetic or environmental variations. This is commonly observed experimentally from "enhancer" screens, although there are instances where loss of a gene activity confers relative system stability against particular perturbations as well (i.e., genetic suppressors). In experimental organisms it's not uncommon for a gene to be essential in one inbred strain, yet dispensable in another [1], possibly due to "enhancing" modifiers in one genetic background or "suppress-

Nutritional Genomics: Discovering the Path to Personalized Nutrition
Edited by Jim Kaput and Raymond L. Rodriguez

ing" modifiers in the other. Similarly in natural populations, a particular mutant allele may contribute to severe disease phenotypes in some individual, yet have relatively little effect on others [7–14]. Genetic buffering provides a conceptual tool for global, quantitative analysis of the genes that modify phenotypes, embracing the idea that genetic "causation" and "modification" of phenotypic traits represent the ends of a spectrum of gene interactions [7, 15–18].

The combinatorial nature and quantitative complexity of gene interaction analysis is perhaps the biggest challenge for understanding genetic buffering networks. If one considers the cell as a networked genetic system, variation in the activity of any one component may not only affect the observed system output directly, but may also alter the buffering capacity of the system against environmental perturbations or variations in activities of other system components. The complexity becomes even greater when considering the networks of phenotypic interaction between cells in multicellular systems, where the robustness of a phenotype against genetic, environmental, and stochastic variations could be attributable to genes acting in different cells, tissues, or organs. Superimposed on the complex topology of interacting components is the quantitative challenge of distinguishing phenotypic effects actually attributable to interaction from individual effects that are expressed independently of combination. Unicellular systems offer the advantage of reduced complexity in determining genetic principles for cellular robustness. Just as gene activities and cellular functions have been conserved over evolution, principles of gene interaction, which provide cellular stability against variations in these activities and functions, should be conserved as well [18].

Old Observations About Natural Selection, New Opportunities for Understanding Phenotypic Buffering

C. H. Waddington, wrote in 1942, ". . . developmental reactions, as they occur in organisms submitted to natural selection, are in general canalized. That is to say, they are adjusted so as to bring about one definite end result regardless of minor variations in conditions during the course of the reaction." Waddington was brought to this conclusion by the emergence of distinct tissue types adjacent to one another during development and the absence of intermediate forms. His metaphor was of a number of well-defined canals down which development could flow. He continued, "the constancy of the wild type must be taken as evidence of the buffering of the genotype against minor variations not only in the environment in which the animals developed but also in its genetic make-up" [19]. Waddington's observations are resurfacing, as the feasibility of bridging theoretical and experimental genetics to understand biological complexity increases [18, 20, 21].

Many of Waddington's observations and theoretical explanations came before it was known that DNA was the genetic material. Now, with the sequencing of entire genomes and the cumulative progress in understanding genetic and molecular mechanisms, cellular pathways, and regulatory processes, sufficient knowledge and appropriate tools are available to conceptualize and experimentally measure "cellular robustness" and "genetic buffering" as properties of "biological systems" [22].

The primary aim of this chapter is to point out how global and quantitative analysis of gene interactions provides a strategy to understand why genetic systems are robust to perturbations [3, 23–25]. The key concept is that genetic interactions themselves underlie genetic buffering [18]. Qualitatively, gene interaction implies that the presence, absence, or variation of a gene's activity has the capacity to alter the phenotypic output of the system in response to a perturbation [23, 25]. Quantitatively, this means that the phenotypic response from combining gene mutation and the perturbation is nonadditive (Figure 5.1) [24]. Comprehensive *and* quantitative analyses of gene interactions reveal a functional network, indicating the genetic requirements for phenotypic stability. Gene annotation databases can then be used to infer the functions of genes that interact, generating testable hypotheses for the molecular basis of experimentally observed genetic interaction networks.

There are two points for emphasis regarding quantitative modeling of buffering. (1) While an additive model is described here as the null hypothesis for gene interaction (Figure 5.1), multiplicative models have also been used [26, 27]. In any

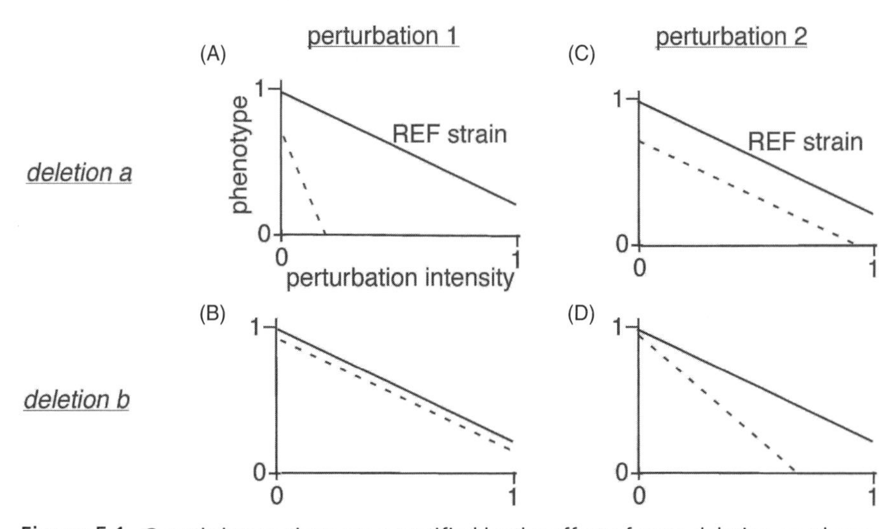

Figure 5.1. Genetic interactions are quantified by the effect of gene deletion on phenotypic response. Gene *deletion strain a* (A, C) and *deletion strain b* (B, D) are used to quantify the interaction between *GENE A* or *GENE B* and perturbation 1 (A, B) or perturbation 2 (C, D), respectively. Differences between "phenotypic slopes" of the reference strain (REF) (solid lines) and the deletion strains (dashed lines) represent the perturbation-specific buffering capacity for each gene. For ease of illustration, the reference strain is shown to have identical phenotypic responses to perturbations 1 and 2 (A–D). The phenotypic effect of gene deletion alone is observed in the absence of perturbation, where it can be seen that deletion of *GENE A* but not *GENE B* has a phenotypic effect. In this illustration, *GENE A* strongly interacts with perturbation 1 (A) but does not interact with perturbation 2 (C). *GENE B* does not interact with perturbation 1 (B) but interacts weakly with perturbation 2 (D). Thus, phenotypic stability (robustness) against perturbation 1 depends on buffering activity of *GENE A*, but not *GENE B*, while the reverse is true for perturbation 2.

case, the null hypothesis provides a common reference point for quantifying whether effects interact, and if so how strongly. (2) Genes interact with both other genes and the environment. For example, genes that buffer a particular genetic mutation ("epistasis") might also be expected to buffer drug perturbations ("chemical–genetic" interaction) that target the same gene (see Figure 5.8) [25].

The experimental platform of using the yeast deletion strain collection to assess genetic interaction networks is an expansion and further integration of classical "reductionist" genetic approaches, where the molecular functions of select genes and pathways that determine phenotypes are ascertained under a controlled environment and genetic background. Other biomolecular networks, such as transcriptional networks [28–30], and protein–protein interaction networks [31] subserve genetic interaction networks, however, it is important to recognize that simple correlations between gene activities over these different networks do not exist. For example, it has been shown that transcriptional regulation of a gene has weak correlation with the phenotypic impact of losing the same gene activity [32–34]. The mechanism by which naturally selected genes buffer the phenotype may or may not involve transcriptional regulation, because cells have a wide array of other gene regulatory mechanisms that may be used for a particular buffering response. Additionally, the cell may have parallel circuits for buffering any particular perturbation. Accordingly, most of the genes that are coregulated as part of the transcriptional "signature" for a buffering response are dispensable for phenotypic stability.

A major goal for systems biology is to utilize knowledge about biomolecular networks to predict human disease phenotypes and manage disease on an individualized basis. Functional overlap between different genes and pathways and the many layers of gene regulation that cells utilize, such as transcriptional activation/repression, splicing, degradation, translational regulation, protein transport, and modification, explains why the genetic appearance of "molecular" networks is so different from "phenotypic" networks [35, 36]. Thus, there is a need to understand these different genetic networks independently in order to understand their relationships to one another. Ultimately, this will lead to integration of genomes, transcriptomes, proteomes, and phenomes [37–40].

Relevance of Genetic Buffering to Nutritional Genomics

The concept of genetic buffering is relevant to how phenotypic stability is maintained, because nutrition itself is an ongoing perturbation: nutritional status is in constant flux. Natural selection acts on biological systems for phenotypic robustness against nutritional environmental variation, as evidenced by genetic interactions between certain loci and dietary factors variably modifying the health of individual people. Genetic buffering provides a framework for identifying such effects systematically and quantitatively because foods are complex mixtures of bioactive substances, exhibiting complex phenotypic interactions across genetically diverse populations. Interactions can be dissected using the "reductive" power of large-scale studies in genetic model systems under controlled but variable environmental conditions, generating hypotheses about sites of functional allelic variation in natural populations. Since many pharmaceutical agents target and modulate pathways that interact with bioactive components of foods, the topology and dynamics of genetic interaction networks should be mutually informative for nutrigenomics and pharmacogenomics.

5.2 EXAMPLES OF BUFFERING

Diploidy, Haplo-Insufficiency, Enhancers, and Suppressors

Diploidy is the most abundant source of genetic buffering. Though a special case, it illustrates well the principle of nonadditive gene interactions. Most genes are "haplo-sufficient," meaning that one functional allele suffices as well as two. This can be viewed as a functional allele buffering a nonfunctional counterpart. This does not require dosage compensation [41], but only that the biological system can absorb the reduced input of gene activity and maintain its same phenotypic output. Haplo-sufficiency is the principle behind recessive inherited diseases and dominant diseases due to somatic loss of a buffering counterpart. In either case, phenotypic selection against organisms with mutant alleles is relaxed. From this view, both heterozygotes and homozygotes, for a disease-contributing allele, have reduced phenotypic buffering capacity, relative to two normal alleles. Consistent with this view, the heterozygous diploid set of yeast deletion strains has been used to identify genetic targets of drugs [42, 43].

This loss of buffering in a homozygote may be sufficient to produce disease, but it does not account for all of the buffering capacity, since even "monogenic" disease in a genetically heterogeneous population produces a broad range of phenotypes [7–14]. In contrast, some loci are haplo-insufficient. Thus, not all genes are buffered by diploidy, and natural selection may act on single alleles even in the presence of a functional counterpart [44]. Molecular explanations for dominance have been considered theoretically and in a limited number of experimental contexts [45–49] but exceed the intended scope of discussion, which is focused on genetic interactions between different loci. Nevertheless, it is worth mentioning that a long-standing question in evolutionary biology is why sexual reproduction is under such strong natural selection [50]. Genetic buffering via diploidy provides a potential explanation, because the constraint of negative selection against genetic mutation, which potentially gives rise to novel beneficial traits, is alleviated.

Epistasis (gene–gene interaction) reveals instances of phenotypic buffering between different genetic loci, which may involve homologous or nonhomologous genes [1, 3, 23, 51]. "Enhancer" and "suppressor" screens exploit this phenomenon to identify genes that act within a common pathway; however, multiple pathways are typically discovered in such screens [52]. In the case of genetic suppressors, *loss* of a gene function may provide relative phenotypic stability against a second perturbation [53, 54]. However, presumably yet other genes buffer loss of suppressor gene functions. An extreme example of phenotypic "enhancement" is synthetic lethality, where the combination of two nonlethal mutations causes lethality in haploid cells.

Synthetic lethal interactions imply that the function of one gene compensates, or buffers, loss of another [18, 23, 55, 56]. In this regard, it is remarkable that over 80% of genes are nonessential for growth under optimal laboratory conditions in haploid yeast [33, 57]. Challenging viable deletion strains with additional genetic or environmental perturbations and measuring the resulting phenotypes provides insights as to why seemingly dispensable genes have been maintained by natural selection. Notably, the number of gene–gene interactions alone has been estimated to be over 100,000 [23]. Thus, phenotypic profiling of gene knockout strains is a focus for investigating buffering networks that confer phenotypic robustness against genetic and environmental perturbations. Before going into more detail about genetic interaction network analysis, other perspectives of buffering are briefly discussed.

Genetic Redundancy: Genome Duplication and Gene Families

Gene duplication provides a strategy for genes and organisms to evolve new functions and traits, via a mechanism of buffering that is less constrained than diploidy. Following duplication, a gene may be mutated with little fitness cost to the organism, presuming the duplication provides no original selective disadvantage or advantage. Mutation in duplicate genes may result in subfunctionalization—for example, new temporal or spatial activity, cell-type specific regulation, modified substrate specificity, product formation, or physical interaction. Once the "new" member of the gene family has achieved a specialized function, it undergoes purifying selection and is maintained independently [58–60]. The evolved gene may have redundant functions that can buffer the original gene that gave rise to it, and vice versa.

Saccharomyces cerevisiae and other organisms have undergone genome duplication relatively recently [61–63]. One study concluded that over 25% of the *S. cerevisiae* gene deletion strains that maintain robust growth were compensated by redundant partners [3]. The lines of evidence supporting this conclusion were: (1) a higher probability of functional compensation for a duplicate gene than for a singleton, (2) a high correlation between the frequency of compensation and the sequence similarity of two duplicates, and (3) a higher probability of a severe fitness effect when the more highly expressed duplicate copy was deleted [3]. Other studies have suggested that alternative pathways and regulatory networks, in addition to redundant/duplicated genes, contribute significantly to phenotypic robustness [1, 2, 23–25, 35, 36, 51, 64–66].

Dedicated Buffering Proteins: Molecular Chaperones

An alternative strategy to genetic redundancy and gene regulatory networks for buffering phenotypes against genetic/environmental variation is molecular chaperoning, which compensates the structural/functional loss of other proteins by restoring the defective protein itself, rather than by substituting a "new" gene activity to buffer the lost activity. HSP90 (heat shock protein) is the prototype for this family of proteins, which, in addition to providing structural stability to proteins during high-temperature stress, has been shown to stabilize morphologic traits by buffering genetic variation in signal transduction proteins in animals and plants [67, 68]. This provides a specific molecular mechanism for strengthening and relaxing the evolvability of traits, and thus HSP90 has been considered as a "capacitor" for evolution [67–70]. Molecular chaperones represent a unique example of genetic buffering in that their activity appears relatively dedicated to particular signaling pathways rather than an emergent property of the overall system function [1, 70]. It will be interesting to learn if other classes of proteins are dedicated to buffering, and to learn more about phenotypic buffering by protein chaperones operating in different pathways, cell types, and organisms.

Quantitative Genetic Interaction Networks

Qualitatively, genetic interaction means that the phenotypic effect of perturbation (e.g., gene mutation or drug treatment) is dependent on allele status at the "interacting" locus (Figures 5.1–5.3). Quantitatively, there must be a definition for "independent" phenotypic responses to perturbation that can serve as a null hypothesis for interaction and a reference point for quantifying strength of different genetic interactions (Figures 5.1–5.8). Additive and multiplicative models have been used in this way to model gene interactions quantitatively [24, 26, 27, 66]. For example, if two perturbations each have a 30% effect on reducing the phenotypic measure, the null hypothesis would be rejected if the combination of phenotype effects was

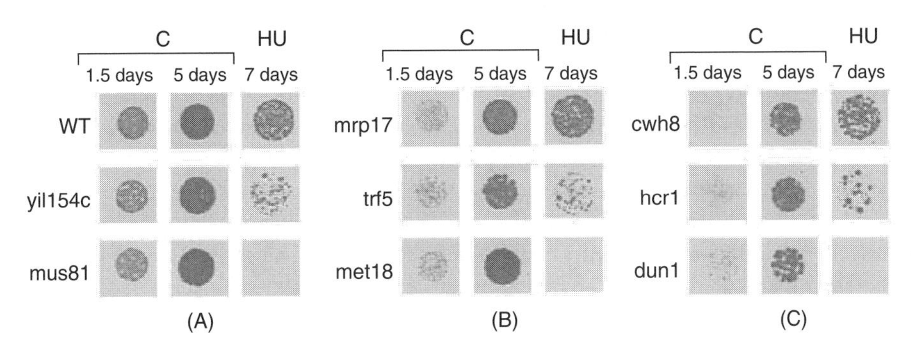

Figure 5.2. Genetic interaction is a quantitative continuum. Each row of data represents growth of a particular strain, either in the absence (columns 1 and 2 of each 3 × 3 block) or presence (column 3) of 150 mM HU. The row labels indicate yeast genes deleted from WT, the isogeneic starting strain, from which all deletion strains were created. In each block, earlier (1.5 days) and later (5 days) time in the control ("C") condition reflect the effect of each deletion alone (no HU). In panel (A), deletion of *YIL154c* or *MUS81* appreciably modifies the phenotypic response to HU but does not modify growth in the absence of HU. Panels (B) and (C) represent strains with moderately and severely reduced growth due to gene deletion alone (compare first columns in each block). In each case a quantitative range of interactions with HU (compare third columns) is also observed. Thus, in order to quantify interactions precisely, the phenotypic effects of gene deletion and perturbation must be considered separately as well as in combination, and a range of perturbation intensities must be tested (see Figs. 5.4, 5.5, 5.6).

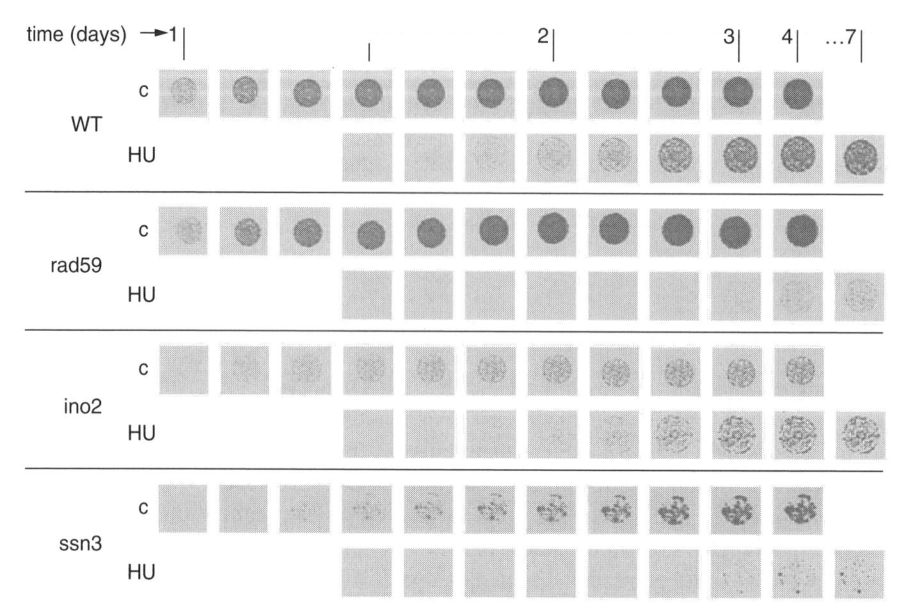

Figure 5.3. Genetic interaction is quantified by nonadditive phenotypic effects. Images of the reference strain (WT) and three different deletion strains, growing in the absence (C) or presence (HU) of 150 mM hydroxyurea, are depicted at several different times over 7 days. Nonadditive (synergistic) interaction between HU treatment and deletion of *RAD59* is evidenced by equivalent growth of the *rad59* deletion strain and the reference strain (WT) without HU exposure, while *RAD59* is required for growth in the presence of HU. In contrast to the *rad59* deletion strain, the *ino2* deletion strain has reduced growth in the absence of HU, which must be accounted for when assessing whether the robustness of growth against HU perturbation depends on the presence/absence of the *INO2* gene. Like the *ino2* deletion strain, the *ssn3* deletion strain also demonstrates slow growth in the absence of HU. However, in contrast to deletion of *INO2*, deletion of *SSN3* appears to have a greater-than-additive effect on growth when combined with HU.

Figure 5.4. Cell proliferation is a dynamic function of perturbation intensity and time. Panels (A) and (B) represent mean values for many replicates of the reference strain. In panel (A), image density of spotted cultures (see Figure 5.3) is plotted against time for different perturbation intensities. In panel (B), image density is plotted against perturbation intensity at discrete times during the experiment. Note that reference strain growth is robust against HU perturbation up to a concentration of 10 mM; beginning with 25 mM the cell's buffering capacity for HU is exceeded by additional increases in concentration (panel A). The calculation of interaction is potentially affected by the time at which the assay is scored, as illustrated in panel (B), where it is shown that HU concentrations as high as 150 mM appear to have little affect on reference strain proliferation by 126 hours, although the same concentrations have large effects at 40 hours. (See insert for color representation.)

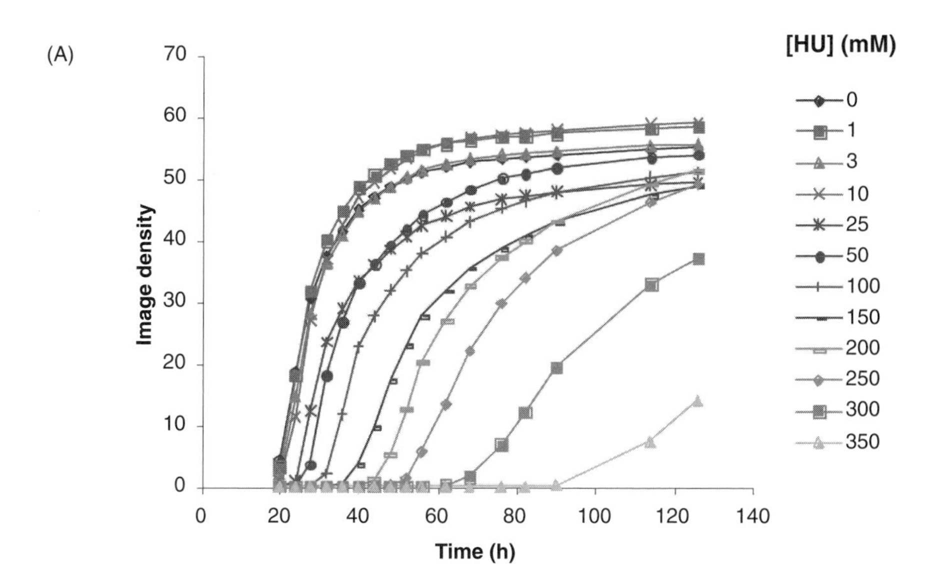
(A)
Image density
Time (h)
[HU] (mM)
0
1
3
10
25
50
100
150
200
250
300
350

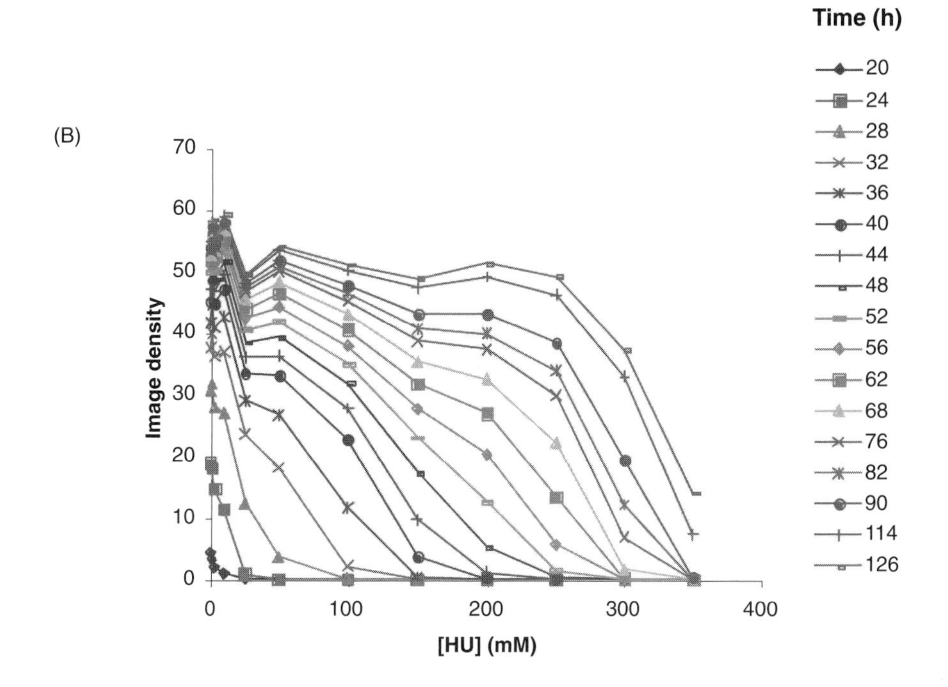
(B)
Image density
[HU] (mM)
Time (h)
20
24
28
32
36
40
44
48
52
56
62
68
76
82
90
114
126

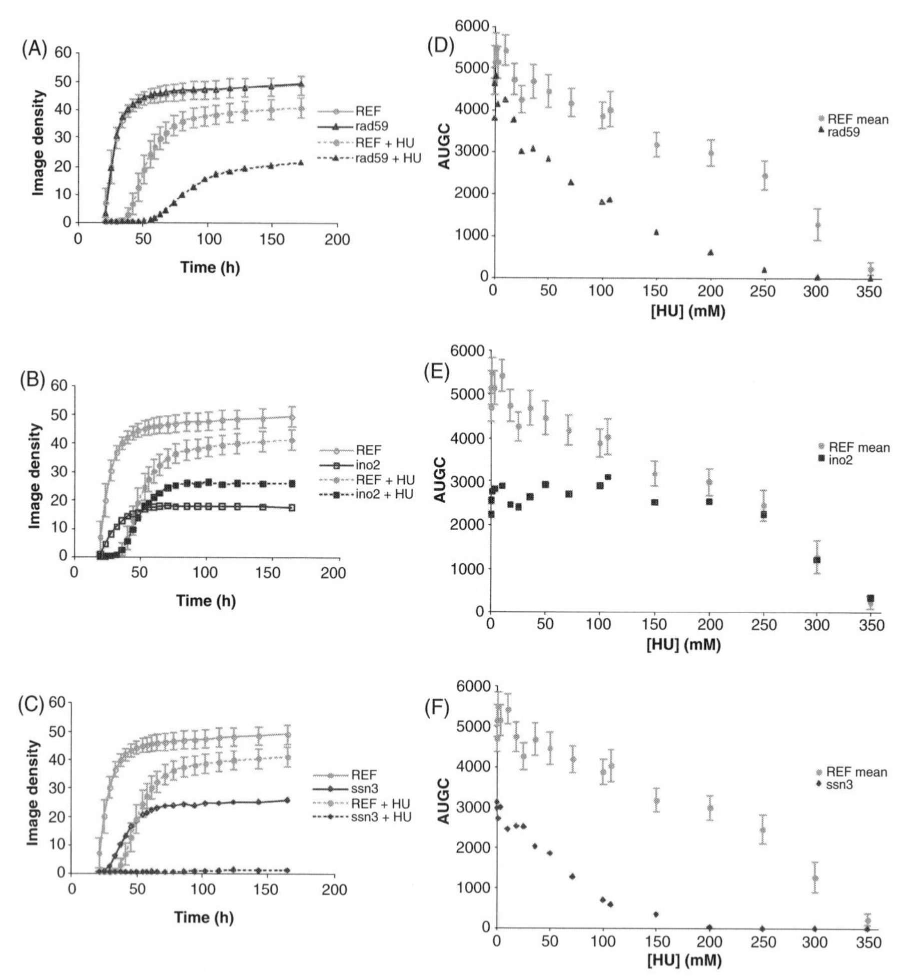

Figure 5.5. Area under the growth curve (AUGC) as a unit measure of cell proliferation for quantifying genetic interactions. Panels (A)–(C) are representations of data derived from spot cultures as in Figures 5.3 and 5.4. In panels (A)–(C), mean growth and standard deviation for replicate cultures of the reference strain in the absence (gray solid lines) an d presence (gray dashed lines) of 150 mM HU are reproduced on each panel. The corresponding growth curves for the *rad59* (panel A), *ino2* (panel B), and *ssn3* (panel C) deletion strains are shown in the absence (solid black lines) and presence (dashed black lines) of perturbation with 150 mM HU. In panels (D)–(F), AUGC is plotted against the perturbation intensity (HU concentration) for replicates of the reference strain (the gray data points represent the AUGCs from Figure 5.4A). The *rad59* and *ssn3* deletion strains demonstrate synergistic (greater than additive) gene interaction (panels D and F), implying that *RAD59* and *SSN3* genes help buffer cells against growth inhibition by HU. The *ino2* deletion strain demonstrates antagonistic (less than additive) gene interaction (panel E), meaning that deletion of *INO2* masks the HU effect, which is observed in the reference strain between 25 mM and 200 mM HU. Beyond 200 mM, no further interaction is observed.

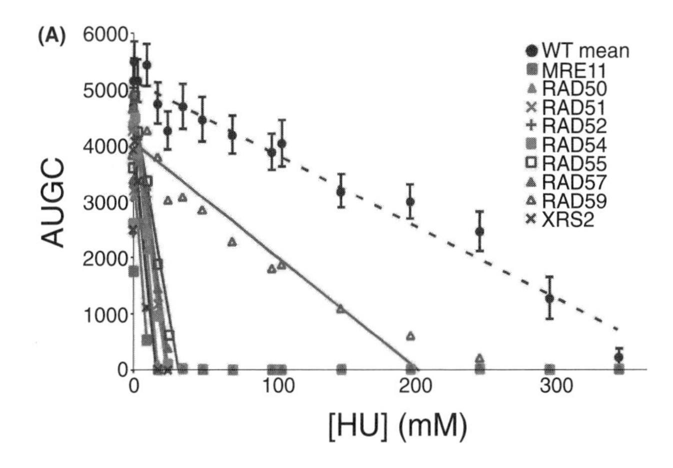

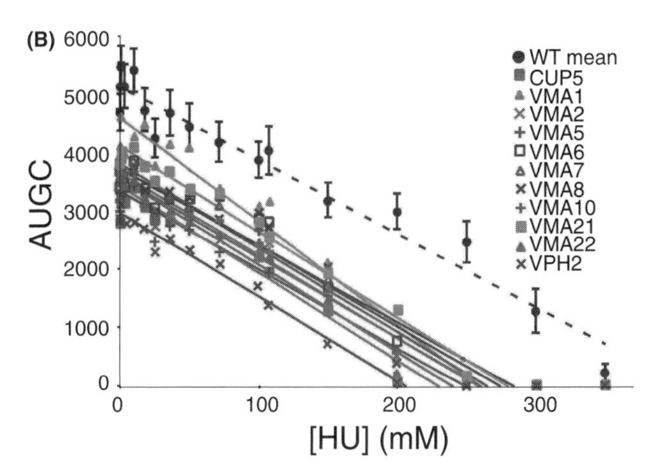

Figure 5.6. Buffering capacity (strength of genetic interaction) aids identification of functional genetic interaction modules. In panels (A) and (B), area under the growth curve (AUGC) is plotted against HU perturbation intensity. Genes functioning in homologous DNA recombination are observed to interact strongly with HU perturbation, indicating a uniform and high buffering capacity (*RAD59* is an exception) for this functional gene group (panel A). In contrast, genes required for the structure and assembly of the vacuolar H^+/ATPase interact weakly with HU, by virtue of essentially additive, and thus independent, phenotypic effects (panel B). (See insert for color representation.)

significantly different than 60% for the additive model, while a 51% reduction would be the expectation with the multiplicative model. To a greater extreme, if each perturbation caused 50% phenotypic reduction, the multiplicative model would predict a 25% phenotypic response to the combined perturbation, while the additive model would predict complete loss of the phenotype.

The figures in this chapter are intended to emphasize how *quantitative* analysis of gene interaction (networks) helps to understand genetic buffering and cellular robustness (Figures 5.1–5.9). In all examples, the primary phenotype is cell proliferation, which is under strong natural selection, the driving force for cellular robustness. In particular, genetic buffering of DNA replication, a cellular process closely tied to cell proliferation phenotypes, is discussed.

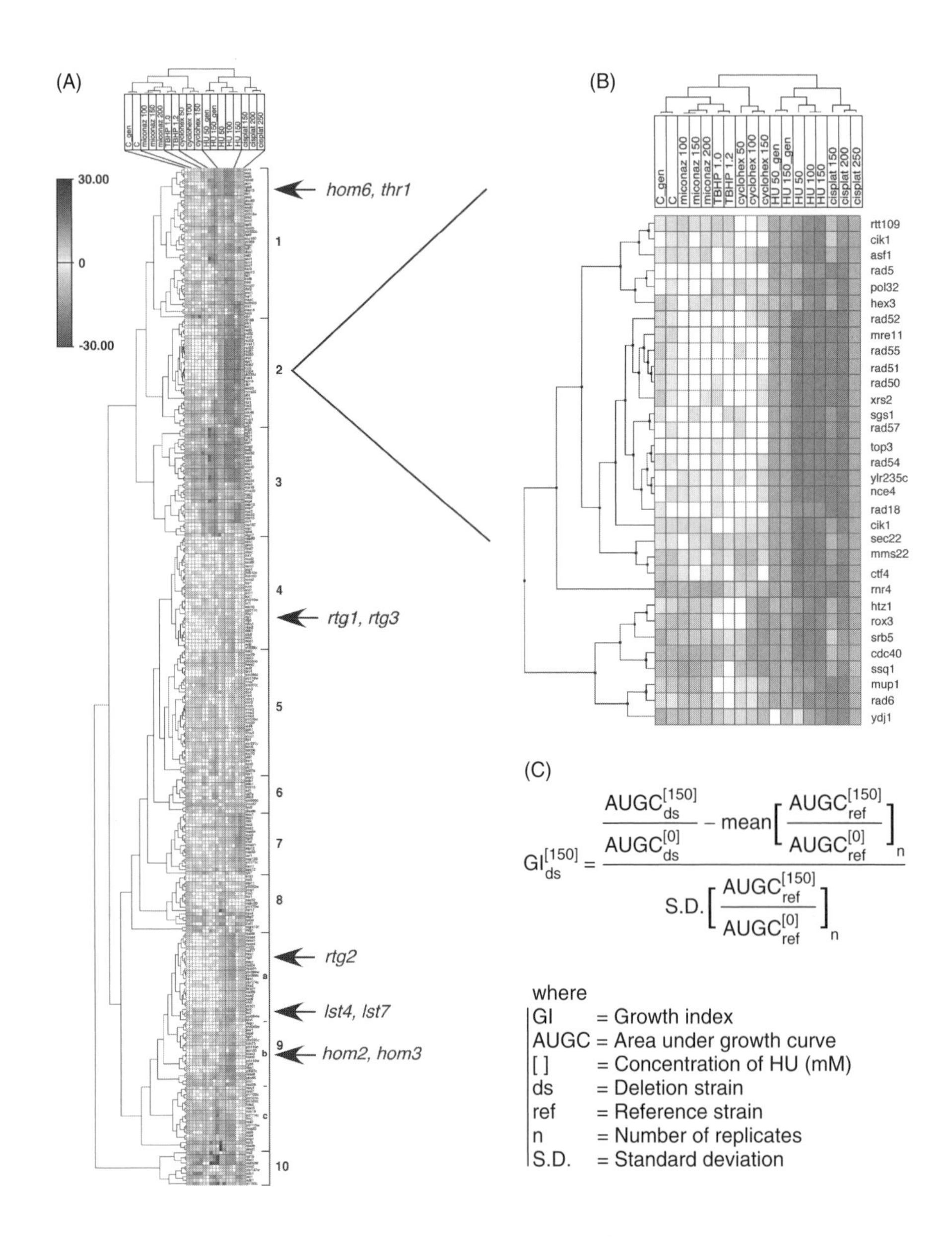

5.3 EXPERIMENTAL CONCEPTS FOR GENETIC BUFFERING ANALYSIS

Gene Interactions Underlie Buffering; Buffering Underlies Robustness

The terms buffering and robustness are more often defined in chemistry and engineering than genetics or biology. It is in part for this reason that they help to frame genetic and biological questions in a new and useful way. Genetic buffering is analogous to pH buffering in the sense that the "activity" of the buffer is to maintain system homeostasis by absorbing the perturbation(s). By this definition, one need not know the mechanism of buffering in order to determine whether one system or another is better buffered against a particular perturbation; one need only measure the perturbation input and the system output in the presence and absence of candidate buffers. In the case of pH buffering, measuring the pH in response to addition of acid or base, with and without a candidate buffering substance, defines the buffering characteristics of the substance. By analogy, one assesses the buffering characteristics of a gene by measuring the phenotypic output of its (cell) system in its presence and absence. In principle, one needs only titratable perturbations, and a method for quantifying the phenotype [24]. By titrating various perturbations, in the presence and absence of the candidate gene buffer, the effect of the gene activity on the phenotypic responses defines the buffering characteristics of the gene (Figures 5.1, 5.5, and 5.6).

Figure 5.7. Hierarchical clustering used to identify genetic interaction modules in complex data sets involving multiple different perturbation types. A set of 298 haploid deletion strains was selected from a genome-wide screen of over 4800 strains based on growth index values for perturbation with 150 mM HU calculated as shown in panel (C) [24]. The growth index for all HU-interacting deletion strains was determined for many perturbations of different type and intensity as indicated by column labels in panels (A) and (B), where C is the no drug control for each strain, miconaz is miconazole (nM), TBHP is *t*-butyl-hydroperoxide (mM), cyclohex is cycloheximide (ng/mL), HU is hydroxyurea (mM), and cisplat is cisplatin (μM); _gen refers to data from the genome-wide screen. The row labels indicate individual gene deletion strains, and the growth index values are indicated by the shading intensity at the intersection of rows (gene deletions) and columns (perturbations). Synergistic interactions have a negative growth index (more details about quantification found in [24]). Clusters are assigned numbers along the right side for ease of reference. Clusters 1–3 are comprised overall of genes with the highest buffering capacities. Cluster 2 contains genes with high buffering selectivity for DNA-damaging perturbations (HU and cisplatin), as compared with Clusters 1 and 3. The inset of Cluster 2 is to emphasize the enrichment of DNA repair genes in this highly specific genetic buffering module, particularly those involved in homologous recombination (rad = radiation sensitive; see also Figure 5.6A). Cluster 7 indicates genes with relative buffering selectivity for cisplatin perturbation, while Cluster 9 is more selective for hydroxyurea. The genes indicated by arrows are further described in Figure 5.9. (See insert for color representation.)

(A)

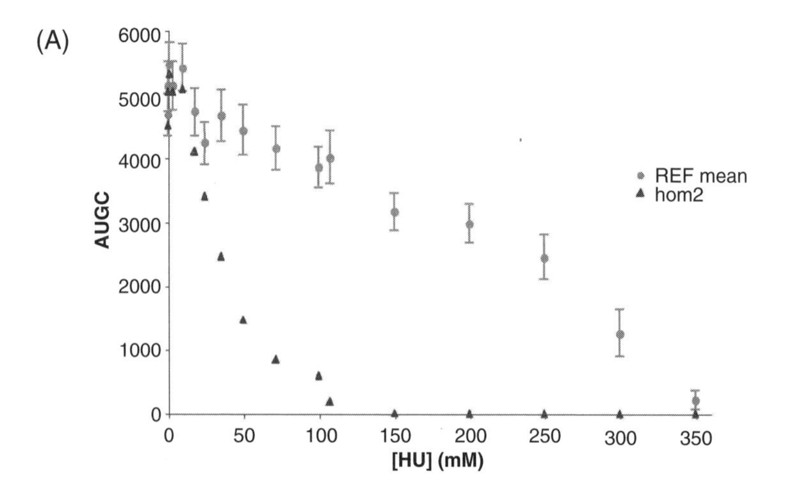
6000
5000
4000
3000
2000
1000
0
AUGC
0
50
100
150
200
250
300
350
[HU] (mM)
REF mean
hom2

(B)

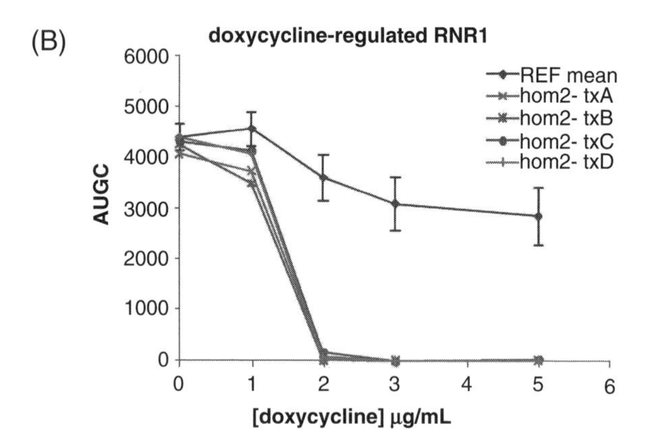
doxycycline-regulated RNR1
6000
5000
4000
3000
2000
1000
0
AUGC
0
1
2
3
4
5
6
[doxycycline] μg/mL
REF mean
hom2- txA
hom2- txB
hom2- txC
hom2- txD

(C)

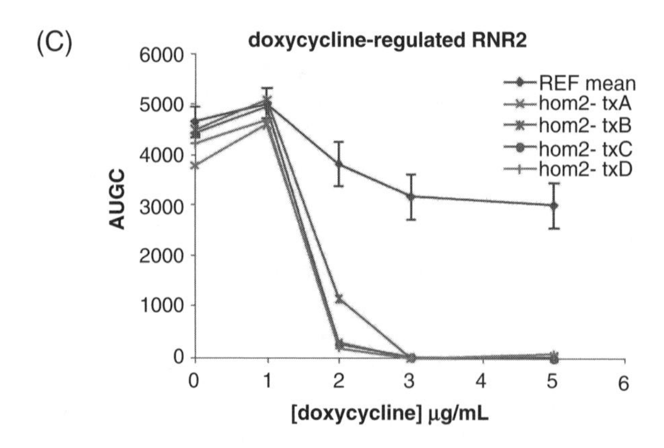
doxycycline-regulated RNR2
6000
5000
4000
3000
2000
1000
0
AUGC
0
1
2
3
4
5
6
[doxycycline] μg/mL
REF mean
hom2- txA
hom2- txB
hom2- txC
hom2- txD

Figure 5.8. Tet-regulatable alleles of genetic drug targets are useful for validating chemical–genetic interactions. Hydroxyurea acts to inhibit ribonucleotide reductase activity, which is required for production of dNTPs and thus for DNA replication. Panel (A) depicts the buffering capacity of *HOM2* (common pathway of threonine and methionine synthesis) in response to HU perturbation. To more directly test whether *HOM2* buffers RNR activity, tetracycline/doxycycline-regulatable alleles of *RNR1* and *RNR2* were constructed by chromosomal integration into the reference strain and the *hom2* deletion strain. Panel (B) depicts the average AUGC from multiple transformants of the reference strain, in addition to four transformants each, for the tet-RNR1(regulatory subunit)/*hom2* deletion strain, each tested at multiple concentrations of doxycycline, designed to repress *RNR1* gene transcription. In panel (C), similar data for tet-RNR2 (catalytic subunit)/*hom2* deletion strain is shown for comparison. (See insert for color representation.)

In contrast to the mechanistic simplicity of a chemically buffered solution, robustness properties of biological systems are more analogous to those of highly engineered machines, such as airplanes [4–6], where system-level organization involves modular architectures, elaborate hierarchies of protocols, functional redundancies, and layers of feedback regulation, which are driven by the demand for system stability in the face of a dynamic and uncertain environment. Csete and Doyle [6] have developed these and other useful analogies and insights from engineering theory, providing a useful perspective on biological complexity.

From a systems perspective, one can envision the genome as the blueprint of an organism, with gene products as the system components and regulatory networks as the functional organization [22]. In this analogy, natural selection would be the "engineer," designing the modular architectures, hierarchical protocols, and layers of feedback that regulate phenotypic responses to perturbation. Experimentally, gene knockout collections provide a way to reverse engineer the gene interaction networks that provide buffering, because they facilitate dissection of the genetic basis for "robustness." Using defined model systems, the activity of each of the components in the system, the perturbation type and intensity, and the phenotypic output of the system can all be manipulated. As with other global, quantitative genetic data, one faces the challenge of partitioning and integrating the information in order to achieve particular biological insights. Interpretation of phenotypic buffering networks is perhaps facilitated relative to other biomolecular networks because genes are commonly annotated with respect to their phenotypic effects.

Buffering Capacity

Buffering capacity is the amount of stability/robustness that a gene or genetic module imparts on the system in response to a particular perturbation input. Thus, buffering capacity reflects the strength of genetic interaction. The concept of buffering capacity is motivated by the observation that gene interaction is ubiquitous and occurs over a quantitative continuum (Figures 5.2–5.7). Buffering capacity remains little explored because the models that quantitative geneticists have used to identify quantitative trait loci have generally lacked the power to quantify epistasis or

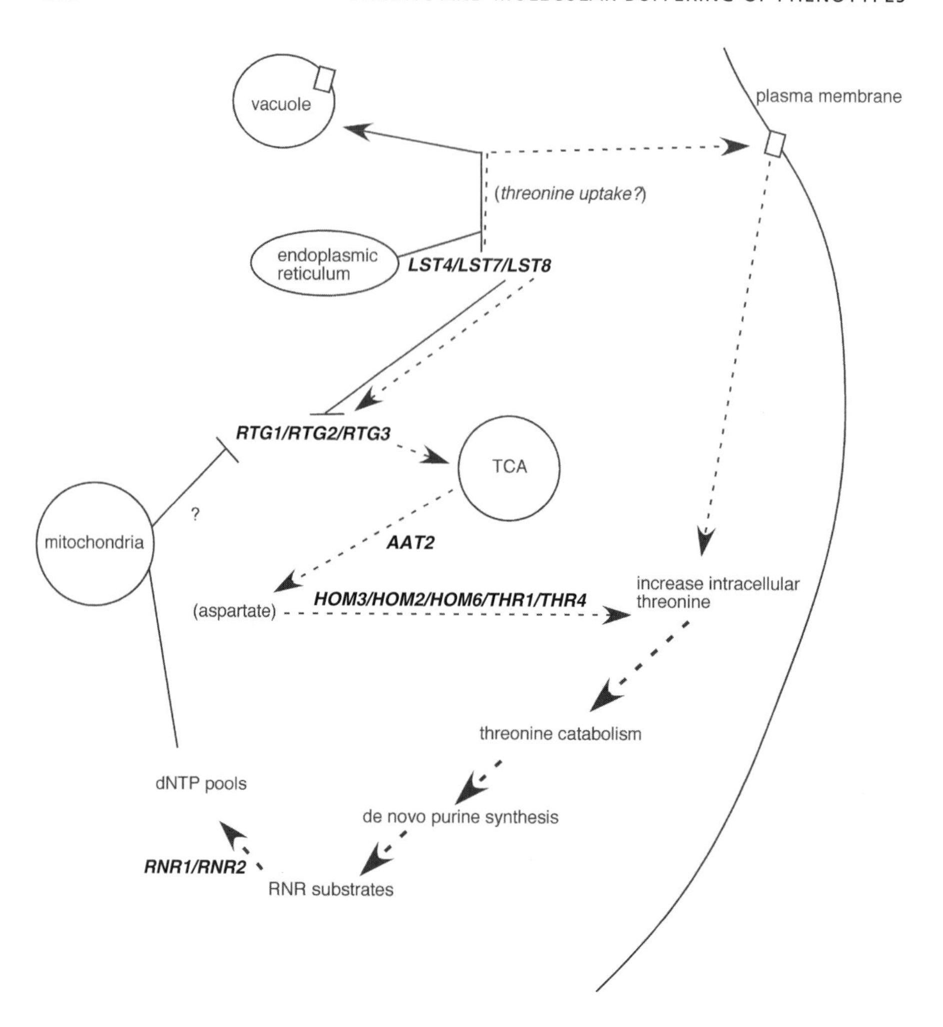

Figure 5.9. A hypothetical model for a buffering protocol that provides cellular robustness against HU perturbation. A buffering "protocol" is the temporal, spatial, and molecular logic of connectivity between different genetic buffering modules. Each genetic buffering module is defined experimentally by genes having shared buffering specificity; that is, patterns of buffering capacity and selectivity (see Figure 5.7). The proposed connectivity between modules is based on collective gene annotations. Protocols can, in theory, involve any number of genes/pathways, acting in any order, and on any variety of biochemical intermediates. For the proposed example, four genetic buffering modules are incorporated as explained by the following annotations. Briefly, computational analysis of the *E. coli* metabolome predicted that a high-flux metabolic backbone acts to reprogram metabolism for optimal growth in response to a changing environment, and that it is likely conserved in eukaryotic cells [90]. Threonine synthesis was on the high-flux backbone. Accordingly, all genes on this pathway were recovered from the HU chemical–genetic

gene–environment interactions [16, 71], while experimental molecular geneticists have more typically sought biochemical/mechanistic information than quantitative genetic information. The concept of buffering capacity put forth here is that strong interactions indicate genes with high buffering capacity, since the activity is required for stability with relatively small degrees of perturbation (Figure 5.6A), while weak interactions indicate genes with low buffering capacity, because loss of activity has little impact on the phenotypic response to perturbation (Figure 5.6B). The precision of gene buffering capacity measurements results from the range and number of perturbation intensities tested—low-intensity perturbations being required to discriminate between genes having the highest capacity (strongest interactions) and high-intensity perturbations being required to discriminate between genes with low buffering capacity (weak interactions) (Figure 5.6). Time, as a variable, also impacts phenotypic measurements and thus buffering capacity calculations (Figures 5.2–5.5).

In order to visualize large matrices of genetic interaction data, the multidimensionality of genetic interactions can be reduced by representing buffering capacity as a single value, derived from many growth curves (Figure 5.5). For example, one could use the difference between areas under each curve in Fig. 5.8A. A caveat to such "reduced" representations of interaction quantities, however, is that biologically important features of the data may be lost to the analysis (Figures 5.5 and 5.8). For this reason, buffering capacity should probably be analyzed with respect to different features of growth curves (e.g., lag time maximum rate, final population density), each representing physiologically distinct biological phenomena.

---

screen (*HOM3, HOM2, HOM6, THR1, THR4*). The computational analysis predicted threonine catabolism to be upstream of de novo purine synthesis on the high-flux backbone, forming the foundation for a second hypothesis that purines are limiting in the context of HU inhibition of RNR, and that the outlined protocol of genetic interactions function to buffer RNR activity by regulation of purine fluxes in yeast. In support of the model, *RTG1/RTG2/RTG3* regulate expression of tricarboxylic acid genes, potentially controlling the provision of aspartate for threonine synthesis (*AAT2*, which converts TCA intermediates to aspartate, the substrate of HOM3p, was also found to buffer HU perturbation). RTG stands for retrograde, because these genes initiate their transcriptional program in response to mitochondrial perturbations. Coincidentally, RNR deficiency is known to cause mitochondrial malfunction and loss of respiratory capacity, potentially indicating activation of such a regulatory loop. Additionally, *LST4/LST7/LST8* may participate in a "second layer" of regulation because *LST8* has been shown to regulate both RTG signaling and amino acid permease sorting. Thus, it is plausible that *LST4/LST7/LST8* "communicates" with the high-flux backbone via interaction with *RTG1/RTG2/RTG3* to help coordinate both threonine uptake and threonine synthesis to boost overall purine catabolism and augment production of dNTPs by RNR. The hypothesis that ADP is the limiting substrate for RNR is consistent with the fact that dATP is the primary allosteric negative feedback product for RNR. Interestingly, it has been observed that several of the genes in the threonine synthesis pathway are also lethal with *SEC13* (LST).

Buffering Selectivity

Buffering selectivity is the qualitative pattern of interaction a given gene or genetic module demonstrates across perturbations affecting different cellular processes [24, 25]. For example, a series of different DNA damaging agents might be more useful for discriminating between the buffering selectivity of different DNA repair genes, while a more diverse set of drug perturbations targeting different cellular processes would be more useful for classifying the buffering selectivity of a set of genes representing all protein activities (Figure 5.7). Buffering selectivity highlights local areas on a gene interaction network that provide robustness in particular perturbation environments. Returning to the DNA repair example, it's important to consider that genes that buffer cell proliferation against various forms of DNA damage comprise many functional classes (transcription factors, polymerases, endonucleases, helicases, ligases, kinases, ubiquitin-modifying proteins, cytoskeletal proteins, etc.). While some of these genes may function to buffer a range of other types of perturbations, others likely interact relatively selectively in the case of DNA damage, and some may be even more selective, acting on only certain types of DNA damage. Thus, buffering selectivity and buffering capacity are complementary for characterizing modularity in genetic interaction networks.

Buffering Specificity and Modular Buffering Networks

Buffering specificity is the combined selectivity and capacity of genetic interactions across a series of different perturbations. To the extent that buffering specificity is defined comprehensively and quantitatively for any single gene, it represents the contribution of a single gene to overall robustness of the cell. Modularity in genetic interaction networks is thus defined by correlation of buffering specificity among all genes for a particular set of perturbations and identifies sets of genes that act together to modify a phenotype in a similar way in various contexts that the cell encounters (Figure 5.7). Thus, the interpretation of modularity depends on the sets of gene–perturbation interactions analyzed, the method for calculating interactions, and the method for assigning similarities within the global data set.

5.4 EXPERIMENTAL PLATFORMS FOR GLOBAL GENETIC INTERACTION ANALYSIS

Use of the Yeast Knockout Collection for Gene Interaction Analysis

Yeast was the first eukaryotic genome sequenced [72]. Following this achievement, a consortium of laboratories knocked out all yeast genes [33, 57]. Pairs of 20 mers (oligonucleotides of 20 bases) flanking a selectable gene marker were designed for each locus and inserted by homologous recombination. Since each pair of oligonucleotides is unique to its deletion locus, relative growth of all strains can be characterized by microarray hybridization analysis—essentially a bar-code system. By this method, pooled deletion strains are grown competitively under different conditions of interest and collected at various times for extraction, labeling, and hybridization of the genomic DNA to oligonucleotide microarrays. Hybridization signal data are filtered for quality and then statistical algorithms are applied to determine whether the changes in relative hybridization signal for each strain indicate loss of fitness for each strain in the perturbed pool [33, 57].

A more common use of the deletion set has been to replicate the strains to agar growth media and score the growth subjectively after a designated time period. Many growth-perturbing conditions have now been tested in this way, but no standards have been adopted that would facilitate pooling the various data sets for quantitative analysis. Many of these studies were designed as more traditional genetic screens, generating a list of candidate genes for a pathway or process, rather than for studying genetic buffering or cellular robustness per se [18, 73].

The Boone laboratory has performed the most comprehensive qualitative analysis of gene–gene and gene–environment interactions [23, 25, 55]. For gene–gene interactions they developed a method for using the deletion set as a starting point to test for synthetic lethality of double mutations [23, 55]. In brief, the synthetic genetic array (SGA) strategy is powerful because, beginning with the deletion set created by the consortium, DNA transformation is not required to introduce a second gene mutation into the entire set. For SGA, haploid double mutants are created by a mating strategy: a "query strain" carrying a mutation of interest, marked with a unique dominant drug resistance marker, is of opposite mating type and contains a mating-type-silenced auxotrophy and a recessive drug resistance marker, but lacks the dominant marker used to create the deletion array. This strain is mated against the deletion collection array and the diploid double heterozygous mutant is selected and then sporulated to obtain all haploid meiotic products. The haploid double mutants can be selected by growth under appropriate selective growth conditions [23, 55]. Synthetic lethality is presumed to occur when the double mutant is not obtained, implying that the combination of two nonlethal mutations causes lethality. In actuality, many of the reported interactions are "synthetic sick," meaning that, rather than lethality per se, there is an observable effect of interaction on growth of the double mutant. It is also possible for a given combination to result in synthetic phenotypes only a percentage of the time, or to express various degrees of "sickness," over multiple tests. Growth is typically scored qualitatively for SGA, and hierarchical clustering is performed on interaction data after it is converted to a binary form [23, 25].

SGA has been used to characterize a large network of interactions, revealing global aspects of gene interaction networks. The analysis concludes that connectivity between interacting genes across the network follows a power-law distribution, providing an example of "small-world networks," which exhibit dense local neighborhoods and hubs such that the shortest path between any two vertices is typically short [23]. From use of the gene ontology resource, genetic interaction networks appeared to strongly reflect functional aspects of cellular organization [23, 74]. Novel functions for some genes were discovered from hypotheses generated for unknown genes, based on connectivity to genes of known function [23]. This work set a standard for genetic interaction network analysis, representing interdisciplinary contributions from several laboratories. Boone and colleagues also used the matrix of gene–gene interactions to show that correlation between gene–gene and chemical–gene interaction profiles is predictive of the genetic targets of drugs [25].

In contrast to the work from the Boone laboratory, most of the work described below emphasizes development of methods for *quantitative* analysis of gene interaction networks recognizing that important biological aspects of genetic interaction networks are lost when interactions are treated as binary rather than continuous phenomena. With the extreme complexity of biological networks being well documented, strategies for improved resolution are needed. To this end, high-throughput methods for quantifying growth of tens of thousands of cultures per experiment will be discussed along with analytical techniques for automating summary measurements of gene interaction, based on the growth of multiple individual cultures [24].

Buffering of DNA Replication in Yeast: A Model for Genome Instability and Cancer

The work reviewed next focuses on how perturbations of ribonucleotide reductase (RNR), an enzyme activity required for producing deoxyribonucleotides from ribonucleotides, is buffered by activities of other genes. The rationale for choosing RNR as a primary perturbation is based on the hypothesis that cell systems have been naturally selected to buffer variation in RNR activity in order to maintain phenotypic stability in the face of perturbed cellular dNTP pools. Genetic buffering of RNR was investigated by measuring chemical–genetic interactions between hydroxyurea (HU), a small molecule inhibitor of RNR, and the set of yeast deletions. Examples from this work illustrate how buffering concepts such as capacity, selectivity, and specificity are used to generate hypotheses regarding the organization of genetic circuitry underlying phenotypic robustness to particular perturbations (Figures 5.6–5.9).

As alluded to previously, cell proliferation is an ideal phenotypic output for these studies. First, it is a quantitative trait that is relatively easy to measure. Second, due to strong natural selection for cell proliferation, it is a robust and genetically complex phenotypic output that represents overall function of the entire genetic system because nearly every gene (in some cellular context) may modify cell proliferation. Third, cell proliferation is a fundamental feature of all cells. Thus, principles learned from genetic buffering of DNA replication, which is closely tied to cell proliferation, will likely be conserved in many cell types [18]. Because DNA replication fidelity is a major determinant of genome stability, a complete understanding of how DNA replication is buffered in yeast can be applied to better understand tumorigenesis in multicellular organisms, where genetic instability nearly always precedes cancer [75].

Genetic Buffering of Ribonucleotide Reductase

RNR activity is required for production of all dNTPs [76, 77] and thus is required for DNA replication and cell proliferation. RNR regulation is elaborate, the complex in yeast being assembled from four different subunits (*RNR1, RNR2, RNR3, RNR4*) and the activity being regulated during the cell cycle and in response to DNA damage [78–80]. Its activity is regulated at the levels of transcription [81], mRNA degradation [82, 83], translation [84], protein degradation [85, 86], protein–protein interaction [87], allosteric feedback regulation [88], and cellular localization [89]. Physiologically, elevation of dNTP pools is mutagenic [78, 90] and pool depletion causes stalled replication forks, producing single and double strand DNA breaks [91]. Additionally, DNA damage responses include transcriptional induction of RNR and other DNA synthesis genes, helping to stabilize stalled forks, repair damaged DNA, and resume replication [78, 81, 92]. The compendium of genes acting to maintain cell proliferation, in spite of abnormal RNR activity, is said to "buffer" RNR [18, 24].

There are multiple ways to perturb RNR activity in yeast. The primary perturbation reviewed here is addition of hydroxyurea (HU), an enzymatic inhibitor that scavenges a hydroxyl radical on the catalytic subunit of RNR (RNR2), which is required for the reduction reaction for all dNDPs. In general, the strategy of perturbation with inhibitors, like HU, for dissecting genetic networks has two significant drawbacks: (1) there are "off-target" effects of drugs that add unrecognized complexity to the genetic interpretation and (2) most genes do not have inhibitors, so the scalability of this approach is limited. A second perturbation strategy uses tet-regulated transcription of the RNR subunits [93, 94]. This approach can be modified to perturb the activity of any gene in a titratable fashion (Figure 5.8).

A stepwise strategy for assessing buffering globally and quantitatively was to first screen the entire set of deletion strains for interactions against a few "sensitizing" concentrations of drug, and then to characterize buffering specificity of selected strains in greater detail [24]. HU chemical–genetic interactions were studied in this way, where it was first determined that HU inhibited growth of the reference strain over a concentration range of 10–300 mM (Figures 5.4–5.6). Based on this finding, HU concentrations, eliciting around a 20–60% reduction in growth of the reference strain (50, 100, and 150 mM), were chosen for genome-wide screening (Figure 5.5). Approximately 200 replicate cultures of the reference strain were used to quantify experimental noise. The phenotypic assay for these studies consists of spotting dilute cultures onto agar slabs as 8 × 12 cellular arrays, maintaining the 96 well configuration used for distribution of the strains (Figure 5.2). Growth is quantified by image analysis of the population growth in each spotted culture (Figure 5.5). Images are collected every few hours over 3–4 days using an optical scanner that images 10 arrays per scan (Figure 5.3). Approximately 30 scans can be collected per hour, making it practical to phenotype 30,000 (8 × 12 array format) to 120,000 (16 × 24 array format) cultures per experiment (Figure 5.7). The image values are quantified by pixel analysis and plotted against time to generate growth curves for each strain (Figures 5.4 and 5.5) [24].

Area under the growth curve (AUGC) was used as the unit of growth for each culture, incorporating all features of growth (lag, proliferation rate, and total yield) into a single measure (Figure 5.5). AUGC is used to calculate the growth index, a z-statistic representing the probability of nonadditive interaction for a single perturbation condition (Figure 5.7). The growth index represents the interacting effects of a gene deletion and the perturbation on the phenotype, normalized by the effects of the deletion or perturbation alone and divided by the experimental noise (standard deviation of reference strain growth) (Figure 5.7C). As expected for a z-statistic, the growth index values for reference strain replicates were distributed normally, with a range of less than ±3 (SD). A growth index threshold of ±6 (SD) was used to select 300 interacting strains from the approximately 4850 strains tested. Use of a z-statistic facilitated statistical interpretation of quantitative results between different perturbation types and intensities (Figure 5.7) [24]. The growth index for a gene deletion strain (at a single concentration of one drug) provides a limited representation of buffering capacity, which is represented more completely by the difference between phenotypic slopes, or difference in areas under the phenotypic response curves (Figure 5.6).

Hierarchical Clustering Reveals Buffering Specificity of Genetic Interaction Modules

Buffering specificity is experimentally defined by both the buffering selectivity of a gene across multiple perturbation contexts and the buffering capacity of the gene in each context. Thus, correlation of buffering specificity, ascertained by hierarchical clustering [95], highlights genetic interaction "modules" of genes that coordinately act to buffer the phenotype against perturbations the cell encounters.

Buffering selectivity for HU was evaluated using qualitatively different, but quantitatively normalized, perturbations including miconazole (an inhibitor of ERG11p, required for ergosterol synthesis), cycloheximide (an inhibitor of RPL28, a subunit of the 60S ribosome), cisplatin (a DNA cross-linking agent), and *t*-butyl hydroperoxide (a nonspecific oxidant). To standardize comparison of growth index values between perturbations with different drugs, the reference strain was used to determine phenotypically equivalent concentrations of each drug (inhibiting growth of the reference strain to the same extent as 50, 100, and 150 mM HU). Three concentrations of each drug were used to test all 300 HU-interacting strains for selectivity of chemical–genetic interactions (Figure 5.7) [24].

To understand the modular organization of genetic interaction networks that buffer cell proliferation against HU perturbation, annotations of gene clusters from the 300 (number of deletion strains) × 18 (number of drug perturbations) matrix of perturbation combinations was assessed (Figure 5.7). Gene clusters represent interaction modules containing genes that have shared buffering specificity, based on both the qualitative (selectivity) and quantitative (capacity) patterns of buffering across all perturbations tested. The particular modules discovered and the literature annotations supporting their classification are discussed elsewhere in greater detail [24]. The general conclusion is that *both buffering capacity and buffering selectivity contributed to the resolution of genetic interaction modules containing sets of genes with shared functional annotations*. Selectivity was useful for distinguishing general processes, involving large numbers of representative genes, such as protein trafficking (found to have low specificity) and DNA repair (found to have high specificity). Buffering capacity was useful for finer distinctions, for example, a specific class of vacuolar protein sorting genes, apart from other vesicular trafficking genes, had greater buffering capacity for DNA damaging perturbations, while DNA repair genes involved in homologous recombination had greater capacity than genes involved in S-phase checkpoint signaling [24].

Dynamic Nature of Genetic Interaction Modules and Molecular Network Connectivity

Modularity in gene interaction networks is a paradigm for understanding how genetic circuitry is functionally organized to confer phenotypic robustness on cells by buffering the various genetic and environmental perturbations that cells encounter. However, it is well known from innumerable examples that individual genetic interactions are themselves dependent on the genetic background and the environment in which they occur. Thus, "the network" itself and the modular properties thereof are also dynamic, and a function of the many variables that comprise "cellular context." The dynamic nature of genetic interaction networks means that to maintain cellular robustness in the face of a constantly changing genetic and environmental context, individual genes may interact with a wide variety of other genes and participate in a large number of different "modules" in buffering phenotypes. The isogenic set of yeast deletion strains permits one to control the context and search for buffering modules in a systematic, comprehensive, and quantitative manner [18, 24]. However, to address the problem of combinatorial explosion in quantitative global genetic interaction network analysis, further advances in high-capacity, high-resolution phenotyping methodologies are needed to resolve the modular organization of buffering networks for all cellular contexts.

It is almost certain that genetic interaction networks are more complex in nature than can be recapitulated in the laboratory. For example, natural genetic variation is not conveniently packaged as deleted open reading frames, and the combinatorial effects of variation across all loci as well as environmental effects are overwhelming by comparison. This increase in complexity is further compounded in multicellular systems compared to the yeast system described here. Insights into genetic buffering and cellular robustness gained from powerful experimental model systems provide testable hypotheses that serve as focal points, and thus indirectly reduce the complexity of analyzing natural genetic variation. In this regard, model systems and comparative genomics should play an even more important role in "systems biology" than previously, because the reduction of biological complexity is inversely combinatorial.

As one progresses in conceptualizing ways to model and test the complex, dynamic, and continuous nature of gene interaction networks, the difficulties in representing these concepts with language and images increase. For example, genetic network models are typically represented as graphs, with balls and sticks, representing genes and their connections, respectively. This is an obviously static representation. How does one represent dynamic changes in buffering networks as perturbation/cellular context is varied? How can genes associated with multiple different modules be simultaneously represented? How might one represent quantitative differences in genetic interaction network activity? How can one simulate activity over the network? These questions all point to the need for increasing interdisciplinary efforts among geneticists, biochemists, theoretical biologists, mathematicians, physicists, and engineers.

Since interdisciplinary science is required for systems biology, it is useful to consider why lines are drawn between disciplines in the first place: each discipline has different constraints imposed by appropriate applications of their respective disciplinary tools (e.g., mathematics or biology), which is closely related to the motivation behind the questions that dictate their use [96]. Many of the ideas presented in this chapter attempt to bridge polarized "local" and "global" perspectives of biology by recognizing the importance of both governing properties of molecular networks and mechanistic details of the component activities [97]. For example, the localist might be interested in genetic buffering and phenotypic robustness to understand molecular mechanisms underlying "key" parts of the network that act in select cellular contexts, while the globalist would instead want to know the overall structure, regardless of the biological/biochemical details of how each component acts. Awareness of these artificial and unnecessary lines between disciplines will help to break them down. The experimental platform described here utilizes localist resources (the well-annotated set of yeast deletion strains) adapted to a globalist design (array-based measurement of interaction quantities). Globalist analytical abstractions (see Glossary) are formulated with the goal of generating localist hypotheses (threonine flux buffers dNTP pools against RNR limitation by augmenting de novo purine synthesis; see Figures 5.7 and 5.9 [24]).

Extrapolating Experimentally Derived Insights to Natural Populations

Ultimately, the convergence of knowledge about individual genes and genetic systems will lead to understanding particular biological problems in far greater detail. The accumulation of knowledge about gene regulation, protein functions, and the organization of biomolecular pathways can now be integrated for understanding organisms as systems. As the complexity of analysis increases, it is evermore true that model organisms provide unique opportunities to understand biological systems. Solving the puzzle of phenotypic robustness for even one simple organism opens the door for comparative analysis in other systems, integrative analysis of other biomolecular networks, and new approaches for genetic analysis of natural populations. Previously inconceivable biomedical advances in diagnosis, management, and prevention of human disease may be possible as a result of understanding how functional natural genetic variation is buffered/accumulated and thus how phenotypic expressions are regulated.

5.5 CONCLUSION

Biological systems are robust to perturbation, meaning that phenotypic outputs remain relatively stable in response to variable genetic and environmental inputs. The process by which a robust system absorbs changing inputs, while maintaining stable outputs, is known as buffering. Genetic interactions reveal sources of buffering and robustness because they identify functional redundancies of cellular organization. Genetic interactions can be studied in a controlled manner on a genome-wide scale using collections of gene knockout strains or mutants that are available in inbred model systems. Gene–gene or gene–environment interactions are assessed by perturbing the entire deletion strain collection; for example, by introducing a second mutation into all strains or by drug treatments. Quantitatively, genetic interaction implies that the phenotypic response to perturbation is dependent on the allele status at a particular genetic locus. This can be determined by a four-way comparison of the phenotypic output between the reference and gene deletion strains, with and without the perturbation. An expectation of the quantitative effect on a phenotypic trait, from the *actual combination* of gene deletion and perturbation (e.g., additive effects or multiplicative effects), provides a null hypothesis for genetic interaction. The amount by which the observed effect of the combination departs from the null hypothesis represents the strength of interaction, also called the buffering capacity for the gene. The network of genetic interactions reflects the dynamic organization of genetic circuitry that provides phenotypic buffering and robustness in those contexts.

Genetic principles for phenotypic buffering and robustness, learned from model systems, could help focus efforts to map functional allelic variation underlying phenotypic diversity in natural populations, where the problem of combinatorial explosion currently severely limits statistically meaningful global analysis. This chapter focuses on efforts to map genetic interaction networks in yeast, with an emphasis on high-capacity, high-resolution cellular phenotyping for quantitative and global analysis. The quantitative emphasis addresses two fundamental needs in high-throughput biological studies: (1) standards for objectivity and consistency in measurement and interpretation; and (2) functional partitioning of the hierarchical organization of complex biological systems. Establishment of genetic principles that underlie buffering of biological systems will provide novel insights for generating and testing hypotheses regarding combinations of genetic and environmental perturbations that produce and/or modify disease phenotypes in natural populations. This will increase opportunities to apply the rapid advances in genome characterization to understand phenotypic diversity so that disease can be managed in a more individualized manner.

ACKNOWLEDGMENTS

I thank Lee Hartwell and Maynard Olson for their encouragement and guidance in using the yeast system to investigate the genetic complexity of cellular phenotypes;

Lue Ping Zhao and colleagues for their many insights into quantitative analysis; and Nic Tippery for innovative technical assistance. I am also grateful for generous funding support from the Howard Hughes Medical Institute and NIH/National Cancer Institute.

REFERENCES

1. A. Wagner (2000). Robustness against mutations in genetic networks of yeast, *Nat. Genet.* **24**:355.
2. K. Wolfe (2000). Robustness—it's not where you think it is, *Nat. Genet.* **25**:3.
3. Z. Gu, L. M. Steinmetz, X. Gu, C. Scharfe, R. W. Davis, and W. H. Li (2003). Role of duplicate genes in genetic robustness against null mutations, *Nature* **421**:63.
4. H. Kitano (2004). Biological robustness, *Nat. Rev. Genet.* **5**:826.
5. J. Stelling, U. Sauer, Z. Szallasi, F. J. Doyle, 3rd, and J. Doyle (2004). Robustness of cellular functions, *Cell* **118**:675.
6. M. E. Csete and J. C. Doyle (2002). Reverse engineering of biological complexity, *Science* **295**:1664.
7. J. L. Badano and N. Katsanis (2002). Beyond Mendel: an evolving view of human genetic disease transmission, *Nat. Rev. Genet.* **3**:779.
8. C. S. Carlson, M. A. Eberle, L. Kruglyak, and D. A. Nickerson (2004). Mapping complex disease loci in whole-genome association studies, *Nature* **429**:446.
9. J. Zielenski and L. C. Tsui (1995). Cystic fibrosis: genotypic and phenotypic variations, *Annu. Rev. Genet.* **29**:777.
10. J. H. Nadeau (2001). Modifier genes in mice and humans, *Nat. Rev. Genet.* **2**:165.
11. R. L. Nagel and M. H. Steinberg (2001). Role of epistatic (modifier) genes in the modulation of the phenotypic diversity of sickle cell anemia, *Pediatr. Pathol. Mol. Med.* **20**:123.
12. F. Salvatore, O. Scudiero, and G. Castaldo (2002). Genotype–phenotype correlation in cystic fibrosis: the role of modifier genes, *Am. J. Med. Genet.* **111**:88.
13. E. Beutler (2003). The HFE Cys282Tyr mutation as a necessary but not sufficient cause of clinical hereditary hemochromatosis, *Blood* **101**:3347.
14. E. Beutler (2001). Discrepancies between genotype and phenotype in hematology: an important frontier, *Blood* **98**:2597.
15. N. H. Barton and P. D. Keightley (2002). Understanding quantitative genetic variation, *Nat. Rev. Genet.* **3**:11.
16. O. Carlborg and C. S. Haley (2004). Epistasis: too often neglected in complex trait studies? *Nat. Rev. Genet.* **5**:618.
17. J. H. Moore (2003). The ubiquitous nature of epistasis in determining susceptibility to common human diseases, *Hum. Hered.* **56**:73.
18. J. L. Hartman, B. Garvik, and L. Hartwell (2001). Principles for the buffering of genetic variation, *Science* **291**:1001.
19. C. H. Waddington (1942). Canalization of development and the inheritance of acquired characters, *Nature* **150**:563.

20. S. L. Rutherford (2000). From genotype to phenotype: buffering mechanisms and the storage of genetic information, *Bioessays* **22**:1095.
21. J. M. Slack (2002). Conrad Hal Waddington: the last Renaissance biologist? *Nat. Rev. Genet.* **3**:889.
22. T. Ideker, T. Galitski, and L. Hood (2001). A new approach to decoding life: systems biology, *Annu. Rev. Genomics Hum. Genet.* **2**:343.
23. A. H. Tong, G. Lesage, G. D. Bader, H. Ding, H. Xu, et al. (2004). Global mapping of the yeast genetic interaction network, *Science* **303**:808.
24. J. L. T. Hartman and N. P. Tippery (2004). Systematic quantification of gene interactions by phenotypic array analysis, *Genome Biol.* **5**:R49.
25. A. B. Parsons, R. L. Brost, H. Ding, Z. Li, C. Zhang, et al. (2004). Integration of chemical–genetic and genetic interaction data links bioactive compounds to cellular target pathways, *Nat. Biotechnol.* **22**(1):62–69.
26. S. F. Elena and R. E. Lenski (1997). Test of synergistic interactions among deleterious mutations in bacteria, *Nature* **390**:395.
27. D. Segre, A. Deluna, G. M. Church, and R. Kishony (2005). Modular epistasis in yeast metabolism, *Nat. Genet.* **37**:77.
28. T. I. Lee, N. J. Rinaldi, F. Robert, D. T. Odom, Z. Bar-Joseph, et al. (2002). Transcriptional regulatory networks in *Saccharomyces cerevisiae*, *Science* **298**:799.
29. T. R. Hughes, M. J. Marton, A. R. Jones, C. J. Roberts, R. Stoughton, et al. (2000). Functional discovery via a compendium of expression profiles, *Cell* **102**:109.
30. G. Chua, M. D. Robinson, Q. Morris, and T. R. Hughes (2004). Transcriptional networks: reverse-engineering gene regulation on a global scale, *Curr. Opin. Microbiol.* **7**: 638.
31. J. D. Han, N. Bertin, T. Hao, D. S. Goldberg, G. F. Berriz, et al. (2004). Evidence for dynamically organized modularity in the yeast protein–protein interaction network, *Nature* **430**:88.
32. G. W. Birrell, J. A. Brown, H. I. Wu, G. Giaever, A. M. Chu, R. W. Davis, and J. M. Brown (2002). Transcriptional response of *Saccharomyces cerevisiae* to DNA-damaging agents does not identify the genes that protect against these agents, *Proc. Natl. Acad. Sci. U.S.A.* **99**:8778.
33. G. Giaever, A. M. Chu, L. Ni, C. Connelly, L. Riles, et al. (2002). Functional profiling of the *Saccharomyces cerevisiae* genome, *Nature* **418**:387.
34. M. Chang, M. Bellaoui, C. Boone, and G. W. Brown (2002). A genome-wide screen for methyl methanesulfonate-sensitive mutants reveals genes required for S phase progression in the presence of DNA damage, *Proc. Natl. Acad. Sci. U.S.A.* **99**:16934.
35. U. Alon, M. G. Surette, N. Barkai, and S. Leibler (1999). Robustness in bacterial chemotaxis, *Nature* **397**:168.
36. T. M. Yi, Y. Huang, M. I. Simon, and J. Doyle (2000). Robust perfect adaptation in bacterial chemotaxis through integral feedback control, *Proc. Natl. Acad. Sci. U.S.A.* **97**:4649.
37. T. Toyoda and A. Wada (2004). Omic space: coordinate-based integration and analysis of genomic phenomic interactions, *Bioinformatics* **20**:1759.
38. A. J. Walhout, J. Reboul, O. Shtanko, N. Bertin, P. Vaglio, et al. (2002). Integrating interactome, phenome, and transcriptome mapping data for the *C. elegans* germline, *Curr. Biol.* **12**:1952.

39. N. Freimer and C. Sabatti (2003). The human phenome project, *Nat. Genet.* **34**:15.
40. G. D. Bader, A. Heilbut, B. Andrews, M. Tyers, T. Hughes, and C. Boone (2003). Functional genomics and proteomics: charting a multidimensional map of the yeast cell, *Trends Cell Biol.* **13**:344.
41. I. Marin, M. L. Siegal, and B. S. Baker (2000). The evolution of dosage-compensation mechanisms, *Bioessays* **22**:1106.
42. G. Giaever, D. D. Shoemaker, T. W. Jones, H. Liang, E. A. Winzeler, A. Astromoff, and R. W. Davis (1999). Genomic profiling of drug sensitivities via induced haploinsufficiency, *Nat. Genet.* **21**:278.
43. G. Giaever, P. Flaherty, J. Kumm, M. Proctor, C. Nislow, et al. (2004). Chemogenomic profiling: identifying the functional interactions of small molecules in yeast, *Proc. Natl. Acad. Sci. U.S.A.* **101**:793.
44. P. J. Mason and M. Bessler (2004). Heterozygous telomerase deficiency in mouse and man: when less is definitely not more, *Cell Cycle* **3**:1127.
45. H. Kacser and J. A. Burns (1981). The molecular basis of dominance, *Genetics* **97**:639.
46. A. Cornish-Bowden (1987). Dominance is not inevitable, *J. Theor. Biol.* **125**:333.
47. H. Kacser and J. A. Burns (1995). The control of flux, *Biochem. Soc. Trans* **23**:341.
48. U. Grossniklaus, M. S. Madhusudhan, and V. Nanjundiah (1996). Nonlinear enzyme kinetics can lead to high metabolic flux control coefficients: implications for the evolution of dominance, *J. Theor. Biol.* **182**:299.
49. H. Bagheri-Chaichian, J. Hermisson, J. R. Vaisnys, and G. P. Wagner (2003). Effects of epistasis on phenotypic robustness in metabolic pathways, *Math. Biosci.* **184**:27.
50. N. H. Barton and B. Charlesworth (1998). Why sex and recombination? *Science* **281**:1986.
51. G. C. Conant and A. Wagner (2004). Duplicate genes and robustness to transient gene knock-downs in *Caenorhabditis elegans*, *Proc. R. Soc. Lond. B Biol. Sci.* **271**:89.
52. E. M. Jorgensen and S. E. Mango (2002). The art and design of genetic screens: *Caenorhabditis elegans*, *Nat. Rev. Genet.* **3**:356.
53. F. Fabre, A. Chan, W. D. Heyer, and S. Gangloff (2002). Alternate pathways involving Sgs1/Top3, Mus81/Mms4, and Srs2 prevent formation of toxic recombination intermediates from single-stranded gaps created by DNA replication, *Proc. Natl. Acad. Sci. U.S.A.* **99**:16887.
54. E. Shor, S. Gangloff, M. Wagner, J. Weinstein, G. Price, and R. Rothstein (2002). Mutations in homologous recombination genes rescue top3 slow growth in *Saccharomyces cerevisiae*, *Genetics* **162**:647.
55. A. H. Tong, M. Evangelista, A. B. Parsons, H. Xu, G. D. Bader, et al. (2001). Systematic genetic analysis with ordered arrays of yeast deletion mutants, *Science* **294**:2364.
56. L. Hartwell (2004). Genetics. Robust interactions, *Science* **303**:774.
57. E. A. Winzeler, D. D. Shoemaker, A. Astromoff, et al. (1999). Functional characterization of the *S. cerevisiae genome* by gene deletion and parallel analysis, *Science* **285**:901.
58. M. Lynch and J. S. Conery (2000). The evolutionary fate and consequences of duplicate genes, *Science* **290**:1151.
59. M. Lynch (2002). Genomics. Gene duplication and evolution, *Science* **297**:945.
60. M. Lynch and J. S. Conery (2003). The origins of genome complexity, *Science* **302**:1401.

61. K. H. Wolfe and D. C. Shields (1997). Molecular evidence for an ancient duplication of the entire yeast genome, *Nature* **387**:708.
62. C. Seoighe and K. H. Wolfe (1999). Updated map of duplicated regions in the yeast genome, *Gene* **238**:253.
63. A. McLysaght, K. Hokamp, and K. H. Wolfe (2002). Extensive genomic duplication during early chordate evolution, *Nat. Genet.* **31**:200.
64. J. M. Cork and M. D. Purugganan (2004). The evolution of molecular genetic pathways and networks, *Bioessays* **26**:479.
65. M. Goulian (2004). Robust control in bacterial regulatory circuits, *Curr. Opin. Microbiol.* **7**:198.
66. J. Warringer, E. Ericson, L. Fernandez, O. Nerman, and A. Blomberg (2003). High-resolution yeast phenomics resolves different physiological features in the saline response, *Proc. Natl. Acad. Sci. U.S.A.* **100**:15724.
67. S. L. Rutherford and S. Lindquist (1998). Hsp90 as a capacitor for morphological evolution, *Nature* **396**:336.
68. C. Queitsch, T. A. Sangster, and S. Lindquist (2002). Hsp90 as a capacitor of phenotypic variation, *Nature* **417**:618.
69. S. L. Rutherford (2003). Between genotype and phenotype: protein chaperones and evolvability, *Nat. Rev. Genet.* **4**:263.
70. C. C. Milton, B. Huynh, P. Batterham, S. L. Rutherford, and A. A. Hoffmann (2003). Quantitative trait symmetry independent of Hsp90 buffering: distinct modes of genetic canalization and developmental stability, *Proc. Natl. Acad. Sci. U.S.A.* **100**:13396.
71. N. Yi and S. Xu (2002). Mapping quantitative trait loci with epistatic effects, *Genet. Res.* **79**:185.
72. A. Goffeau, B. G. Barrell, H. Bussey, R. W. Davis, B. Dujon, et al. (1996). Life with 6000 genes, *Science* **274**:546.
73. B. Scherens and A. Goffeau (2004). The uses of genome-wide yeast mutant collections, *Genome Biol.* **5**:229.
74. M. A. Harris, J. Clark, A. Ireland, J. Lomax, M. Ashburner, et al. (2004). The Gene Ontology (GO) database and informatics resource, *Nucleic Acids Res* **32**(Database issue): D258.
75. B. Vogelstein and K. W. Kinzler (2004). Cancer genes and the pathways they control, *Nat. Med.* **10**:789.
76. A. Jordan and P. Reichard (1998). Ribonucleotide reductases, *Annu. Rev. Biochem.* **67**:71.
77. L. Thelander and P. Reichard (1979). Reduction of ribonucleotides, *Annu. Rev. Biochem.* **48**:133.
78. A. Chabes, B. Georgieva, V. Domkin, X. Zhao, R. Rothstein, and L. Thelander (2003). Survival of DNA damage in yeast directly depends on increased dNTP levels allowed by relaxed feedback inhibition of ribonucleotide reductase, *Cell* **112**:391.
79. S. P. Angus, L. J. Wheeler, S. A. Ranmal, X. Zhang, M. P. Markey, C. K. Mathews, and E. S. Knudsen (2002). Retinoblastoma tumor suppressor targets dNTP metabolism to regulate DNA replication, *J. Biol. Chem.* **277**:44376.
80. H. Tanaka, H. Arakawa, T. Yamaguchi, K. Shiraishi, S. Fukuda, K. Matsui, Y. Takei, and Y. Nakamura (2000). A ribonucleotide reductase gene involved in a p53-dependent cell-cycle checkpoint for DNA damage, *Nature* **404**:42.

81. S. J. Elledge, Z. Zhou, J. B. Allen, and T. A. Navas (1993). DNA damage and cell cycle regulation of ribonucleotide reductase, *Bioessays* **15**:333.
82. F. M. Amara, R. A. Hurta, A. Huang, and J. A. Wright (1995). Altered regulation of message stability and tumor promoter-responsive cis–trans interactions of ribonucleotide reductase R1 and R2 messenger RNAs in hydroxyurea-resistant cells, *Cancer Res.* **55**:4503.
83. S. Saitoh, A. Chabes, W. H. McDonald, L. Thelander, J. R. Yates, and P. Russell (2002). Cid13 is a cytoplasmic poly(A) polymerase that regulates ribonucleotide reductase mRNA, *Cell* **109**:563.
84. M. R. Abid, Y. Li, C. Anthony, and A. De Benedetti (1999). Translational regulation of ribonucleotide reductase by eukaryotic initiation factor. 4E links protein synthesis to the control of DNA replication, *J. Biol. Chem.* **274**:35991.
85. A. L. Chabes, C. M. Pfleger, M. W. Kirschner, and L. Thelander (2003). Mouse ribonucleotide reductase R2 protein: a new target for anaphase-promoting complex-Cdh1-mediated proteolysis, *Proc. Natl. Acad. Sci. U.S.A.* **100**:3925.
86. A. Chabes and L. Thelander (2000). Controlled protein degradation regulates ribonucleotide reductase activity in proliferating mammalian cells during the normal cell cycle and in response to DNA damage and replication blocks, *J. Biol. Chem.* **275**:17747.
87. X. Zhao and R. Rothstein (2002). The Dun1 checkpoint kinase phosphorylates and regulates the ribonucleotide reductase inhibitor Sml1, *Proc. Natl. Acad. Sci. U.S.A.* **99**:3746.
88. P. Reichard (2002). Ribonucleotide reductases: the evolution of allosteric regulation, *Arch. Biochem. Biophys.* **397**:149.
89. R. Yao, Z. Zhang, X. An, B. Bucci, D. L. Perlstein, J. Stubbe, and M. Huang (2003). Subcellular localization of yeast ribonucleotide reductase regulated by the DNA replication and damage checkpoint pathways, *Proc. Natl. Acad. Sci. U.S.A.* **100**:6628.
90. B. A. Kunz, S. E. Kohalmi, T. A. Kunkel, C. K. Mathews, E. M. McIntosh, and J. A. Reidy (1994). International Commission for Protection Against Environmental Mutagens and Carcinogens. Deoxyribonucleoside triphosphate levels: a critical factor in the maintenance of genetic stability, *Mutat. Res.* **318**:1.
91. J. A. Tercero, M. P. Longhese, and J. F. Diffley (2003). A central role for DNA replication forks in checkpoint activation and response, *Mol. Cell. Biol.* **11**:1323.
92. X. Wang, G. Ira, J. A. Tercero, A. M. Holmes, J. F. Diffley, and J. E. Haber (2004). Role of DNA replication proteins in double-strand break-induced recombination in *Saccharomyces cerevisiae*, *Mol. Cell. Biol.* **24**:6891.
93. G. Belli, E. Gari, L. Piedrafita, M. Aldea, and E. Herrero (1998). An activator/repressor dual system allows tight tetracycline-regulated gene expression in budding yeast, *Nucleic Acids Res.* **26**:942.
94. S. Mnaimneh, A. P. Davierwala, J. Haynes, J. Moffat, W. T. Peng, et al. (2004). Exploration of essential gene functions via titratable promoter alleles, *Cell* **118**:31.
95. M. B. Eisen, P. T. Spellman, P. O. Brown, and D. Botstein (1998). Cluster analysis and display of genome-wide expression patterns, *Proc. Natl. Acad. Sci. U.S.A.* **95**:14863.
96. S. Huang (2004). Back to the biology in systems biology: What can we learn from biomolecular networks? *Brief Funct. Genomic Proteomic* **2**:279.
97. E. Almaas, B. Kovacs, T. Vicsek, Z. N. Oltvai, and A. L. Barabasi (2004). Global organization of metabolic fluxes in the bacterium *Escherichia coli*, *Nature* **427**:839.

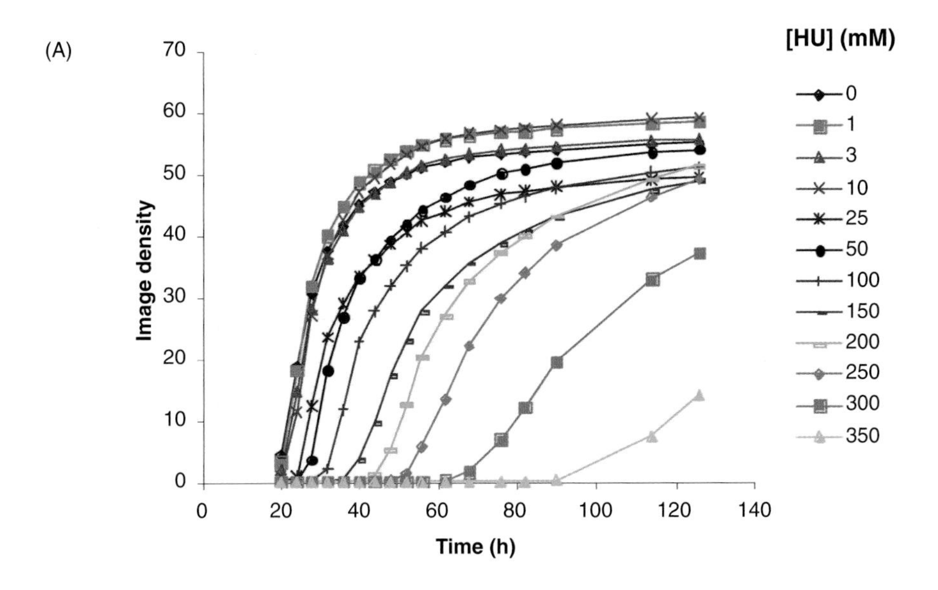

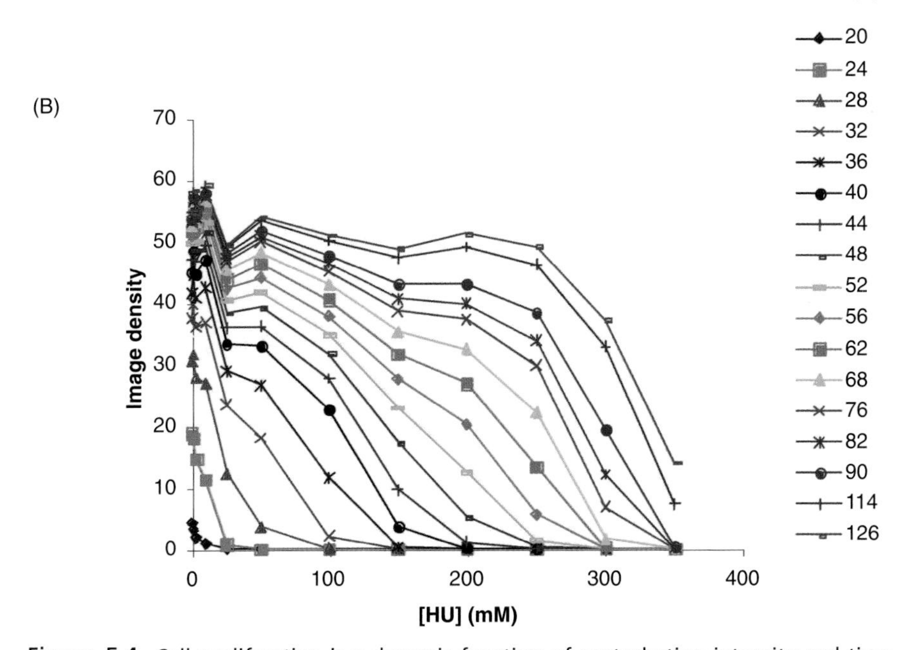

Figure 5.4. Cell proliferation is a dynamic function of perturbation intensity and time. Panel A, image density of spotted cultures (from Fig 5.3) is plotted for different perturbation intensities. Panel B, image density is plotted against perturbation intensity at discrete times during the experiment. See text page 112 for complete figure legend.

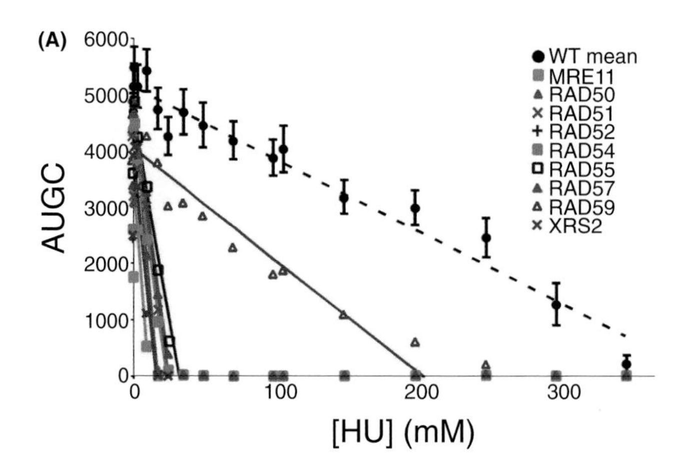

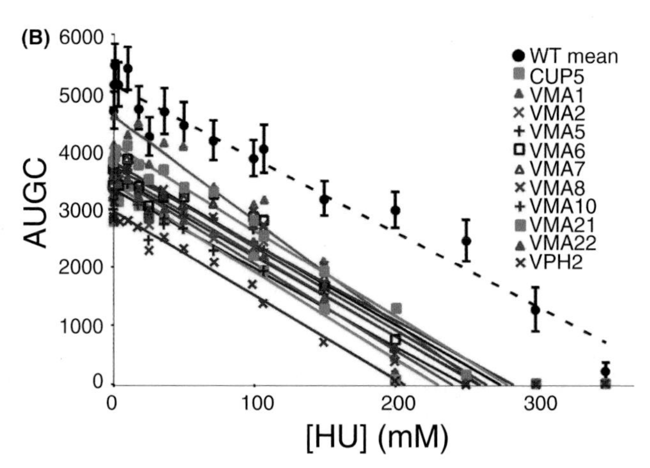

Figure 5.6. Buffering capacity (strength of genetic interaction) aids identification of functional interaction modules. Panel A, genes function in homologous DNA recombination. Panel B, genes involved in H+/ATPase structure and assembly. See text page 115 for complete figure legend.

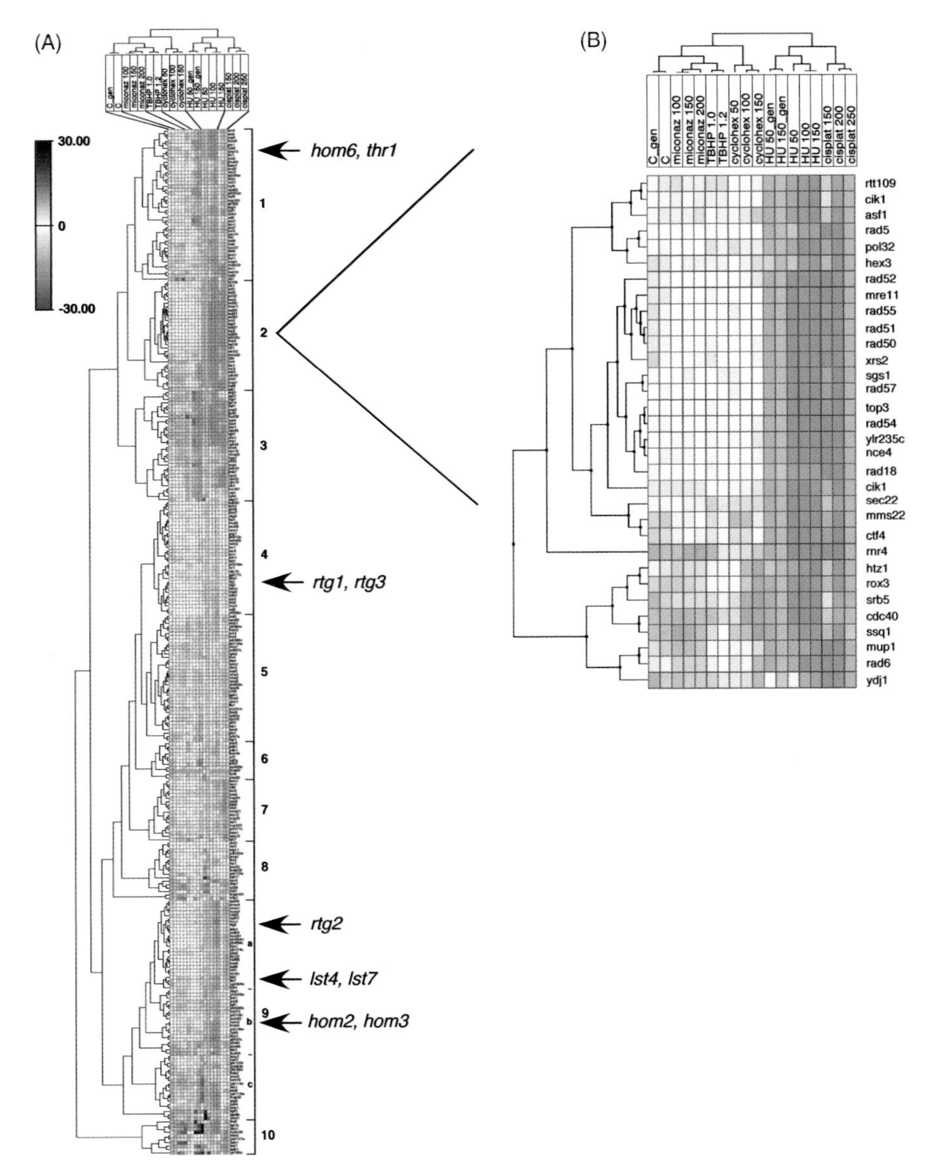

Figure 5.7. Hierarchical clustering used to identify genetic interaction modules in complex data sets involving multiple different perturbation types. See text page 117 for complete figure legend.

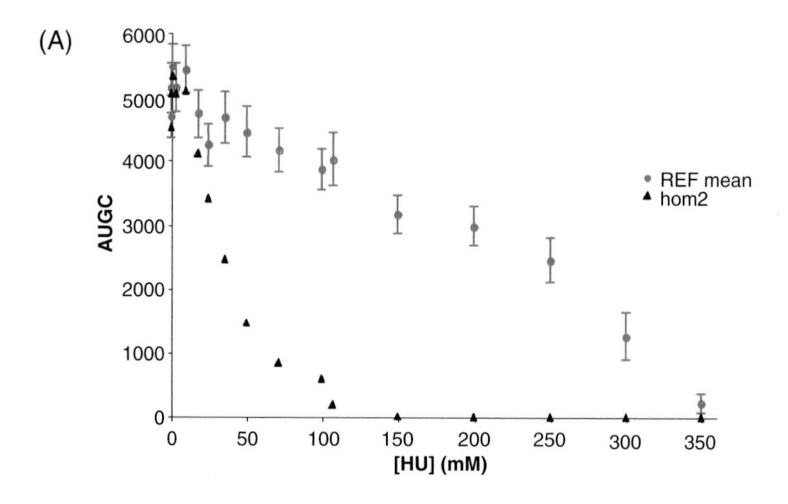

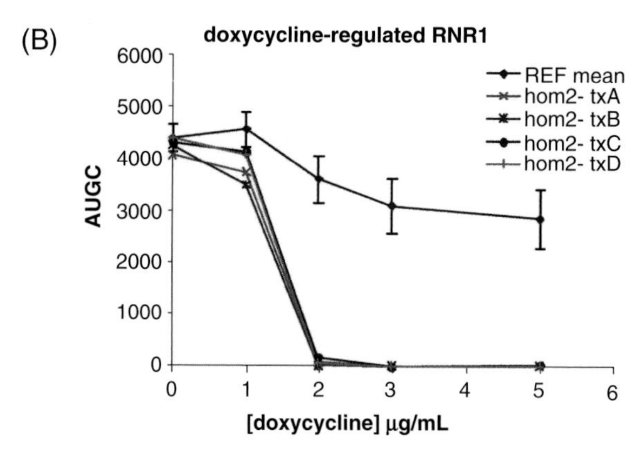

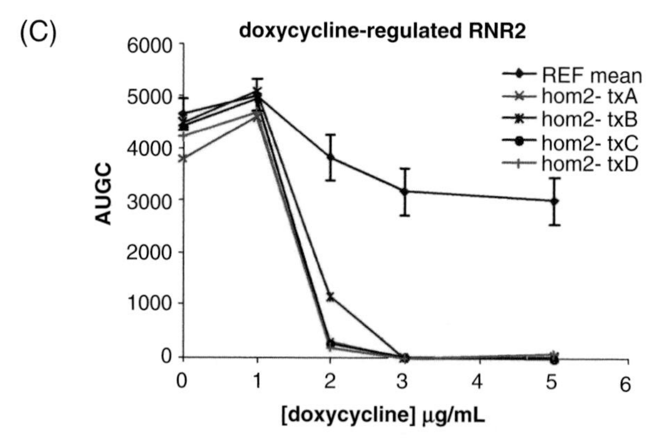

Figure 5.8. Tet-regulatable alleles of genetic drug targets are useful for validating chemical-genetic interactions. Panel A, buffering capacity of hom2 in response to HU perturbation. Panel B, depicts the average AUGC from multiple transformants of the reference strain. Panel C, similar data for tet-rnr2 deletion strain. See text page 119 for complete figure legend.

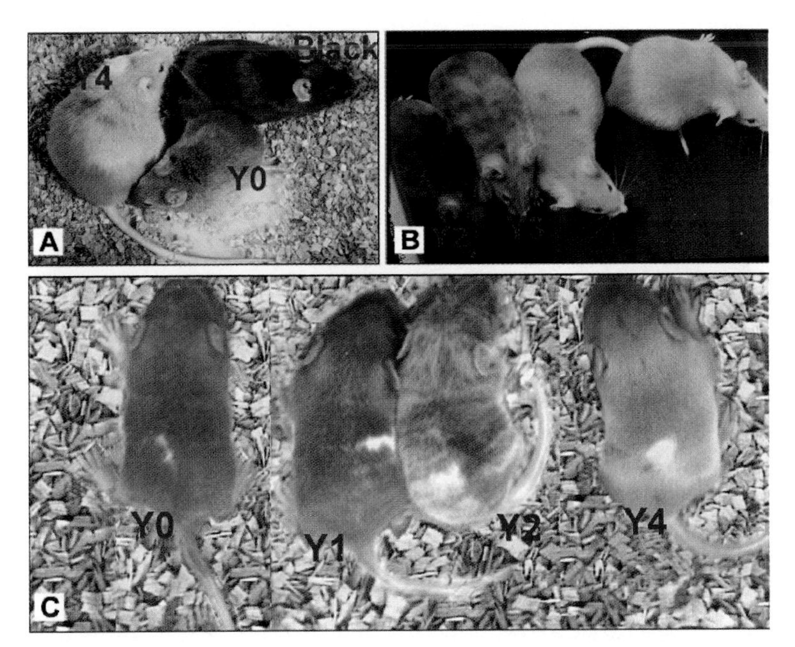

Figure 10.3. Examples of mice from viable yellow mouse model. (A) VY mice, (B) strain VY A^{vy}/a, (C) strain YS A^{vy}/a. See text page 228 for complete figure legend.

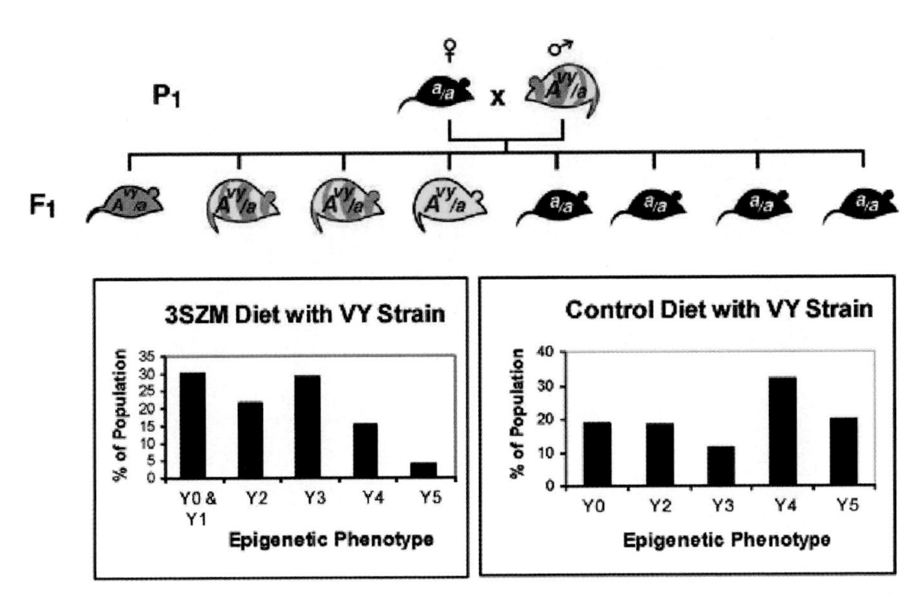

Figure 10.4. Mice mated a/a dam X A^{vy}/a sire. Dam is on 3SZM diet resulting in epigenetic phenotype shown in left panel or on control diet resulting in epigenetic phenotype shown in right panel. See text page 232 for complete figure legend.

Figure 17.8. Isomap visualization of normal (red) and clear cell carcinoma (blue) cells isolated from kidney tissues. See text page 390 for complete figure legend.

Figure 17.9. Principal component analysis (PCA) identifies three clusters in single nucleotide polymorphism (SNP) dataset. Asians (magenta), Yoruba Africans (dark blue), African-Americans (red) and European-Americans (light blue). See text page 391 for complete figure legend.

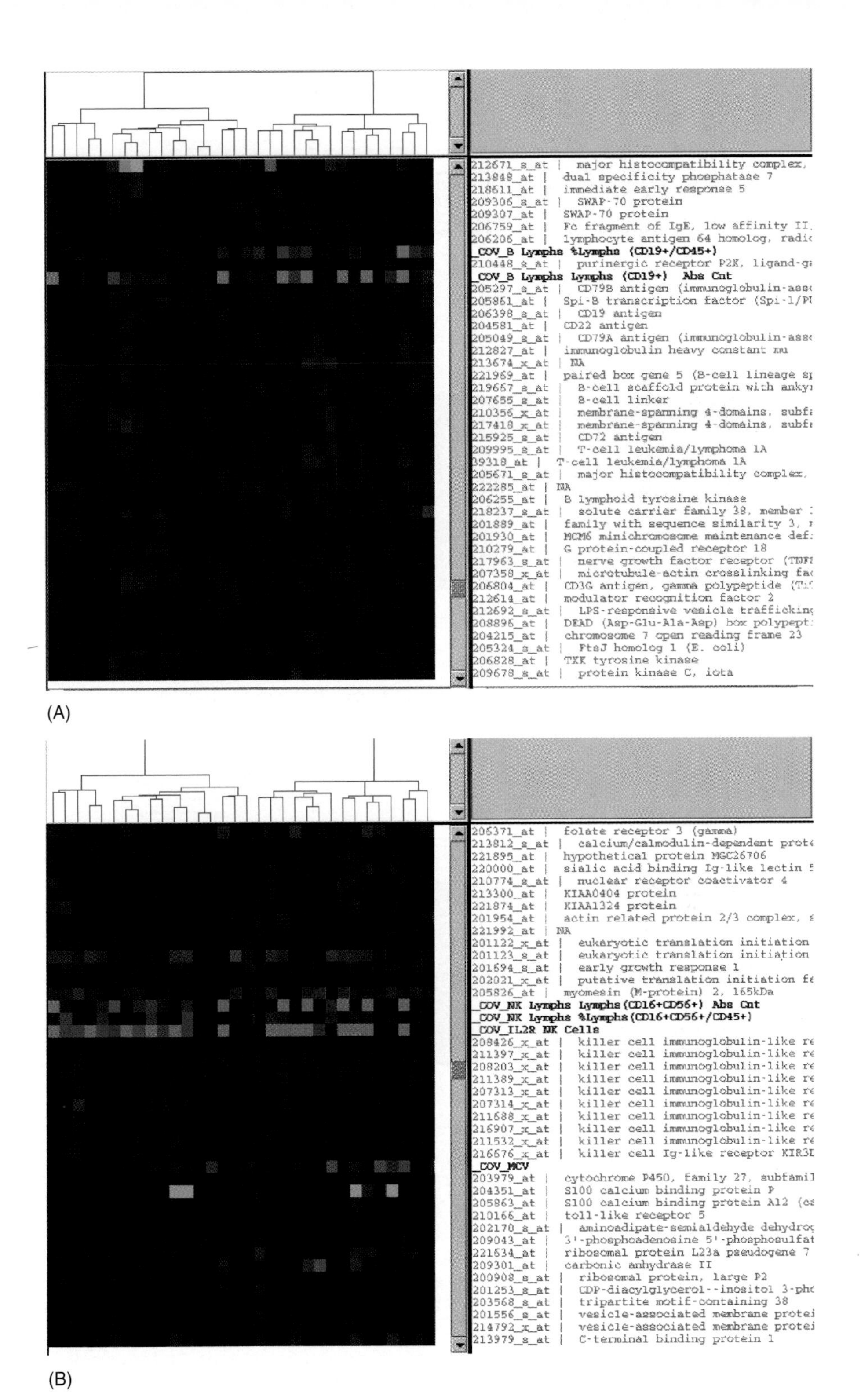

Figure 17.11. Expression values of genes expressed in particular cell types. (A) CD19+/CD45+, (B) CD16+/CD56+. See text page 397 for complete figure legend.

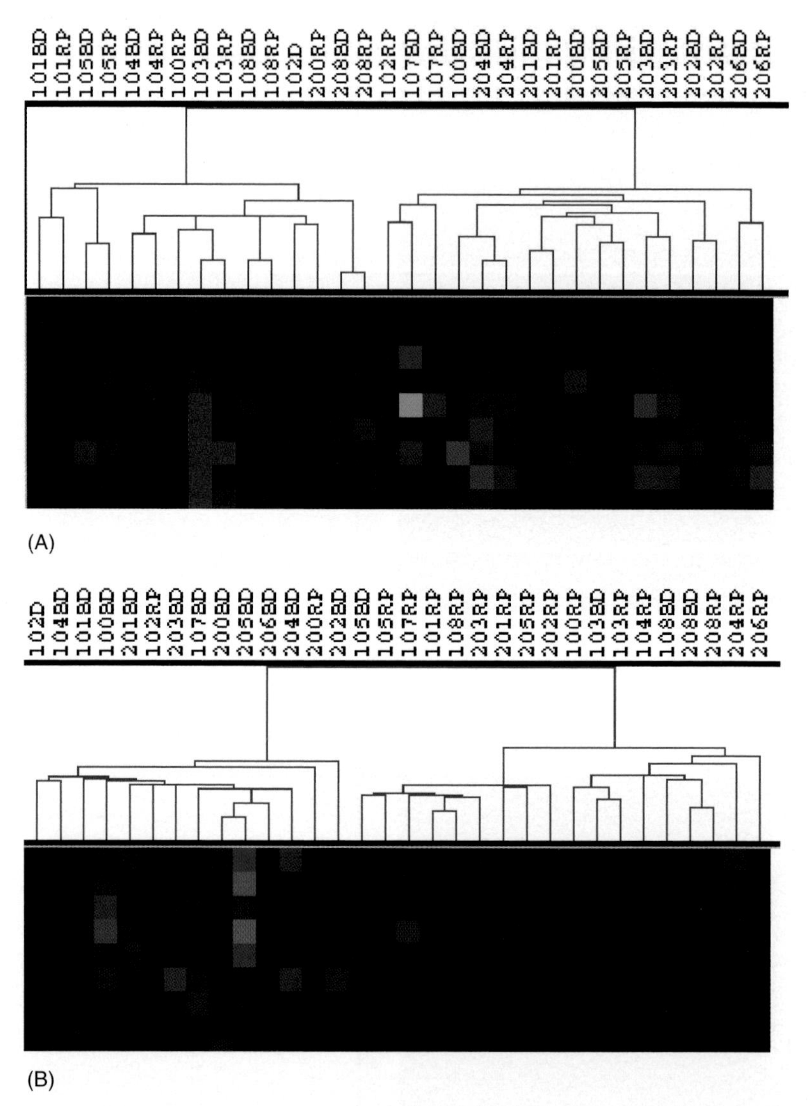

Figure 17.12. Inpatient variance of gene expression is usually higher than the effect of dietary intervention. (A) Hierarchical clustering of gene expression values of 16 subjects before and after dietary intervention. (B) ANOVA separates patient specific variance from treatment specific variance. See text page 398 for complete figure legend.

6

GENE–GENE EPISTASIS AND GENE–ENVIRONMENT INTERACTIONS INFLUENCE DIABETES AND OBESITY

Sally Chiu,[1] Adam L. Diament,[1] Janis S. Fisler,[2] and Craig H. Warden[1,3]

[1] *Rowe Program in Genetics and Department of Pediatrics, University of California–Davis, Davis, California*
[2] *Department of Nutrition, University of California–Davis, Davis, California*
[3] *Section of Neurobiology, Physiology, and Behavior, University of California–Davis, Davis, California*

6.1 GENE–GENE AND GENE–ENVIRONMENT INTERACTIONS

Epistasis

Epistasis, in its strictest classical genetic definition, is the interaction of genes where one gene (or locus) masks the effect of another [1]. To comprehend epistasis, an understanding of basic genetic terms, such as *gene*, *allele*, *dominant*, and *recessive* is first necessary. The definition of a gene has evolved over time as our knowledge of genetic mechanisms has increased, but for the purpose of this chapter, a gene is defined as a stretch of continuous DNA that occupies a specific location on a chromosome and that codes for a protein or functional RNA product. This stretch of DNA can have minor changes in sequence and still lead to a product, though this

Nutritional Genomics: Discovering the Path to Personalized Nutrition
Edited by Jim Kaput and Raymond L. Rodriguez

product may be slightly, or significantly, different as a consequence of the altered sequence. These alternate sequences of DNA at a specific locus (region on a chromosome) are called alleles. Each gene can have many alleles throughout a population, since mutations happen at different times and different locations on a chromosome through evolution, but each individual can only have two alleles of a gene (one allele of a gene per chromosome). Allele therefore can be defined as a form of a gene. Some alleles are deleterious and cause disease, others are harmless, and some may be beneficial. If a person has a harmless allele and a disease allele, what determines the outcome of whether this person will have a disease is whether the disease allele is dominant or recessive. A dominant allele needs only one copy of the allele to show its effects. Examples of diseases that are dominant are Huntington's disease, MODY (mature-onset diabetes of the young) diabetes, and certain forms of polycystic kidney disease. If two copies of the deleterious allele are necessary for the disease to manifest, the disease allele is said to be recessive, or masked by having one copy of a normal allele. In a recessive disease, the normal unmutated allele masks the effects of the mutated allele. What is important to remember about dominance and recessiveness is that these terms describe the effect that a *single* gene has on the measurable trait, not the effect of two different genes at different chromosomal loci.

In contrast to dominance and recessiveness, the classical definition of epistasis refers to at least *two* genes that are not alleles of each other, interacting to have a measurable effect on a trait. An example in the animal world is that of coat color of Labrador retrievers, which can be yellow, black, or chocolate. Coat color in this example is not caused by alleles of a single gene, but rather by the interaction of alleles of two different genes. In this respect, epistasis can be defined in a manner similar to dominant and recessive, except that instead of one allele being dominant to another allele of that same gene, alleles of one gene can be dominant to alleles of a different gene. In other words, when a gene has epistatic effects, the sum of the parts do not equal the whole: one cannot add up the effects of the yellow coat color gene, and the effects of a black coat color gene, to predict the outcome of the color of the dog. One cannot do this because the genes interact with each other, and the allelic form of one gene influences the effect of the second gene. While the term epistasis has been defined various ways throughout time and scientific field, the current geneticists' definition, and described in its broadest sense, is that gene–gene interactions show a nonlinearity of the effects of the two different alleles [2].

Gene–Environment Interactions

Many common, complex diseases are influenced not only by gene–gene epistasis, but by gene–environment interactions as well. The effects of certain genes are modified by the context of the environment, which may include factors such as diet, exercise, and life style. This chapter illustrates the general principles of both complex processes by discussing published data on their influence on obesity and diabetes. More complex interactions are also possible, such as gene–gene–environment interactions where gene–gene epistasis is modified by environment. Although this is a potentially important contributor to obesity and diabetes, such complex relationships cannot yet form the basis for a review because the literature is too limited.

6.2 EPISTASIS AND GENE–ENVIRONMENT INTERACTIONS IN OBESITY AND DIABETES

While the example of coat color is easy to visualize, other traits such as body fat percent or plasma glucose are harder to understand intuitively because such traits are quantitative and cannot be put into discrete categories. However, the principle of epistasis remains the same. The presence of alleles of one gene that increase body fat may mask the effects of alleles of a second gene that decrease body fat so that the overall effect cannot be predicted by adding up the effects of individual genes; that is, the effect of one gene is epistatic to other genes. In Figure 6.1A, heterozygous alleles for both gene X and gene Y are needed to increase the phenotype, whereas each heterozygous allele alone is not sufficient to produce an effect. Likewise, diet and/or exercise may interact with genes to produce unexpected outcomes. This concept is shown in Figure 6.1B, where a certain phenotype depends on both the alleles one possesses and the diet one consumes, in this case amount of fat in the diet. The effect of the low-fat diet masks the effect of the B allele to increase body weight. Since individuals have different alleles for many genes, how the body handles food or exercise will not be the same in all individuals. Metabolic rate, calorie partitioning, and other body composition effects will differ depending on how nutrients or exercise interact with different genes.

The rates of obesity and its comorbidity, type 2 diabetes mellitus (T2DM), are on the rise and an understanding of the molecular mechanisms is needed in order to

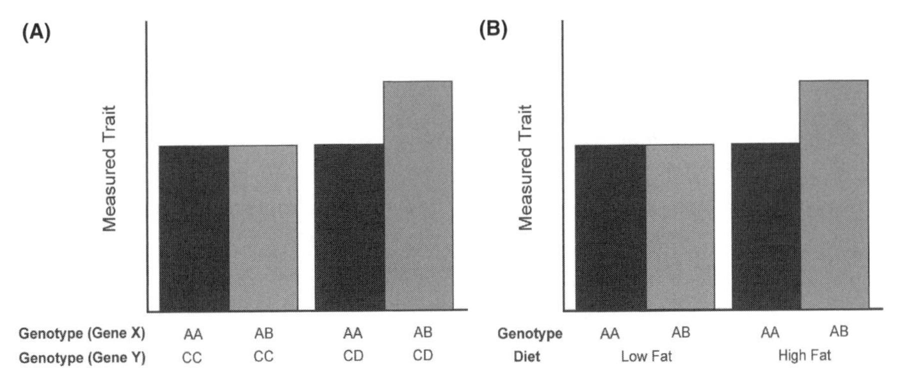

Figure 6.1. Gene-gene and gene-diet interactions. (A) A hypothetical gene-gene interaction. Both heterozygous alleles for genes X and Y are needed to increase the phenotype; one alone has no effect. (B) A hypothetical gene-diet interaction. On a low-fat diet, both genotypes AA and AB respond equally. However, on a high-fat diet, the presence of the B allele yields a greater response to the high-fat diet.

develop effective and safe treatments. Both obesity and T2DM are complex traits and increasing evidence suggests that both gene–gene interactions and gene–diet interactions contribute to the pathogenesis of both diseases. Identifying these interactions is important to the dissection of both the pathways involved as well as the genetic and environmental contributions to the diseases. Gene–diet interactions will allow better dietary recommendations to be made to individuals based on their genetic makeup and subsequent response to dietary factors.

In this chapter, we review models used to detect both gene–gene epistasis and gene–diet effects as well as the specific interactions identified thus far that contribute to obesity and diabetes.

6.3 ANIMAL MODELS FOR DETECTING GENE INTERACTIONS

Genetic studies in animals are faster and allow more detailed and accurate phenotyping than human studies. Mice will frequently be the animal model of choice due to the availability of hundreds of inbred and congenic strains as well as a completed mouse genome project. Since the homologous regions of mouse and human chromosomes are well-defined, mapping and identification of a gene in mice allows immediate examination in humans. Ability to manipulate diet and obtain any tissues at any time point are critical advantages of animal studies. In the past, rats, due to their larger size, were more frequently used than mice for physiological studies. However, newer tools and techniques have now made studies of energy balance and glucose metabolism feasible in mice.

Use of experimental crosses in inbred animal models is an efficient method for dissection of complex (quantitative) traits into discrete genetic factors. This method, called quantitative trait locus (QTL) mapping, is an ideal model for whole genome searches where environmental factors can be controlled. QTL analysis identifies the chromosomal location of a gene causing the phenotype of interest without needing any biochemical or physiological knowledge of the gene's function [3]. QTL mapping utilizes genetic linkage, or recombination maps, measured in centimorgans (cM), that can be constructed for loci that occur in two or more heritable forms or alleles. A QTL may contain hundreds of genes, one or more of which may influence the target trait. The gene needs both to participate in the target pathway and to have functionally different alleles in the two inbred strains being studied, that is, to have sequence differences that alter protein function or mRNA expression for that gene.

To accomplish QTL mapping, two inbred strains that differ in the trait of interest are bred to produce F1 mice. The F1 mice, which are all genetically identical, are then intercrossed to produce F2 progeny having three possible genotypes at each marker (Figure 6.2A), or backcrossed to either parent to produce progeny having two possible genotypes at each marker (Figure 6.2B). Genotyping identifies which of the two parental inbred strains contributed the allele at a specific locus in each of the F2 or backcross progeny. A genetic linkage map of ordered markers is constructed for the cross and a linear regression analysis of phenotype on genotype, such as analysis of variance, is done to determine if a genetic locus significantly affects a trait. However, linear regression can be used only to detect association at each genotyped marker. Therefore, more powerful statistics that utilize multiple markers and that can model multiple QTLs were developed. These statistics can locate genes that have independent effects on a trait and genes that interact with other genes to modify the trait [4, 5].

Many inbred mouse strain crosses have been used productively to map QTLs for obesity associated traits. The Human Obesity Gene Map: The 2004 Update [6] lists 89 QTLs for adiposity, measured as percent body fat, adiposity index, or the weight of specific fat pad depots in various mouse crosses, plus numerous QTLs for

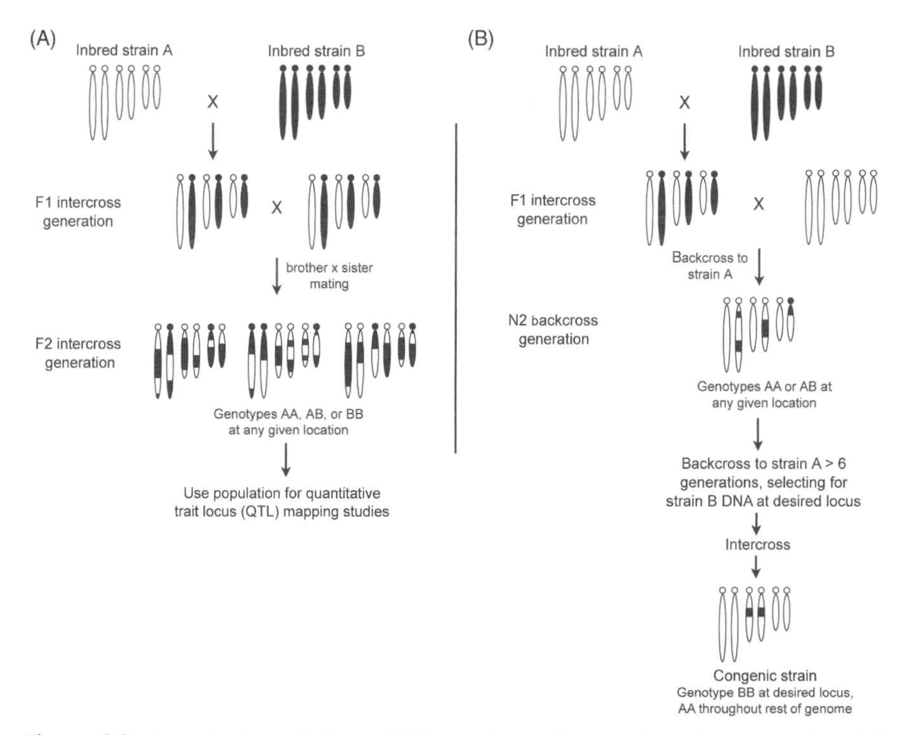

Figure 6.2. Quantitative trait locus (QTL) mapping and congenic strain construction. (A) General design of a QTL study. Two inbred strains, A and B, are crossed to produce F1 mice that are heterozygous for A and B alleles at all loci. F1 mice are intercrossed to produce F2 mice that are of genotype AA, AB, or BB at a given locus and are genetically distinct from each other. Genotype along the genome and phenotype are determined for this population and used to detect QTLs. (B) Congenic strains begin in a similar manner, with the production of F1 mice from strains A and B. F1 mice are backcrossed to the designated background strain (strain A) to produce N2 mice. An N2 mouse with the donor strain DNA (strain B) at a desired locus is backcrossed more than six times to the background strain, and at each generation, the B allele at the desired locus is selected for. Resulting mice are intercrossed to produce mice with the BB genotype at the desired locus and >99% AA genotypes throughout the rest of the genome.

body weight. Including human and all animal studies, a total of 38 genomic regions contain adiposity QTLs that have been replicated in two or more studies [6].

Once the QTL's location is identified, isolation of the locus into a congenic strain can confirm and fine map the trait. Congenic strains are produced by successive backcrosses to one parental (background) strain, selecting for donor strain genotype in the region of interest and deselecting for donor strain DNA elsewhere in the genome (Figure 6.2B) [7]. Subcongenic strains can be made that further subdivide the congenic region, and so on, until the QTL is located within a region containing very few genes [8]. Although congenic strains are frequently produced after QTL mapping, their use for the isolation of complex trait genes is an independent tool; that is, congenic strains with interesting phenotypes can be used whether or not the corresponding chromosomal region was first identified as a QTL. Advantages of the congenic approach are that it separates a gene affecting the phenotype from other genes that also affect that phenotype and allows the assessment of phenotype in many genetically identical individuals, greatly increasing statistical power [9].

6.4 GENE–GENE INTERACTION IN OBESITY AND DIABETES

The existence of gene–gene interactions has been known for many years. Genes targeted for knockout on one strain often show different phenotypes when on the background of another strain. The *ob* mutation of leptin (Lep^{ob}) causes leptin deficiency and on a C57BK/KsJ background results in obesity, hyperglycemia, and severe diabetes with degeneration of the islets of Langerhans. On a C57BL/6J background, the mutation also results in obesity; however, the diabetes phenotype is only mild with hypertrophy and hyperplasia of the islet cells [10]. An almost identical situation occurs with the $Lepr^{db}$ mutation, where the diabetes phenotype is exaggerated on a C57BL/KsJ background in comparison to a C57BL/6J background [11]. These studies show not only the importance of genetic background, but also the concept of gene interaction as different alleles from different strains interact with a mutation to produce variation in phenotype.

Double knockout mouse models, where two genes are disrupted, or double transgenics, where two genes are inserted, can often reveal interactions between genes. $Irs1^{-/-}/Irs3^{-/-}$ double knockouts are hyperglycemic and have severe lipoatrophy, while the $Irs1^{-/-}$ mice have only mild glucose intolerance and $Irs3^{-/-}$ mice have no detectable phenotype. Presence of only one knockout produces mice with normal serum glucose levels [12], suggesting that *Irs1* and *Irs3* compensate for each other. Mice with null mutations of either neuropeptide Y (NPY) or galanin have no marked obesity phenotypes, but when combined, the mutations in males result in a 30% increase in body weight relative to wild type males [13]. In another double mutant, the melanin-concentrating hormone knockout ($MCH^{-/-}$) was crossed onto an *ob/ob* mouse [14]. The authors expected that the $MCH^{-/-}$ would attenuate obesity in the *ob/ob* mouse by decreasing food intake since previous studies showed that $MCH^{-/-}$ mice ate less. However, in the presence of the *ob/ob* mutation, $MCH^{-/-}$ promoted weight loss, not through a decrease in hyperphagia, but entirely through an increase in energy expenditure. These results indicate that MCH lies downstream of leptin and is involved in leptin-mediated energy expenditure.

Recent statistical methods have been developed to detect gene–gene epistasis in QTL studies [5]. In our B6–*Mus spretus* backcross (BSB) model, lean C57BL/6J × lean *Mus spretus* progeny were backcrossed to C57BL/6J, producing backcross progeny varying in body fat from 1% to 50% [15, 16]. Several interacting chromosomal regions were found where the genotype at one locus affected obesity depending on the genotype at another locus [17]. A locus on chromosome 12 had only weak independent effects, but interacted with a locus on chromosome 2 to account for a greater amount of phenotypic variation. The same locus on chromosome 12 also interacted with one on chromosome 15 to affect body fat. In a cross using the TallyHo (TH) mouse, a model of noninsulin-dependent diabetes mellitus [18], TH mice were mated to either C57BL/6J or CAST/Ei mice. The resulting progeny were backcrossed to TH mice. These backcross progeny were used to search for multiple and interacting QTLs. Two pairs of interacting loci were found, one interaction between chromosomes 18 and 19, and a second between chromosomes 16 and 13, both of which increased hyperglycemia. The loci on chromosomes 18 and 16 had no individual effects on blood glucose. In a LG × SM cross, eight loci were found to influence adiposity with all loci participating in at least one epistatic interaction. With epistasis, twice the variance in females and 1.3× the variance in males was explained by the eight QTLs as compared to the nonepistatic model [19]. While these studies do not identify the underlying genes, they direct further studies aimed at isolating the genes involved and characterizing their relationships to each other.

6.5 DIETARY FAT IN OBESITY AND DIABETES

Evidence of Gene–Diet Interactions in Mouse Models

Though somewhat controversial, a large body of evidence associates dietary fat intake with the development of obesity [20, 21] and insulin resistance [22]. This relationship may be attenuated by interaction with genetic variants. Numerous inbred mouse strains exhibit different subphenotypes of obesity and T2DM when fed high-fat or high-sucrose diets, while genetic crosses have revealed chromosomal regions involved in obesity and T2DM that are sensitive to diet. These models provide information about the organs and pathways that contribute to complex phenotypes and can be used to isolate and identify specific genes that interact with dietary factors. Some of these gene–diet models that identified chromosomal loci controlling obesity or diabetes phenotypes are outlined in Table 6.1 and described below.

C57BL/6J mice fed a high-fat, high-sucrose diet are one model of "diabesity" [23]. High-fat diets promote more weight gain, hyperglycemia, and hyperinsulinemia in C57BL/6J mice than in the diet resistant A/J strain. Reducing the energy

TABLE 6.1. Quantitative Trait Loci (QTLs) for Diet–Gene Interaction

Strain	Phenotype	Interaction	Diet	QTL Name and Location	Reference
(C57BL/6J × CAST/Ei) F2	Body fat	QTL found on high-fat diet, no low-fat control diet for comparison	*High-fat diet*: 75% chow + 7.5% cocoa butter (total 30% kcal as fat)	*Mob5:* Chr 2 *Mob6:* Chr 2 *Mob7:* Chr 2 *Mob8:* Chr 9	[26]
(C57BL/6J × CAST/Ei) F2	Adiposity, mesenteric fat depot	QTL found on high-fat diet, no low-fat control diet for comparison	*High-fat condensed milk diet:* 32% kcal from fat	Chr 15	[27]
CAST/Ei congenic donor on C57Bl/6J background	Adiposity index, fat depot weights	Reduction of body fat on high-fat diet, not seen on low-fat diet	*High-fat condensed milk diet:* 32.6% kcal from fat *Low-fat chow diet:* 12% kcal from fat	Chr 7	[29]
(AKR/J × SWR/J) F2	Adiposity index	Authors state F2 mice do not become obese in absence of a high-fat diet	*High-fat diet:* 32.6% kcal from fat	*Do2:* Chr 9 *Do3:* Chr 15	[33]
NZO × (SJLxNZO) F1	Hyperglycemia	Hyperglycemia associated with *Nidd/SJL* locus exacerbated by *Nob1* locus only on high-fat diet, not low-fat diet	*Standard rodent chow*: 5/48/22.5% fat/CHO/protein *High fat rodent chow:* 16/46.8/17.1% fat/CHO/protein	*Nidd/SLJ:* Chr 4 *Nob1:* Chr. 5	[39]
(C57BL/6J × CAST/Ei) F2	Carbohydrate preference	Preference for 78%/22% of energy from CHO/protein	*High CHO:* 78%/22% of energy from CHO/protein, 3.6 kcal/g *High fat:* 78%/22% of energy from fat/protein, 6.0 kcal/g	*Mnic1:* Chr 17 *Mnic2:* Chr 6 *Mnic3:* Chr X	[42]
(C57BL/6J × CAST/Ei) F2	Dietary fat preference	Preference for 78%/22% of energy from fat/protein	*High CHO:* 78%/22% of energy from CHO/protein, 3.6 kcal/g *High fat:* 78%/22% of energy from fat/protein, 6.0 kcal/g	*Mnif1:* Chr 8 *Mnif2:* Chr 18 *Mnif3:* Chr X	[42]
(C57BL/6J × 129S1/SvImJ) F2	Body fat mass, body mass index	*Obq18* had no effect on percentage of fat itself, but had a large combined effect with *Obq16*	*High-fat diet:* 15% dairy fat	*Obq16:* Chr 8 *Obq18*: Chr 9	[28]

CHO = carbohydrate; Chr = chromosome. Diet compositions given are as published.

consumed from a high-fat diet attenuates but does not prevent the development of T2DM and obesity in C57BL/6J mice [24]. Linear relationships of body weight increases and deterioration of glucose tolerance were observed when C57BL/6J mice were fed diets varying from 10% to 60% of kilocalories from safflower oil [25].

Crosses utilizing C57BL/6J mice have been used to identify dietary fat responsive genetic loci. Mehrabian and colleagues [26] found that CAST/Ei mice are unusually lean and reported three QTLs on mouse chromosome 2 and one on chromosome 9 in an F2 cross of C57BL/6J and CAST/Ei mice maintained on a high-fat (30% of kcal) diet. York et al. [27] found a locus on chromosome 15 in a similar cross of C57BL/6J and CAST/Ei mice. A cross of C57BL/6J and 129S1/SvImJ mice discovered loci on chromosome 8 and chromosome 9 that affect body fat mass on a high-fat diet of 15% dairy fat [28]. Unfortunately, none of these studies included a low-fat diet comparison and it is not clear that the loci identified were a result of the fat content of the diet.

York et al. [29], in a study of a chromosome 7 CAST/Ei congenic on the C57BL/6J background found that fat depot weights were approximately 50% lower in congenic mice than in control C57BL/6J mice fed a 33% fat diet. This difference was not seen in mice fed a low-fat (12%) diet, suggesting that a gene(s) within the congenic region of approximately 25 cM responded to the increased dietary fat, although no differences in food intake or energy expenditure could be demonstrated between congenic and control mice [29].

Numerous other inbred mouse strains have been examined for susceptibility to dietary fat induced obesity. Salmon and Flatt [30] found that the incidence of obesity, defined as a third or more of body weight as fat, increased progressively from 0% to 35% in adult female CDI albino mice as the diet's fat content varied from 1% to 64% of total calories. West and colleagues [31] identified nine inbred mouse strains that were differentially sensitive to dietary fat, with AKR/J being the most sensitive and SWR/J being the least sensitive to dietary fat [32]. These authors then identified obesity QTLs on chromosomes 9 and 15 in F2 or backcrosses between AKR/J and SWR/J mice on a high-fat diet. Since these mice were not fed a low-fat diet, it is not known whether these QTLs contain genes that respond to changes in dietary fat [33]. Cheverud and colleagues [34] compared response to both low- and high-fat diets in mouse models bred for low (SM/J) and high (LG/J) growth. Although LG/J mice consumed more total food, SM/J mice consumed more food per unit body weight, and SM/J mice responded more strongly to dietary fat than the LG/J strain for obesity related traits [35].

The New Zealand obese (NZO) mouse is a model of obesity and diabetes with insulin and leptin resistance [36]. The lean Swiss/Jackson Laboratory (SJL) mouse carries a T2DM susceptibility allele (*Nidd/SJL*) [37]. A locus for body mass index (BMI), *Nob1*, identified in an NZO × SJL cross [38] also contributes to dietary induced obesity. In backcross mice of these two strains (NZO × F1(SJL × NZO)) raised on a high-fat diet, the presence of the *Nidd/SJL* locus resulted in hyperglycemia, hypoinsulinemia, reduced islet cell volume, and loss of beta cells of the pancreas [39]. Mice that also carried the *Nob1* locus were responsible for 90% of the diabetes in the backcross population. When raised on a low-fat diet, backcross mice carrying the *Nidd/SJL* locus showed an increased incidence of diabetes but the *Nob1* locus had no additional effect. Thus, the effect of *Nob1* in this backcross model to enhance diabetes is dependent on the fat content of the diet [39].

A few studies have examined inbred mouse strain differences in macronutrient selection. Smith et al. [40] found that AKR/J mice ate 30% more kilocalories and selected a significantly higher proportion of energy as fat (62% vs 28%) compared to SWR/J mice, suggesting that the increased susceptibility to dietary fat of the AKR/J strain was due to a preference for fat and to increased energy intake. These authors extended studies of macronutrient selection to a total of 13 mouse strains, including C57BL/6J, A/J, CAST/Ei, AKR/J, and SWR/J, and found that nine strains preferentially consumed more calories from fat over carbohydrate, while two strains preferentially consumed carbohydrate over fat [41]. Fat consumption varied from 26% to 83% of total energy and epididymal fat pad weight correlated with fat consumption in strains that consumed higher levels of dietary fat. The authors [41] concluded that SWR/J and CAST/Ei strains are highly sensitive to negative feedback from dietary fat, whereas AKR/J and C57BL/6J strains are not. Smith and colleagues [42] continued studies of diet selection where mice of a C57BL/6J and CAST/Ei F2 intercross were given the choice of two macronutrient diets containing 78%/22% of energy as either fat/protein or carbohydrate/protein. QTL analysis identified three loci, on chromosomes 8, 18, and X, for fat intake and three loci, on chromosomes 6, 17, and X, for carbohydrate intake. Loci for energy intake adjusted for body weight were found on chromosomes 17 and 18. All loci, with the exception of those on chromosomes 8 and X, overlapped loci for obesity in this F2 population [42].

Studies in rats and humans show that fats with differing chain length and saturation have different effects on energy metabolism and/or obesity. However, only one study in mice looked at whether the type of dietary fat could influence fat storage. Bell et al. [43] placed ARC Swiss albino mice on either a low-fat diet, or the equivalent diet where 41% of the carbohydrate was replaced with either beef fat or canola oil, and found that the amount of body fat of mice fed the monounsaturated canola oil diet was significantly less than that of mice fed the beef fat diet, suggesting that the type as well as amount of fat alters dietary fat induced obesity in mice.

Not all obesity QTLs discovered in studies of high-fat-fed strains or backcrosses encode genes influenced by diet since many studies fail to feed the F2 or backcross generation mice low-fat diets. Comparing the response of F2 mice fed high-fat versus low-fat diet controls for those QTLs exerting effect in all environments rather than those sensitive to changes in dietary fat. Since some obesity QTL analyses use high-fat-fed mice and others use low-fat-fed mice, caution should be used when comparing across studies. In addition, changes in dietary fat levels mean that other macronutrient levels, usually carbohydrate, also change. Hence, a QTL linked with dietary fat may actually be a QTL for carbohydrate intake. Finally, the response may be a result of increased caloric density, not type of nutrient, if the diets were not carefully made. Given these caveats, specific gene–diet interactions should be analyzed by mechanistic molecular and biochemical studies.

Evidence of Gene–Diet Interactions in Humans

Detecting novel gene–diet interactions in humans is difficult due to genetic diversity (heterogeneity), variable diets, and other uncontrollable environmental factors, leading to studies of reduced statistical power. However, several studies investigated gene–diet interactions by testing the association of dietary and environment factors with specific gene polymorphisms that had previously been associated with obesity or T2DM. The Pro12Ala polymorphism in the peroxisome proliferators activated receptor γ (PPARγ) gene has been associated with a decreased risk of T2DM [44–46]. PPARγ is activated by specific fatty acids and is a master regulator of adipocyte differentiation [47]. One study reported an interaction with the Pro12Ala polymorphism and the polyunsaturated fat to saturated fat (P:S) ratio of the diet. The authors found that in carriers of the Ala allele (both heterozygotes and homozygotes), there was a negative association with the P:S ratio and fasting insulin, but only in physically active subjects and not sedentary subjects [48]. Another study reported an inverse relationship between BMI and monosaturated fat intake in Ala carriers, but not Pro homozygotes [49]. The interaction between dietary fat, both total and saturated fat, and the L162 carriers of the L162V polymorphism of PPARγ was shown to explain slightly more variance in waist circumference than either factor alone. These studies need to be replicated since the introduction of multiple testing increases statistical errors [50].

Two common polymorphisms in the β_2-adrenergic receptor gene are associated with obesity by altering receptor function [51, 52]. In one case–control study, intake of carbohydrates at greater than 50% total kilocalories is associated with increased insulin levels in women carrying the Gln27Glu polymorphism [53]. A study of 12 pairs of monozygotic twins showed that long-term overfeeding led to a greater increase in obesity in Gln27 homozygotes than in Glu27 carriers [54]. In the same twin population, the HincII polymorphism of adipsin was associated with body fat and the Gln223Arg polymorphism of the leptin receptor was associated with insulin parameters [55]. Women carrying the Trp64Arg polymorphism of the β_3-adrenergic receptor had more difficulty losing weight in response to diet and exercise intervention than wild-type carriers [56].

Nieters and others [57] looked at polymorphisms in several candidate obesity genes and their interaction with dietary fatty acids in a case–control study. Of those, the Ala variant of the Pro12Ala polymorphism of PPARγ interacted with a high intake of arachidonic acid, resulting in an increased risk of obesity. An interaction was also found between intake of fatty acids and increased risk of obesity for carriers of the A alleles of the −2548G/A polymorphism of the leptin gene and A alleles of the −307G/A polymorphism of the tumor necrosis factor α (TNFα) gene.

6.6 MATERNAL EFFECTS

Gene interactions can also occur as a result of a maternal effect. The general concept of maternal effects refers to a mother's influence, either in utero or prior to weaning, on the phenotype of its progeny. Examples may include nutrients in the milk, uterine factors, or maternal health status that have a long-lasting effect on the offspring. This differs from maternal imprinting, where progeny selectively express the maternally inherited allele. There can be both maternal environmental effects, such as poor nutrition, or maternal genetic effects, such as genes controlling milk composition.

Evidence from rodent models shows that maternal genotype can affect the growth and development of obesity of a mother's progeny. Cowley et al. [58] performed embryo transfer experiments and showed that embryos, regardless of genetic background, had increased postnatal growth and adult body weight when transferred into C3H dams compared to SWR dams. Because all mice were then nursed by CB6F1/J females, the experiment showed that the maternal uterine environment, as determined by the genotype, was able to significantly influence adult body weight. In a cross between SM/J and LG/J mice, cross fostering experiments revealed three maternal effects QTLs affecting early growth in mice [59]. These QTLs accounted for almost a third of the among litter variance in growth. In a cross of NZO and NON mice, several QTLs were detected that controlled diabesity. Cross fostering showed that some of these QTLs were dependent on, or enhanced by, the maternal postparturitional environment [60].

Though identifying maternal effect genes influencing obesity or diabetes is difficult, increasing evidence indicates that maternal effects are important in the pathogenesis of some chronic diseases. Recent methods have been developed to detect QTL interactions of maternal and offspring genomes that affect offspring traits [61].

6.7 FUTURE DIRECTIONS AND CONCLUSION

Although statistical evidence suggests that genetics has a large influence on obesity, on weight gain in response to diet, on weight loss in response to exercise, and on development and severity of T2DM, only a few genes have been shown to influence obesity or T2DM in relatively small numbers of people and in some studies. Thus, either the statistical evidence is highly flawed or many more genes and pathways influencing weight homeostasis remain to be discovered. We have focused on studies in mice because mouse T2DM and obesity models have been the single most productive resource to identify human obesity and some T2DM genes. Investigators who read both the mouse and human literatures on obesity are often struck by the observation that there are many lessons from the mouse models literature that have not been applied to humans. Some of these conclusions are reproducibly found in many papers in the mouse models literature, but there are almost no comparable studies or acknowledgments of epistatic or gene–environment interactions in the human literature.

There have been literally dozens of QTL mapping studies of obesity and diet-responsive obesity in mice and rats. In this chapter, we reviewed some of these papers. We also reviewed the mouse models literature, which demonstrates that gene–gene epistasis for control of obesity is virtually universal, and papers that suggest that maternal genetic effects influence a substantial proportion of individual weight variance in some mice. Thus, we believe that it is now possible to begin looking at the big picture and asking about patterns that are evident from an analysis of the whole.

1. *Epistasis is common.* Studies in mice have demonstrated that gene–gene epistasis is universal and has a statistically significant impact on overall obesity. Nevertheless, very few human studies have searched for gene–gene epistasis in obesity. Almost all human obesity genetics studies are too underpowered for this analysis.
2. *Diet–genotype interactions are common.* Studies have demonstrated that some mice or rats gain weight on diets where other strains remain lean. The obvious implications are that some people will gain weight on diets where others remain lean and that gene–diet interactions influence initiation, development, and severity of chronic diseases.
3. *Maternal effects are modified by genetics.* Studies in mice have demonstrated that maternal effects do not just depend on the environment (i.e., what diet the mother is fed) but also depend on maternal genotype. Although there have been many studies of maternal environmental effects on obesity in humans, investigations of the effects of maternal genotype are virtually nonexistent.

In this chapter we discussed some of the complex interactions of diets and genotypes that influence obesity in mice and humans. Studies of three-way interactions of gene–gene–diet are almost nonexistent; thus, we cannot assess the importance of even more complex models for weight regulation. Nevertheless, we believe it is essential to perform human genetic studies in carefully phenotyped populations with sufficient size to detect effects that are dependent on gene–diet, gene–gene interaction, and/or maternal effects. These studies are necessary if we are to discover the underlying causes of human obesity because these mechanisms all are clearly reproducibly present in mice and are untested in humans.

REFERENCES

1. W. Bateson (1909). *Mendel's Principles of Heredity.* Cambridge University Press, Cambridge, U.K.
2. R. A. Fisher (1918). The correlation between relatives on the supposition of Mendelian inheritance, *Trans. R. Soc. Edin.* **52**:1127–1136.
3. E. S. Lander and D. Botstein (1989). Mapping Mendelian factors underlying quantitative traits using RFLP linkage maps, *Genetics* **121**:185–199.
4. S. Sen and G. A. Churchill (2001). A statistical framework for quantitative trait mapping, *Genetics* **159**:371–387.

5. N. Yi and S. Xu (2002). Mapping quantitative trait loci with epistatic effects, *Genet. Res.* **79**:185–198.
6. L. Pérusse, T. Rankinen, A. Zuber, Y. C. Chagnon, S. J. Weisnagel, G. Argyropoulos, B. Walts, E. E. Snyder, and C. Bouchard (2005). The human obesity gene map: The 2004 update, *Obes. Res.* **13**:381–490.
7. L. M. Silver (1995). *Mouse Genetics*. Oxford University Press, New York, p. 362.
8. A. L. Diament, P. Farahani, S. Chiu, J. Fisler, and C. H. Warden (2004). A novel mouse chromosome 2 congenic strain with obesity phenotypes, *Mamm. Genome* **15**:452–459.
9. O. Abiola, J. M. Angel, P. Avner, A. A. Bachmanov, J. K. Belknap, et al. (2003). The nature and identification of quantitative trait loci: a community's view, *Nat. Rev. Genet.* **4**:911–916.
10. D. L. Coleman and K. P. Hummel (1973). The influence of genetic background on the expression of the obese (Ob) gene in the mouse, *Diabetologia* **9**:287–293.
11. K. P. Hummel, D. L. Coleman, and P. W. Lane (1972). The influence of genetic background on expression of mutations at the diabetes locus in the mouse. I. C57BL-KsJ and C57BL-6J strains, *Biochem. Genet.* **7**:1–13.
12. Y. Terauchi, J. Matsui, R. Suzuki, N. Kubota, K. Komeda, et al. (2003). Impact of genetic background and ablation of insulin receptor substrate (IRS)-3 on IRS-2 knock-out mice, *J. Biol. Chem.* **278**:14284–14290.
13. J. G. Hohmann, D. N. Teklemichael, D. Weinshenker, D. Wynick, D. K. Clifton, and R. A. Steiner (2004). Obesity and endocrine dysfunction in mice with deletions of both neuropeptide Y and galanin, *Mol. Cell. Biol.* **24**:2978–2985.
14. G. Segal-Lieberman, R. L. Bradley, E. Kokkotou, M. Carlson, D. J. Trombly, et al. (2003). Melanin-concentrating hormone is a critical mediator of the leptin-deficient phenotype, *Proc. Natl. Acad. Sci. U.S.A.* **100**:10085–10090.
15. C. H. Warden, J. S. Fisler, S. M. Shoemaker, P. Z. Wen, K. L. Svenson, M. J. Pace, and A. J. Lusis (1995). Identification of four chromosomal loci determining obesity in a multifactorial mouse model, *J. Clin. Invest.* **95**:1545–1552.
16. J. S. Fisler, C. H. Warden, M. J. Pace, and A. J. Lusis (1993). BSB: a new mouse model of multigenic obesity, *Obes. Res.* **1**:271–280.
17. N. Yi, A. Diament, S. Chiu, K. Kim, D. B. Allison, J. S. Fisler, and C. H. Warden (2004). Characterization of epistasis influencing complex spontaneous obesity in the BSB model, *Genetics* **167**:399–409.
18. J. H. Kim, S. Sen, C. S. Avery, E. Simpson, P. Chandler, P. M. Nishina, G. A. Churchill, and J. K. Naggert (2001). Genetic analysis of a new mouse model for non-insulin-dependent diabetes, *Genomics* **74**:273–286.
19. J. M. Cheverud, T. T. Vaughn, L. S. Pletscher, A. C. Peripato, E. S. Adams, C. F. Erikson, and K. J. King-Ellison (2001). Genetic architecture of adiposity in the cross of LG/J and SM/J inbred mice, *Mamm. Genome* **12**:3–12.
20. G. A. Bray and B. M. Popkin (1998). Dietary fat intake does affect obesity! *Am. J. Clin. Nutr.* **68**:1157–1173.
21. J. C. Peters (2003). Dietary fat and body weight control, *Lipids* **38**:123–127.
22. J. C. Lovejoy (2002). The influence of dietary fat on insulin resistance, *Curr. Diab. Rep.* **2**:435–440.
23. R. S. Surwit, M. N. Feinglos, J. Rodin, A. Sutherland, A. E. Petro, E. C. Opara, C. M. Kuhn, and M. Rebuffe-Scrive (1995). Differential effects of fat and sucrose on the development of obesity and diabetes in C57BL/6J and A/J mice, *Metabolism* **44**:645–651.

24. A. E. Petro, J. Cotter, D. A. Cooper, J. C. Peters, S. J. Surwit, and R. S. Surwit (2004). Fat, carbohydrate, and calories in the development of diabetes and obesity in the C57BL/6J mouse, *Metabolism* **53**:454–457.

25. M. Takahashi, S. Ikemoto, and O. Ezaki (1999). Effect of the fat/carbohydrate ratio in the diet on obesity and oral glucose tolerance in C57BL/6J mice, *J. Nutr. Sci. Vitaminol. (Tokyo)* **45**:583–593.

26. M. Mehrabian, P. Z. Wen, J. Fisler, R. C. Davis, and A. J. Lusis (1998). Genetic loci controlling body fat, lipoprotein metabolism, and insulin levels in a multifactorial mouse model, *J. Clin. Invest.* **101**:2485–2496.

27. B. York, K. Lei, and D. B. West (1996). Sensitivity to dietary obesity linked to a locus on chromosome 15 in a CAST/Ei x C57BL/6J F2 intercross, *Mamm. Genome* **7**:677–681.

28. N. Ishimori, R. Li, P. M. Kelmenson, R. Korstanje, K. A. Walsh, G. A. Churchill, K. Forsman-Semb, and B. Paigen (2004). Quantitative trait loci that determine plasma lipids and obesity in C57BL/6J and 129S1/SvImJ inbred mice, *J. Lipid Res.* **45**: 1624–1632.

29. B. York, A. A. Truett, M. P. Monteiro, S. J. Barry, C. H. Warden, J. K. Naggert, T. P. Maddatu, and D. B. West (1999). Gene–environment interaction: a significant diet-dependent obesity locus demonstrated in a congenic segment on mouse chromosome 7, *Mamm. Genome* **10**:457–462.

30. D. M. Salmon and J. P. Flatt (1985). Effect of dietary fat content on the incidence of obesity among ad libitum fed mice, *Int. J. Obes.* **9**:443–449.

31. D. B. West, C. N. Boozer, D. L. Moody, and R. L. Atkinson (1992). Dietary obesity in nine inbred mouse strains, *Am. J. Physiol.* **262**:R1025–R1032.

32. D. B. West, J. Waguespack, and S. McCollister (1995). Dietary obesity in the mouse: interaction of strain with diet composition, *Am. J. Physiol.* **268**:R658–R665.

33. D. B. West, J. Goudey-Lefevre, B. York, and G. E. Truett (1994). Dietary obesity linked to genetic loci on chromosomes 9 and 15 in a polygenic mouse model, *J. Clin. Invest.* **94**:1410–1416.

34. J. M. Cheverud, L. S. Pletscher, T. T. Vaughn, and B. Marshall (1999). Differential response to dietary fat in large (LG/J) and small (SM/J) inbred mouse strains, *Physiol. Genomics* **1**:33–39.

35. T. H. Ehrich, J. P. Kenney, T. T. Vaughn, L. S. Pletscher, and J. M. Cheverud (2003). Diet, obesity, and hyperglycemia in LG/J and SM/J mice, *Obes. Res.* **11**:1400–1410.

36. J. L. Halaas, C. Boozer, J. Blair-West, N. Fidahusein, D. A. Denton, and J. M. Friedman (1997). Physiological response to long-term peripheral and central leptin infusion in lean and obese mice, *Proc. Natl. Acad. Sci. U.S.A.* **94**:8878–8883.

37. L. Plum, R. Kluge, K. Giesen, J. Altmuller, J. R. Ortlepp, and H. G. Joost (2000). Type 2 diabetes-like hyperglycemia in a backcross model of NZO and SJL mice: characterization of a susceptibility locus on chromosome 4 and its relation with obesity, *Diabetes* **49**:1590–1596.

38. R. Kluge, K. Giesen, G. Bahrenberg, L. Plum, J. R. Ortlepp, and H. G. Joost (2000). Quantitative trait loci for obesity and insulin resistance (Nob1, Nob2) and their interaction with the leptin receptor allele (LeprA720T/T1044I) in New Zealand obese mice, *Diabetologia* **43**:1565–1572.

39. L. Plum, K. Giesen, R. Kluge, E. Junger, K. Linnartz, A. Schurmann, W. Becker, and H. G. Joost (2002). Characterisation of the mouse diabetes susceptibilty locus Nidd/SJL:

islet cell destruction, interaction with the obesity QTL Nob1, and effect of dietary fat, *Diabetologia* **45**:823–830.

40. B. K. Smith, D. B. West, and D. A. York (1997). Carbohydrate versus fat intake: differing patterns of macronutrient selection in two inbred mouse strains, *Am. J. Physiol.* **272**: R357–R362.
41. B. K. Smith, P. K. Andrews, and D. B. West (2000). Macronutrient diet selection in thirteen mouse strains, *Am. J. Physiol. Regul. Integr. Comp. Physiol.* **278**: R797–R805.
42. B. K. Smith Richards, B. N. Belton, A. C. Poole, J. J. Mancuso, G. A. Churchill, R. Li, J. Volaufova, A. Zuberi, and B. York (2002). QTL analysis of self-selected macronutrient diet intake: fat, carbohydrate, and total kilocalories, *Physiol. Genomics* **11**:205–217.
43. R. R. Bell, M. J. Spencer, and J. L. Sherriff (1997). Voluntary exercise and monounsaturated canola oil reduce fat gain in mice fed diets high in fat, *J. Nutr.* **127**: 2006–2010.
44. D. Altshuler, J. N. Hirschhorn, M. Klannemark, C. M. Lindgren, M. C. Vohl, et al. (2000). The common PPARgamma Pro12Ala polymorphism is associated with decreased risk of type 2 diabetes, *Nat. Genet.* **26**:76–80.
45. S. S. Deeb, L. Fajas, M. Nemoto, J. Pihlajamaki, L. Mykkanen, J. Kuusisto, M. Laakso, W. Fujimoto, and J. Auwerx (1998). A Pro12Ala substitution in PPARgamma2 associated with decreased receptor activity, lower body mass index and improved insulin sensitivity, *Nat. Genet.* **20**:284–287.
46. S. Masud and S. Ye (2003). Effect of the peroxisome proliferator activated receptor-gamma gene Pro12Ala variant on body mass index: a meta-analysis, *J. Med. Genet.* **40**:773–780.
47. R. M. Evans, G. D. Barish, and Y. X. Wang (2004). PPARs and the complex journey to obesity, *Nat. Med.* **10**:355–361.
48. P. W. Franks, J. Luan, P. O. Browne, A. H. Harding, S. O'Rahilly, V. K. Chatterjee, and N. J. Wareham (2004). Does peroxisome proliferator-activated receptor gamma genotype (Pro12ala) modify the association of physical activity and dietary fat with fasting insulin level? *Metabolism* **53**:11–16.
49. A. Memisoglu, F. B. Hu, S. E. Hankinson, J. E. Manson, I. De Vivo, W. C. Willett, and D. J. Hunter (2003). Interaction between a peroxisome proliferator-activated receptor gamma gene polymorphism and dietary fat intake in relation to body mass, *Hum. Mol. Genet.* **12**:2923–2929.
50. J. Robitaille, C. Brouillette, A. Houde, S. Lemieux, L. Perusse, A. Tchernof, D. Gaudet, and M. C. Vohl (2004). Association between the PPARalpha-L162V polymorphism and components of the metabolic syndrome, *J. Hum. Genet.* **49**:482–489.
51. V. Large, L. Hellstrom, S. Reynisdottir, F. Lonnqvist, P. Eriksson, L. Lannfelt, and P. Arner (1997). Human beta-2 adrenoceptor gene polymorphisms are highly frequent in obesity and associate with altered adipocyte beta-2 adrenoceptor function, *J. Clin. Invest.* **100**:3005–3013.
52. K. Leineweber and O. E. Brodde (2004). Beta2-adrenoceptor polymorphisms: relation between in vitro and in vivo phenotypes, *Life Sci.* **74**:2803–2814.
53. J. A. Martinez, M. S. Corbalan, A. Sanchez-Villegas, L. Forga, A. Marti, and M. A. Martinez-Gonzalez (2003). Obesity risk is associated with carbohydrate intake in women carrying the Gln27Glu beta2-adrenoceptor polymorphism, *J. Nutr.* **133**:2549–2554.

54. O. Ukkola, A. Tremblay, and C. Bouchard (2001). Beta-2 adrenergic receptor variants are associated with subcutaneous fat accumulation in response to long-term overfeeding, *Int. J. Obes. Relat. Metab. Disord.* **25**:1604–1608.

55. O. Ukkola and C. Bouchard (2004). Role of candidate genes in the responses to long-term overfeeding: review of findings, *Obes. Rev.* **5**:3–12.

56. K. Shiwaku, A. Nogi, E. Anuurad, K. Kitajima, B. Enkhmaa, K. Shimono, and Y. Yamane (2003). Difficulty in losing weight by behavioral intervention for women with Trp64Arg polymorphism of the beta3-adrenergic receptor gene, *Int. J. Obes. Relat. Metab. Disord.* **27**:1028–1036.

57. A. Nieters, N. Becker, and J. Linseisen (2002). Polymorphisms in candidate obesity genes and their interaction with dietary intake of n-6 polyunsaturated fatty acids affect obesity risk in a sub-sample of the EPIC-Heidelberg cohort, *Eur. J. Nutr.* **41**:210–221.

58. D. E. Cowley, D. Pomp, W. R. Atchley, E. J. Eisen, and D. Hawkins-Brown (1989). The impact of maternal uterine genotype on postnatal growth and adult body size in mice, *Genetics* **122**:193–203.

59. J. B. Wolf, T. T. Vaughn, L. S. Pletscher, and J. M. Cheverud (2002). Contribution of maternal effect QTL to genetic architecture of early growth in mice, *Heredity* **89**:300–310.

60. P. C. Reifsnyder, G. Churchill, and E. H. Leiter (2000). Maternal environment and genotype interact to establish diabesity in mice, *Genome Res.* **10**:1568–1578.

61. Y. Cui, G. Casella, and R. Wu (2004). Mapping quantitative trait loci interactions from the maternal and offspring genomes, *Genetics* **167**:1017–1026.

7

NUTRIENTS AND GENE EXPRESSION

Gertrud U. Schuster

Nutrition Department, University of California-Davis, Davis, California

7.1 INTRODUCTION

The last decade has provided evidence that major and minor dietary constituents can regulate gene expression in a hormonally independent manner. Fatty acids, cholesterol, and glucose, including their metabolites, and lipid soluble vitamins mediate their functions via transcription factors. Sterol regulatory element binding proteins (SREBPs), carbohydrate sensitive response element binding protein (ChREBP), and nuclear receptors are nutrient sensing factors, which are pivotal players in the regulation of diverse biological processes including cancer and inflammatory response, and in metabolic pathways such as carbohydrate, lipid, and cholesterol homeostasis (Table 7.1).

7.2 SREBPs AND ChREBP: TRANSCRIPTION FACTORS INFLUENCED BY DIETARY LIPIDS AND GLUCOSE

Several dietary studies have implicated dietary polyunsaturated fatty acids (PUFAs) in the inhibition of hepatic expression of several genes involved in fatty acid syn-

Nutritional Genomics: Discovering the Path to Personalized Nutrition
Edited by Jim Kaput and Raymond L. Rodriguez

TABLE 7.1. Nutrient and Gene Expression: Transcription Factor Mediated Metabolic Pathways [92]

Nutrient	Component	Transcription Faction
	Macronutrients	
Fats	Fatty acids	SREBPs, ChREBP, HNF4, PPARs, LXR
	Cholesterol	LXR, FXR, PXR
Carbohydrates	Glucose	SREBPs, ChREBP, USFs
Proteins	Amino acids	C/EBPs
	Micronutrients	
Vitamins	Vitamin A	RAR, RXR
	Vitamin D	VDR
	Vitamin E	PXR
Minerals	Calcium	Calcineurin/NF-ATs
	Iron	IRP1, IRP2
	Zinc	MTF1
	Other Food Components	
	Flavonoids/phytoestrogens	ERs, NF-κB, AP1
	Xenobiotics	CAR/PXR

Abbreviations used: AP-1, activating protein-1; CAR, constitutively active receptor; C/EBP, CAAT/enhancer binding protein; ChREBP, carbohydrate response element binding protein; ER, estrogen receptor; FXR, farnesoid X receptor; HNF4, hepatocyte nuclear factor; IRP, iron regulatory protein; LXR, liver X receptor; NF-AT, nuclear factor of activated T cells; NF-κB, nuclear factor-κB; MTF, metal-responsive transcription factor; PPAR, peroxisome proliferator activated receptor; PXR, pregnane X receptor; RAR, retinoid acid receptor; RXR, retinoid X receptor; SREBP, sterol regulatory element binding protein; USF, upstream stimulatory factor; VDR, vitamin D receptor.

thesis [1, 2]. However, until the discoveries that fatty acid induced gene expression was mediated by transcription factors and the cloning and characterization of the previously so-called orphan receptors, the underlying molecular mechanisms were poorly understood. PUFAs have dramatic effects on gene expression by regulating the activity or abundance of several transcription factors. Except for SREBPs and ChREBP, all other transcription factors are members of the superfamily of nuclear receptors.

SREBPs

SREBPs regulate the expression of over 30 genes connected with endogenous cholesterol, fatty acids, triacylglycerol, and phospholipid synthesis. SREBPs can be separated into three isoforms: SREBP-2, SREBP-1a, and SREBP-1c (also known as ADD1), of which SREBP-1c appears to be the most physiologically relevant (see below) [3]. *SREBP-1a* and *SREBP-1c* are derived from a single gene through alternative splicing and the use of alternative transcription start sites at their first exon [3]. They belong to the family of basic helix–loop–helix (bHLH) leucine zipper transcription factors. But unique within the bHLH-Zip family members, they are synthesized as inactive precursors bound to endoplasmic reticulum [3]. Each SREBP precursor protein has three functional domains: an amino terminal segment with the bHLH domain, a central region with two hydrophobic transmembrane-spanning segments, linked to each other by a short region located in the lumen of the endoplasmic reticulum or Golgi, and the regulatory carboxy terminal part [3]. Their release and activation requires transport from the endoplasmic reticulum to the Golgi complex by the sterol sensor SREBP cleavage activating protein (SCAP) followed by a two-step cleavage by Golgi located proteases (S1P and S2P) [3]. The formation of mature SREBPs is controlled by cellular levels of cholesterol, insulin/glucose, and PUFAs through feedback inhibition of their proteolytic cleavage [4]. SREBPs activate the transcription of their target genes by binding as homodimers at sterol response elements (SREs) with the consensus sequence YCAYnYCAY, which are typically located close to binding sites for stimulating protein-1 (Sp1) and/or NF-Y[5].

Although they differ in their potency, SREBP-1a and SREBP-1c induce the expression of genes facilitating the synthesis of mono- and polyunsaturated fatty acids as well as their incorporation in triglycerides and phospholids [6, 7]. Among those genes are ATP-citrate lyase, acetyl-CoA synthetase, acetyl-CoA carboxylase, fatty acid synthase, stearoyl CoA desaturase, and other desaturases [6, 7]. SREBP-2 preferentially regulates the expression of genes involved in uptake of cholesterol from lipoprotein particles and the entire cascade of de novo biosynthesis of cholesterol, among them low-density lipoprotein receptor (LDLR) or farnesyl pyrophosphate synthase, HMG-CoA synthase and HMG-CoA reductase [5–7].

ChREBP

The carbohydrate sensitive response element binding protein (ChREBP) is required for both basal as well as carbohydrate-induced expression of certain lipogenic genes, which are essential for the conversion of dietary carbohydrates into triglycerides, thereby promoting long-term storage of carbohydrates as triglycerides [8]. These studies showed that nutrients themselves appeared to induce the expression of several regulatory enzymes of glycolysis and lipogenesis without the presence of insulin in primary hepatocytes [8]. These effects appear to be mediated by ChREBP. ChREBP was first isolated from primary hepatocytes and later found to be highly expressed in brown and white adipose tissue, kidney, the intestine, and skeletal muscle as well [9]. Many lipogenic genes that mediate glucose responsiveness contain glucose/carbohydrate response elements (ChREs) within their promoters [8]. These glucose response elements consist of two imperfect E boxes (CACGGG and CCCGTG), spaced by 5 nucleotides. ChREBP is a large protein consisting of a N terminal nuclear localization signal (NLS), polyproline domains, a basic helix–loop–helix–leucinezipper (bHLH/Zip), and a leucine zipper-like (Zip-like) domain as well as several potential phosphorylation sites for AMP-activated protein kinase and cAMP-dependent protein kinase. High carbohydrate diets induce the transcription of more than 15 genes involved in the metabolic conversion of glucose to fat, including liver specific pyruvate kinase, a regulatory enzyme in the pathway of liver glycolysis, fatty acid synthase, which uses acetyl-CoA and malonyl-CoA to form long-chain fatty acids, acetyl-CoA carboxylase, whose activity provides malonyl-CoA, and Spot14, a nuclear protein thought to be involved in stimulating lipogenesis (reviewed in [10]).

Crosstalk Between SREBPs and ChREBP

Fatty acids, cholesterol, or carbohydrates apparently do not bind directly to SREBPs or ChREBP, which differentiates them from nuclear receptors. Nevertheless, SREBP and ChREBP are a pair of important transcription factors involved in the nutrient and hormonal regulation of genes encoding enzymes of glucose metabolism and lipogenesis. Key lipogenic genes are regulated synergistically by both SREBPs and ChREBP [11]. High carbohydrate diets lead to insulin secretion from the pancreatic beta cells, which mediate proteolytic cleavage and activation of SREBP. In addition, ChREBP is present in the nucleus at high glucose levels but is localized in the cytosolic compartment of the cell during fasting. It appears that glucose regulates its dephosphorylation, which enables the activation of the import of ChREBPs into the nucleus, where it induces the expression of genes containing glucose response elements. Dietary PUFA (1) can efficiently decrease both the mRNA levels as well as the mature nuclear form of SREBP-1c, suggesting that in the latter case PUFAs may influence proteolytic processing; (2) can inhibit the transcriptional effects of SREBPs and ChREBP and subsequently glucose utilization [8] in the presence of glucagon; and (3) can inactivate ChREBP by phosphorylation [12].

7.3 SUPERFAMILY OF NUCLEAR RECEPTORS

Receptors in the nucleus can directly or indirectly regulate gene expression in response to lipid-soluble nutrients and their metabolites. The human genome project has revealed 48 members of nuclear transcription factors that comprise a superfamily of related genes and functions [13]. The classical steroid receptors, glucocorticoid, estrogen, androgen, progesterone, and mineralocorticoid receptor (GR, ER, AR, PR, and MR) respond primarily to endogenous hormonal lipids. Steroid hormones bind to their receptors with high affinity (dissociation constant K_d in nanomolar range) [14]. Other nuclear receptors can mediate the effects of fat-soluble vitamins, fatty acids, or cholesterol metabolites. These nuclear receptors act as lipid sensor due their ability to bind dietary lipids or lipids that are intermediates in metabolic pathways. Lipid sensing receptors were cloned about a decade ago because of their sequence homology to steroid receptors [15]. Since their natural ligands, target genes, and physiological importance were initially unknown, they were named orphan nuclear receptors. The hunt for their ligands and target genes led to the identification of their natural ligands and they are placed now in the group of adopted orphan receptors (reviewed in [16]).

7.4 NUCLEAR RECEPTORS: STRUCTURE AND FUNCTION

The nuclear receptors share common structures [14, 17] (Figure 7.1A). The amino terminal A/B domain is highly variable in length and sequence in the different family members. It contains the activation function-1 (AF-1), which is responsible for ligand-independent transcriptional activation by mediating the coordinated interaction of coregulatory proteins. A highly conserved C domain or DNA binding domain (DBD) is found 3′ to the AF-1 domain and structurally consists of two zinc finger motifs with affinity to specific target DNA sequences called response elements (REs, Figure 7.1C). Adjacent to the DBD is the short D domain, also known as the hinge region, which harbors a putative nuclear localization signal (NLS) and residues important for the interaction with transcriptional corepressor proteins that mediate the transcriptional repression function of unliganded receptors. The E domain includes the ligand binding domain (LBD), a NLS, and AF-2. This carboxy terminal region controls ligand binding, transactivation, nuclear localization, and dimerization. The hinge region serves as a highly flexible link between the DBD and LBD that allows for simultaneous receptor dimerization and DNA binding. This flexibility ensures a high degree of rotational freedom, so that the nuclear receptors can bind to a variety of response elements (see below). In response to ligand binding, nuclear receptors undergo transformational changes, allowing dimerization, protein–DNA interaction, and recruiting of cofactors and other transcription factors [14, 18–21]. In the absence of ligands, however, certain nuclear receptors, like the retinoid sensitive RAR/RXR heterodimer, are located in the nucleus and bound to the response elements in the regulatory region of their target genes. Unliganded RAR/RXR is associated with histone deacetylase-containing (HDAC) complexes, tethered through corepressors, which results in repression of gene transcription. Upon ligand binding, the corepressor dissociates, enabling recruitment of coactivators, which are associated in complexes with histone acetylase (HAT), methyltransferase, kinase, or ATP-dependent remodeling (SWI/SNF complex) activities that decompact chromatin. Finally, the general transcription machinery can initiate transcription [19, 22]. The regulation of gene transcription by nuclear receptors extends to their ability to transactivate specific target genes in an agonist-dependent manner. Many members of the nuclear hormone receptor superfamily, once activated by an agonist, can interact physically with other types of transcription factors influencing their functional properties. This interaction can result in either an inhibition or an enhancement of the transcriptional activities of the individual interacting proteins [23]. Most of the nutrient sensing nuclear receptors discussed here belong to the group of adopted orphan receptors that have low affinities for dietary lipids (dissociation constant K_d in micromolar range) [14, 16]. These receptors require heterodimerization with RXR in order to mediate DNA binding and regulation of transcription, with the exception of RXR, which acts as a homodimer (Figure 7.1B) [15]. Their response elements are composed of two degenerated hexameric AGGTCA sequences, oriented either as direct repeats in tandem (DR), inverted (IR), or everted, and separated by 1 to 8 nucleotides, depending on the receptor. Hence, a direct repeat, spaced by 1 nucleotide, is referred to as DR1. There can be high variability in the AGGTCA hexamer

(A)

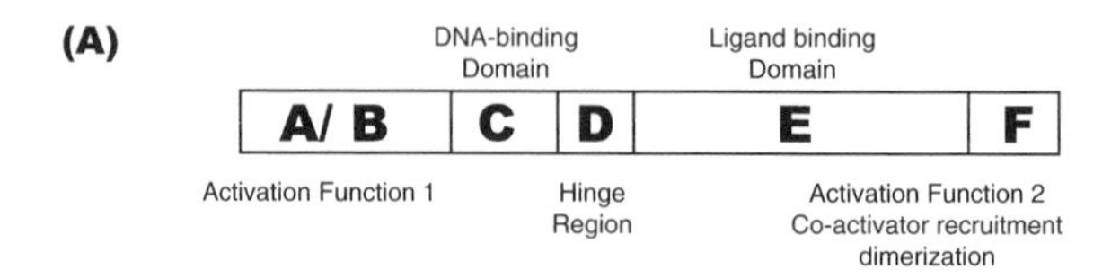

(B)

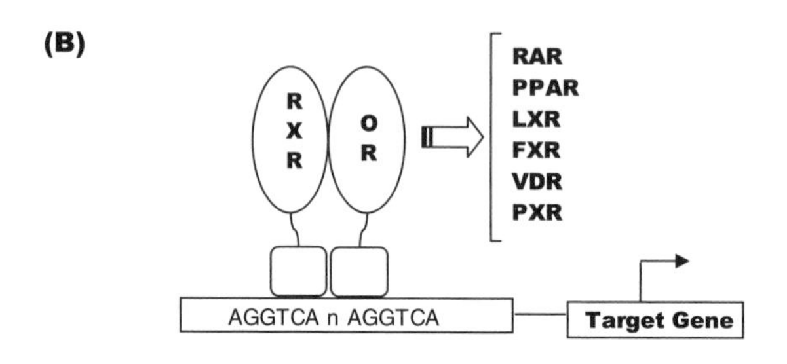

(C)

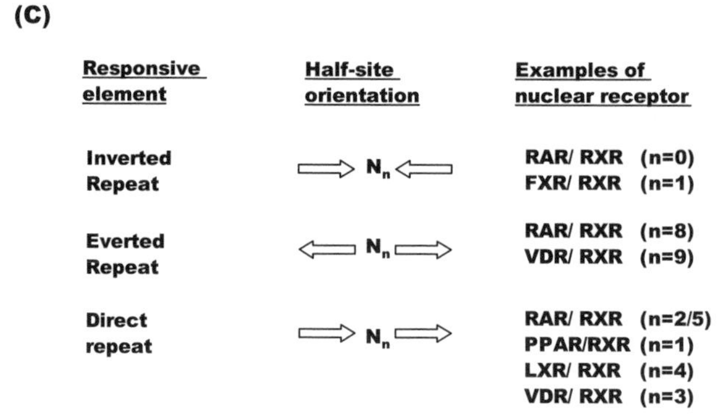

Figure 7.1. Overview of structural and functional properties of nutrient sensing nuclear receptors. (A) Common structure and organizations of nuclear receptors. Nuclear receptors contain five distinct regions. The most conserved region is the C domain, which contains the highly conserved domains for their DNA binding (DBD) and dimerization functions. The D domain consists of the variable hinge region. The moderately conserved and multifunctional E domain harbors a region to interact with their ligands (LBD) and a putative NLS. The ligand-independent AF-1 activation domain resides in the less conserved A/B part, the ligand-dependent AF-2 activation domain in the E part. (B) The RXR heterodimerization partners. RARα, RARβ, RARγ, and VDR bind high-affinity lipids; RXRα, RXRβ, RXRγ, PPARα, PPARβ/δ, PPARγ, LXRα, LXRβ, FXR, and PXR bind low-affinity lipids. These nuclear receptors function as heterodimers with RXR and mediate the effects of their ligands, of RXR's ligands, or synergistically of both receptors' ligands. (C) Different types of response elements. DNA response elements that are recognized by RXR heterodimers are composed of two hexamers (6 bp elements) with the consensus sequence 5′-AGGTCA-3′. The hexamer motifs are separated by 0 to 8 nucleotides and can be arranged as direct, inverse, or everse repeats. Abbreviations used: AF, activation function; CAR, constitutively active receptor; DBD, DNA binding domain; FXR, farnesoid X receptor; LBD, ligand binding domain; LXR, liver X receptor; N, nucleotide; n = spacing nucleotide between two hexamers (0–8); NLS, nuclear localization signal; PPAR, peroxisome proliferator activated receptor; PXR, pregnane X receptor; RAR, retinoid acid receptor; RXR, retinoid X receptor; VDR, vitamin D receptor.

sequence. The surrounding and intervening DNA sequences may significantly affect the binding affinity and function of RXR heterodimers.

7.5 NUCLEAR RECEPTORS AS METABOLIC SENSOR

The Hepatocyte Nuclear Factor 4

Fatty acids constitutively bind to the LBD and are activating the transcriptional activity of HNF4 (NR2A1) [24]. In this respect the fatty acid/HNF4 complex differs significantly from the activation of other nuclear factors by fatty acids. HNF4 binds as homodimer to DR1 motifs, regulates the expression of genes encoding apolipoproteins (apoC-II, C-III, A-II, A-IV), enzymes regulating carbohydrate and bile acid metabolism, insulin secretion, and certain cytochrome P450s [1, 24]. Its transcripts are mainly found in liver and pancreatic islets, befitting its pivotal function in glucose and triglyceride metabolism.

The Peroxisome Proliferator Activated Receptors

The peroxisome proliferator activated receptors (PPARs) were first cloned as nuclear receptors mediating the effects of synthetic compounds called peroxisome proliferators on gene expression. The ligands, which are activating PPARs and therefore gene expression, are several PUFAs, among them α-linoleic (C18:3), γ-linoleic (C18:3), arachidonic (C20:4), and eicosapentaenoic acid (C20:5), including their derivatives. The three PPAR isotypes (PPARα, PPARβ or δ, PPARγ1, and PPARγ2, also named NR1C1, NR1C2, and NR1C3, respectively) have distinct expression patterns and biological functions.

PPARα is a global regulator of energy homeostasis and is expressed in liver, kidney, heart, and muscle [25]. PPARα directly induces the expression of genes facilitating fatty acid uptake and intracellular transport of fatty acids into peroxisomes and mitochondria for their fatty acid oxidation, like fatty acid binding protein (FATP) in the liver, the major organ of fatty acid metabolism [26]. PPARα also regulates the expression of several catabolic enzymes, utilizing mitochondrial β-oxidation and peroxisome ω-oxidation, including acyl-CoA oxidase or cytochrome P450 4A [26]. Fibrates, potent synthetic PPARα agonists, can lower serum triglyceride levels and increase HDL cholesterol in patients with hyperlipidemia [27].

PPARγ, considered the master switch of adipocyte differentiation [28], regulates adipogenesis as well as diverse physiological processes, among them cellular differentiation and insulin sensitization [26, 29–31]. *PPARγ* has two splice variants (Figure 7.2A), *PPARγ1* and *PPARγ2*. PPARγ1 is more widely expressed, but highest in adipose tissue, macrophages, and endothelial cells. PPARγ2 appears to be expressed exclusively in adipocytes. PPARγ further promotes the storage of fat by increasing adipocyte differentiation and enhancing the transcription of genes that are important for lipogenesis [26, 31]. Several in vivo and in vitro studies reported that PPAR ligands have pronounced insulin-sensitizing effects through their influ-

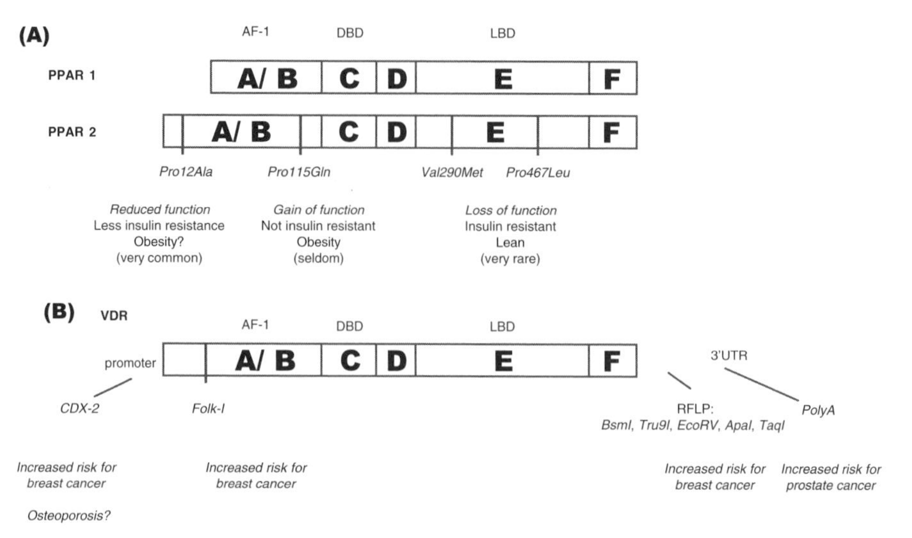

Figure 7.2. Polymorphic sides on the *PPARγ* and *VDR* genes. (A) Polymorphisms of the *PPARγ2* isoform and their localization linked to insulin sensitivity. The relatively common Pro12Ala polymorphism is linked to reduced insulin sensitivity. The gain-of-function missense mutation Pro115Gln is associated with obesity, but not with insulin resistance. The loss-of-function mutations of PPARγ (Val290Met, Pro467Leu) lead to pronounced insulin resistance with early onset of NIDDM. (B) Common polymorphisms of the *VDR* gene and their localizations that are associated with increased risk for breast or prostate cancer. An increased risk for breast cancer is linked to (1) the polymorphism at the site for the transcription factor CDX-2 in the promoter of the *VDR* gene; (2) the *FolkI* RFLP that leads to two VDR isoforms, which differ in length; (3) the polymorphisms at the restriction sites of *BsmI*, *Tru9I*, *EcoRV*, *ApaI*, and *TaqI*. The *L* variant of the length polymorphism in the 3′-UTR region that is classified into long (*L*) and short (*S*) variants was suggested to reduce the risk for breast cancer, but with increased risk to develop prostate cancer. Abbreviations used: AF, activation function; CDX-2, caudal-related homeobox-2; DBD, DNA binding domain; LBD, ligand binding domain; RFLP, restriction fragment length polymorphism; polyA, polyadenyl A; PPAR, peroxisome proliferator activated receptor; UTR, untranslated region; VDR, vitamin D receptor; Pro12Ala, exchange of proline with alanine at codon 12; Pro115Gln, exchange of proline with glycine at codon 115; Val290Met, exchange of valine with methionine at codon 290; Pro467Leu, exchange of proline with leucine at codon 467; NIDDM, non-insulin-dependent diabetes mellitus.

ence on glucose and lipid homeostasis in muscle, liver, and adipose tissue (reviewed in [30]). They affect the expression of adipocyte-secreted hormones such as leptin, tumor necrosis factor-α (TNF-α), plasminogen activator inhibitor-1, interleukin-6 (IL-6), and adiponectin. In addition, PPARγ ligands reduce the levels of circulating free fatty acids (FFAs), which are associated with insulin sensitivity, by increasing the net flux into adipose tissue and by preventing their release. PPARγ stimulates lipolysis of circulating triglyceride and the subsequent uptake of fatty acids into the adipose cell through induction of the expression of lipoprotein lipase (LPL), FATP, and the class B scavenger receptor that binds long-chain fatty acids and modified LDL. Increased expression of phospho*enol*pyruvate carboxykinase (PEPCK) and glycerol kinase GyK9 enables triglyceride synthesis and promotes the storage of fatty acid. In addition, other adipogenic factors involved in glucose and lipid homeostasis, like insulin-responsive glucose transporter GLUT4 or 11-β-hydroxysteroid dehydrogenase 1 (11β-HSD-1) are modulated by PPARγ.

The prevalent isoform PPARβ/δ is expressed ubiquitously and is like the other PPAR isoforms involved in lipid and lipoprotein metabolism [32]. Selective agonists for PPARβ/δ appear to have beneficial effect on reverse cholesterol transport from peripheral tissue by increasing serum high-density lipoprotein (HDL) cholesterol, while decreasing LDL cholesterol, and thus can decrease the risk of cardiovascular disease associated with metabolic syndrome X [32]. The importance of PPARβ/δ-mediated functions in muscle is debatable, since selective PPARβ/δ agonists can influence the expression of genes important for fatty acid catabolism, like malonyl-CoA decarboxylase or uncoupling protein-1 (UCP-1), and those involved in fatty acid oxidation [33]. In addition to metabolic pathways, PPARβ/δ appears to be an essential transcription regulator for early steps of cell differentiation in various adipocytes, keratinoycytes, and oligodendrocytes, rather than being involved in terminal differentiation and maintenance of the differentiated state [34].

The Liver X Receptors

Liver X receptor α (LXRα) (NR1H3) is abundant in the liver, but also in other tissues associated with lipid and cholesterol metabolism, among them adipose tissue, kidney, intestine, adrenals, and macrophages. LXRβ (NR1H2) is found in almost every tissue. Both LXRs, which bind as heterodimers with RXR to DR4, can be activated by certain intermediates in the cholesterol de novo biosynthesis and its metabolites, including oxidized derivatives of cholesterol such as 24,25(*S*)-epoxycholesterol and 22(*R*)- and 24(*S*)-hydroxycholesterol, which are the most potent endogenous ligands [16, 35]. Derivatives of phytosterols can also activate LXRα, even more effectively than the endogenous oxysterols [36]. Geranylgeranyl pyrophosphate, one of the end products in the mevalonate pathway, auto-oxidized cholesterol sulfates, and certain unsaturated fatty acids antagonize LXR activation [16]. LXRs are key players in regulating cholesterol homeostasis at the cellular levels as well as through the regulation of circulating lipoproteins. LXR regulates the complex process of reverse cholesterol transport at various levels, including catabolism and elimination of cholesterol through regulation of genes involved in these processes [16, 37]. High intracellular cholesterol content triggers cholesterol efflux, which is considered to be the first and rate-limiting step in reverse cholesterol transport (RCT). This process is mediated primarily by activated LXRs through the induction of the expression of ATP-binding cassette transporter genes (*ABCA1*, *ABCG1*, *ABCG5*, and *ABCG8*) and the major gene clusters encoding apolipoproteins (*apoE,C-I/C-IV/C-II*). This pathway appears to be very efficient, particularly in macrophages. LXRs have positive impact on the remodeling of lipoproteins by regulating the expression of LPL, cholesterol ester transfer protein (CETP), and phospholipid transfer protein (PLTP), which all participate in the exchange of various lipids between lipoproteins. This results in the clearance of triglyceride-rich lipoproteins and the receptor-mediated uptake and eventual elimination of lipoproteins. A functional LXRE in the promoter region of the scavenger receptor class B, type I (SR-BI) gene suggested that an additional mediator of reverse cholesterol transport, the biological relevant receptor for HDL, is regulated by LXRs [38]. SR-BI facilitates the selective uptake of free and esterified cholesterol from HDL particles into cells (reviewed in [39]). Excessive circulating cholesterol, taken up by the hepatocytes, can be eliminated after conversion into bile acids and after efflux into the intestinal lumen via ABCA5/A8 within the bile acid flow [37]. In addition, LXR regulates several genes involved in lipogenesis, like SREBP-1c, fatty acid synthase, and acetyl-CoA carboxylase [16].

Farnesoid X Receptor

The bile acid receptor (BAR, FXR) was initially shown to be moderately activated by a large variety of endogenous isoprenoids, including farnesol [35]. Subsequent research showed that certain bile acids, among them chenodeoxycholic acid, lithocholic acid, and deoxycholic acid, and their metabolites are more relevant endogenous agonists [16, 35]. In mouse, two different isoforms (*Fxrα/Nr1h4* and *Fxrβ/Nr1h5*) result through alternative splicing. In humans, *FXRβ* does not encode a functional receptor [14]. The FXR/RXR heterodimers bind to DNA sequences consisting of an inverted repeat spaced by one nucleotide (IR1). FXR is mainly expressed in the liver, intestine, kidney, and adrenal cortex [35]. FXR appears to have an important impact in the regulation of the enterohepatic circulation of bile acids, which is critical for the intestinal absorption of nutrients, the biotransformation and excretion of cholesterol, and the clearance of potentially toxic xenobiotics. Binding of bile acids to FXR stimulates not only the export of bile acids from the liver into the bile, but also the reuptake of bile acids by the gut through the induction of genes encoding the bile acid/salt export pump (BSEP) and the cytosolic ileal bile acid binding protein (IBABP), respectively [16, 35]. FXR also protects against the cytotoxic effects of bile acids indirectly by inhibiting the expression of cholesterol 7α-hydroxylase (CYP7A1) and sterol 12α-hydroxylase (CYP8B1), critical enzymes for bile acid synthesis, as well as of basolateral Na^+-taurocholate cotransporting polypeptide (NTCP) and organic anion transporting polypeptide-1 (OARP-1). FXR affects these processes partly by regulating transcription of a transcriptional repressor called the small heterodimerization partner (SHP, NR0B2) [16, 40]. Whether or not SHP is directly involved in the bile acid-induced repression of CYP8B1 is currently not known. However, the main role of FXR is not only in bile acid metabolism but also in directly regulating expression of a variety of genes involved in cholesterol, lipoprotein, and triglyceride metabolism. FXR positively regulates the expression of PLTP, but negatively regulates CETP and apoA-I, the major apolipoprotein component of mature HDL [40]. FXR also affects triglyceride metabolism by inducing the expression of apoC-II, an obligate cofactor for lipoprotein lipase activity [40], and gluconeogenesis via SHP-mediated repression of HNF4 and forkhead transcription factor-1 (FOXO1) [41]. Under certain conditions FXR can counteract the transcriptional activities of other nuclear receptors such as LXR [42] or PPAR, by inducing the expression of SHP, a corepressor of many nuclear receptors.

Nuclear Receptors in Inflammatory Response

All three PPAR isoforms are involved in the regulation of various aspects of inflammatory responses and immunomodulatory properties (reviewed in [23]), although PPARα and PPARγ are the primary mediators. They mediate their anti-inflammatory effects primarily through repression of gene transcription (transrepression), although activation (transactivation) of certain target genes might be involved as well. They are expressed in macrophages and dendritic, B, and T cells. Agonist-activated PPARs compete successfully for rate-limiting amounts of coactivators, among them steroid receptor coactivator-1 (SRC1) or cAMP response element binding (CREBP) binding protein (CBP). Subsequently, the transcriptional activities of other transcription factors depending on these specific coactivators are inhibited. Ligand-induced PPAR/RXR heterodimer can also interact directly and form complexes with other types of transcription factors such as nuclear factor-κB (NF-κB), activating protein-1 (AP-1), signal transducers and activators of transcriptions (STATs), and nuclear factor of activated T cells (NF-AT). One additional mechanism of transrepression is possibly through inhibition of phosphorylation and activation of certain members of the mitogen-activated protein kinase (MAP kinase) cascade. PPAR agonists can repress the activation of both c-Jun N terminal kinase (JNK) and p38 MAP kinase. In addition, the PPARγ agonist 15-deoxy-$\Delta^{12,14}$-prostaglandin (15dPGJ2) can mediate anti-inflammatory effects through a nongenomic or PPARγ-independent pathway. Activation of NF-κB mainly occurs via phosphorylation of inhibitor of NF-κB (IκB) by IκB kinase. 15dPGJ2 directly inhibits NF-κB-dependent gene transcription through covalent modification of critical cysteine residues in IκB kinase and/or in the DBD of NF-κB. PPARs can repress, by inhibiting NF-κB and AP-1, the expression of several genes involved in inflammatory response, including cytokines, cell adhesion molecules, and other proinflammatory signal mediators, such as inducible nitric oxide synthase (iNOS). In dendritic cells (DCs) or T cells, activated PPARγ inhibits the synthesis of important cytokines for T_H1 cell differentiation; IL-12 in DCs as well as both interferon-γ and IL-2 in T cells after T cell activation.

More recently, LXRs have been implicated in inflammation and immune responses. LXR activation in macrophage has been shown to inhibit the expression of a cluster of genes involved in inflammation and the innate immune response [43–45]. Similar to what has been described for PPARs, LXRs can antagonize the NF-κB signaling pathway in response to bacterial pathogens or proinflammatory cytokine, among them lipopolysaccharide (LPS), IL-1β, or TNF-α, through a mechanism that is not completely understood. These processes are associated with inhibited expression of inflammatory mediators such as iNOS, cyclooxygenase-2 (COX2), IL-1b, IL-6, monocyte chemoattractant protein-1 (MCP-1), and metalloproteinase MMP-9 [43, 46–48]. In cultured macrophages and in aortic tissue, activation of toll-like receptors (TLRs) 3 and 4 by microbial ligands or LPS, respectively, can block the induced expression of certain LXR target genes, including ABCA1, ABCG1, or apoE [46]. This crosstalk between LXR and TLR signaling is apparently mediated by IRF3 (interferon-β enhanceosome), a viral response transcription factor activated specifically by TLR 3/4, which inhibits the transcriptional activity of LXRs on its target promoters independently of NF-kB [46]. In addition, LXRα but not LXRβ, was identified as an important regulator of macrophage antimicrobial activity and apoptotic control [45]. LXRα was shown to be involved in the innate immune response to bacterial pathogens, like *Listeria monocytogenes*, a gram-positive bacteria that has been implicated in serious foodborne human illness, among them gastroenteritis, and meningitis, by directly influencing the transcription of the anti-apoptotic factor SPα (scavenger receptor cysteine-rich repeat protein). This seems to be independent of the TLR pathway but may be regulated by NODs (nucleotide binding oligomerization domain proteins), a class of intracellular mediators of immunity and apoptosis. This is the first example of combined functions of a nutrient sensing nuclear receptor in cholesterol, lipid, and lipoprotein metabolism with innate immunity. The role of LXRα in immune and antimicrobial response may be unique or may only be the first example within the class of nuclear receptors forming heterodimers with RXR.

PXR: Mediator of Naturally Occurring Steroids

The pregnane X receptor (PXR, NR1I2) functions as a hepatic sensor for many xenobiotics, such as rifampicin and phenobarbital, and certain bile acids, among them the secondary bile acid lithocholic acid and ursodeoxycholic acid [49]. In addition, a number of naturally occurring steroids, including pregnenolone, progesterone, androstanol, hyperforin (a component of St. John's wort), dexamethasone (a synthetic glucocorticoid), and phytoestrogens have been shown to function as a ligand for PXR [49]. PXR binds to DNA as a heterodimer with RXR to a DR3 and IR6. PXR is mainly expressed in the liver and at low levels in the intestine, tissues that are routinely exposed to environmental chemicals [49]. Activated PXR induces the expression of genes that are involved in hepatic uptake and excretion of toxic substances, among them members of the cytochrome P450-3a (CYP3A) subfamilies. CYP3A enzymes are particularly relevant to xenobiotic metabolism due to their broad substrate specifity in liver and intestine, where they are abundantly present.

7.6 VITAMINS

Vitamin A

Certain lipophilic derivatives of vitamin A, the retinoids, can influence a variety of metabolic signaling pathways, including carbohydrate, fatty acid, and cholesterol metabolism. In addition, retinoids play an important role during embryonic development and in cell proliferation, differentiation, and cell death/apoptosis. Their effects are mediated by the retinoid acid receptor (RAR) and the retinoid X receptor (RXR). All-*trans* retinoid acid can activate only RAR; its 9-*cis* isomer binds to both RAR and RXR [50]. However, certain naturally occurring and synthetic RXR ligands, also referred to as rexinoids, do not interact with RARs. Retinoids activate RAR and bind as RAR/RXR heterodimer to DR5, but also to DR1 and DR2 [22, 50]. RXR also binds to DR1s as RXR/RXR homodimers or as heterodimeric partner with other nuclear receptors to corresponding response elements [15]. Each family consists of three isotypes (RARα, β, and γ, also named NR1B1, NR1B2, and NR1B3; RXRα, β, and γ or NR2B1, NR2B2, and NR2B3, respectively), which are encoded by distinct genes [15]. RXRs are expressed in a very tissue specific pattern [50]. RXRβ is expressed ubiquitously, while transcripts of RXRα are mainly found in liver, kidney, spleen, placenta, and epidermis, and those of RXRγ in skeletal and cardiac muscle, and to a lesser degree in the brain. Retinoids control a wide variety of hormone-responsive genes [15, 51], because RXR forms heterodimers with RARs and several other nuclear receptors (PPARs, LXRs, FXR, PXRs) [22]. Moreover, the promoters of retinoid target genes contain not only cognate response elements (RREs), but other regulatory sequences that bind to other transcription activators and enhanceosomes [22].

Vitamin D

The biological active form of vitamin D, 1,25-dihydroxyvitamin D_3 (1,25$(OH)_2D_3$), is an important regulator of calcium homeostasis through its interaction with the vitamin D receptor (VDR, NR1I1). VDR (Figure 7.2B) heterodimerizes with RXR in order to interact with the vitamin D response elements (VDREs) of its target genes [52, 53]. Several of the VDREs, which have been characterized, are DR3s. However, neither the sequence of the half-sites nor the number of the spacing nucleotides between them has been well conserved. Vitamin D can be obtained through dietary intake (fish oil or vitamin D fortified food, such as milk or margarine) or through the endogenous pathway. The precursor 7-dehydrocholesterol is converted in the skin by the action of sunlight into pre-vitamin D, which is hydroxylated in liver and subsequently converted in a variety of organs such as kidney, prostate, breast, colon, or skin into its biological active metabolite (1,25$(OH)_2D_3$) [54]. VDR was not only found in bone, kidney, intestine, and parathyroid glands, but also in the brain, heart, pancreas, skin, lymphocytes, and gonads [55]. Vitamin D is essential for an intact mineral metabolism by increasing intestinal calcium absorption or mobilizing it from the bone [52, 53]. Besides these well-known functions on classical target tissues (bone, kidney, intestine, parathyroid gland), 1,25$(OH)_2D_3$ has been shown to inhibit the proliferation of a number of malignant cells and to affect the differentiation and function of cells in the immune system (reviewed in [52]). Vitamin D promotes monocyte differentiation and inhibits lymphocyte proliferation as well as secretion of cytokines, such as IL-2, interferon-γ, and IL-12 [56]. Vitamin D has been shown to have antiproliferative effects in several different cancer cell types. Even though there are many ideas about the pleiotropic effects of 1,25$(OH)_2D_3$, the number of VDR target genes identified so far is limited.

Among their major transcriptionally regulated genes are calcium binding proteins calbindin-D_{28K} and calbindin-D_{9K}, which are mainly present in kidney and intestine, respectively, and the bone matrix proteins osteocalcin and osteopontin, secreted by osteoblasts [57]. Vitamin D 24-hydroxylase is also a target gene and it is involved in the breakdown of $1,25(OH)_2D_3$ [57]. CCAAT enhancer binding protein-β (C/EBP-β), epithelial calcium channels (EcaCs) (reviewed in [53]), or c-fos, an early response gene, and transforming growth factor-β2 (TGF-β2) are directly regulated by VDR [57]. VDR is also activated by lithocholic acid and induces the expression of CYP3A, which detoxifies various xenobiotics and lithocholic acid itself in the liver and intestine [58].

7.7 PHYTOESTROGENS: NUTRIENTS MIMICKING ESTROGENS

Beginning in the 1940s it was first realized that some plant-derived components can mimic estrogenic effects [59]. Animals grazing on pastures containing red clover showed signs of infertility and females that were pregnant often had miscarriages. In addition, immature animals developed signs of estrus. Twenty years later it was discovered that clover in these pastures had a high content of the isoflavones formononetin and biochanin A, which are among the first identified phytoestrogens. Meanwhile, several plant compounds are classified as phytoestrogens, among them isoflavonoids, stilbenes, lignans, and coumestans [59–61]. Their structural similarity to the mammalian estrogen 17β-estradiol enables phytosterols to bind to the two types of estrogen receptors (ERs): ERα (NR3A1) and ERβ (NR3A2). Both receptor subtypes are encoded by two different genes [21, 62]. They share high amino acid identity and bind to estrogen response elements (EREs) with similar specifit and affinity, even though differences in a subset of natural EREs have been reported [62, 63]. The consensus sequence of EREs has been determined to be 5′-AGGTCAnnnTGACCT-3′ [64]. ERs are able to regulate gene expression either by direct DNA binding (as homo- or heterodimers) or by forming complexes with other DNA-bound transcription factors, like AP-1 and Sp1. ERα and ERβ form functional heterodimers in vitro and in vivo, although the physiological relevance of heterodimerization is not yet known [21, 64]. These receptors exhibit a cell- and tissue-specific distribution and display both functional differences as well as similarities [62, 65]. Both ERs can be found not only in reproductive organs but also in a variety of tissues, including endothelial and vascular smooth muscle cells, breast, brain, bone, liver, kidney, and the cardiovascular system. ERβ is primarily expressed in ovary, prostate, lung, gastrointestinal tract, and hematopoietic and central nervous systems [21, 62, 64, 65]. Both ERs function in breast tissue and normal ovarian follicular development, whereby ERα is more important in maintaining follicle stimulating and luteinizing hormone concentrations in blood. ERα is involved in bone maturation in both females and males; however, only ERβ plays a role in maintaining bone density by regulating the formation and resorption of bone in females [65, 66]. In certain tissues, where both receptors are present, among them mammary gland, adrenals, bone, and distinct areas of the brain, ERβ can inhibit or even oppose the effects of ERα [21, 64].

Drugs targeted to the ER, such as raloxifene and tamoxifen, act either as ER agonist or antagonist. They may exhibit different receptor specificity, depending on the tissue and the dose administered [67]. Certain ER antagonists are very efficient agents for treating breast cancer; the treatment of postmenopausal women with hormone and estrogen replacement therapies (HRT and ERT) has exhibited undesirable side effects, like breast and endometrial cancer or irregular bleeding [59–61]. Many women ingest phytoestrogens as a reasonable alternative for HRT or ERT, since these plant-derived compounds do not produce deleterious side effects.

Phytoestrogens have considerable beneficial impact on cardiovascular disease, osteoporosis, menopausal syndromes, cancers, and coronary heart disease [59–61, 66, 67]. Their great diversity and complex biological actions make it difficult to draw general conclusions about their beneficial properties, since different classes of phytoestrogens may have different activities, pharmacokinetic properties, and metabolic fates. Phytoestrogens, depending on their subset, bind with different affinity to either ERα or ERβ and act as estrogen agonists and antagonists [67]. However, their diverse biological activity is only due in part to their influence on transcriptional ability. Phytoestrogens can also inhibit enzymatic activity of enzymes involved in the synthesis of steroid hormones, including aromatase and 17β-hydroxysteroid dehydrogenase, can inhibit protein tyrosine kinases, DNA topoisomerases, and angiogenesis, and can exert antioxidant effects [60]. Even with this considerable progress, we are far from understanding all the potential health effects of phytoestrogens.

7.8 POLYMORPHISMS

Molecular and nutrition research during the last decade has shown the importance of understanding the effects of nutrients on gene expression and subsequent physiology. A parallel effort in genetics demonstrated that variations in the genomic sequence (polymorphisms) affect an individual's response to diet and susceptibility to disease. The most common form of polymorphism is the single nucleotide polymorphism (SNP) and in combination (i.e., sets of gene variants) can have the largest impact on complex traits including responses to diet and disease development. Over 30,000 genes are encoded within 3 billion base pairs in the human genome; 5 SNPs are found per 10 kb of genome sequence [68]. Hence, over 1.5 million SNPs are predicted in the complete human genome or more than 1 SNP per 1 kb of gene sequence [69]. A subset of SNPs will occur in coding regions leading to functional alterations of regulating proteins or modifying enzymes at the top of biological cascades or at rate-limiting steps in intermediary metabolism. Among the most extensively studied polymorphisms within nuclear receptors are those in the genomic regions of *PPARs*, *VDR*, and *ERs*. Polymorphisms in these genes may be linked with altered risk to develop obesity, osteoporosis, and cancer.

Polymorphisms in the *PPARγ* Gene

Several naturally occurring genetic receptor variants of *PPARγ* have been identified and associated with changes in insulin sensitivity, insulin secretion, or susceptibility for obesity, all of which influence the risk for non-insulin-dependent diabetes mellitus (NIDDM) (Figure 7.2A) [70]. A relatively common polymorphism within the amino terminus of the PPARγ2 isoform is an alanine for proline change at codon 12 (Pro12Ala), which appears to reduce the risk of developing NIDDM. This allele, which occurs with a frequency of almost 15% among Caucasians, was originally reported to be associated with a lower body mass index (BMI) and insulin sensitivity. A rare gain-of-function missense mutation at codon 115 (Pro115Gln) in PPARγ2 was found to be associated with severe obesity, but not with insulin resistance. The PPARγ2 (Pro115Gln) variant is more susceptible to ligand-independent activation, since the phosphorylation properties of the serine residue at position 114 seem to be affected. This serine is part of a typical phosphorylation site for MAP kinase that may be activated by insulin and growth factors. The reference (i.e., wild-type) PPARγ2 has reduced sensitivity to ligands following MAP kinase activation. Individuals with one or two heterozygous loss-of-function mutations within the LBD of PPARγ (Val290Met, Pro467Leu), exhibit marked insulin resistance with early onset of NIDDM, symptoms associated with the metabolic syndrome including dyslipidemia (high triglyceride and low HDL cholesterol) and hypertension. At a molecular level, both mutant receptors revealed impaired transcriptional activity due to attenuated ligand binding and failure to recruit transcriptional coactivators. Moreover, the function of the other naturally occurring variant of PPARγ, which is expressed from the unaffected allele, is inhibited by its mutant counterpart in a dominant negative way.

Polymorphism in the *VDR* Gene

Several allelic variants in the *VDR* gene were identified over a decade ago (Figure 7.2B)[71, 72]. These *VDR* polymorphisms were studied in relation to bone mass, but conflicting results were reported. The restriction fragment length polymorphism (RFLP) at the *FolkI* restriction site in exon 2 created an alternative translational start site, resulting in two isoforms of the VDR protein, which differ in length [73]. Studies addressing their effects on function and disease symptoms have yielded conflicting results. A cluster of polymorphisms, located between exons 8 and 9, can be discriminated by the restriction enzymes *BsmI*, *Tru9I*, *EcoRV*, *ApaI*, and *TaqI*. These polymorphisms do not alter the amino acid sequence of VDR and the mechanisms by which the VDR function is affected are unclear. Further evidence has suggested that the association between *VDR* alleles and bone mineral density (BMD) may be dependent on calcium and vitamin D intake, although these studies had small sample size [73].

Certain polymorphisms in the *VDR* gene may be associated with cancer of the prostate and breast [74]. Polymorphisms at the *BsmI* and the *FolkI* restriction sites appear to be associated with increased risk for breast cancer, whereas the *FF* genotype was associated with a 50% decreased risk of breast cancer in African-Americans. A length polymorphism in the 3′ untranslated region, which alters a polyadenyl A (polyA) microsatellite, was classified into long (*L*) and short (*S*) variants [75]. The results from several studies suggest that a long polyA microsatellite variant might be associated with decreased mRNA stability and protein levels, causing a reduction in the mediating effect of VDR on differentiation of adipocytes and muscle cells [76]. In breast cancer, the presence of *L* alleles was also associated with about 50% reduction in risk. However, in prostate cancer, the presence of the *L* variant was associated with a three- to fivefold increased risk to develop cancer [74]. Another polymorphism has been identified in the promoter of *VDR* at the site for the transcription factor CDX-2 (caudal-related homeobox-2), and this polymorphism has been associated with BMD in Japanese subjects [73]. The *CDX-2* polymorphism appears to be functional because it influences DNA protein binding and modulates gene expression in reporter assays.

Despite extensive research, no consensus has yet been reached regarding the association of VDR genotypes and BMD. One possible reason for the controversy is the differences in the confounders in different genetic epidemiology studies that include ethnic background, age, calcium intake, caffeine intake, and other complex and genetic–environmental interactions.

Polymorphisms in the *ERα* and *ERβ* Genes

Osteoporosis is a common disease characterized by reduced bone mass and an increased risk for fracture. An association has been reported between a TA repeat polymorphism in the *ERα* promoter with bone mass in Japanese, American, and European populations [73]. Three common sequence variations and several additional SNPs and variable-number tandem repeat polymorphisms have been described in the *ERα* gene [77]. The polymorphisms defined by the restriction enzymes *PvuII* (also known as 454–397T > C; IVS1–397 T/C; rs2234693) and *XbaI* (also known as 454–351 A > G; IVS1–351 A/G; rs9340799) of the *ERα* gene are located in the first intron. This intron seems to contain regulatory elements important for gene transcription. The *PvuII* polymorphism interferes with a potential Mybl2 (v-myb avian myeloblastosis viral oncogene homolog-like 2) binding site with subsequent effects on gene transcription [77]. So far their functional consequences have not been precisely defined. The GENOMOS (Genetic Markers for Osteoporosis) project, which included almost 20,000 participants from eight different European countries, led to the conclusion that these three common polymorphisms within the *ERα* gene had either no or only minor effects on BMD. However, a significant reduction of fracture risk was associated with women homozygous for the absence of an *XbaI* recognition site. These effects were independent of BMD. The other polymorphisms were not associated with increased fracture risk. The Rotterdam Study, which investigated a *PvuII* and *XbaI* polymorphic allele in the *ERα* gene in over 6000 Dutch persons of similar age, revealed that about 30% of postmenopausal women and men are homozygous carriers, 50% are heterozygous, and only 20% are non carriers [78]. Results from the Rotterdam Study revealed further that postmenopausal Dutch women, who are heterozygous or homozygous carriers of haplotype 1, have a twofold increased risk of myocardial infarction, compared with noncarriers [78]. In men, no association with myocardial infarction or ischemic heart disease was observed for this polymorphism. The results from this study also support the hypothesis that the 454–397T allele leads to lower estrogen action. Evidence suggests that genetic variations in the *ERα* gene might influence the risk for ischemic or coronary heart disease. The sequence variations affecting the *ERα* gene are undefined. Other polymorphisms in the *ERα* gene, among them the 792C > T or 975C > G polymorphisms, are linked to various forms of breast or endometrial cancer [79–83], menstrual disorders, and Alzheimer disease [84, 85]. Fewer sequence variations have been identified for the *ERβ* gene. Several studies suggested a possible link between polymorphisms and ovulatory dysfunction, anorexia nervosa, hypertension, bone density, and androgen levels [84, 86–90]. Previously, several additional variants of *ERβ* in the African population were identified [91]. In two cases, this results in changes of the amino acid sequences. Whereas the substitution of the third amino acid, isoleucine, by a valine (I3V) did not lead to any functional changes, the V320G variant of *ERβ*, has reduced transcriptional activity, as shown in reporter assays [91]. However, the effect of these polymorphisms on disease symptoms has not yet been analyzed.

7.9 CONCLUSION

Epidemiological research examining the association among polymorphisms and multifactorial diseases, such as obesity, NIDDM, osteoporosis, and cancer, has been inconsistent or conflicting with other studies. Among the most important factors is the limited number of individuals within one study or differences in ethnic background, country, or living standard. It is becoming evident that more parameters from different areas within the field of nutrition have to be accounted for in such studies. The investigations on how nutrients regulate gene expression on molecular levels are only one piece of the complex puzzle in our understanding of how each individual responds to certain nutrients.

ACKNOWLEDGMENTS

I would like to thank Dr. Kaput and Dr. Stephensen for critical reading of the manuscript. This work was supported by NIH grant R01-AI050863 to Charles Stephensen.

REFERENCES

1. D. B. Jump (2002). Dietary polyunsaturated fatty acids and regulation of gene transcription, *Curr. Opin. Lipidol.* **13**:155–164.
2. D. B. Jump and S. D. Clarke (1999). Regulation of gene expression by dietary fat, *Annu. Rev. Nutr.* **19**:63–90.

3. M. S. Brown and J. L. Goldstein (1997). The SREBP pathway: regulation of cholesterol metabolism by proteolysis of a membrane-bound transcription factor, *Cell* **89**: 331–340.
4. C. J. Loewen and T. P. Levine (2002). Cholesterol homeostasis: Not until the SCAP lady INSIGs, *Curr. Biol.* **12**:R779–R781.
5. P. A. Edwards, D. Tabor, H. R. Kast, and A. Venkateswaran (2000). Regulation of gene expression by SREBP and SCAP, *Biochim. Biophys. Acta* **1529**:103–113.
6. J. D. Horton, N. A. Shah, J. A. Warrington, N. N. Anderson, S. W. Park, M. S. Brown, and J. L. Goldstein (2003). Combined analysis of oligonucleotide microarray data from transgenic and knockout mice identifies direct SREBP target genes, *Proc. Natl. Acad. Sci. U.S.A.* **100**:12027–12032.
7. J. D. Horton, J. L. Goldstein, and M. S. Brown (2002). SREBPs: Activators of the complete program of cholesterol and fatty acid synthesis in the liver, *J. Clin. Invest.* **109**:1125–1131.
8. K. Uyeda, H. Yamashita, and T. Kawaguchi (2002). Carbohydrate responsive element-binding protein (ChREBP): A key regulator of glucose metabolism and fat storage, *Biochem. Pharmacol.* **63**:2075–2080.
9. K. Iizuka, R. K. Bruick, G. Liang, J. D. Horton, and K. Uyeda (2004). Deficiency of carbohydrate response element-binding protein (ChREBP) reduces lipogenesis as well as glycolysis, *Proc. Natl. Acad. Sci. U.S.A.* **101**:7281–7286.
10. H. C. Towle, E. N. Kaytor, and H. M. Shih (1997). Regulation of the expression of lipogenic enzyme genes by carbohydrate, *Annu. Rev. Nutr.* **17**:405–433.
11. H. C. Towle (2001). Glucose and cAMP: Adversaries in the regulation of hepatic gene expression, *Proc. Natl. Acad. Sci. U.S.A.* **98**:13476–13478.
12. S. Kersten (2002). Effects of fatty acids on gene expression: Role of peroxisome proliferator-activated receptor alpha, liver X receptor alpha and sterol regulatory element-binding protein-1c, *Proc. Nutr. Soc.* **61**:371–374.
13. J. M. Maglich, A. Sluder, X. Guan, Y. Shi, D. D. McKee, K. Carrick, K. Kamdar, T. M. Willson, and J. T. Moore (2001). Comparison of complete nuclear receptor sets from the human, Caenorhabditis elegans and Drosophila genomes, *Genome Biol.* **2**: RESEARCH0029.
14. H. Gronemeyer, J. A. Gustafsson, and V. Laudet (2004). Principles for modulation of the nuclear receptor superfamily, *Nat. Rev. Drug Discov.* **3**:950–964.
15. D. J. Mangelsdorf and R. M. Evans (1995). The RXR heterodimers and orphan receptors, *Cell* **83**:841–850.
16. A. Chawla, J. J. Repa, R. M. Evans, and D. J. Mangelsdorf (2001). Nuclear receptors and lipid physiology: Opening the X-files, *Science* **294**:1866–1870.
17. R. M. Evans (1988). The steroid and thyroid hormone receptor superfamily, *Science* **240**:889–895.
18. A. Warnmark, E. Treuter, A. P. Wright, and J. A. Gustafsson (2003). Activation functions 1 and 2 of nuclear receptors: Molecular strategies for transcriptional activation, *Mol. Endocrinol.* **17**:1901–1909.
19. L. Nagy and J. W. Schwabe (2004). Mechanism of the nuclear receptor molecular switch, *Trends Biochem. Sci.* **29**:317–324.
20. C. K. Glass and M. G. Rosenfeld (2000). The coregulator exchange in transcriptional functions of nuclear receptors, *Genes Dev.* **14**:121–141.

21. J. Matthews and J. A. Gustafsson (2003). Estrogen signaling: A subtle balance between ER alpha and ER beta, *Mol. Interv.* **3**:281–292.
22. J. Bastien and C. Rochette-Egly (2004). Nuclear retinoid receptors and the transcription of retinoid-target genes, *Gene* **328**:1–16.
23. R. A. Daynes and D. C. Jones (2002). Emerging roles of PPARs in inflammation and immunity, *Nat. Rev. Immunol.* **2**:748–759.
24. G. B. Wisely, A. B. Miller, R. G. Davis, A. D. Thornquest, Jr., R. Johnson, T. Spitzer, A. Sefler, B. Shearer, J. T. Moore, T. M. Willson, and S. P. Williams (2002). Hepatocyte nuclear factor 4 is a transcription factor that constitutively binds fatty acids, *Structure (Camb)* **10**:1225–1234.
25. W. Wahli, P. R. Devchand, A. IJpendberg, and B. Desvergne (1999). Fatty acids, eicosanoids, and hypolipidemic agents regulate gene expression through direct binding to peroxisome proliferator-activated receptors, *Adv. Exp. Med. Biol.* **447**:199–209.
26. B. Desvergne and W. Wahli (1999). Peroxisome proliferator-activated receptors: Nuclear control of metabolism, *Endocr. Rev.* **20**:649–688.
27. G. F. Watts and S. B. Dimmitt (1999). Fibrates, dyslipoproteinaemia and cardiovascular disease, *Curr. Opin. Lipidol.* **10**:561–574.
28. P. Tontonoz, E. Hu, and B. M. Spiegelman (1994). Stimulation of adipogenesis in fibroblasts by PPAR gamma 2, a lipid-activated transcription factor, *Cell* **79**:1147–1156.
29. S. Kersten, B. Desvergne, and W. Wahli (2000). Roles of PPARs in health and disease, *Nature* **405**:421–424.
30. S. M. Rangwala and M. A. Lazar (2004). Peroxisome proliferator-activated receptor gamma in diabetes and metabolism, *Trends Pharmacol. Sci.* **25**:331–336.
31. E. D. Rosen, C. J. Walkey, P. Puigserver, and B. M. Spiegelman (2000). Transcriptional regulation of adipogenesis, *Genes Dev.* **14**:1293–1307.
32. L. Michalik, B. Desvergne, and W. Wahli (2003). Peroxisome proliferator-activated receptors beta/delta: Emerging roles for a previously neglected third family member, *Curr. Opin. Lipidol.* **14**:129–135.
33. D. M. Muoio, P. S. MacLean, D. B. Lang, S. Li, J. A. Houmard, J. M. Way, D. A. Winegar, J. C. Corton, G. L. Dohm, and W. E. Kraus (2002). Fatty acid homeostasis and induction of lipid regulatory genes in skeletal muscles of peroxisome proliferator-activated receptor (PPAR) alpha knock-out mice. Evidence for compensatory regulation by PPAR delta, *J. Biol. Chem.* **277**:26089–26097.
34. L. Michalik, B. Desvergne, and W. Wahli (2004). Peroxisome-proliferator-activated receptors and cancers: Complex stories, *Nat. Rev. Cancer* **4**:61–70.
35. G. A. Francis, E. Fayard, F. Picard, and J. Auwerx (2003). Nuclear receptors and the control of metabolism, *Annu. Rev. Physiol.* **65**:261–311.
36. E. Kaneko, M. Matsuda, Y. Yamada, Y. Tachibana, I. Shimomura, and M. Makishima (2003). Induction of intestinal ATP-binding cassette transporters by a phytosterol-derived liver X receptor agonist, *J. Biol. Chem.* **278**:36091–36098.
37. J. J. Repa and D. J. Mangelsdorf (2002). The liver X receptor gene team: Potential new players in atherosclerosis, *Nat. Med.* **8**:1243–1248.
38. L. Malerod, L. K. Juvet, T. Gjoen, and T. Berg (2002). The expression of scavenger receptor class B, type I (SR-BI) and caveolin-1 in parenchymal and nonparenchymal liver cells, *Cell Tissue Res.* **307**:173–180.

39. M. Krieger (1999). Charting the fate of the "good cholesterol": Identification and characterization of the high-density lipoprotein receptor SR-BI, *Annu. Rev. Biochem.* **68**:523–558.

40. A. Sirvent, T. Claudel, G. Martin, J. Brozek, V. Kosykh, R. Darteil, D. W. Hum, J. C. Fruchart, and B. Staels (2004). The farnesoid X receptor induces very low density lipoprotein receptor gene expression, *FEBS Lett.* **566**:173–177.

41. K. R. Stayrook, K. S. Bramlett, R. S. Savkur, J. Ficorilli, T. Cook, M. E. Christe, L. F. Michael, and T. P. Burris (2005). Regulation of carbohydrate metabolism by the farnesoid X receptor, *Endocrinology* **146**:984–991.

42. C. Brendel, K. Schoonjans, O. A. Botrugno, E. Treuter, and J. Auwerx (2002). The small heterodimer partner interacts with the liver X receptor alpha and represses its transcriptional activity, *Mol. Endocrinol.* **16**:2065–2076.

43. A. Castrillo, S. B. Joseph, C. Marathe, D. J. Mangelsdorf, and P. Tontonoz (2003). Liver X receptor-dependent repression of matrix metalloproteinase-9 expression in macrophages. *J. Biol. Chem.* **278**:10443–10449.

44. A. J. Fowler, M. Y. Sheu, M. Schmuth, J. Kao, J. W. Fluhr, L. Rhein, J. L. Collins, T. M. Willson, D. J. Mangelsdorf, P. M. Elias, and K. R. Feingold (2003). Liver X receptor activators display anti-inflammatory activity in irritant and allergic contact dermatitis models: Liver-X-receptor-specific inhibition of inflammation and primary cytokine production. *J. Invest. Dermatol.* **120**:246–255.

45. S. B. Joseph, M. N. Bradley, A. Castrillo, K. W. Bruhn, P. A. Mak, L. Pei, J. Hogenesch, M. O'Connell, G. Cheng, E. Saez, J. F. Miller, and P. Tontonoz (2004). LXR-dependent gene expression is important for macrophage survival and the innate immune response, *Cell* **119**:299–309.

46. A. Castrillo, S. B. Joseph, S. A. Vaidya, M. Haberland, A. M. Fogelman, G. Cheng, and P. Tontonoz (2003). Crosstalk between LXR and toll-like receptor signaling mediates bacterial and viral antagonism of cholesterol metabolism, *Mol. Cell* **12**:805–816.

47. S. B. Joseph, A. Castrillo, B. A. Laffitte, D. J. Mangelsdorf, and P. Tontonoz (2003). Reciprocal regulation of inflammation and lipid metabolism by liver X receptors, *Nat. Med.* **9**:213–219.

48. B. A. Laffitte, L. C. Chao, J. Li, R. Walczak, S. Hummasti, S. B. Joseph, A. Castrillo, D. C. Wilpitz, D. J. Mangelsdorf, J. L. Collins, E. Saez, and P. Tontonoz (2003). Activation of liver X receptor improves glucose tolerance through coordinate regulation of glucose metabolism in liver and adipose tissue, *Proc. Natl. Acad. Sci. U.S.A.* **100**:5419–5424.

49. S. A. Kliewer (2003). The nuclear pregnane X receptor regulates xenobiotic detoxification, *J. Nutr.* **133**:2444S–2447S.

50. H. S. Ahuja, A. Szanto, L. Nagy, and P. J. Davies (2003). The retinoid X receptor and its ligands: Versatile regulators of metabolic function, cell differentiation and cell death, *J. Biol. Regul. Homeost. Agents* **17**:29–45.

51. P. Chambon (1996). A decade of molecular biology of retinoic acid receptors, *Faseb J.* **10**:940–954.

52. S. Christakos, P. Dhawan, Y. Liu, X. Peng, and A. Porta (2003). New insights into the mechanisms of vitamin D action, *J. Cell Biochem.* **88**:695–705.

53. S. Christakos, F. Barletta, M. Huening, P. Dhawan, Y. Liu, A. Porta, and X. Peng (2003). Vitamin D target proteins: Function and regulation, *J. Cell Biochem.* **88**:238–244.

54. M. Holick (2002). Vitamin D: The underappreciated D-lightful hormone that is important for skeletal and cellular health, *Curr. Opin. Endocrinol. Diabets.* **9**:87–98.
55. R. G. Erben, D. W. Soegiarto, K. Weber, U. Zeitz, M. Lieberherr, R. Gniadecki, G. Moller, J. Adamski, and R. Balling (2002). Deletion of deoxyribonucleic acid binding domain of the vitamin D receptor abrogates genomic and nongenomic functions of vitamin D, *Mol. Endocrinol.* **16**:1524–1537.
56. M. R. Haussler, G. K. Whitfield, C. A. Haussler, J. C. Hsieh, P. D. Thompson, S. H. Selznick, C. E. Dominguez, and P. W. Jurutka (1998). The nuclear vitamin D receptor: Biological and molecular regulatory properties revealed, *J. Bone Miner Res.* **13**:325–349.
57. P. Bortman, M. A. Folgueira, M. L. Katayama, I. M. Snitcovsky, and M. M. Brentani (2002). Antiproliferative effects of 1,25-dihydroxyvitamin D3 on breast cells: A mini review, *Braz. J. Med. Biol. Res.* **35**:1–9.
58. M. Makishima, T. T. Lu, W. Xie, G. K. Whitfield, H. Domoto, R. M. Evans, M. R. Haussler, and D. J. Mangelsdorf (2002). Vitamin D receptor as an intestinal bile acid sensor, *Science* **296**:1313–1316.
59. T. Cornwell, W. Cohick, and I. Raskin (2004). Dietary phytoestrogens and health, *Phytochemistry* **65**:995–1016.
60. A. M. Duncan, W. R. Phipps, and M. S. Kurzer (2003). Phyto-oestrogens, *Best Pract. Res. Clin. Endocrinol. Metab.* **17**:253–271.
61. A. L. Ososki and E. J. Kennelly (2003). Phytoestrogens: A review of the present state of research, *Phytother. Res.* **17**:845–869.
62. J. A. Gustafsson (1999). Estrogen receptor beta—a new dimension in estrogen mechanism of action, *J. Endocrinol.* **163**:379–383.
63. K. Pettersson, K. Grandien, G. G. Kuiper, and J. A. Gustafsson (1997). Mouse estrogen receptor beta forms estrogen response element-binding heterodimers with estrogen receptor alpha, *Mol. Endocrinol.* **11**:1486–1496.
64. P. J. Kushner (2003). Estrogen receptor action through target genes with classical and alternative response elements, *Pure Appl. Chem.* **75**:1757–1769.
65. S. Nilsson and J. A. Gustafsson (2002). Biological role of estrogen and estrogen receptors, *Crit. Rev. Biochem. Mol. Biol.* **37**:1–28.
66. S. H. Windahl, G. Andersson, and J. A. Gustafsson (2002). Elucidation of estrogen receptor function in bone with the use of mouse models, *Trends Endocrinol. Metab.* **13**:195–200.
67. J. A. Gustafsson (2003). What pharmacologists can learn from recent advances in estrogen signaling, *Trends Pharmacol. Sci.* **24**:479–485.
68. Z. Zhao, Y. X. Fu, D. Hewett-Emmett, and E. Boerwinkle (2003). Investigating single nucleotide polymorphism (SNP) density in the human genome and its implications for molecular evolution, *Gene* **312**:207–213.
69. A. Chakravarti (2001). To a future of genetic medicine, *Nature* **409**:822–823.
70. M. Stumvoll and H. Haring (2002). The peroxisome proliferator-activated receptor-gamma2 Pro12Ala polymorphism, *Diabetes* **51**:2341–2347.
71. A. G. Uitterlinden, Y. Fang, J. B. van Meurs, H. A. Pols, and J. P. van Leeuwen (2004). Genetics and biology of vitamin D receptor polymorphisms, *Gene* **338**:143–156.
72. A. G. Uitterlinden, Y. Fang, J. B. van Meurs, J. P. van Leeuwen, and H. A. Pols (2004). Vitamin D receptor gene polymorphisms in relation to Vitamin D related disease states, *J. Steroid Biochem. Mol. Biol.* **89-90** :187–193.

73. S. H. Ralston (2002). Genetic control of susceptibility to osteoporosis, *J. Clin. Endocrinol. Metab.* **87**:2460–2466.

74. J. E. Osborne and P. E. Hutchinson (2002). Vitamin D and systemic cancer: Is this relevant to malignant melanoma? *Br. J. Dermatol.* **147**:197–213.

75. E. Grundberg, H. Brandstrom, E. L. Ribom, O. Ljunggren, A. Kindmark, and H. Mallmin (2003). A poly adenosine repeat in the human vitamin D receptor gene is associated with bone mineral density in young Swedish women, *Calcif. Tissue Int.* **73**:455–462.

76. E. Grundberg, H. Brandstrom, E. L. Ribom, O. Ljunggren, H. Mallmin, and A. Kindmark (2004). Genetic variation in the human vitamin D receptor is associated with muscle strength, fat mass and body weight in Swedish women, *Eur. J. Endocrinol.* **150**: 323–328.

77. J. P. Ioannidis, S. H. Ralston, S. T. Bennett, M. L. Brandi, D. Grinberg, F. B. Karassa, B. Langdahl, J. B. van Meurs, L. Mosekilde, S. Scollen, O. M. Albagha, M. Bustamante, A. H. Carey, A. M. Dunning, A. Enjuanes, J. P. van Leeuwen, C. Mavilia, L. Masi, F. E. McGuigan, X. Nogues, H. A. Pols, D. M. Reid, S. C. Schuit, R. E. Sherlock, and A. G. Uitterlinden (2004). Differential genetic effects of ESR1 gene polymorphisms on osteoporosis outcomes, *JAMA* **292**:2105–2114.

78. S. C. Schuit, H. H. Oei, J. C. Witteman, C. H. Geurts van Kessel, J. B. van Meurs, R. L. Nijhuis, J. P. van Leeuwen, F. H. de Jong, M. C. Zillikens, A. Hofman, H. A. Pols, and A. G. Uitterlinden (2004). Estrogen receptor alpha gene polymorphisms and risk of myocardial infarction, *JAMA* **291**:2969–2977.

79. T. I. Andersen, K. R. Heimdal, M. Skrede, K. Tveit, K. Berg, and A. L. Borresen (1994). Oestrogen receptor (ESR) polymorphisms and breast cancer susceptibility, *Hum. Genet.* **94**:665–670.

80. H. J. Kang, S. W. Kim, H. J. Kim, S. J. Ahn, J. Y. Bae, S. K. Park, D. Kang, A. Hirvonen, K. J. Choe, and D. Y. Noh (2002). Polymorphisms in the estrogen receptor-alpha gene and breast cancer risk, *Cancer Lett.* **178**:175–180.

81. E. L. Schubert, M. K. Lee, B. Newman, and M. C. King (1999). Single nucleotide polymorphisms (SNPs) in the estrogen receptor gene and breast cancer susceptibility. *J. Steroid Biochem. Mol. Biol.* **71**:21–27.

82. E. Weiderpass, I. Persson, H. Melhus, S. Wedren, A. Kindmark, and J. A. Baron (2000). Estrogen receptor alpha gene polymorphisms and endometrial cancer risk, *Carcinogenesis* **21**:623–627.

83. S. Wedren, L. Lovmar, K. Humphreys, C. Magnusson, H. Melhus, A. C. Syvanen, A. Kindmark, U. Landegren, M. L. Fermer, F. Stiger, I. Persson, J. Baron, and E. Weiderpass (2004). Oestrogen receptor alpha gene haplotype and postmenopausal breast cancer risk: A case control study, *Breast Cancer Res.* **6**:R437–R449.

84. H. Lu, T. Higashikata, A. Inazu, A. Nohara, W. Yu, M. Shimizu, and H. Mabuchi (2002). Association of estrogen receptor-alpha gene polymorphisms with coronary artery disease in patients with familial hypercholesterolemia, *Arterioscler. Thromb. Vasc. Biol.* **22**:817–823.

85. M. L. Brandi, L. Becherini, L. Gennari, M. Racchi, A. Bianchetti, B. Nacmias, S. Sorbi, P. Mecocci, U. Senin, and S. Govoni (1999). Association of the estrogen receptor alpha gene polymorphisms with sporadic Alzheimer's disease, *Biochem. Biophys. Res. Commun.* **265**:335–338.

86. M. Nilsson, S. Naessen, I. Dahlman, A. Linden Hirschberg, J. A. Gustafsson, and K. Dahlman-Wright (2004). Association of estrogen receptor beta gene polymorphisms with bulimic disease in women, *Mol. Psychiatry* **9**:28–34.

87. S. Ogawa, T. Hosoi, M. Shiraki, H. Orimo, M. Emi, M. Muramatsu, Y. Ouchi, and S. Inoue (2000). Association of estrogen receptor beta gene polymorphism with bone mineral density, *Biochem. Biophys. Res. Commun.* **269**:537–541.

88. S. Ogawa, M. Emi, M. Shiraki, T. Hosoi, Y. Ouchi, and S. Inoue (2000). Association of estrogen receptor beta (ESR2) gene polymorphism with blood pressure, *J. Hum. Genet.* **45**:327–330.

89. H. Eastwood, K. M. Brown, D. Markovic, and L. F. Pieri (2002). Variation in the ESR1 and ESR2 genes and genetic susceptibility to anorexia nervosa, *Mol. Psychiatry* **7**:86–89.

90. L. Westberg, F. Baghaei, R. Rosmond, M. Hellstrand, M. Landen, M. Jansson, G. Holm, P. Bjorntorp, and E. Eriksson (2001). Polymorphisms of the androgen receptor gene and the estrogen receptor beta gene are associated with androgen levels in women, *J. Clin. Endocrinol. Metab.* **86**:2562–2568.

91. C. Zhao, L. Xu, M. Otsuki, G. Toresson, K. Koehler, Q. Pan-Hammarstrom, L. Hammarstrom, S. Nilsson, J. A. Gustafsson, and K. Dahlman-Wright (2004). Identification of a functional variant of estrogen receptor beta in an African population, *Carcinogenesis* **25**:2067–2073.

92. M. Muller and S. Kersten (2003). Nutrigenomics: Goals and strategies, *Nat. Rev. Genet.* **4**:315–322.

8

GREEN TEA POLYPHENOLS AND CANCER PREVENTION

Shangqin Guo and Gail Sonenshein

Department of Biochemistry and Women's Health Interdisciplinary Research Center, Boston University School of Medicine, Boston, Massachusetts

8.1 INTRODUCTION

Consumption of tea around the world is believed second only to water, although the levels vary widely by country. There are three major types of tea produced: black tea (78%), green tea (20%), and oolong tea (2%) [1]. Green tea is predominant in China, Japan, Korea, India, and a few countries in North Africa and the Middle East, while black tea is predominant in Western countries and some Asian countries.

All major types of tea are derived from the leaves of the *Camellia sinensis* plant. The methods of harvesting and processing of the tea leaves determines the type of tea and the characteristic flavonoid content. Flavonoids are polyphenolic compounds that can be categorized according to chemical structure into flavonols, flavones, flavanones, isoflavones, catechins, and so on. Of the various flavonoids in tea, flavonol content is less affected by processing and is thus present in comparable quantities in green, oolong, and black teas. On the other hand, processing associated with preparation of black tea reduces the amounts of catechins via release of polyphenol oxidizing enzymes, which are in separate compartments from catechins in the fresh

Nutritional Genomics: Discovering the Path to Personalized Nutrition
Edited by Jim Kaput and Raymond L. Rodriguez

tea leaves. When the tea leaves are rolled or broken during industry manufacture of black tea, catechins come into contact with polyphenol oxidase, which joins the monomeric flavonoids into oligomeric and polymeric polyphenols (converting primarily catechins and gallocatechins to theaflavins and thearubigins) (Figure 8.1). Furthermore, black tea is "fermented" and highly oxidized, resulting in increased concentrations of theaflavins and thearubigins and relatively low catechin concentrations. In contrast, fresh tea leaves meant for green tea are steamed or heated immediately after harvest, leading to the rapid inactivation of the oxidases. As a result, green tea contains high concentrations of catechins; the major compounds include epigallocatechin-3-gallate (EGCG), epicatechin (EC), epicatechin-3-gallate (ECG), and epigallocatechin (EGC). EGCG is the most abundant polyphenol present in green tea, accounting for approximately one-third of all the polyphenols, and has been the subject of extensive studies. Oolong tea is partially oxidized and has a mixture of catechin and theaflavin components. In this chapter, we focus on green tea and EGCG, and their effects on cancer and cancer prevention. We first present epidemiological evidence, and then data from animal models and cell culture studies identifying affected signaling pathways that are linked to cancer development and progression.

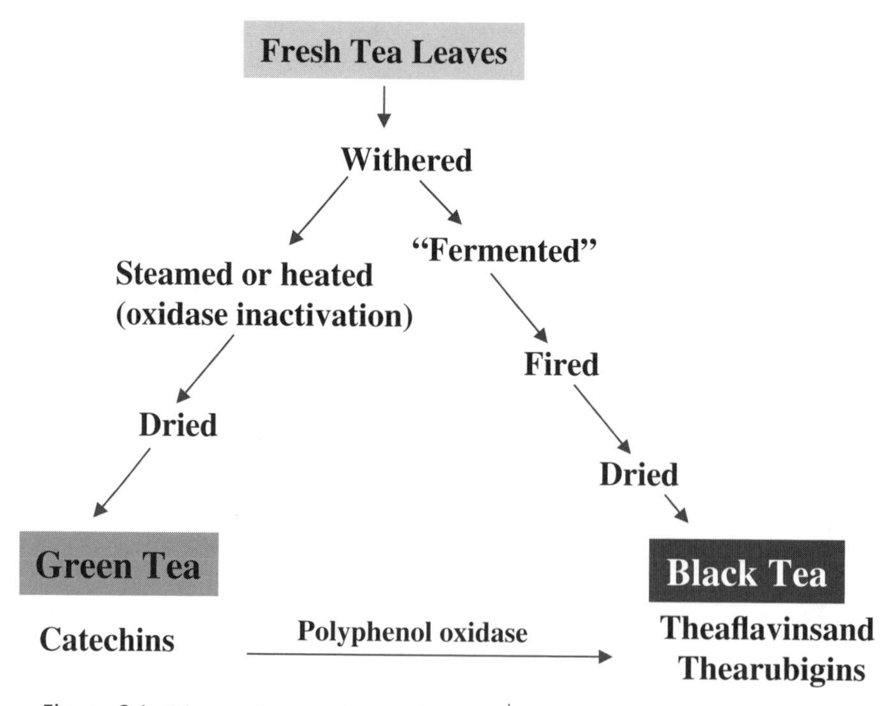

Figure 8.1. Scheme of processing tea leaves to produce either green or black tea.

8.2 GREEN TEA AND CANCER EPIDEMIOLOGY

Since the incidence of different cancers varies widely by geographical region, substantial efforts have been made to identify dietary links. Many of these studies have shown an inverse association between green tea consumption and cancer incidence, although there are some inconsistencies in the findings. Comprehensive reviews of the epidemiological literature have been performed [2–5]. Epidemiological studies on breast cancer risk and tea consumption are among the best characterized. Consumption of green tea is associated with improved prognosis of patients with breast cancer. Increased consumption of green tea was closely associated with decreased numbers of axillary lymph node metastases among premenopausal patients with stage I and II breast cancer and with increased expression of PR and ER among postmenopausal women [6]. An inverse correlation between regular green tea consumption prior to breast cancer diagnosis and the subsequent risk of recurrence has also been demonstrated [7]. When a study was done to identify the specific contribution of black tea and green tea to breast cancer risk, green tea drinkers showed a significantly reduced risk of breast cancer, while black tea drinkers did not enjoy this benefit [8]. When the mechanisms of these effects were examined, high intake of green tea correlated directly with increasing sex hormone-binding globulin and decreasing free estradiol on day 11 of the menstrual cycle. Thus, green tea protection appears to function, in part, by reducing hormone levels associated with the risk of developing breast cancer [9].

In addition to breast cancer, tea consumption has been associated with reduced risk of developing malignancies in a wide variety of organ sites. For example, studies from many different groups have repeatedly presented a lower risk of cancer of the digestive tract, including esophageal, stomach, colorectal, and pancreatic cancer, in association with green tea consumption [10–15]. Fraumeni and co-workers found that drinking green tea provided a protective effect on esophageal cancer in women, and that the risk decreased further as tea consumption increased [10]. Matsuyama and co-workers demonstrated a decreased risk of cancer in people consuming over 10 cups of green tea daily, compared to those consuming 3 cups or less. In addition, they found a significant delay in the onset of cancer with increased green tea consumption [16, 17]. An inverse association was also observed between green tea drinking and chronic gastritis and gastric cancers [14, 18, 19]. The association with the risk of chronic atrophic gastritis was significant even after adjusting for *Helicobacter pylori* infection [20]. In particular, when urinary tea polyphenols, including EGC and EC, and their respective metabolites 5-(3′,4′,5′-trihydroxyphenyl)-gamma-valerolactone (M4) and 5-(3′,4′-dihydroxyphenyl)-gamma-valerolactone (M6), were used as biomarkers for tea consumption, instead of a conventional questionnaire based study, EGC positivity showed a statistically significant inverse association with gastric cancer after adjustment for *Helicobactor pylori* seropositivity and other confounding factors. Similar tea polyphenol–cancer risk associations were observed when gastric cancer and esophageal cancer sites were combined [21].

The risk of developing ovarian cancer has also been shown to be decreased with an increasing frequency and duration of green tea drinking [22]. Interestingly, regular consumption of green tea started even after diagnosis seemed to enhance the survival of epithelial ovarian cancer patients [23]. Tea consumption has also been reported to associate negatively with prostate cancer risk in Japan [24, 25]. Other studies have shown that regular drinking of green tea may protect smokers from oxidative damage, as judged by urinary levels of 8-hydroxydeoxyguanosine as a measure of oxidative DNA damage, and could thereby reduce the risk of developing cancers that are caused by free radicals associated with smoking [26]. However, consumption of green tea was associated with a reduced risk of lung cancer among nonsmoking, but not smoking, women in a small case-controlled study of the association between past consumption of green tea and the risk of lung cancer in women in Shanghai (China) [27].

While most of the epidemiological evidence indicates a protective effect against cancer, there have been studies reporting the absence of any effect of green tea. For example, in a pooled analysis of two prospective studies with 35,004 Japanese women, green tea intake was not associated with a lower risk of breast cancer [28]. Gastric cancer risk reduction also showed mixed results, and these studies have been reviewed elsewhere [29, 30].

Several potential reasons for the inconsistencies between these studies have been proposed. Differences in the flavonol content might account at least partially for the variable results in epidemiological studies, as large variations exist in the flavonol contents of different brands of teas [31]. When 12 brands of tea products from six provinces in China were studied, the tea quality grades inversely correlated with fluoride level [32]. Furthermore, while catechins derived primarily from fruits tended to be inversely associated with upper digestive tract cancer, catechins derived from tea were inversely associated with rectal cancer [33]. Thus, the overall results of catechin consumption will be a composite of the effects caused by the individual catechins from the different dietary sources. Although most of studies are presented after adjustment for many possible confounding factors (e.g., age, alcohol use, body weight, and menarche), additional factors may need to be considered such as smoking. The protection from drinking green tea may be organ site specific. Bladder cancer risk was not reduced by green tea drinking [34, 35].

Importantly, recent evidence has demonstrated the genetic background of the subjects alters metabolism or oxidative stress and thereby can affect the conclusions of these epidemiological studies. Catechol-*O*-methyltransferase (COMT) is involved in the rapid biotransformation and elimination of tea catechins after ingestion [36]. In human populations, the *COMT* allele is polymorphic, with an allele encoding an enzyme with higher activity. Thus, an individual can inherit either none, one, or two copies of this polymorphic allele. In a case-controlled study of Chinese-American, Japanese-American, and Filipino-American women in Los Angeles County, an association between tea consumption and breast cancer risk reduction could be observed only among individuals who possessed at least one copy of the low-activity *COMT* allele [37]. In these individuals, even black tea drinking could reduce breast cancer risk. In contrast, in individuals that are homozygous for the high-activity *COMT* allele, no difference could be observed in cancer risk between tea drinkers and nondrinkers [37].

Another genetic locus shown to affect the responsiveness to green tea is that for glutathione *S*-transferase (GST), a member of the phase II group of xenobiotic metabolizing enzymes. The *GST* $\mu 1$ (*GSTM1*) and *GST* $\theta 1$ (*GSTT1*) genes have a null-allele variant in which the entire gene is absent. When the effects of high consumption (4 cups/day) of green tea or black tea on oxidative DNA damage caused by cigarette smoking were studied, *GSTM1* and *GSTT1* genotypes showed different responses. DNA damage was assessed by urinary level of 8-hydroxydeoxyguanosine (8-OHdG). Heavy smokers were recruited to the study and were given green tea, black tea, or water to drink over a 4 month period. Although the baseline levels of 8-OHdG were the same before the experiment, only the *GSTM1*-positive and *GSTT1*-positive smokers showed a statistically significant decrease in urinary 8-OHdG levels after 4 months of drinking green, but not black, tea [38].

Overall, the findings suggest that in people with certain genetic makeups, green tea can play a potent role in prevention of certain types of malignancies.

8.3 ANIMAL MODELS

While epidemiological evidence provides a correlative link between reduced cancer incidence and green tea consumption, studies using animal models have convincingly shown a protective effect of tea against a wide variety of either spontaneous or induced (chemical or irradiation) malignancies. Frequent consumption of green tea by mice leads to a high level of green tea polyphenols (GTPs) in a wide array of organs [39, 40]. Protection against cancer has been achieved by using polyphenolic fraction of green tea, a water extract of green tea, or individual polyphenols present in green tea delivered either through the oral route (in diet or drinking water), topical application, gavage, or injection. The findings of these studies are summarized below and have recently been reviewed elsewhere [41, 42].

Skin Cancer Models

Treatment of *sensitive to carcinogenesis* (SENCAR) mice with (+/–)-7 beta,8 alpha-dihydroxy-9 alpha,10 alpha-epoxy-7,8,9,10-tetrahydrobenzo[*a*]pyrene (BPDE-2) and 12-*O*-tetradecanoyl phorbol-13-acetate (TPA) induces skin tumor formation. Pretreatment of the skin of SENCAR mice with tannic acid and GTP topically afforded significant protection against skin tumor induction. The latency of tumor formation was prolonged and tumor numbers were reduced [43]. Later, similar protective effects by GTP were reported for skin tumors induced by other carcinogenic polycyclic aromatic hydrocarbons (PAHs) in BALB/c mice and by 7,12-dimethylbenz[*a*]anthracene (DMBA) in SENCAR mice. In addition to topical application, oral feeding of GTPs in drinking water was effective [44]. SKH-1 hairless mice were used to study ultraviolet B (UVB)-induced skin tumor formation. Continuous oral feeding or topical application of GTP resulted in significantly lower tumor yield (percent of animals with tumors and number of tumors per mouse)—with oral feeding being more effective than topical application [45]. Oral feeding of

GTP was also effective in protecting against skin tumors induced by different combinations of carcinogens and UV irradiation [46]; in BALB/cAnNHsd mice, topical application of purified EGCG, but not oral administration, was effective in inhibiting skin tumors induced by UV [47]. Studies with individual polyphenolic compounds of green tea indicated that topical application of EGCG, EGC, and ECG all inhibited TPA-induced epidermal inflammation (a promoter of tumor development) in a dose-dependent manner, with EGCG showing the greatest inhibition [48, 49]. Green tea was also effective in causing regression of established malignancies. Administration of green tea as sole drinking source inhibited the growth and caused the regression of established experimentally induced skin papillomas [50].

Digestive Tract Cancer Models

A wide spectrum of chemical carcinogens, including *N*-methyl-*N'*-nitro-*N*-nitrosoguanidine (MNNG) [51, 52], *N*-methyl-*N*-nitrosourea (MNU) [53], *N*-nitrosodiethylamine (NDEA) [54], and 2-dimethylhydrazine (1,2-DMH) [55], induce tumors of the digestive tract. Water extracts of green tea, green tea catechins, low molecular weight tannin fraction of green tea extracts, and EGCG have been found to inhibit tumor formation in several models. For example, treatment of A/J mice with 10 mg/kg NDEA once a week for 8 weeks resulted in forestomach (and lung) tumor formation in more than 90% of the animals. With 0.63 or 1.25% green tea (12.5 g green tea leaves brewed with 1 liter of boiling water) as the sole drinking source, green tea treatment reduced forestomach tumor incidence and multiplicity [54]. Protection was also achieved when GTP was given by gavage in this model [56]. Recently, it was shown that green tea inhibited aberrant crypt focus formation (a precursor of colon tumors) in mice on a high corn oil diet [57]. Rats given *N*-nitrosomethylbenzylamine (NMBzA) orally develop esophageal tumors. Feeding rats either green, jasmine, black, or oolong tea dramatically protected against NMBzA-induced esophageal tumor formation [58]. Interestingly, the protection was effective when the green tea was given in the drinking water either before or during the carcinogen treatment [59]. In a multiorgan carcinogenesis model in male F344 rats, the animals were treated with a combination of multiple carcinogens and then fed a diet rich in green tea catechins. The numbers of small intestinal tumors (adenomas and carcinomas) per rat were significantly reduced in the groups treated with green tea catechins during and after carcinogen treatment [60]. In Fisher rats, GTP given in drinking water inhibited the development of azoxymethane (AOM)-induced colon carcinogenesis and EGCG inhibited the cellular kinetics of the gastric mucosa during the promotion stage of MNNG-induced gastric carcinogenesis [61, 62]. Further studies from the same group revealed that green tea was well tolerated by the animals without any obvious toxicity [63].

Prostate Tumor Models

Early studies were performed using a transplantable tumor model formed by subcutaneous inoculation of human prostate cancer cell lines PC-3 or LNCaP 104-R into nude mice. Intraperitoneal injection of EGCG in this model inhibited the growth and rapidly reduced the size of transplanted prostate tumors in nude mice [64]. Ornithine decarboxylase (ODC), a rate-controlling enzyme in the polyamine biosynthetic pathway, is overexpressed in prostate cancers and prostatic fluid in humans. Oral feeding of GTPs in drinking water decreased the testosterone-mediated induction of ODC activity in castrated rats and C57BL/6 mice [65]. More recently, the ability of green tea to prevent prostate cancer formation has also been studied in a molecularly induced model: the Transgenic Adenocarcinoma of Mouse Prostate (TRAMP) mouse. TRAMP mice, which express the oncogene SV40 T antigen in the prostate epithelium, develop spontaneous autochthonous prostate cancer and are considered a model for prostate cancer initiation and progression [66]. Oral infusion of a polyphenolic fraction isolated from green tea significantly inhibited prostate cancer development and increased survival of TRAMP mice [67].

Lung Tumor Models

Treatment of A/J mice with a single dose (103 mg/kg) of the tobacco-specific 4-(methylnitrosamino)-1-(3-pyridyl)-1-butanone (NNK) resulted in the formation of pulmonary adenomas in almost all of the animals. Extracts of green tea decreased both pulmonary tumor incidence and multiplicity [54, 68]. Green tea given in drinking water prior to, during, and after the carcinogen treatment similarly protected C3H mice from diethylnitrosamine-induced lung tumors [69]. In hamsters, green tea significantly inhibited lung tumor multiplicity initiated with the tobacco carcinogen NHK in a neuroendocrine model, although it increased adenocarcinomas, consistent with differerential effects on cell signaling pathways (see below) [70].

Mammary Gland Tumor Models

About 90% of female Sprague-Dawley rats treated with DMBA develop mammary tumors [71]. When DMBA-treated rats were fed diets containing green tea catechins, the catechin-consuming animals showed significantly higher survival than those given the regular chow diet [72]. However, when the incidence, multiplicity, and size of the tumors were compared, no significant differences were seen between the green tea-fed and control groups. Thus, green tea catechins fed in diet do not appear to be effective at inhibiting progression of rat mammary carcinogenesis [73]. In a separate report, green tea catechins were shown to have a weak inhibition on mammary tumors induced by DMBA, but the effect was not dose-dependent [74]. In work from our laboratory, Sprague-Dawley rats were given 0.3% green tea or water as their sole fluid source and then treated with DMBA. A significant increase in mean latency to first tumor (84 versus 66 days), an approximate 70% decrease in tumor burden (2.5 ± 4.5 g versus 8.3 ± 6.9 g), and an 87% reduction in the number of invasive tumors per tumor-bearing animal were observed [75]. Although water-fed rats had 3.0 tumors per tumor-bearing animal, and the green tea-fed rats had 1.9; this difference did not reach statistical significance [75].

Several transplantable breast tumor models have also been used. Human breast cancer MCF-7 cells were inoculated into nude mice to form tumors. EGCG rapidly inhibited the MCF-7 cell formed tumor growth [64]. In another model, precancerous mouse mammary epithelial cells RIII/MG were inoculated into mice, which led to the formation of tumors, and this tumor formation in the animals was inhibited in a dose-dependent manner by EGCG [76].

Miscellaneous Other Tumor Models

Diethylnitrosamine-treated C3H mice develop liver tumors, in addition to lung tumors. Green tea or black tea given in drinking water reduced the liver tumor numbers, although there was no apparent relationship between EGCG concentration and liver tumor response [69]. Green tea was also reported to inhibit hepatocarcinogenesis induced by either aflatoxin B1,2-nitropropane or pentachlorophenol [40, 77, 78]. Green tea significantly inhibited tumor growth in NOD/SCID mice transplanted intraperitoneally with human non-Hodgkin's lymphoma cell lines [79]. Furthermore, green tea and its major GTP components given in drinking fluid ameliorated immune dysfunction in mice bearing Lewis lung carcinoma and treated with the carcinogen NNK [80]. Green tea preparations protected against DMBA-induced oral carcinogenesis in hamsters [81], as well as urinary bladder tumor formation in rats induced by *N*-butyl-*N*-(4-hydroxybutyl)-nitrosamine [82, 83]. Intravesical instillation of EGCG inhibited bladder tumor growth in a rat model [84]. Green tea extracts also inhibited pancreatic carcinogenesis and tumor promotion of transplanted pancreatic cancer in hamsters [85].

In summary, green tea extracts and polyphenol constituents display a strong protective effect on carcinogen-induced and genetically induced cancers in many different animal models. These findings lead to questions regarding the mechanisms of action of green tea polyphenols, which have been addressed using cell culture systems.

8.4 MECHANISMS OF GREEN TEA ACTION: MOLECULAR SIGNALING PATHWAYS AND GENE TARGETS

Tumor malignancy is manifested in various aspects of the physiology and metabolic status of the cell [86]. The malignant cell typically displays enhanced potential for growth, proliferation, survival, and lifespan and a reduced requirement for cell-matrix adhesion as compared to a normal cell. In recent years, it has been recognized that sustained tumor growth in animals requires angiogenesis to maintain the blood supply. Studies on cultured cells have demonstrated the power of green tea polyphenols to combat virtually all aspects of tumor cell physiology. It is not surprising then to see that green tea or EGCG affects many key signaling molecules and pathways.

The Receptor for EGCG

The 67 kDa laminin receptor (67LR) has recently been identified as the major receptor for EGCG [88]. All *trans*-retinoic acid (RA) induced 67LR expression and enhanced cell surface binding of EGCG. The 67LR was found on many tumor cells and its expression correlated with invasion and metastasis [87]. Human lung cancer cells A549 ectopically expressing 67LR showed increased cell surface binding of EGCG and slowed growth in response to EGCG treatment. The receptor appears to be specific for EGCG as other catechins present in green tea did not show any binding to 67LR-expressing cells, nor did they inhibit the growth [88]. In our own studies, we have found that EGCG treatment of breast cancer cells leads to a down regulation of the 67LR protein. In cells that are resistant to EGCG inhibition, the 67LR levels were low. However, when a retrovirus expressing the 67LR was introduced into the EGCG resistant cells, it did not substantially restore the sensitivity (S.G. and G.E.S., unpublished data).

Cell Cycle Machinery

Green tea polyphenols have been shown to affect several key regulators of cell cycle progression (Table 8.1). GTPs induced the tumor suppressor p53 and its target gene, the cyclin-dependent kinase inhibitor (CKI) $p21^{WAF1/CIP1}$, in both in vivo and in vitro experimental systems. It is known that p53 is mutated in more than half of human tumors. Pretreatment of SKH-1 mice with orally administered green tea for 2 weeks enhanced the number of p53/p21-positive and apoptosis-positive cells in the epider-

TABLE 8.1. EGCG Effects on Cell Cycle Genes in Cells in Culture

Cells	Protein	Effect	Function	References
HepG2	p53 p21	↑ ↑	Tumor suppressor Cyclin-dependent kinase inhibitor (CKI). Blocks cell cycle progression.	[89]
LNCaP	p53 p21	Stabilized ↑	Tumor suppressor CKI	[90]
Normal human keratinocytes	$p57^{Kip1}$	↑	CKI and apoptosis inhibitor	[91]
Breast cancer	$p27^{Kip1}$ FOXO3a	↑ ↑	CKI, blocks G1/S Forkhead box O transcription factor. Induces cell cycle blockade.	[75, 92] [100]
A431	pRb E2F p130 p107	Hypo-phosphorylated ↓ ↓ ↓	Tumor suppressor Transcription factor required for cell cycle progression Tumor suppressor family Tumor suppressor family	[93]
SKOV-3 OVACR-3	p21 pRb	↑ ↓	CKI Tumor suppressor	[94]

mis following UV treatment and sunburn [95]. EGCG treatment of human liver cancer HepG2 cells led to increased p53 and p21 protein levels, leading to cell cycle blockade and apoptosis [89]. EGCG induced p53 stabilization and upregulation of p21 and Bax in human prostate carcinoma LNCaP cells [90]. While the concomitant increase in p53 and p21 can be attributed in most cases to the transcriptional activation of p21 by p53, p53-independent induction of p21 has also been reported [92]. Green tea has also been shown to affect UVB-induced mutagenesis of p53. UVB irradiation of mouse skin induces mutations clustering in exon 5 and exon 8. However, when the irradiated mice were treated with green tea, the UVB-induced mutations now also occurred in exon 6, presumably leading to a mutant p53 protein of different activity [96]. Interestingly, Hsu and co-workers [91] have shown that in untransformed cells, EGCG induced another CKI $p57^{Kip2}$ and that the p57 prevented caspase-3 induction and protected the normal cell from apaf-1-induced cell death.

We and others have found that $p27^{Kip1}$, a CKI that promotes G1/S phase growth arrest, is induced in breast cancer cells by treatment with EGCG [75, 92]. The regulation of p27 occurs at multiple levels, including transcription as well as protein and mRNA stability. Members of the forkhead box O family of proteins (FOXO) have been shown to transactivate *p27* gene expression [97–99]. Recently, we have shown that EGCG treatment leads to the activation and nuclear accumulation of FOXO3a [100]. Preliminary studies in breast cancer cells indicate that the induction of active FOXO3a may account, at least in part, for the increase in p27 protein levels (unpublished data).

In addition to p27, the G1/S phase regulator and tumor suppressor pRb was shown to be regulated by EGCG in A431 cells. EGCG induced an increase in the ratio of hypo/hyperphosphorylated pRb and downregulation of E2F factors. EGCG was found to downregulate the protein expression of other members of the pRb family, for example, p130 and p107, in a dose- and time-dependent manner as cells accumulated in the late G1 [93]. Interestingly, EGCG treatment led to the induction of p21 and a drop in Rb levels in ovarian cancer SKOV-3 and OVCAR-3 cells, which lack normal p53 protein [94].

Apoptotic Machinery

Proper integration and execution of apoptotic signals are crucial for development, homeostasis, and immune defense. Death receptor–ligand binding or induction of p53 can activate caspases leading to apoptosis. Poly(ADP-ribose)polymerase (PARP) cleavage and DNA fragmentation are frequently used indicators of apoptotic death. The Bcl-2 family of proteins plays an important role in control of apoptosis via regulating mitochondrial permeability and release of cytochrome *c*, which activates the caspase cascade. Green tea polyphenols have been shown to induce the apoptotic machinery in a number of cancer cells (Table 8.2). EGCG decreased the viability of HTB-94 human chondrosarcoma cells, as judged by enhanced DNA fragmentation, induction of caspase-3/CPP32 activity, and cleavage of PARP. Pretreatment of cells with either a synthetic pan-caspase inhibitor (Z-VAD-FMK) or a caspase-3-specific inhibitor (DEVD-CHO) prevented EGCG-induced PARP cleavage [101]. In Ehrlich ascites tumor cells, green tea extract caused a dose- and time-dependent increase in caspase-3-like protease activation, preceded by a release of cytochrome *c* from the mitochondria [102]. In prostate cancer cells, treatment with tea polyphenols or EGCG blocked expression of the hyperphosphorylated, but not hypophosphorylated, Bcl-xl in mitochondria, which was accompanied by cytochrome *c* release, caspase activation, and apoptosis [103]. In HepG2 human liver cancer cells, EGCG treatment increased the levels of Fas ligand and pro-apoptotic Bax proteins [89]. Monomeric flavonoids EGCG, GCG, ECG, and CG, and polymeric flavonoids theaflavin, theaflavanin, and theaflavin-3′ gallate were also found to promote apoptosis via binding tightly to Bcl-2 family proteins (i.e., K_i in the nanomolar range) and suppressing their activity [104].

Growth and Proliferation Signals

Cell Surface Tyrosine Receptor Kinases. Gene amplification, overexpression, or mutation of cell surface tyrosine kinase growth factor receptors, which has frequently been found in cancer cells, lead to autophosphorylation or to a reduced growth factor requirement [86]. Several groups have found signaling by these receptors is inhibited by green tea polyphenols. EGCG treatment of human epidermoid carcinoma A431 cells reduced the autophosphorylation of the epithelial growth factor receptor (EGFR) activated by EGF. Using an in vitro kinase assay,

TABLE 8.2. EGCG or Green Tea Effects on Apoptotic Genes in Cells in Culture

Cells	Protein	Cellular Effects	Function	References
HTB-94		↑ DNA fragmentation PARP cleavage	Indicates apoptosis	[101]
	Caspase-3/CPP32	↑ activity	Activates apoptotic pathway	
Erlich ascites		Cytochrome c release	Indicates loss of mitochondria integrity and activates apoptotic pathway	[102]
	Caspase-3 like	↑	Activates apoptotic pathway	
Prostate cancer	Bcl-xl	↓ Hyperphosphorylated Bcl-xl	Induce apoptosis	[103]
		Cytochrome c release and caspase activation	As above	
HepG2	Fas ligand	↑	Death inducer	[89]
	Bax proteins	↑	Pro-apoptotic	

EGCG strongly inhibited the tyrosine kinase activity of EGFR, platelet derived growth factor receptor (PDGFR), and fibroblast growth factor receptor (FGFR), but only modestly affected kinase activity of pp60v-src, PKC, or PKA [105]. EGCG similarly inhibited PDGF beta-receptor activity in vivo [106, 107]. Another tea polyphenol, theaflavin-3,3′-digallate, was found to more potently inhibit the autophosphorylation of EGFR and PDGFR induced by their respective ligands than EGCG [110]. Very recently, tyrosine phosphorylation of the vascular epithelial growth factor receptor (VEGFR) was found to be inhibited by EGCG in chronic lymphocytic leukemia cells [108]. Inhibition of VEGF signaling is further discussed below with regard to tumor angiogenesis.

The second member of the EGFR family Her-2/neu receptor has a critical role in mammary gland malignancies. It is overexpressed in about 30% of human breast cancers and indicates poor prognosis. While the humanized monoclonal Her-2/neu antibody Herceptin in adjuvant treatments has proven effective in a portion of the Her-2/neu overexpressing patients [111], many patients fail to respond and several complications occur. We have shown that EGCG inhibited the autophosphorylation induced by Her-2/neu receptor overexpression, as well as the downstream signaling events mediated via activation of the PI3K/Akt to NF-κB pathway [109]. EGCG also inhibited the malignant growth of tumor cells in soft agar. During our study, we have established EGCG resistant cells from the original Her-2/neu overexpressing population. Interestingly, these resistant cells lose the constitutive tyrosine phosphorylation on the Her-2/neu receptor, without any obvious change in the level of protein expression (unpublished data). Similar inhibition on the Her-2/neu receptor by tea polyphenols has been reported in head and neck and ovarian cancers [112, 113].

NF-κB Transcription Factors. NF-κB is a family of heterodimeric transcription factors, consisting of subunits p65/RelA, RelB, c-Rel, p50, and p52. NF-κB is ubiquitously expressed, but it is sequestered in the cytoplasm by a family of inhibitor proteins, IκBs, in normal resting non-B cells. The canonical pathway to activate NF-κB involves the phosphorylation of IκB by upstream IκB kinases (IKK), and their subsequent proteasome-mediated degradation. The freed NF-κB dimers then translocate into the nucleus to regulate target gene expression. NF-κB-regulated genes play important roles in proliferation, survival, angiogenesis, malignant transformation, and inflammation.

As discussed earlier, we demonstrated that EGCG blocked the activation of NF-κB mediated by Her-2/neu signaling in breast cancer cells. Other mechanisms of EGCG-mediated inhibition of NF-κB activation have also been proposed. In murine peritoneal macrophages, EGCG, EGC, and theaflavin-3,3′-digallate were found to inhibit lipopolysaccharide-induced NF-κB activity, by inhibiting IKK activity [114, 115]. EGCG inhibited activation of NF-κB by IL-1β or TNF via preventing IKK activation and reducing phosphorylation of p65 [116, 117]. EGCG also blocks the activation of TNF itself, preventing paracrine activation of NF-κB by this factor [118, 119]. In addition, EGCG led to degradation of NF-κB p65 subunits in human epidermoid carcinoma A431 cells via activation of a caspase cascade [120]. Since green tea polyphenols have been shown as inhibitors of the proteasome, NF-κB inhibition has also been proposed to be the result of prolonged IκBα turnover [121]. No matter which is the predominant mechanism of inhibition in each experimental system, NF-κB appears to be an important mediator of EGCG inhibitory action. We have observed that EGCG-resistant NF639 breast cancer cells display an increase in NF-κB activity, and ectopic expression of IκBα or cotreatment with the NF-κB inhibitor dexamethasone restores sensitivity of the cells to EGCG (unpublished data).

AP-1 Transcription Factors. AP-1 is another family of dimeric transcription factors that promotes increased cell proliferation through the control of important cell cycle regulators such as cyclin D1 and p21 [122]. EGCG and theaflavins inhibited AP-1-dependent transcriptional activity and DNA binding activity in JB6 mouse epidermal cells [123]. Similar inhibition of AP-1 was observed with most of the tea polyphenols [124]. The inhibition of AP-1 in these cases appeared to be mediated through upstream kinases c-jun NH2-terminal kinase or extracellular signal-regulated protein kinase. In addition to the basal activity, UVB induced AP-1 gene expression and activity could be blocked by tea polyphenols, with theaflavins being somewhat more effective [125, 126]. In contrast, in normal human keratinocytes, EGCG markedly increased AP-1 associated responses [127].

The inhibition of NF-κB and AP-1 should have profound effects on multiple aspects of tumor biology, as the genes encoding many key molecules in tumor invasion, metastasis, and angiogenesis are direct targets of these transcription factors.

Forkhead Transcription Factor FOXO3a. There are three forkhead box O (FOXO) transcription factors—which are mammalian orthologues of *C. elegans daf-16*—FOXO1a, FOXO3a, and FOXO4 (previously known as FKHR, FKHR-L1, and AFX, respectively). FOXO proteins are phosphorylated and inactivated by Akt/protein kinase B [128]. Specifically, upon phosphorylation by Akt, FOXO3a is bound by 14-3-3 and exported from the nucleus. Activation of FOXO3a protein upon inhibition of Akt proceeds via dephosphorylation and nuclear translocation. FOXO proteins induce transcription of genes that inhibit proliferation (e.g., the CKI p27 and GADD45) and induce apoptosis (i.e., Bim). For example, FOXO4 controlled cell cycle progression by activating *p27* gene expression [99]. FOXO3a induced *p27* transcription, which leads to apoptosis of Ba/F3 pro-B and WEHI 231 immature B cells [97, 98].

Since we had previously shown that EGCG inhibited Her-2/neu receptor autophosphorylation and downstream PI3K/Akt activation [109], we examined the effects of EGCG on FOXO3a expression. EGCG treatment of breast cancer cell lines led to a substantial increase in functional levels of FOXO3a [100]. Two functional FOXO3a elements were identified upstream of the ERα B promoter and we demonstrated FOXO3a induces ERα expression. Furthermore, EGCG induced ERα expression in breast cancer cells [100]. Importantly, about 60% of breast cancers express ERα; these patients have a better prognosis than those with ERα-negative cancers. The ERα-expressing tumors respond readily to antiestrogen therapies, while the ERα-negative tumors do not. The identification of Her-2/neu/Akt/FOXO3a/ERα axis suggests the possibility of combinatorial treatments for ERα-negative/low, antiestrogen-resistant tumors. It has been reported that the combination of tamoxifen and Herceptin (antibody targeting the Her-2/neu receptor) synergistically inhibited tumor cell growth and enhanced cell cycle arrest [129]. EGCG might function in a similar manner, namely, inhibiting Her-2/neu, while increasing the ERα content. Of note, it was reported that ERα-negative human breast cancer cells are more sensitive to EGCG, and the combination of EGCG and 4-hydroxy tamoxifen are synergistically toxic to ERα-negative MDA-MB-231 cells [130]. By enhancing the expression of ERα through the inhibition of the Her-2/neu/Akt pathway, EGCG enhances the effect of antiestrogens.

Invasion and Metastasis: The Extracellular Matrix

Tumor cell invasion and metastasis requires extensive extracellular matrix remodeling. As a result, invasive and metastatic tumors usually overexpress matrix-degrading proteinases [131]. Matrix metalloproteinases (MMPs) are a family of endopeptidases requiring Zn^{2+} for their enzymatic activity and are enzymatically activated by cleavage of the propeptide. Green tea polyphenols have been found to be potent inhibitors of many matrix proteinases. Theaflavin, theaflavin digallate, and EGCG inhibited type IV collagenases [132]. In animal models, oral feeding of GTP inhibited the expression levels of MMP-2, MMP-9, VEGF, and urokinase type plasminogen activator (uPA). Inhibition appears to be mediated via multiple mechanisms. EGCG was an inhibitor of MMP-2 and MMP-9 in in vitro zymogen assays [133]. Tissue type plasminogen activator (tPA) and uPA are the principal enzymes releasing active plasmin that activates MMP-1 and MMP-3, which in turn cleaves pro-MMP-9. Molecular modeling studies showed EGCG binds to uPA, blocking His 57 and Ser 195 of the uPA catalytic triad and extending toward Arg 35 from a positively charged loop of uPA, which leads to enzymatic inhibition [134].

Inhibition of MMPs by GTPs can also occur secondary to the inhibition of NF-κB and AP-1 transcription factors. Tumor cells produce large amounts of reactive oxygen species (ROS), which leads to the induction of MMPs [135, 136]. NF-κB and AP-1 are both redox-sensitive transcription factors [122]. The promoters of many MMP genes contain NF-κB or AP-1 elements [137–139]. Green tea polyphenols, especially EGCG, has strong antioxidative capacity [140]. The antioxidant function of GTPs has been proposed as one of the major mechanisms of the chemopreventive effects.

DNA breaks, mutations, lipid peroxidation, and changes in gene expression are other effects of ROS damage. Pretreatment of cultured human lung cells with GTP prevented cigarette smoke solution-induced DNA breakage [141]. Green tea consumption decreased oxidative DNA damage, lipid peroxidation, and free radical in blood and urine in smokers [142]. GTPs also cause a significant increase in plasma antioxidant status, which, in turn, lowered oxidative damage to DNA [143]. Thus, through antioxidative properties, green tea polyphenols have an additional mechanism of inhibiting the induction of genes that promote the invasive and metastatic features associated with advanced malignancies.

Angiogenesis

Angiogenesis is regulated by a balance of a battery of proangiogenic vs antiangiogenic factors. Basic fibroblast growth factor (bFGF), interleukin-8 (IL-8), VEGF, TNFα, and prostaglandins are all proangiogenic, while thrombospondin is antiangiogenic [144].

Given the proper cues, endothelial cells can form tubes, similar to the capillary vasculature surrounding tumors, and are commonly used as an in vitro model for angiogenesis. EGCG caused a dose-dependent inhibition of in vitro capillary endothelial cell growth and chorioallantoic membrane angiogenesis that had been stimulated by the FGF-2 (also called bFGF). Tea consumed in drinking water significantly prevented VEGF-induced corneal neovascularization in mice [145]. EGCG inhibited VEGF-induced tube formation, in a dose-dependent fashion [146]. H_2O_2-induced IL-8 production and tube formation were also prevented by green tea catechins [147]. In a transplanted tumor model, EGCG inhibited inoculated human colon tumor cell growth in nude mice by inhibiting VEGF induction [148]. Green tea extract inhibited human umbilical cord endothelial cell (HUVEC) angiogenesis through inhibition of VEGF and VEGF receptors [149], while EGCG decreased the mRNA and protein levels of bFGF [150]. These findings are not surprising given that many of the proangiogenic factors are direct targets of NF-κB and AP-1 (e.g., VEGF and IL-8). Thus, EGCG reduced VEGF mRNA levels through inhibition of AP-1 [151] and NF-κB transcription factor activity [152].

Other mechanisms of antiangiogenesis activity of green tea polyphenols have also been reported. For example, tea polyphenols (including EGCG) inhibited VEGF-dependent tyrosine phosphorylation of VEGFR-2 [153]. EGCG also inhibited VEGF binding to its receptors [154]. In a phase I/II trial, green tea ingestion rapidly decreased prostaglandin E_2 levels in rectal mucosa in humans [155]. Taken together, the concept of angiogenesis prevention has been proposed and tea polyphenols may play an important mediator in this process [156].

Curbing the Proliferation Potential: Inhibition of the Telomerase

Telomeres, the protein–DNA complexes that cap the chromosomal ends, are replicated by telomerase. Telomerase has reverse transcriptase activity and is responsible for maintaining the integrity of chromosome ends due to the inability of DNA polymerase to initiate DNA synthesis in the 3′ to 5′ direction. Telomeres also prevent chromosomal ends from being recognized as DNA double strand breaks by the cells. In most normal somatic cells, telomerase activity is extremely low and the telomeric DNA sequence shortens over cell cycle divisions due to incomplete replication by DNA polymerase. Shortened telomeres are ultimately detected by cell cycle checkpoint pathways, resulting in irreversible cell cycle arrest (senescence) or apoptosis.

This mechanism is crucial for maintaining limited proliferative capacity in normal cells [157]. However, more than 85% of cancers and cancer cells express a high level of telomerase activity. The telomerase activity endows the tumor cells with unlimited proliferative potential without causing overt chromosome loss and abnormality. Ectopic expression of the telomerase enzyme together with other oncogenes transforms primary cells into tumor cells [158].

EGCG directly inhibited telomerase in a cell-free system as well as in living cells. The growth of U937 monoblastoid leukemia cells and HT29 colon adenocarcinoma cells was inhibited along with telomere shortening, chromosomal abnormalities, and expression of the senescence-associated beta-galactosidase [159]. In a nude mouse model when both telomerase-dependent and telomerase-independent xenograft tumors were treated with EGCG, only the telomerase-dependent tumors responded to prolonged oral administration of EGCG [160]. The inhibition also happened at the mRNA level [161], which may have resulted from the effects of EGCG on the NF-κB to c-*myc* pathway [162, 163], as telomerase is a target of these factors [164–167].

Microarray Analyses Identifies Additional Target Genes

Additional targets for green tea polyphenols may exist besides the previously discussed ones, such as the DNA methyltransferase [168] and topoisomerase I [169]. To discover additional targets of green tea polyphenols, four recent studies have taken advantage of microarray analysis. (1) Human lung cancer cell line PC-9 cells were treated with 200 μM of EGCG for 7 hours and gene expression was profiled using Atlas Human Cancer cDNA Expression Array containing 588 genes [170]. (2) Using a Micromax Direct System, human prostate cancer cells LNCaP were treated with 12 μM EGCG for 12 hours and analyzed [171, 172]. (3) Human papillomavirus-16 (HPV-16) associated cervical cancer cell line CaSki was treated with 35 μM of EGCG for 12, 24, and 48 hours, and gene expression was profiled using a 384 cDNA chip from Macrogen, Seoul, Korea [173]. (4) HUVEC was exposed to green tea extracts for 6 and 48 hours, and gene expression was profiled using an Affymetrix chip containing 12,625 genes [174].

These studies reported genes up- or down regulated by more than twofold. These included gene categories involved in proliferation, cell cycle control, and apoptosis, confirming the findings before with conventional molecular and cellular biology studies. There were novel genes identified as well, for example, RhoB [170] and cyclin A2 [174]. However, the differences between the cell types, transformation status, and duration and dose of treatment and the chips used for these experiments make a direct comparison almost impossible. Our group has been profiling gene expression changes induced by EGCG in two types of breast cancer cells: the oncogene Her-2/neu-transformed and the carcinogen DMBA-transformed breast cancer cells. Additionally, we are comparing established derivative breast cancer cells that are more resistant to EGCG with the parental sensitive populations. These studies should contribute to the understanding of the molecular action of EGCG.

8.5 CLINICAL STUDIES AND THE PROMISE OF TEA IN COMBINATORIAL THERAPY

Clinical Trials

A phase I clinical trial has studied the maximum tolerated dose, toxicity, and pharmacology of oral green tea extract and demonstrated that a dose equivalent to 7 to 8 Japanese cups (120 mL) of green tea three times daily was safe for at least 6 months [175]. In a recent phase I clinical trial using green tea extract in advanced lung cancer patients, green tea extract has been shown to have limited activity as a cytotoxic agent in established malignancies [176]. In another phase II clinical trial, green tea was found to have limited activity among patients with androgen-independent prostate cancers [171, 177]. These studies suggest that tea may not be useful against advanced malignancies. In contrast, green tea has been shown to be effective in fighting some viral-related diseases that are precursors of cancers. Green tea extracts in a form of ointment or a 200 mg capsule taken for 8-12 weeks was effective in treatment of HPV-infected cervical lesions [178]. Daily intake of 9 (27.3 mg EGCG) capsules of green tea (equivalent of 10 cups of tea) for 5 months significantly diminished the HTLV-1 provirus load in the peripheral blood lymphocytes of HTLV-1 carriers [179]. Thus, green tea might be useful in preventing these virus-related malignancies. Furthermore, the tea polyphenols may function in chemoprevention or potentially at the early stages of tumorigenesis in adjuvant therapy (see below). More clinical studies with early stage malignancies are needed to address this issue.

Green Tea in Adjuvant Therapy

Green tea extracts (GTEs) have been used effectively in animal and cell culture models in combination with antimetabolites in cancer chemotherapy. GTEs inhibited thymidine and uridine transport in mouse leukemia cells, blocked the rescue effect of exogenous nucleosides, and enhanced the cytotoxicity of AraC and MTX [180]. Combination of EGCG and dacarbazine was more effective than either EGCG or dacarbazine alone in reducing the number of pulmonary metastases and primary tumor growths, and increased the survival rate of melanoma-bearing mice (181). EGCG induced tumor cell apoptosis synergistically with sulindac and tamoxifen in cultured human lung cancer cell line PC-9 and in the intestinal tumors in multiple intestinal neoplasia (Min) mice [182, 183].

Tea polyphenols may increase the effective concentrations of other chemotherapeutic agents. The activity of P-glycoprotein (P-gp), which is involved in the multidrug resistance phenotype of cancer cells, is inhibited by EGCG [184]. In M5076 tumor-bearing mice, the injection of adriamycin alone did not inhibit tumor growth, whereas the combination of theanine and adriamycin significantly reduced the tumor weight. When combined with theanine, effective antitumor activity of adriamycin was observed without an increase in the dosage. In vitro experiments proved that theanine inhibited the efflux of adriamycin from tumor cells. Furthermore, the oral administration of theanine or green tea similarly enhanced the antitumor activity of adriamycin [185, 186]. Powdered green tea significantly enhanced mitomycin C uptake in Ehrlich ascites carcinoma cells in a dose-dependent fashion [187]. The increase in effectiveness may also lead to dose reduction, eliminating some side effects associated with high doses of pharmacological interventions. For example, the combination of EGCG and curcumin allowed for a dose reduction of 4.4–8.5-fold for EGCG and 2.2–2.8-fold for curcumin [188]. These findings suggest that additional trials are needed to test the efficacy of the use of EGCG or other tea polyphenols in enhancing the effectiveness of therapeutic regimens currently in use in the clinic.

8.6 FUTURE DIRECTIONS AND CONCLUSION

Epidemiology, animal models, and cell and molecular biology studies have all established a role for green tea polyphenols, especially EGCG, in cancer chemoprevention. Green tea prevents multiple organ site malignancies that either appear spontaneously or are induced via genetic manipulations or carcinogen treatment. Green tea was shown to prevent transformation by reducing mitogenic signals, inducing cell cycle arrest and apoptosis, preventing angiogenesis and metastasis, and limiting the proliferation potential. Mechanistic studies in cell culture systems suggest that different signaling pathways show variable susceptibility to EGCG or green tea. The ability of EGCG and other tea polyphenols to inhibit multiple tyrosine kinase receptors makes it a good template to derive small molecule tyrosine kinase inhibitor drugs. Modifications in structure may improve the pharmacokinetics and effectiveness. Although green tea was not effective in treating established advanced malignancies in humans, it holds promise for early stage disease, especially when combined with other chemotherapeutic agents. New target identification with gene expression profiling may further help design new effective drug combinatorial treatments for cancer. Lastly, it will also be of interest to test specific transgene-driven animal tumor models for susceptibility to a green tea regimen. Studies from these animals could help elucidate a genetic basis of green tea responsiveness in human populations at high risk, and identify those who would enjoy particular benefit from drinking green tea to reduce their risk of developing cancer.

ACKNOWLEDGMENTS

We thank Dr. Adrianne Rogers for helpful comments on this chapter. This work was supported by grants from the NIH PO1 ES11624 and RO1 CA82742.

REFERENCES

1. H. Mukhtar and N. Ahmad (1999). Green tea in chemoprevention of cancer, *Toxicol Sci.* **52**:111–117.

2. L. Arab and D. Il'yasova (2003). The epidemiology of tea consumption and colorectal cancer incidence, *J. Nutr.* **133**:3310S–3318S.
3. W. J. Blot, J. K. McLaughlin, and W. H. Chow (1997). Cancer rates among drinkers of black tea, *Crit. Rev. Food Sci. Nutr.* **37**:739–760.
4. J. L. Bushman (1998). Green tea and cancer in humans: a review of the literature, *Nutr. Cancer* **31**:151–159.
5. C. S. Yang and J. M. Landau (2000). Effects of tea consumption on nutrition and health, *J. Nutr.* **130**:2409–2412.
6. K. Nakachi, K. Suemasu, K. Suga, T. Takeo, K. Imai, and Y. Higashi (1998). Influence of drinking green tea on breast cancer malignancy among Japanese patients, *Jpn. J. Cancer Res.* **89**:254–261.
7. M. Inoue, K. Tajima, M. Mizutani, H. Iwata, T. Iwase, et al. (2001). Regular consumption of green tea and the risk of breast cancer recurrence: Follow-up study from the Hospital-based Epidemiologic Research Program at Aichi Cancer Center (HERPACC), Japan, *Cancer Lett.* **167**:175–182.
8. A. H. Wu, M. C. Yu, C. C. Tseng, J. Hankin, and M. C. Pike (2003). Green tea and risk of breast cancer in Asian Americans, *Int. J. Cancer* **106**:574–579.
9. C. Nagata, M. Kabuto, and H. Shimizu (1998). Association of coffee, green tea, and caffeine intakes with serum concentrations of estradiol and sex hormone-binding globulin in premenopausal Japanese women, *Nutr. Cancer* **30**:21–24.
10. Y. T. Gao, J. K. McLaughlin, W. J. Blot, B. T. Ji, Q. Dai, and J. F. Fraumeni, Jr. (1994). Reduced risk of esophageal cancer associated with green tea consumption, *J. Natl. Cancer Inst.* **86**:855–858.
11. M. Inoue, K. Tajima, K. Hirose, N. Hamajima, T. Takezaki, T. Kuroishi, and S. Tominaga (1998). Tea and coffee consumption and the risk of digestive tract cancers: Data from a comparative case-referent study in Japan, *Cancer Causes Control* **9**:209–216.
12. B. T. Ji, W. H. Chow, A. W. Hsing, J. K. McLaughlin, Q. Dai, Y. T. Gao, W. J. Blot, and J. F. Fraumeni, Jr. (1997). Green tea consumption and the risk of pancreatic and colorectal cancers, *Int. J. Cancer* **70**:255–258.
13. B. T. Ji, W. H. Chow, G. Yang, J. K. McLaughlin, R. N. Gao, et al. (1996). The influence of cigarette smoking, alcohol, and green tea consumption on the risk of carcinoma of the cardia and distal stomach in Shanghai, China, *Cancer* **77**:2449–2457.
14. V. W. Setiawan, Z. F. Zhang, G. P. Yu, Q. Y. Lu, Y. L. Li, et al. (2001). Protective effect of green tea on the risks of chronic gastritis and stomach cancer, *Int. J. Cancer* **92**:600–604.
15. G. P. Yu, C. C. Hsieh, L. Y. Wang, S. Z. Yu, X. L. Li, and T. H. Jin (1995). Green-tea consumption and risk of stomach cancer: A population-based case–control study in Shanghai, China, *Cancer Causes Control* **6**:532–538.
16. K. Imai, K. Suga, and K. Nakachi (1997). Cancer-preventive effects of drinking green tea among a Japanese population, *Prev. Med.* **26**:769–775.
17. K. Nakachi, S. Matsuyama, S. Miyake, M. Suganuma, and K. Imai (2000). Preventive effects of drinking green tea on cancer and cardiovascular disease: Epidemiological evidence for multiple targeting prevention, *Biofactors* **13**:49–54.
18. S. Kono, M. Ikeda, S. Tokudome, and M. Kuratsune (1988). A case–control study of gastric cancer and diet in northern Kyushu, Japan, *Jpn. J. Cancer Res.* **79**:1067–1074.

19. S. Sasazuki, M. Inoue, T. Hanaoka, S. Yamamoto, T. Sobue, and S. Tsugane (2004). Green tea consumption and subsequent risk of gastric cancer by subsite: The JPHC Study, *Cancer Causes Control* **15**:483–491.
20. K. Shibata, M. Moriyama, T. Fukushima, A. Kaetsu, M. Miyazaki, and H. Une (2000). Green tea consumption and chronic atrophic gastritis: A cross-sectional study in a green tea production village, *J. Epidemiol.* **10**:310–316.
21. C. L. Sun, J. M. Yuan, M. J. Lee, C. S. Yang, Y. T. Gao, R. K. Ross, and M. C. Yu (2002). Urinary tea polyphenols in relation to gastric and esophageal cancers: A prospective study of men in Shanghai, China, *Carcinogenesis* **23**:1497–1503.
22. M. Zhang, C. W. Binns, and A. H. Lee (2002). Tea consumption and ovarian cancer risk: A case–control study in China, *Cancer Epidemiol. Biomarkers Prev.* **11**:713–718.
23. C. W. Binns, M. Zhang, A. H. Lee, and C. X. Xie (2004). Green tea consumption enhances survival of epithelial ovarian cancer patients, *Asia Pac. J. Clin. Nutr.* **13**:116.
24. L. K. Heilbrun, A. Nomura, and G. N. Stemmermann (1986). Black tea consumption and cancer risk: A prospective study, *Br. J. Cancer* **54**:677–683.
25. L. J. Kinlen, A. N. Willows, P. Goldblatt, and J. Yudkin (1988). Tea consumption and cancer, *Br. J. Cancer* **58**:397–401.
26. I. A. Hakim, R. B. Harris, S. Brown, H. H. Chow, S. Wiseman, S. Agarwal, and W. Talbot (2003). Effect of increased tea consumption on oxidative DNA damage among smokers: A randomized controlled study, *J. Nutr.* **133**:3303S–3309S.
27. L. Zhong, M. S. Goldberg, Y. T. Gao, J. A. Hanley, M. E. Parent, and F. Jin (2001). A population-based case–control study of lung cancer and green tea consumption among women living in Shanghai, China, *Epidemiology* **12**:695–700.
28. Y. Suzuki, Y. Tsubono, N. Nakaya, Y. Koizumi, and I. Tsuji (2004). Green tea and the risk of breast cancer: Pooled analysis of two prospective studies in Japan, *Br. J. Cancer* **90**:1361–1363.
29. W. J. Blot, W. H. Chow, and J. K. McLaughlin (1996). Tea and cancer: A review of the epidemiological evidence, *Eur. J. Cancer Prev.* **5**:425–438.
30. L. Kohlmeier, K. G. Weterings, S. Steck, and F. J. Kok (1997). Tea and cancer prevention: An evaluation of the epidemiologic literature, *Nutr. Cancer* **27**:1–13.
31. S. M. Henning, C. Fajardo-Lira, H. W. Lee, A. A. Youssefian, V. L. Go, and D. Heber (2003). Catechin content of 18 teas and a green tea extract supplement correlates with the antioxidant capacity, *Nutr. Cancer* **45**:226–235.
32. Y. Lu, W. F. Guo, and X. Q. Yang (2004). Fluoride content in tea and its relationship with tea quality, *J. Agric. Food Chem.* **52**:4472–4476.
33. I. C. Arts, D. R. Jacobs, Jr., M. Gross, L. J. Harnack, and A. R. Folsom (2002). Dietary catechins and cancer incidence among postmenopausal women: The Iowa Women's Health Study (United States), *Cancer Causes Control* **13**:373–382.
34. K. Wakai, K. Hirose, T. Takezaki, N. Hamajima, Y. Ogura, S. Nakamura, N. Hayashi, and K. Tajima (2004). Foods and beverages in relation to urothelial cancer: Case–control study in Japan, *Int. J. Urol.* **11**:11–19.
35. K. Wakai, Y. Ohno, K. Obata, and K. Aoki (1993). Prognostic significance of selected lifestyle factors in urinary bladder cancer, *Jpn. J. Cancer Res.* **84**:1223–1229.
36. H. Lu, X. Meng, and C. S. Yang (2003). Enzymology of methylation of tea catechins and inhibition of catechol-*O*-methyltransferase by (–)-epigallocatechin gallate, *Drug Metab. Dispos.* **31**:572–579.

37. A. H. Wu, C. C. Tseng, D. Van Den Berg, and M. C. Yu (2003). Tea intake, COMT genotype, and breast cancer in Asian-American women, *Cancer Res.* **63**:7526–7529.
38. I. A. Hakim, R. B. Harris, H. H. Chow, M. Dean, S. Brown, and I. U. Ali (2004). Effect of a 4-month tea intervention on oxidative DNA damage among heavy smokers: Role of glutathione *S*-transferase genotypes, *Cancer Epidemiol. Biomarkers Prev.* **13**:242–249.
39. S. G. Khan, S. K. Katiyar, R. Agarwal, and H. Mukhtar (1992). Enhancement of antioxidant and phase II enzymes by oral feeding of green tea polyphenols in drinking water to SKH-1 hairless mice: Possible role in cancer chemoprevention, *Cancer Res.* **52**:4050–4052.
40. T. Umemura, S. Kai, R. Hasegawa, K. Kanki, Y. Kitamura, A. Nishikawa, and M. Hirose (2003). Prevention of dual promoting effects of pentachlorophenol, an environmental pollutant, on diethylnitrosamine-induced hepato- and cholangiocarcinogenesis in mice by green tea infusion, *Carcinogenesis* **24**:1105–1109.
41. D. Chen, K. G. Daniel, D. J. Kuhn, A. Kazi, M. Bhuiyan, et al. (2004). Green tea and tea polyphenols in cancer prevention, *Front. Biosci.* **9**:2618–2631.
42. F. L. Chung, J. Schwartz, C. R. Herzog, and Y. M. Yang (2003). Tea and cancer prevention: studies in animals and humans, *J. Nutr.* **133**:3268S–3274S.
43. W. A. Khan, Z. Y. Wang, M. Athar, D. R. Bickers, and H. Mukhtar (1988). Inhibition of the skin tumorigenicity of (+/–)-7 beta,8 alpha-dihydroxy-9 alpha,10 alpha-epoxy-7,8,9,10-tetrahydrobenzo[*a*]pyrene by tannic acid, green tea polyphenols and quercetin in Sencar mice, *Cancer Lett.* **42**:7–12.
44. Z. Y. Wang, W. A. Khan, D. R. Bickers, and H. Mukhtar (1989). Protection against polycyclic aromatic hydrocarbon-induced skin tumor initiation in mice by green tea polyphenols, *Carcinogenesis* **10**:411–415.
45. Z. Y. Wang, R. Agarwal, D. R. Bickers, and H. Mukhtar (1991). Protection against ultraviolet B radiation-induced photocarcinogenesis in hairless mice by green tea polyphenols, *Carcinogenesis* **12**:1527–1530.
46. Z. Y. Wang, M. T. Huang, T. Ferraro, C. Q. Wong, Y. R. Lou, et al. (1992). Inhibitory effect of green tea in the drinking water on tumorigenesis by ultraviolet light and 12-*O*-tetradecanoylphorbol-13-acetate in the skin of SKH-1 mice, *Cancer Res.* **52**:1162–1170.
47. H. L. Gensler, B. N. Timmermann, S. Valcic, G. A. Wachter, R. Dorr, K. Dvorakova, and D. S. Alberts (1996). Prevention of photocarcinogenesis by topical administration of pure epigallocatechin gallate isolated from green tea, *Nutr. Cancer* **26**:325–335.
48. R. Agarwal, S. K. Katiyar, S. I. Zaidi, and H. Mukhtar (1992). Inhibition of skin tumor promoter-caused induction of epidermal ornithine decarboxylase in SENCAR mice by polyphenolic fraction isolated from green tea and its individual epicatechin derivatives, *Cancer Res.* **52**:3582–3588.
49. M. T. Huang, C. T. Ho, Z. Y. Wang, T. Ferraro, T. Finnegan-Olive, et al. (1992). Inhibitory effect of topical application of a green tea polyphenol fraction on tumor initiation and promotion in mouse skin, *Carcinogenesis* **13**:947–954.
50. Z. Y. Wang, M. T. Huang, C. T. Ho, R. Chang, W. Ma, et al. (1992). Inhibitory effect of green tea on the growth of established skin papillomas in mice, *Cancer Res.* **52**:6657–6665.
51. Y. Fujita, T. Yamane, M. Tanaka, K. Kuwata, J. Okuzumi, T. Takahashi, H. Fujiki, and T. Okuda (1989). Inhibitory effect of (–)-epigallocatechin gallate on carcinogenesis

with *N*-ethyl-*N*′-nitro-*N*-nitrosoguanidine in mouse duodenum, *Jpn. J. Cancer Res.* **80**:503–505.

52. A. K. Jain, K. Shimoi, Y. Nakamura, T. Kada, Y. Hara, and I. Tomita (1989). Crude tea extracts decrease the mutagenic activity of *N*-methyl-*N*′-nitro-*N*-nitrosoguanidine in vitro and in intragastric tract of rats, *Mutat. Res.* **210**:1–8.
53. T. Narisawa and Y. Fukaura (1993). A very low dose of green tea polyphenols in drinking water prevents *N*-methyl-*N*-nitrosourea-induced colon carcinogenesis in F344 rats, *Jpn. J. Cancer Res.* **84**:1007–1009.
54. Z. Y. Wang, J. Y. Hong, M. T. Huang, K. R. Reuhl, A. H. Conney, and C. S. Yang (1992). Inhibition of *N*-nitrosodiethylamine- and 4-(methylnitrosamino)-1-(3-pyridyl)-1-butanone-induced tumorigenesis in A/J mice by green tea and black tea, *Cancer Res.* **52**:1943–1947.
55. P. Yin, J. Zhao, S. Cheng, Q. Zhu, Z. Liu, and L. Zhengguo (1994). Experimental studies of the inhibitory effects of green tea catechin on mice large intestinal cancers induced by 1,2-dimethylhydrazine, *Cancer Lett.* **79**:33–38.
56. S. K. Katiyar, R. Agarwal, and H. Mukhtar (1993). Protective effects of green tea polyphenols administered by oral intubation against chemical carcinogen-induced forestomach and pulmonary neoplasia in A/J mice, *Cancer Lett.* **73**:167–172.
57. J. Ju, Y. Liu, J. Hong, M. T. Huang, A. H. Conney, and C. S. Yang (2003). Effects of green tea and high-fat diet on arachidonic acid metabolism and aberrant crypt foci formation in an azoxymethane-induced colon carcinogenesis mouse model, *Nutr. Cancer* **46**:172–178.
58. J. Chen (1992). The effects of Chinese tea on the occurrence of esophageal tumors induced by *N*-nitrosomethylbenzylamine in rats, *Prev. Med.* **21**:385–391.
59. Z. Y. Wang, L. D. Wang, M. J. Lee, C. T. Ho, M. T. Huang, A. H. Conney, and C. S. Yang (1995). Inhibition of *N*-nitrosomethylbenzylamine-induced esophageal tumorigenesis in rats by green and black tea, *Carcinogenesis* **16**:2143–2148.
60. M. Hirose, T. Hoshiya, K. Akagi, S. Takahashi, Y. Hara, and N. Ito (1993). Effects of green tea catechins in a rat multi-organ carcinogenesis model, *Carcinogenesis* **14**:1549–1553.
61. T. Yamane, N. Hagiwara, M. Tateishi, S. Akachi, M. Kim, et al. (1991). Inhibition of azoxymethane-induced colon carcinogenesis in rat by green tea polyphenol fraction, *Jpn. J. Cancer Res.* **82**:1336–1339.
62. T. Yamane, T. Takahashi, K. Kuwata, K. Oya, M. Inagake, Y. Kitao, M. Suganuma, and H. Fujiki (1995). Inhibition of *N*-methyl-*N*′-nitro-*N*-nitrosoguanidine-induced carcinogenesis by (−)-epigallocatechin gallate in the rat glandular stomach, *Cancer Res.* **55**:2081–2084.
63. T. Yamane, H. Nakatani, N. Kikuoka, H. Matsumoto, Y. Iwata, Y. Kitao, K. Oya, and T. Takahashi (1996). Inhibitory effects and toxicity of green tea polyphenols for gastrointestinal carcinogenesis, *Cancer* **77**:1662–1667.
64. S. Liao, Y. Umekita, J. Guo, J. M. Kokontis, and R. A. Hiipakka (1995). Growth inhibition and regression of human prostate and breast tumors in athymic mice by tea epigallocatechin gallate, *Cancer Lett.* **96**:239–243.
65. S. Gupta, N. Ahmad, R. R. Mohan, M. M. Husain, and H. Mukhtar (1999). Prostate cancer chemoprevention by green tea: in vitro and in vivo inhibition of testosterone-mediated induction of ornithine decarboxylase, *Cancer Res.* **59**:2115–2120.

66. B. A. Foster, P. J. Kaplan, and N. M. Greenberg (1998). Peptide growth factors and prostate cancer: New models, new opportunities, *Cancer Metastasis Rev.* **17**:317–324.

67. S. Gupta, K. Hastak, N. Ahmad, J. S. Lewin, and H. Mukhtar (2001). Inhibition of prostate carcinogenesis in TRAMP mice by oral infusion of green tea polyphenols, *Proc. Natl. Acad. Sci. U.S.A.* **98**:10350–10355.

68. Y. Xu, C. T. Ho, S. G. Amin, C. Han, and F. L. Chung (1992). Inhibition of tobacco-specific nitrosamine-induced lung tumorigenesis in A/J mice by green tea and its major polyphenol as antioxidants, *Cancer Res.* **52**:3875–3879.

69. J. Cao, Y. Xu, J. Chen, and J. E. Klaunig (1996). Chemopreventive effects of green and black tea on pulmonary and hepatic carcinogenesis, *Fundam. Appl. Toxicol.* **29**:244–250.

70. H. M. Schuller, B. Porter, A. Riechert, K. Walker, and R. Schmoyer (2004). Neuroendocrine lung carcinogenesis in hamsters is inhibited by green tea or theophylline while the development of adenocarcinomas is promoted: Implications for chemoprevention in smokers, *Lung Cancer* **45**:11–18.

71. A. E. Rogers (1990). Dimethylbenzanthracene-induced mammary tumorigenesis in ethanol-fed rats, *Nutr. Res.* **10**:915–928.

72. M. Hirose, T. Hoshiya, K. Akagi, M. Futakuchi, and N. Ito (1994). Inhibition of mammary gland carcinogenesis by green tea catechins and other naturally occurring antioxidants in female Sprague-Dawley rats pretreated with 7,12-dimethylbenz[alpha]-anthracene, *Cancer Lett.* **83**:149–156.

73. M. Hirose, Y. Mizoguchi, M. Yaono, H. Tanaka, T. Yamaguchi, and T. Shirai (1997). Effects of green tea catechins on the progression or late promotion stage of mammary gland carcinogenesis in female Sprague-Dawley rats pretreated with 7,12-dimethylbenz-[*a*]anthracene, *Cancer Lett.* **112**:141–147.

74. H. Tanaka, M. Hirose, M. Kawabe, M. Sano, Y. Takesada, A. Hagiwara, and T. Shirai (1997). Post-initiation inhibitory effects of green tea catechins on 7,12-dimethylbenz-[*a*]anthracene-induced mammary gland carcinogenesis in female Sprague-Dawley rats, *Cancer Lett.* **116**:47–52.

75. K. T. Kavanagh, L. J. Hafer, D. W. Kim, K. K. Mann, D. H. Sherr, A. E. Rogers, and G. E. Sonenshein (2001). Green tea extracts decrease carcinogen-induced mammary tumor burden in rats and rate of breast cancer cell proliferation in culture, *J. Cell. Biochem.* **82**:387–398.

76. H. Yanaga, T. Fujii, T. Koga, R. Araki, and K. Shirouzu (2002). Prevention of carcinogenesis of mouse mammary epithelial cells RIII/MG by epigallocatechin gallate, *Int. J. Mol. Med.* **10**:311–315.

77. G. Qin, P. Gopalan-Kriczky, J. Su, Y. Ning, and P. D. Lotlikar (1997). Inhibition of aflatoxin B1-induced initiation of hepatocarcinogenesis in the rat by green tea, *Cancer Lett.* **112**:149–154.

78. K. Sai, S. Kai, T. Umemura, A. Tanimura, R. Hasegawa, T. Inoue, and Y. Kurokawa (1998). Protective effects of green tea on hepatotoxicity, oxidative DNA damage and cell proliferation in the rat liver induced by repeated oral administration of 2-nitropropane, *Food Chem. Toxicol.* **36**:1043–1051.

79. F. Bertolini, L. Fusetti, C. Rabascio, S. Cinieri, G. Martinelli, and G. Pruneri (2000). Inhibition of angiogenesis and induction of endothelial and tumor cell apoptosis by green tea in animal models of human high-grade non-Hodgkin's lymphoma, *Leukemia* **14**:1477–1482.

80. M. Zhu, Y. Gong, Z. Yang, G. Ge, C. Han, and J. Chen (1999). Green tea and its major components ameliorate immune dysfunction in mice bearing Lewis lung carcinoma and treated with the carcinogen NNK, *Nutr. Cancer* **35**:64–72.

81. N. Li, C. Han, and J. Chen (1999). Tea preparations protect against DMBA-induced oral carcinogenesis in hamsters, *Nutr. Cancer* **35**:73–79.

82. D. Sato (1999). Inhibition of urinary bladder tumors induced by *N*-butyl-*N*-(4-hydroxybutyl)-nitrosamine in rats by green tea, *Int. J. Urol.* **6**:93–99.

83. D. Sato and M. Matsushima (2003). Preventive effects of urinary bladder tumors induced by *N*-butyl-*N*-(4-hydroxybutyl)-nitrosamine in rat by green tea leaves, *Int. J. Urol.* **10**:160–166.

84. J. K. Kemberling, J. A. Hampton, R. W. Keck, M. A. Gomez, and S. H. Selman (2003). Inhibition of bladder tumor growth by the green tea derivative epigallocatechin-3-gallate, *J. Urol.* **170**:773–776.

85. A. Hiura, M. Tsutsumi, and K. Satake (1997). Inhibitory effect of green tea extract on the process of pancreatic carcinogenesis induced by *N*-nitrosobis-(2-oxypropyl) amine (BOP) and on tumor promotion after transplantation of *N*-nitrosobis-(2-hydroxypropyl)amine (BHP)-induced pancreatic cancer in Syrian hamsters, *Pancreas* **15**:272–277.

86. D. Hanahan and R. A. Weinberg (2000). The hallmarks of cancer, *Cell* **100**:57–70.

87. S. Menard, E. Tagliabue, and M. I. Colnaghi (1998). The 67 kDa laminin receptor as a prognostic factor in human cancer, *Breast Cancer Res. Treat.* **52**:137–145.

88. H. Tachibana, K. Koga, Y. Fujimura, and K. Yamada (2004). A receptor for green tea polyphenol EGCG, *Nat. Struct. Mol. Biol.* **11**:380–381.

89. P. L. Kuo and C. C. Lin (2003). Green tea constituent (–)-epigallocatechin-3-gallate inhibits Hep G2 cell proliferation and induces apoptosis through p53-dependent and Fas-mediated pathways, *J. Biomed. Sci.* **10**:219–227.

90. K. Hastak, S. Gupta, N. Ahmad, M. K. Agarwal, M. L. Agarwal, and H. Mukhtar (2003). Role of p53 and NF-kappaB in epigallocatechin-3-gallate-induced apoptosis of LNCaP cells, *Oncogene* **22**:4851–4859.

91. S. Hsu, J. B. Lewis, J. L. Borke, B. Singh, D. P. Dickinson, et al. (2001). Chemopreventive effects of green tea polyphenols correlate with reversible induction of p57 expression, *Anticancer Res.* **21**:3743–3748.

92. Y. C. Liang, S. Y. Lin-Shiau, C. F. Chen, and J. K. Lin (1999). Inhibition of cyclin-dependent kinases 2 and 4 activities as well as induction of Cdk inhibitors p21 and p27 during growth arrest of human breast carcinoma cells by (–)-epigallocatechin-3-gallate, *J. Cell. Biochem.* **75**:1–12.

93. N. Ahmad, V. M. Adhami, S. Gupta, P. Cheng, and H. Mukhtar (2002). Role of the retinoblastoma (pRb)-E2F/DP pathway in cancer chemopreventive effects of green tea polyphenol epigallocatechin-3-gallate, *Arch. Biochem. Biophys.* **398**:125–131.

94. S. W. Huh, S. M. Bae, Y. W. Kim, J. M. Lee, S. E. Namkoong, et al. (2004). Anticancer effects of (–)-epigallocatechin-3-gallate on ovarian carcinoma cell lines. *Gynecol. Oncol.* **94**:760–768.

95. Y. P. Lu, Y. R. Lou, X. H. Li, J. G. Xie, D. Brash, M. T. Huang, and A. H. Conney (2000). Stimulatory effect of oral administration of green tea or caffeine on ultraviolet light-induced increases in epidermal wild-type p53, p21(WAF1/CIP1), and apoptotic sunburn cells in SKH-1 mice, *Cancer Res.* **60**:4785–4791.

96. Q. Liu, Y. Wang, K. A. Crist, Z. Y. Wang, Y. R. Lou, M. T. Huang, A. H. Conney, and M. You (1998). Effect of green tea on p53 mutation distribution in ultraviolet B radiation-induced mouse skin tumors, *Carcinogenesis* **19**:1257–1262.
97. V. Chandramohan, S. Jeay, S. Pianetti, and G. E. Sonenshein (2004). Reciprocal control of forkhead box O 3a and c-Myc via the phosphatidylinositol 3-kinase pathway coordinately regulates p27Kip1 levels, *J. Immunol.* **172**:5522–5527.
98. P. F. Dijkers, R. H. Medema, C. Pals, L. Banerji, N. S. Thomas, et al. (2000). Forkhead transcription factor FKHR-L1 modulates cytokine-dependent transcriptional regulation of p27(KIP1), *Mol. Cell. Biol.* **20**:9138–9148.
99. R. H. Medema, G. J. Kops, J. L. Bos, and B. M. Burgering (2000). AFX-like Forkhead transcription factors mediate cell-cycle regulation by Ras and PKB through p27kip1, *Nature* **404**:782–787.
100. S. Guo and G. E. Sonenshein (2004). Forkhead box transcription factor FOXO3a regulates estrogen receptor alpha expression and is repressed by the Her-2/neu/phosphatidylinositol 3-kinase/Akt signaling pathway, *Mol. Cell Biol.* **24**:8681–8690.
101. S. Islam, N. Islam, T. Kermode, B. Johnstone, H. Mukhtar, et al. (2000). Involvement of caspase-3 in epigallocatechin-3-gallate-mediated apoptosis of human chondrosarcoma cells, *Biochem. Biophys. Res. Commun.* **270**:793–797.
102. D. O. Kennedy, A. Kojima, Y. Yano, T. Hasuma, S. Otani, and I. Matsui-Yuasa (2001). Growth inhibitory effect of green tea extract in Ehrlich ascites tumor cells involves cytochrome *c* release and caspase activation, *Cancer Lett.* **166**:9–15.
103. A. Kazi, D. M. Smith, Q. Zhong, and Q. P. Dou (2002). Inhibition of bcl-x(l) phosphorylation by tea polyphenols or epigallocatechin-3-gallate is associated with prostate cancer cell apoptosis, *Mol. Pharmacol.* **62**:765–771.
104. M. Leone, D. Zhai, S. Sareth, S. Kitada, J. C. Reed, and M. Pellecchia (2003). Cancer prevention by tea polyphenols is linked to their direct inhibition of antiapoptotic Bcl-2-family proteins, *Cancer Res.* **63**:8118–8121.
105. Y. C. Liang, S. Y. Lin-shiau, C. F. Chen, and J. K. Lin (1997). Suppression of extracellular signals and cell proliferation through EGF receptor binding by (−)-epigallocatechin gallate in human A431 epidermoid carcinoma cells, *J. Cell. Biochem.* **67**:55–65.
106. H. Y. Ahn, K. R. Hadizadeh, C. Seul, Y. P. Yun, H. Vetter, and A. Sachinidis (1999). Epigallocathechin-3 gallate selectively inhibits the PDGF-BB-induced intracellular signaling transduction pathway in vascular smooth muscle cells and inhibits transformation of sis-transfected NIH 3T3 fibroblasts and human glioblastoma cells (A172), *Mol. Biol. Cell* **10**:1093–1104.
107. A. Sachinidis, C. Seul, S. Seewald, H. Ahn, Y. Ko, and H. Vetter (2000). Green tea compounds inhibit tyrosine phosphorylation of PDGF beta-receptor and transformation of A172 human glioblastoma. *FEBS Lett.* **471**:51–55.
108. Y. K. Lee, N. D. Bone, A. K. Strege, T. D. Shanafelt, D. F. Jelinek, and N. E. Kay (2004). VEGF receptor phosphorylation status and apoptosis is modulated by a green tea component, epigallocatechin-3-gallate (EGCG), in B-cell chronic lymphocytic leukemia, *Blood* **104**:788–794.
109. S. Pianetti, S. Guo, K. T. Kavanagh, and G. E. Sonenshein (2002). Green tea polyphenol epigallocatechin-3 gallate inhibits Her-2/neu signaling, proliferation, and transformed phenotype of breast cancer cells, *Cancer Res.* **62**:652–655.

110. Y. C. Liang, Y. C. Chen, Y. L. Lin, S. Y. Lin-Shiau, C. T. Ho, and J. K. Lin (1999). Suppression of extracellular signals and cell proliferation by the black tea polyphenol, theaflavin-3,3′-digallate, *Carcinogenesis* **20**:733–736.

111. J. Baselga (2001). Clinical trials of Herceptin(R) (trastuzumab), *Eur. J. Cancer* **37**(Suppl 1):18–24.

112. M. Masuda, M. Suzui, J. T. Lim, and I. B. Weinstein (2003). Epigallocatechin-3-gallate inhibits activation of HER-2/neu and downstream signaling pathways in human head and neck and breast carcinoma cells, *Clin. Cancer Res.* **9**:3486–3491.

113. T. D. Way, H. H. Lee, M. C. Kao, and J. K. Lin (2004). Black tea polyphenol theaflavins inhibit aromatase activity and attenuate tamoxifen resistance in HER2/neu-transfected human breast cancer cells through tyrosine kinase suppression, *Eur. J. Cancer* **40**:2165–2174.

114. Y. L. Lin and J. K. Lin (1997). (–)-Epigallocatechin-3-gallate blocks the induction of nitric oxide synthase by down-regulating lipopolysaccharide-induced activity of transcription factor nuclear factor-kappaB, *Mol. Pharmacol.* **52**:465–472.

115. M. H. Pan, S. Y. Lin-Shiau, C. T. Ho, J. H. Lin, and J. K. Lin (2000). Suppression of lipopolysaccharide-induced nuclear factor-kappaB activity by theaflavin-3,3′-digallate from black tea and other polyphenols through down-regulation of IkappaB kinase activity in macrophages, *Biochem. Pharmacol.* **59**:357–367.

116. N. Ahmad, S. Gupta, and H. Mukhtar (2000). Green tea polyphenol epigallocatechin-3-gallate differentially modulates nuclear factor kappa B in cancer cells versus normal cells, *Arch. Biochem. Biophys.* **376**:338–346.

117. D. S. Wheeler, J. D. Catravas, K. Odoms, A. Denenberg, V. Malhotra, and H. R. Wong (2004). Epigallocatechin-3-gallate, a green tea-derived polyphenol, inhibits IL-1 beta-dependent proinflammatory signal transduction in cultured respiratory epithelial cells, *J. Nutr.* **134**:1039–1044.

118. H. Fujiki, M. Suganuma, S. Okabe, E. Sueoka, K. Suga, K. Imai, K. Nakachi, and S. Kimura (1999). Mechanistic findings of green tea as cancer preventive for humans, *Proc. Soc. Exp. Biol. Med.* **220**:225–228.

119. F. Yang, W. J. de Villiers, C. J. McClain, and G. W. Varilek (1998). Green tea polyphenols block endotoxin-induced tumor necrosis factor-production and lethality in a murine model, *J. Nutr.* **128**:2334–2340.

120. S. Gupta, K. Hastak, F. Afaq, N. Ahmad, and H. Mukhtar (2004). Essential role of caspases in epigallocatechin-3-gallate-mediated inhibition of nuclear factor kappa B and induction of apoptosis, *Oncogene* **23**:2507–2522.

121. S. Nam, D. M. Smith, and Q. P. Dou (2001). Ester bond-containing tea polyphenols potently inhibit proteasome activity in vitro and in vivo, *J. Biol. Chem.* **276**:13322–13330.

122. J. E. Klaunig and L. M. Kamendulis (2004). The role of oxidative stress in carcinogenesis, *Annu. Rev. Pharmacol. Toxicol.* **44**:239–267.

123. Z. Dong, W. Ma, C. Huang, and C. S. Yang (1997). Inhibition of tumor promoter-induced activator protein 1 activation and cell transformation by tea polyphenols, (–)-epigallocatechin gallate, and theaflavins, *Cancer Res.* **57**:4414–4419.

124. J. Y. Chung, C. Huang, X. Meng, Z. Dong, and C. S. Yang (1999). Inhibition of activator protein 1 activity and cell growth by purified green tea and black tea polyphenols in H-ras-transformed cells: Structure–activity relationship and mechanisms involved, *Cancer Res.* **59**:4610–4617.

125. W. Chen, Z. Dong, S. Valcic, B. N. Timmermann, and G. T. Bowden (1999). Inhibition of ultraviolet B-induced c-fos gene expression and p38 mitogen-activated protein kinase activation by (–)-epigallocatechin gallate in a human keratinocyte cell line, *Mol. Carcinog.* **24**:79–84.

126. M. Nomura, W. Y. Ma, C. Huang, C. S. Yang, G. T. Bowden, K. Miyamoto, and Z. Dong (2000). Inhibition of ultraviolet B-induced AP-1 activation by theaflavins from black tea, *Mol. Carcinog.* **28**:148–155.

127. S. Balasubramanian, T. Efimova, and R. L. Eckert (2002). Green tea polyphenol stimulates a Ras, MEKK1, MEK3, and p38 cascade to increase activator protein 1 factor-dependent involucrin gene expression in normal human keratinocytes, *J. Biol. Chem.* **277**:1828–1836.

128. A. Brunet, A. Bonni, M. J. Zigmond, M. Z. Lin, P. Juo, et al. (1999). Akt promotes cell survival by phosphorylating and inhibiting a forkhead transcription factor, *Cell* **96**:857–868.

129. A. Argiris, C. X. Wang, S. G. Whalen, and M. P. DiGiovanna (2004). Synergistic interactions between tamoxifen and trastuzumab (Herceptin), *Clin. Cancer Res.* **10**:1409–1420.

130. K. Chisholm, B. J. Bray, and R. J. Rosengren (2004). Tamoxifen and epigallocatechin gallate are synergistically cytotoxic to MDA-MB-231 human breast cancer cells, *Anticancer Drugs* **15**:889–897.

131. M. Egeblad and Z. Werb (2002). New functions for the matrix metalloproteinases in cancer progression, *Nat. Rev. Cancer* **2**:161–174.

132. M. Sazuka, H. Imazawa, Y. Shoji, T. Mita, Y. Hara, and M. Isemura (1997). Inhibition of collagenases from mouse lung carcinoma cells by green tea catechins and black tea theaflavins, *Biosci. Biotechnol. Biochem.* **61**:1504–1506.

133. S. Garbisa, L. Sartor, S. Biggin, B. Salvato, R. Benelli, and A. Albini (2001). Tumor gelatinases and invasion inhibited by the green tea flavonol epigallocatechin-3-gallate, *Cancer* **91**:822–832.

134. J. Jankun, S. H. Selman, R. Swiercz, and E. Skrzypczak-Jankun (1997). Why drinking green tea could prevent cancer, *Nature* **387**:561.

135. P. Brenneisen, K. Briviba, M. Wlaschek, J. Wenk, and K. Scharffetter-Kochanek (1997). Hydrogen peroxide (H_2O_2) increases the steady-state mRNA levels of collagenase/MMP-1 in human dermal fibroblasts, *Free Radic. Biol. Med.* **22**:515–524.

136. Y. Nonaka, H. Iwagaki, T. Kimura, S. Fuchimoto, and K. Orita (1993). Effect of reactive oxygen intermediates on the in vitro invasive capacity of tumor cells and liver metastasis in mice, *Int. J. Cancer* **54**:983–986.

137. H. S. Kim, M. H. Kim, M. Jeong, Y. S. Hwang, S. H. Lim, B. A. Shin, B. W. Ahn, and Y. D. Jung (2004). EGCG blocks tumor promoter-induced MMP-9 expression via suppression of MAPK and AP-1 activation in human gastric AGS cells, *Anticancer Res.* **24**:747–753.

138. M. Maeda-Yamamoto, N. Suzuki, Y. Sawai, T. Miyase, M. Sano, A. Hashimoto-Ohta, and M. Isemura (2003). Association of suppression of extracellular signal-regulated kinase phosphorylation by epigallocatechin gallate with the reduction of matrix metalloproteinase activities in human fibrosarcoma HT1080 cells, *J. Agric. Food Chem.* **51**:1858–1863.

139. M. Seiki and I. Yana (2003). Roles of pericellular proteolysis by membrane type-1 matrix metalloproteinase in cancer invasion and angiogenesis, *Cancer Sci.* **94**:569–574.

140. Y. D. Jung and L. M. Ellis (2001). Inhibition of tumour invasion and angiogenesis by epigallocatechin gallate (EGCG), a major component of green tea, *Int. J. Exp. Pathol.* **82**:309–316.

141. P. Leanderson, A. O. Faresjo, and C. Tagesson (1997). Green tea polyphenols inhibit oxidant-induced DNA strand breakage in cultured lung cells. *Free Radic. Biol. Med.* **23**:235–242.

142. J. E. Klaunig, Y. Xu, C. Han, L. M. Kamendulis, J. Chen, C. Heiser, M. S. Gordon, and E. R. Mohler, 3rd. (1999). The effect of tea consumption on oxidative stress in smokers and nonsmokers, *Proc. Soc. Exp. Biol. Med.* **220**:249–254.

143. I. F. Benzie, Y. T. Szeto, J. J. Strain, and B. Tomlinson (1999). Consumption of green tea causes rapid increase in plasma antioxidant power in humans, *Nutr. Cancer* **34**:83–87.

144. J. Folkman and Y. Shing (1992). Angiogenesis, *J. Biol. Chem.* **267**:10931–10934.

145. Y. Cao and R. Cao (1999). Angiogenesis inhibited by drinking tea, *Nature* **398**:381.

146. F. Y. Tang, N. Nguyen, and M. Meydani (2003). Green tea catechins inhibit VEGF-induced angiogenesis in vitro through suppression of VE-cadherin phosphorylation and inactivation of Akt molecule, *Int. J. Cancer* **106**:871–878.

147. F. Y. Tang and M. Meydani (2001). Green tea catechins and vitamin E inhibit angiogenesis of human microvascular endothelial cells through suppression of IL-8 production, *Nutr. Cancer* **41**:119–125.

148. Y. D. Jung, M. S. Kim, B. A. Shin, K. O. Chay, B. W. Ahn, W. Liu, C. D. Bucana, G. E. Gallick, and L. M. Ellis (2001). EGCG, a major component of green tea, inhibits tumour growth by inhibiting VEGF induction in human colon carcinoma cells, *Br. J. Cancer* **84**:844–850.

149. A. Kojima-Yuasa, J. J. Hua, D. O. Kennedy, and I. Matsui-Yuasa (2003). Green tea extract inhibits angiogenesis of human umbilical vein endothelial cells through reduction of expression of VEGF receptors, *Life Sci.* **73**:1299–1313.

150. M. R. Sartippour, D. Heber, L. Zhang, P. Beatty, D. Elashoff, R. Elashoff, V. L. Go, and M. N. Brooks (2002). Inhibition of fibroblast growth factors by green tea, *Int. J. Oncol.* **21**:487–491.

151. M. R. Sartippour, Z. M. Shao, D. Heber, P. Beatty, L. Zhang, et al. (2002). Green tea inhibits vascular endothelial growth factor (VEGF) induction in human breast cancer cells, *J. Nutr.* **132**:2307–2311.

152. M. Masuda, M. Suzui, J. T. Lim, A. Deguchi, J. W. Soh, and I. B. Weinstein (2002). Epigallocatechin-3-gallate decreases VEGF production in head and neck and breast carcinoma cells by inhibiting EGFR-related pathways of signal transduction, *J. Exp. Ther. Oncol.* **2**:350–359.

153. S. Lamy, D. Gingras, and R. Beliveau (2002). Green tea catechins inhibit vascular endothelial growth factor receptor phosphorylation, *Cancer Res.* **62**:381–385.

154. T. Kondo, T. Ohta, K. Igura, Y. Hara, and K. Kaji (2002). Tea catechins inhibit angiogenesis in vitro, measured by human endothelial cell growth, migration and tube formation, through inhibition of VEGF receptor binding, *Cancer Lett.* **180**:139–144.

155. D. A. August, J. Landau, D. Caputo, J. Hong, M. J. Lee, and C. S. Yang (1999). Ingestion of green tea rapidly decreases prostaglandin E_2 levels in rectal mucosa in humans, *Cancer Epidemiol. Biomarkers Prev.* **8**:709–713.

156. F. Tosetti, N. Ferrari, S. De Flora, and A. Albini (2002). Angioprevention: Angiogenesis is a common and key target for cancer chemopreventive agents, *FASEB J.* **16**:2–14.

157. J. Y. Chang (2004). Telomerase: A potential molecular marker and therapeutic target for cancer, *J. Surg. Oncol.* **87**:1–3.

158. B. Elenbaas, L. Spirio, F. Koerner, M. D. Fleming, D. B. Zimonjic, et al. (2001). Human breast cancer cells generated by oncogenic transformation of primary mammary epithelial cells, *Genes Dev.* **15**:50–65.

159. I. Naasani, H. Seimiya, and T. Tsuruo (1998). Telomerase inhibition, telomere shortening, and senescence of cancer cells by tea catechins, *Biochem. Biophys. Res. Commun.* **249**:391–396.

160. I. Naasani, F. Oh-Hashi, T. Oh-Hara, W. Y. Feng, J. Johnston, K. Chan, and T. Tsuruo (2003). Blocking telomerase by dietary polyphenols is a major mechanism for limiting the growth of human cancer cells in vitro and in vivo, *Cancer Res.* **63**:824–830.

161. A. Mittal, M. S. Pate, R. C. Wylie, T. O. Tollefsbol, and S. K. Katiyar (2004). EGCG down-regulates telomerase in human breast carcinoma MCF-7 cells, leading to suppression of cell viability and induction of apoptosis, *Int. J. Oncol.* **24**:703–710.

162. H. Lee, M. Arsura, M. Wu, M. Duyao, A. J. Buckler, and G. E. Sonenshein (1995). Role of Rel-related factors in control of c-myc gene transcription in receptor-mediated apoptosis of the murine B cell WEHI 231 line, *J. Exp. Med.* **181**:1169–1177.

163. M. Wu, H. Lee, R. E. Bellas, S. L. Schauer, M. Arsura, et al. (1996). Inhibition of NF-kappaB/Rel induces apoptosis of murine B cells, *EMBO J.* **15**:4682–4690.

164. R. A. Greenberg, R. C. O'Hagan, H. Deng, Q. Xiao, S. R. Hann, et al. (1999). Telomerase reverse transcriptase gene is a direct target of c-Myc but is not functionally equivalent in cellular transformation, *Oncogene* **18**:1219–1226.

165. Y. Satoh, I. Matsumura, H. Tanaka, S. Ezoe, H. Sugahara, et al. (2004). Roles for c-Myc in self-renewal of hematopoietic stem cells. *J. Biol. Chem.* **279**:24986–24993.

166. U. Sinha-Datta, I. Horikawa, E. Michishita, A. Datta, J. C. Sigler-Nicot, et al. (2004). Transcriptional activation of hTERT through the NF-{kappa}B pathway in HTLV-I-transformed cells, *Blood* **104**:2523–2531.

167. K. J. Wu, C. Grandori, M. Amacker, N. Simon-Vermot, A. Polack, J. Lingner, and R. Dalla-Favera (1999). Direct activation of TERT transcription by c-MYC, *Nat. Genet.* **21**:220–224.

168. M. Z. Fang, Y. Wang, N. Ai, Z. Hou, Y. Sun, H. Lu, W. Welsh, and C. S. Yang (2003). Tea polyphenol (–)-epigallocatechin-3-gallate inhibits DNA methyltransferase and reactivates methylation-silenced genes in cancer cell lines, *Cancer Res.* **63**:7563–7570.

169. S. J. Berger, S. Gupta, C. A. Belfi, D. M. Gosky, and H. Mukhtar (2001). Green tea constituent (–)-epigallocatechin-3-gallate inhibits topoisomerase I activity in human colon carcinoma cells, *Biochem. Biophys. Res. Commun.* **288**:101–105.

170. S. Okabe, N. Fujimoto, N. Sueoka, M. Suganuma, and H. Fujiki (2001). Modulation of gene expression by (–)-epigallocatechin gallate in PC-9 cells using a cDNA expression array, *Biol. Pharm. Bull.* **24**:883–886.

171. V. M. Adhami, N. Ahmad, and H. Mukhtar (2003). Molecular targets for green tea in prostate cancer prevention, *J. Nutr.* **133**:2417S–2424S.

172. S. I. Wang and H. Mukhtar (2002). Gene expression profile in human prostate LNCaP cancer cells by (–) epigallocatechin-3-gallate, *Cancer Lett.* **182**:43–51.

173. W. S. Ahn, S. W. Huh, S. M. Bae, I. P. Lee, J. M. Lee, S. E. Namkoong, C. K. Kim, and J. I. Sin (2003). A major constituent of green tea, EGCG, inhibits the growth of a human cervical cancer cell line, CaSki cells, through apoptosis, G(1) arrest, and regulation of gene expression, *DNA Cell Biol.* **22**:217–224.

174. M. R. Sartippour, D. Heber, S. Henning, D. Elashoff, R. Elashoff, et al. (2004). cDNA microarray analysis of endothelial cells in response to green tea reveals a suppressive phenotype, *Int. J. Oncol.* **25**:193–202.

175. K. M. Pisters, R. A. Newman, B. Coldman, D. M. Shin, F. R. Khuri, W. K. Hong, B. S. Glisson, and J. S. Lee (2001). Phase I trial of oral green tea extract in adult patients with solid tumors, *J. Clin. Oncol.* **19**:1830–1838.

176. S. A. Laurie, V. A. Miller, S. C. Grant, M. G. Kris, and K. K. Ng (2005). Phase I study of green tea extract in patients with advanced lung cancer, *Cancer Chemother. Pharmacol.* **55**:33–38.

177. A. Jatoi, N. Ellison, P. A. Burch, J. A. Sloan, S. R. Dakhil, et al. (2003). A phase II trial of green tea in the treatment of patients with androgen independent metastatic prostate carcinoma, *Cancer* **97**:1442–1446.

178. W. S. Ahn, J. Yoo, S. W. Huh, C. K. Kim, J. M. Lee, S. E. Namkoong, S. M. Bae, and I. P. Lee (2003). Protective effects of green tea extracts (polyphenol E and EGCG) on human cervical lesions, *Eur. J. Cancer Prev.* **12**:383–390.

179. J. Sonoda, C. Koriyama, S. Yamamoto, T. Kozako, H. C. Li, et al. (2004). HTLV-1 provirus load in peripheral blood lymphocytes of HTLV-1 carriers is diminished by green tea drinking, *Cancer Sci.* **95**:596–601.

180. Y. Zhen, S. Cao, Y. Xue, and S. Wu (1991). Green tea extract inhibits nucleoside transport and potentiates the antitumor effect of antimetabolites, *Chin. Med. Sci. J.* **6**:1–5.

181. J. D. Liu, S. H. Chen, C. L. Lin, S. H. Tsai, and Y. C. Liang (2001). Inhibition of melanoma growth and metastasis by combination with (–)-epigallocatechin-3-gallate and dacarbazine in mice, *J. Cell. Biochem.* **83**:631–642.

182. M. Suganuma, Y. Ohkura, S. Okabe, and H. Fujiki (2001). Combination cancer chemoprevention with green tea extract and sulindac shown in intestinal tumor formation in Min mice, *J. Cancer Res. Clin. Oncol.* **127**:69–72.

183. M. Suganuma, S. Okabe, Y. Kai, N. Sueoka, E. Sueoka, and H. Fujiki (1999). Synergistic effects of (–)-epigallocatechin gallate with (–)-epicatechin, sulindac, or tamoxifen on cancer-preventive activity in the human lung cancer cell line PC-9, *Cancer Res.* **59**:44–47.

184. J. Jodoin, M. Demeule, and R. Beliveau (2002). Inhibition of the multidrug resistance P-glycoprotein activity by green tea polyphenols, *Biochim. Biophys. Acta* **1542**: 149–159.

185. T. Sugiyama and Y. Sadzuka (1998). Enhancing effects of green tea components on the antitumor activity of adriamycin against M5076 ovarian sarcoma, *Cancer Lett.* **133**: 19–26.

186. Y. Sadzuka, T. Sugiyama, and S. Hirota (1998). Modulation of cancer chemotherapy by green tea, *Clin. Cancer Res.* **4**:153–156.

187. T. Kurita, A. Miyagishima, Y. Nozawa, Y. Sadzuka, and T. Sonobe (2004). A dosage design of mitomycin C tablets containing finely powdered green tea, *Int. J. Pharmacol.* **275**:279–283.

188. A. Khafif, S. P. Schantz, T. C. Chou, D. Edelstein, and P. G. Sacks (1998). Quantitation of chemopreventive synergism between (–)-epigallocatechin-3-gallate and curcumin in normal, premalignant and malignant human oral epithelial cells, *Carcinogenesis* **19**:419–424.

9

MOLECULAR MECHANISMS OF LONGEVITY REGULATION AND CALORIE RESTRICTION

Su-Ju Lin

Section of Microbiology, College of Biological Sciences, University of California–Davis, Davis, California

9.1 INTRODUCTION

Aging is a complex process that influences many aspects of our lives, yet little is known about the molecular pathways that regulate its progression. To understand the mechanisms of aging, many researchers have employed genetically tractable model systems including yeast, worms, flies, and mice to study aging at the molecular and genetic levels. These studies have demonstrated that longevity can be modulated by single gene mutations [1–9]. Manipulations of these longevity genes not only extend life span but also delay many age-associated phenotypes. All these studies suggest that aging is a regulated process and longevity assurance genes govern the rate of aging [2, 7–9].

9.2 A CONSERVED LONGEVITY FACTOR, Sir2

Recent studies in *Saccharomyces cerevisiae* have led to the identification of a highly conserved longevity regulator, Sir2 (silent information regulator 2). Sir2 proteins

Nutritional Genomics: Discovering the Path to Personalized Nutrition
Edited by Jim Kaput and Raymond L. Rodriguez

exhibit an NAD-dependent histone deacetylase activity that is conserved in the Sir2 family members and is required for chromatin silencing and life span extension in yeast [10–13]. Both human and mouse Sir2 (also known as sirtuins) have been shown to function as NAD-dependent p53 deacetylases [14, 15]. Deacetylation of p53 via Sir2 promotes cell survival under stress [14, 15] and perhaps also plays a role in aging [16]. Interestingly, worms carrying extra copies of the *SIR2* orthologue, Sir-2.1, also exhibit a longer life span [17].

How does the Sir2 family regulate life span? It has been suggested that yeast Sir2 extends life span by increasing chromatin silencing at specific genomic loci (such as the highly repetitive ribosomal DNA and telomere), thereby decreasing genome instability and inappropriate gene expression [18]. A recent study suggests that Sir2 may also increase the fitness of newborn cells by asymmetric partitioning of the oxidatively damaged proteins to mother cells [19]. In worm, Sir-2.1 appears to function in the DAF-2 insulin pathway that senses the nutrient availability and regulates dauer formation and longevity [17]. It is currently unknown whether the Sir2 family in mammals affects life span.

9.3 MOLECULAR MECHANISMS OF CALORIE RESTRICTION

Calorie restriction (CR) is the most effective intervention known to extend life span in a variety of species including mammals (reviewed in [20, 21]). CR has also been shown to delay the onset or reduce the incidence of many age-related diseases [20, 21]. For example, CR suppresses the carcinogenic effect of several classes of chemicals and several forms of radiation-induced cancers in rodents [22–24]. CR also inhibits a variety of spontaneous neoplasias in experimental model systems, including tumors arising in several knockout and transgenic mouse models such as p53-deficient mice [22]. Although it has been suggested that CR may work by reducing the levels of reactive oxygen species due to a slowing in metabolism [20, 21], the mechanism by which CR extends longevity and prevents cancer is still uncertain.

Recent studies in yeast have provided insight into the mechanisms underlying CR-mediated life span extension [5, 25–28]. Calorie restriction (CR) can be imposed in yeast by reducing the glucose concentration from 2% to 0.5% in rich media [5, 26–29] or by downregulating the glucose sensing cyclic-AMP/protein kinase A (PKA) pathway [5]. Under these CR conditions, yeast mother cells show an extended replicative life span (division potential) of about 30%.

CR has recently been linked to the conserved longevity factor, Sir2. In yeast, the benefit of CR requires NAD (nicotinamide adenine dinucleotide, oxidized form) and Sir2 [5, 26]. The requirement of NAD for Sir2 deacetylase activity suggests CR may activate Sir2 by increasing the available NAD pool for Sir2 [18]. In fact, increasing the activity of the NAD salvage pathway appears to function in the same pathway as CR and Sir2 to extend life span [27]. Increasing the NAD salvage activity may also work by decreasing the concentration of nicotinamide, an inhibitor of Sir2 activity [27, 30]. Consistent with this hypothesis, overexpressing Pnc1, a nicotinamidase that converts nicotinamide to nicotinic acid in the salvage pathway, extends life span [28]. Another recent study indicates that CR induces the shunting of carbon metabolism toward the mitochondrial TCA cycle. The concomitant increase in respiration was necessary and sufficient for the activation of Sir2-mediated silencing and extension in life span [26]. Since mitochondrial respiration plays a major role in reoxidizing NAD from NADH [31, 32], it is possible that CR increases the NAD pool and/or NAD/NADH ratio to activate Sir2 by increasing respiration.

How does this metabolic shift toward respiration during CR activate Sir2? Under CR, glucose is low and respiration is preferred. The higher yield of adenosine triphosphate (ATP) per input carbon slows the rate of glycolysis. This may increase the cytosolic/nuclear pool of NAD, because it is less utilized by the glycolytic enzymes such as glyceraldehyde-3-phosphate dehydrogenase (Figure 9.1). Alternatively but not mutually excluded, a higher rate of electron transport may increase the reoxidation of NADH to NAD in the mitochondria, which can be transmitted to the cytosol/nuclear pool by a shuttle that moves redox equivalents across the mitochondrial membrane [31]. Therefore, under CR, a lower rate of glycolysis or an increase in the NADH reoxidation to NAD or both increases NAD/NADH ratio, which then activates Sir2 to extend life span (Figure 9.2).

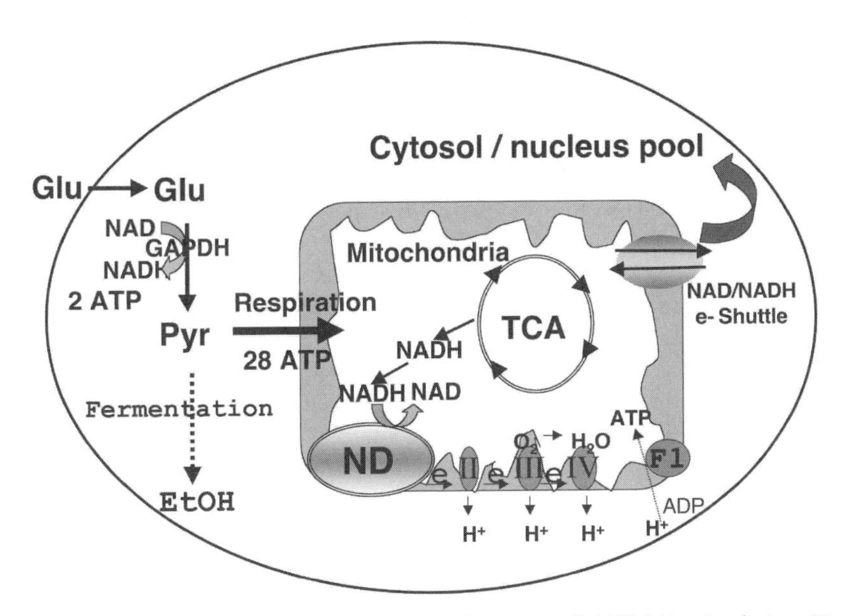

Figure 9.1. Metabolic shift toward respiration increases NAD/NADH ratio during CR. In yeast cells, glucose is metabolized via glycolysis to pyruvate, at which point the pathway bifurcates into fermentation and respiration. Fermentation generates only 2 ATP molecules per molecule of glucose, whereas respiration generates 28 ATP molecules per molecule of glucose. When glucose levels are high, energy is in excess and fermentation is preferred. When glucose is limiting, respiration is preferred. Glu, glucose; GAPDH, glyceraldehyde-3-phosphate dehydrogenase; Pyr, pyruvate; EtOH, ethanol; ND, NADH dehydrogenase; TCA, tricarboxylic (Krebs) cycle.

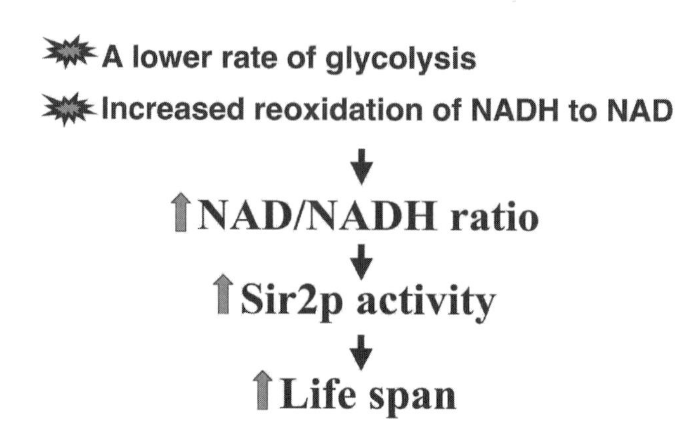

Figure 9.2. A proposed model for calorie restriction in yeast.

Although it has also been suggested that CR activates Sir2 to extend life span by decreasing intracellular levels of nicotinamide, a potent noncompetitive inhibitor of Sir2 [28, 30], it is currently unknown whether the levels of nicotinamide are indeed regulated by CR. On the contrary, it has been shown that, under CR, the ratio of NAD/NADH increases by approximately twofold [33]. This increase in NAD/NADH ratio is manifested by a decrease in the NADH levels since, surprisingly, the levels of NAD remain unchanged under CR. Based on these data, CR may activate Sir2 via NADH. Indeed, NADH can function as a weak competitive inhibitor of Sir2 deacetylase activity. Therefore, our preferred model is that the increase in respiration triggered by CR raises the NAD/NADH ratio and this change activates Sir2 and extends the life span [5, 18, 26].

In summary, recent studies have shown that the longevity factor Sir2 is directly regulated by NAD, NADH, and nicotinamide both in vitro and in vivo. NAD is required for Sir2 activity, whereas NADH and nicotinamide inhibit Sir2 activity. CR appears to activate Sir2 by decreasing the NADH levels. Whether the level of nicotinamide or the NAD/NADH ratio or both play a more important role under CR is still an ongoing debate [33, 34]. This metabolic regulation of Sir2 may underlie the mechanism of CR. However, further studies are required to determine whether this regulation is physiologically relevant. For example, total intracellular levels of NAD and NADH are within the mM range [31, 33], whereas the K_m of Sir2 for NAD is in the low μM range [10, 33, 35]. It is likely that, since most NAD and/or NADH are bound [36], only the free NAD/NADH ratio plays an important role in regulating longevity. Intracellular compartmentalization may also affect the ratio of free NAD/NADH. It is also possible that the NAD/NADH homeostasis/trafficking factors create "virtual" intracellular compartments with a wide range of NAD/NADH ratio. To study how NAD/NADH ratio mediates the metabolic signaling under CR in vivo, an understanding of factors that regulate NAD/NADH trafficking/homeostasis and whether they play a role in longevity regulation is thus essential.

9.4 ROLE OF NAD/NADH RATIO IN AGING AND HUMAN DISEASES

NAD participates in many biological processes including the regulation of energy metabolism [37], DNA repair [38, 39], and transcription [27, 40–42]. In addition to serving as a coenzyme, NAD is also utilized as a substrate. Enzymes that utilize NAD as a substrate include some NAD-dependent DNA ligases [39], NAD-dependent oxidoreductases [37], poly ADP–ribose polymerase (PARP) [38], as well as the Sir2 family, NAD-dependent deacetylases [10–12, 14, 15]. The reduced form of NAD, NADH, is the substrate of the NADH dehydrogenase of the mitochondrial respiratory chain [37]. A derivative of NAD, the NADP(H) coenzyme, is involved in many assimilatory pathways as well as maintaining the intracellular redox state together with glutathione [37].

The NAD/NADH ratio plays an important role in regulating the intracellular redox state and is often considered as a read out of the metabolic state. Several age-related diseases have been directly or indirectly associated with a change in NAD/NADH ratio. For example, the NAD/NADH redox state regulates CtBP corepressor activity and, therefore, plays a role in carcinogenesis [42]. NAD/NADH may also regulate the tumor suppressor p53 via SIRT1 [14, 15]. Furthermore, the mitochondrial NAD/NADH shuttle systems have been shown to play a role in mediating the glucose sensing of beta cells in the pancreas, suggesting a role of NAD/NADH homeostasis in diabetes [43]. Calorie restriction has been shown to decrease the incidence or delay the onset of some of these diseases (reviewed in [20, 44]). Studies in yeast suggest CR may function by increasing the NAD/NADH ratio [5, 26, 33]. CR may ameliorate these human diseases by a similar mechanism. In yeast, a key longevity factor Sir2 appeared to be a downstream target of this NAD/NADH-mediated metabolic regulation. This metabolic regulation probably affects other longevity factors (in addition to Sir2) especially in higher eukaryotes. Further studies to detail the mechanisms underlying these NAD/NADH-mediated metabolic regulations of longevity are thus essential.

9.5 POSSIBLE CR MIMETICS–SMALL MOLECULES THAT REGULATE Sir2 ACTIVITY

Besides genetic modifications, CR is the only way to extend life span in mammals. CR triggers a number of physiological responses including reduced body temperature and plasma insulin levels [20, 21]. Microarray studies on calorie restricted animals have also shown that CR induces a number of genes involved in metabolic and stress response pathways [45]. Studies to identify compounds that mimic CR effects have focused on these pathways [46]. Among these compounds, 2-deoxyglucose and iodoacetate (both inhibit glycolysis) induce similar physiological responses as CR and therefore are promising candidates [46]. However, life span data from these studies still need to be completed to assess the potential of these compounds as real CR mimetics.

Studies suggest that small molecules that regulate Sir2 activity could be possible CR mimetics. In yeast, Sir2 is required for life span extension by CR [5]. A recent study further supports the idea that Sir2 is activated under CR. The expression of mammalian Sir2 is indeed increased in calorie restricted rat and human cells treated with serum isolated from calorie restricted animals [47].

A number of small molecules that can increase human Sir activity have been identified [48]. Among these compounds, resveratrol exhibited the greatest stimulatory activity. Resveratrol is a polyphenol that is found in many plant species including grapes [48]. Resveratrol has been shown to extend life span in yeast as well as metazoans including worms and flies. Life span extension by resveratrol requires a functional *SIR2* gene [48, 49]. In addition, resveratrol did not further extend the life span of calorie restricted cells/animals, suggesting resveratrol functions in the same pathway as CR to extend life span [48, 49].

Several Sir2 inhibitors have also been identified including nicotinamide, sirtuinol [50], and splitomicin [51]. All these chemicals have been shown to inhibit Sir2 deacetylase activity in vitro and Sir2 mediated gene silencing in vivo [30, 50, 51]. These sirtuin inhibitors may be valuable in preventing or treating certain diseases [30]. For example, nicotinamide has been used as a nutritional supplement to prevent type 1 diabetes in genetically predisposed individuals [52].

9.6 MOLECULAR TARGETS OF Sir2 PROTEINS IN MAMMALS

Recent studies have identified several SIRT1 (mammalian Sir2 homologue) interacting proteins, suggesting Sir2 may also regulate longevity in mammals. Both human and mouse SIRT1 have been shown to function as NAD-dependent p53 deacetylases [14, 15]. Deacetylation of p53 via SIRT1 promotes cell survival under stress [14, 15] and may play a role in aging [16]. It has also been shown that mammalian SIRT1 attenuates Bax-mediated apoptosis by deacetylating the DNA repair factor Ku70 [47]. This study suggests CR could extend life span by inducing SIRT1 expression and could promote the long-term survival of irreplaceable postmitotic cells. Activating SIRT1 has also been shown to prevent neurodegeneration. Increasing NAD concentrations in the culture media or overexpressing an NAD biosynthetic enzyme Nmnat1 (nicotinamide mononucleotide adenylyl transferase 1) protects neuronal cells from axonal degeneration by activating the SIRT1 proteins [53].

Another group of studies directly linked SIRT1 to the longevity regulating insulin/IGF pathway. The insulin/IGF pathway contributes to the regulation of life span in worms, flies, and mice [7] (Figure 9.3). One downstream target of the insulin/IGF pathway in worms is the forkhead transcription factor DAF-16 [7]. DAF-16 is a homologue of the mammalian forkhead transcription factors (FOXO) family of forkhead transcription factors, FOXO1 (FKHL), FOXO3 (FKHRL1), and FOXO4 (AFX), that transmits insulin signaling downstream of protein kinase B (Akt kinase) [54, 55]. It has been shown that worm Sir2 extended life span requires a functional DAF-16 [17], suggesting SIRT1 may also interact with the FOXO family in mammalian cells. Indeed, SIRT1 was able to interact with FOXO1 and FOXO3

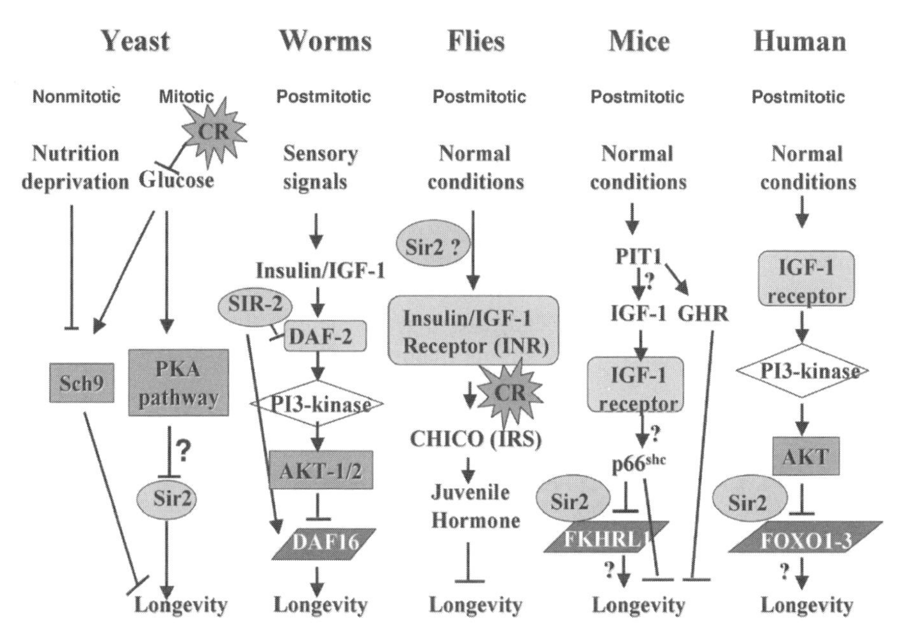

Figure 9.3. A putative conserved longevity regulating pathway. Genes demonstrated to regulate life span in different species are shown. Homologues are highlighted with shading and shapes. The Sir2p family regulates life span in both yeast and worms. In yeast and flies, CR appears to function in these pathways to extend life span. It is still unknown whether CR acts in these pathways in worms, mice, or human. It is also unknown whether the Sir2p family regulates longevity in flies, mice, or human.

and prevent FOXO-induced cell death in a deacetylase-dependent manner [56–58]. SIRT1 also increased the ability of FOXO to activate antioxidant gene expression [56–58]. SIRT1 may therefore increase longevity by preventing apoptosis and increasing stress resistance in a FOXO-dependent manner.

Fat mass is also a primary factor in regulating longevity. Mice lacking the insulin receptor in white adipose tissue showed reduced fat mass and extended longevity [59]. SIRT1 has recently been shown to promote fat mobilization in white adipose tissue by repressing the fat regulator PPARγ (peroxisome proliferator activated receptor γ) [60], which in turn attenuates adipogenesis. These studies suggest that SIRT1-mediated adipogenesis is a possible molecular pathway connecting calorie restriction to life span extension in mammals [60].

9.7 A POSSIBLY CONSERVED LONGEVITY PATHWAY

Identification of the Sir2 family in species ranging from *S. cerevisiae* to *Homo sapiens* suggests that the key longevity regulators are likely to be conserved. In fact, recent studies in yeast have further supported that a conserved longevity pathway may exist. Yeast cells with mutations in Cdc35 (the adenylate cyclase in the cyclic adenosine monophosphate–protein kinase A (cAMP-PKA) pathway) and Sch9 (a kinase that functions in parallel to the PKA pathway) exhibit up to 70% increase in chronological life span [6]. The Sch9 kinase is highly homologous to the AKT kinase family that also regulates longevity in higher eukaryotes [6] (Figure 9.3). In addition, calorie restriction (CR), the only regimen known to extend life span in mammals, appears to extend yeast replicative life span through the PKA pathway [5] (Figure 9.3). An understanding of the CR/PKA pathway may thus provide insight into the conserved longevity pathways. Therefore, understanding the mechanism by which CR extends life span will help to elucidate the molecular pathways of aging and, perhaps, the mechanisms underlying age-related diseases in human.

9.8 APPLICATIONS TO NUTRITIONAL GENOMICS

Studies in model systems have helped in understanding the molecular mechanisms underlying longevity regulation and calorie restriction. Many longevity genes have been found in various model systems, but they have not been systematically studied in humans. One recent study found an association of a G477T marker in SIRT3 (also a mitochondrial histone deacetylase), with longevity in elderly males—the TT genotype increases and the GT genotype decreases ($P = 0.0391$) survival [61]. Association studies, however, require replication. Since aging is a complex trait, it is likely that many genes contribute to its regulation and progression. Studies of gene polymorphisms among these longevity genes and the possible regulation of these genes by diet and other environmental factors are likely to provide valuable insights into the molecular mechanisms of aging and age-related diseases.

Studies using model systems have also provided the basis for the development of nutritional, physiological, or pharmacological mimic of CR, which can be more easily adopted by the public, especially those at high risk for diseases that are associated with aging and aberrant metabolism such as cancer and diabetes. For example, compounds that activate Sir2 activity (e.g., resveratrol) have been shown to mimic CR to extend life span in yeast, worms, and flies [48, 49]. Dietary supplementation of these compounds may generate similar transcriptional profile induced by CR in animal models and may extend life span or ameliorate age-associated diseases in humans.

REFERENCES

1. S. Kim, A. Benguria, C. Y. Lai, and S. M. Jazwinski (1999). Modulation of life-span by histone deacetylase genes in *Saccharomyces cerevisiae*, *Mol. Biol. Cell* **10**:3125–3136.
2. S. M. Jazwinski (2000). Aging and longevity genes, *Acta Biochim. Pol.* **47**:269–279.
3. N. Roy and K. W. Runge (2000). Two paralogs involved in transcriptional silencing that antagonistically control yeast life span, *Curr. Biol.* **10**:111–114.

4. K. Ashrafi, S. S. Lin, J. K. Manchester, and J. I. Gordon (2000). Sip2p and its partner snf1p kinase affect aging in *S. cerevisiae*, *Genes Dev.* **14**:1872–1885.
5. S. J. Lin, P. A. Defossez, and L. Guarente (2000). Requirement of NAD and SIR2 for life-span extension by calorie restriction in *Saccharomyces cerevisiae*, *Science* **289**:2126–2128.
6. P. Fabrizio, F. Pozza, S. D. Pletcher, C. M. Gendron, and V. D. Longo (2001). Regulation of longevity and stress resistance by Sch9 in yeast, *Science* **292**:288–290.
7. C. Kenyon (2001). A conserved regulatory system for aging, *Cell* **105**:165–168.
8. H. A. Tissenbaum and L. Guarente (2002). Model organisms as a guide to mammalian aging, *Dev. Cell* **2**:9–19.
9. K. J. Bitterman, O. Medvedik, and D. A. Sinclair (2003). Longevity regulation in *Saccharomyces cerevisiae*: linking metabolism, genome stability, and heterochromatin, *Microbiol. Mol. Biol. Rev.* **67**:376–399.
10. S. Imai, C. M. Armstrong, M. Kaeberlein, and L. Guarente (2000). Transcriptional silencing and longevity protein Sir2 is an NAD-dependent histone deacetylase, *Nature* **403**:795–800.
11. J. Landry, A. Sutton, S. T. Tafrov, R. C. Heller, J. Stebbins, L. Pillus, and R. Sternglanz (2000). The silencing protein SIR2 and its homologs are NAD-dependent protein deacetylases, *Proc. Natl. Acad. Sci. U.S.A.* **97**:5807–5811.
12. J. S. Smith, C. B. Brachmann, I. Celic, M. A. Kenna, S. Muhammad, et al. (2000). A phylogenetically conserved NAD^+-dependent protein deacetylase activity in the Sir2 protein family, *Proc. Natl. Acad. Sci. U.S.A.* **97**:6658–6663.
13. M. Kaeberlein, M. McVey, and L. Guarente (1999). The SIR2/3/4 complex and SIR2 alone promote longevity in *Saccharomyces cerevisiae* by two different mechanisms, *Genes Dev.* **13**:2570–2580.
14. J. Luo, A. Y. Nikolaev, S. Imai, D. Chen, F. Su, A. Shiloh, L. Guarente, and W. Gu (2001). Negative control of p53 by Sir2alpha promotes cell survival under stress, *Cell* **107**:137–148.
15. H. Vaziri, S. K. Dessain, E. Ng Eaton, S. I. Imai, R. A. Frye, T. K. Pandita, L. Guarente, and R. A. Weinberg (2001). hSIR2(SIRT1) functions as an NAD-dependent p53 deacetylase, *Cell* **107**:149–159.
16. S. D. Tyner, S. Venkatachalam, J. Choi, S. Jones, N. Ghebranious, et al. (2002). p53 mutant mice that display early ageing-associated phenotypes, *Nature* **415**:45–53.
17. H. A. Tissenbaum and L. Guarente (2001). Increased dosage of a sir-2 gene extends lifespan in *Caenorhabditis elegans*, *Nature* **410**:227–230.
18. L. Guarente (2000). Sir2 links chromatin silencing, metabolism, and aging, *Genes Dev.* **14**:1021–1026.
19. H. Aguilaniu, L. Gustafsson, M. Rigoulet, and T. Nystrom (2003). Asymmetric inheritance of oxidatively damaged proteins during cytokinesis, *Science* **299**:1751–1753.
20. W. Weindruch and R. L. Walford (1998). *The Retardation of Aging and Diseases by Dietary Restriction*. Charles C. Thomas, Springfield, IL.
21. G. S. Roth, D. K. Ingram, and M. A. Lane (2001). Caloric restriction in primates and relevance to humans, *Ann. N.Y. Acad. Sci.* **928**:305–315.
22. S. D. Hursting, J. A. Lavigne, D. Berrigan, S. N. Perkins, and J. C. Barrett (2003). Calorie restriction, aging, and cancer prevention: Mechanisms of action and applicability to humans, *Annu. Rev. Med.* **54**:131–152.

23. L. Gross and Y. Dreyfuss (1990). Prevention of spontaneous and radiation-induced tumors in rats by reduction of food intake, *Proc. Natl. Acad. Sci. U.S.A.* **87**:6795–6797.

24. D. Kritchevsky (2003). Diet and cancer: What's next? *J. Nutr.* **133**:3827S–3829S.

25. J. C. Jiang, E. Jaruga, M. V. Repnevskaya, and S. M. Jazwinski (2000). An intervention resembling caloric restriction prolongs life span and retards aging in yeast, *FASEB J.* **14**:2135–2137.

26. S.-J. Lin, M. Kaeberlein, A. A. Andalis, L. A. Sturtz, P.-A. Defossez, V. C. Culotta, G. R. Fink, and L. Guarente (2002). Calorie restriction extends life span by shifting carbon toward respiration, *Nature* **418**:344–348.

27. R. M. Anderson, K. J. Bitterman, J. G. Wood, O. Medvedik, H. Cohen, et al. (2002). Manipulation of a nuclear NAD^+ salvage pathway delays aging without altering steady-state NAD^+ levels, *J. Biol. Chem.* **277**:18881–18890.

28. R. M. Anderson, K. J. Bitterman, J. G. Wood, O. Medvedik, and D. A. Sinclair (2003). Nicotinamide and PNC1 govern lifespan extension by calorie restriction in *Saccharomyces cerevisiae*, *Nature* **423**:181–185.

29. M. Kaeberlein, A. A. Andalis, G. R. Fink, and L. Guarente (2002). High osmolarity extends life span in *Saccharomyces cerevisiae* by a mechanism related to calorie restriction, *Mol. Cell. Biol.* **22**:8056–8066.

30. K. J. Bitterman, R. M. Anderson, H. Y. Cohen, M. Latorre-Esteves, and D. A. Sinclair (2002). Inhibition of silencing and accelerated aging by nicotinamide, a putative negative regulator of yeast Sir2 and human SIRT1, *J. Biol. Chem.* **277**:45099–45107.

31. B. M. Bakker, K. M. Overkamp, A. J. van Maris, P. Kotter, M. A. Luttik, J. P. van Dijken, and J. T. Pronk (2001). Stoichiometry and compartmentation of NADH metabolism in *Saccharomyces cerevisiae*, *FEMS Microbiol. Rev.* **25**:15–37.

32. F. Cruz, M. Villalba, M. A. Garcia-Espinosa, P. Ballesteros, E. Bogonez, J. Satrustegui, and S. Cerdan (2001). Intracellular compartmentation of pyruvate in primary cultures of cortical neurons as detected by (13)C NMR spectroscopy with multiple (13)C labels, *J. Neurosci. Res.* **66**:771–781.

33. S. J. Lin, E. Ford, M. Haigis, G. Liszt, and L. Guarente (2004). Calorie restriction extends yeast life span by lowering the level of NADH, *Genes Dev.* **18**:12–16.

34. R. M. Anderson, M. Latorre-Esteves, A. R. Neves, S. Lavu, O. Medvedik, et al. (2003). Yeast life-span extension by calorie restriction is independent of NAD fluctuation, *Science* **302**:2124–2126.

35. K. G. Tanner, J. Landry, R. Sternglanz, and J. M. Denu (2000). Silent information regulator 2 family of NAD-dependent histone/protein deacetylases generates a unique product, 1-*O*-acetyl-ADP-ribose, *Proc. Natl. Acad. Sci. U.S.A.* **97**:14178–14182.

36. H. Sies (1982). *Metabolic Compartmentation*. Academic Press, New York.

37. C. K. Matthew, K. E. Van Holde, and K. G. Ahern (2000). *Biochemistry*, 3rd ed. Addison-Wesley, Reading, MA.

38. A. Burkle (2001). Physiology and pathophysiology of poly (ADP-ribosyl)ation, *Bioessays* **23**:795–806.

39. A. Wilkinson, J. Day, and R. Bowater (2001). Bacterial DNA ligases, *Mol. Microbiol.* **40**:1241–1248.

40. J. S. Smith and J. D. Boeke (1997). An unusual form of transcriptional silencing in yeast ribosomal DNA, *Genes Dev.* **11**:241–254.

41. J. Rutter, M. Reick, L. C. Wu, and S. L. McKnight (2001). Regulation of clock and NPAS2 DNA binding by the redox state of NAD cofactors, *Science* **293**:510–514.
42. Q. Zhang, D. W. Piston, and R. H. Goodman (2002). Regulation of corepressor function by nuclear NADH, *Science* **295**:1895–1897.
43. K. Eto, Y. Tsubamoto, Y. Terauchi, T. Sugiyama, T. Kishimoto, et al. (1999). Role of NADH shuttle system in glucose-induced activation of mitochondrial metabolism and insulin secretion, *Science* **283**:981–985.
44. M. A. Lane, A. Black, A. Handy, E. M. Tilmont, D. K. Ingram, and G. S. Roth (2001). Caloric restriction in primates, *Ann. N.Y. Acad. Sci.* **928**:287–295.
45. R. Weindruch, T. Kayo, C. K. Lee, and T. A. Prolla (2001). Microarray profiling of gene expression in aging and its alteration by caloric restriction in mice, *J. Nutr.* **131**: 918S–923S.
46. D. K. Ingram, R. M. Anson, R. de Cabo, J. Mamczarz, M. Zhu, J. Mattison, M. A. Lane, and G. S. Roth (2004). Development of calorie restriction mimetics as a prolongevity strategy, *Ann. N.Y. Acad. Sci.* **1019**:412–423.
47. H. Y. Cohen, C. Miller, K. J. Bitterman, N. R. Wall, B. Hekking, et al. (2004). Calorie restriction promotes mammalian cell survival by inducing the SIRT1 deacetylase, *Science* **305**:390–392.
48. K. T. Howitz, K. J. Bitterman, H. Y. Cohen, D. W. Lamming, S. Lavu, et al. (2003). Small molecule activators of sirtuins extend *Saccharomyces cerevisiae* lifespan, *Nature* **425**:191–196.
49. J. G. Wood, B. Rogina, S. Lavu, K. Howitz, S. L. Helfand, M. Tatar, and D. Sinclair (2004). Sirtuin activators mimic caloric restriction and delay ageing in metazoans, *Nature* **430**:686–689.
50. C. M. Grozinger, E. D. Chao, H. E. Blackwell, D. Moazed, and S. L. Schreiber (2001). Identification of a class of small molecule inhibitors of the sirtuin family of NAD-dependent deacetylases by phenotypic screening, *J. Biol. Chem.* **276**:38837–38843.
51. A. Bedalov, T. Gatbonton, W. P. Irvine, D. E. Gottschling, and J. A. Simon (2001). Identification of a small molecule inhibitor of Sir2p, *Proc. Natl. Acad. Sci. U.S.A.* **98**:15113–15118.
52. S. M. Virtanen and M. Knip (2003). Nutritional risk predictors of beta cell autoimmunity and type 1 diabetes at a young age, *Am. J. Clin. Nutr.* **78**:1053–1067.
53. T. Araki, Y. Sasaki, and J. Milbrandt (2004). Increased nuclear NAD biosynthesis and SIRT1 activation prevent axonal degeneration, *Science* **305**:1010–1013.
54. S. Ogg, S. Paradis, S. Gottlieb, G. I. Patterson, L. Lee, H. A. Tissenbaum, and G. Ruvkun (1997). The forkhead transcription factor DAF-16 transduces insulin-like metabolic and longevity signals in *C. elegans*, *Nature* **389**:994–999.
55. K. Lin, J. B. Dorman, A. Rodan, and C. Kenyon (1997). Daf-16: an HNF-3/forkhead family member that can function to double the life-span of *Caenorhabditis elegans*, *Science* **278**:1319–1322.
56. M. C. Motta, N. Divecha, M. Lemieux, C. Kamel, D. Chen, W. Gu, Y. Bultsma, M. McBurney, and L. Guarente (2004). Mammalian SIRT1 represses forkhead transcription factors, *Cell* **116**:551–563.
57. A. Brunet, L. B. Sweeney, J. F. Sturgill, K. F. Chua, P. L. Greer, et al. (2004). Stress-dependent regulation of FOXO transcription factors by the SIRT1 deacetylase, *Science* **303**:2011–2015.

58. H. Daitoku, M. Hatta, H. Matsuzaki, S. Aratani, T. Ohshima, M. Miyagishi, T. Nakajima, and A. Fukamizu (2004). Silent information regulator 2 potentiates Foxo1-mediated transcription through its deacetylase activity, *Proc. Natl. Acad. Sci. U.S.A.* **101**:10042–10047.
59. M. Bluher, B. B. Kahn, and C. R. Kahn (2003). Extended longevity in mice lacking the insulin receptor in adipose tissue, *Science* **299**:572–574.
60. F. Picard, M. Kurtev, N. Chung, A. Topark-Ngarm, T. Senawong, R. Machado De Oliveira, M. Leid, M. W. McBurney, and L. Guarente (2004). Sirt1 promotes fat mobilization in white adipocytes by repressing PPAR-gamma, *Nature* **429**:771–776.
61. G. Rose, S. Dato, K. Altomare, D. Bellizzi, S. Garasto, et al. (2003). Variability of the *SIRT3* gene, human silent information regulator Sir2 homologue, and survivorship in the elderly, *Exp. Gerontol.* **38**:1065–1070.

10

MATERNAL NUTRITION: NUTRIENTS AND CONTROL OF EXPRESSION

Craig A. Cooney

Department of Biochemistry and Molecular Biology, University of Arkansas for Medical Sciences, Little Rock, Arkansas

10.1 INTRODUCTION

In the short term, the environment almost never improves genes by changing the DNA sequence; however, the control of genes can be changed, and potentially improved, by modulating epigenetics and early development through diet. Epigenetic changes are those heritable changes in gene expression that do not require changes in DNA sequence. These usually involve enzymatic DNA methylation and concomitant changes in histone modifications, such as methylation and acetylation, and in other modifications of chromatin structure. Many effects are considered epigenetic if, without mutation, they change the long-term physiological, metabolic, or anatomical state of plants or animals even if the specific gene expression has not been identified.

It is likely that all nutrients have some effects on gene expression and, if used over time, may have persistent, epigenetic, effects. The effects of many nutrients may be much greater during embryonic and fetal development than during adult life. Throughout this chapter, various models of maternal effects on epigenetics, or persistent changes in offspring that are probably epigenetic, will illustrate diverse

Nutritional Genomics: Discovering the Path to Personalized Nutrition
Edited by Jim Kaput and Raymond L. Rodriguez

effects of maternal nutrition on long-term, even multigenerational, health. Some studies aim to determine means by which maternal nutrition might be improved or optimized, whereas others seek to induce a disease state in order to study essential factors and mechanisms.

10.2 METHYL METABOLISM

Enzymatic methylation of DNA and histones is a major component of epigenetic regulation. The methyl groups are either newly synthesized in one-carbon metabolism or are preformed in the diet or by prior metabolism. Methyl metabolism relies on dietary folates or folic acid, dietary methionine, and dietary or endogenous betaine and choline (preformed methyl groups). Folate, methionine, zinc, and vitamin B_{12} (cobalamin) are all dietary essentials and are used as intermediates and enzymatic cofactors to transport and transfer methyl groups in methyl metabolism (Figure 10.1) [1–4]. Choline (also essential) and betaine are widespread in foods and are important sources of preformed methyl groups from the diet [5].

The main methyl donor in mammals is *S*-adenosylmethionine (SAM), which is a product of methyl metabolism. DNA methyltransferases (Dnmts) and histone methyltransferases use SAM to methylate cytosines in DNA and lysines and arginines in histones. Dnmt1 is inhibited by the reaction product *S*-adenosylhomocysteine (SAH) and is a zinc-finger enzyme [6–8]. Histone methyltransferases also use SAM as their methyl donor and are likely inhibited by SAH [9].

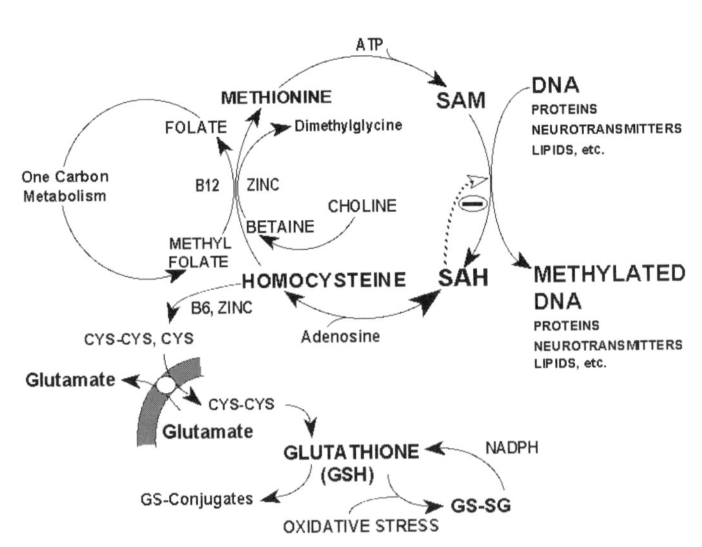

Figure 10.1. Methyl metabolism. Subset of major metabolic intermediates, cofactors, and dietary sources of methyl groups and an intersection with antioxidant metabolism through homocysteine, cysteine, and glutathione.

Methyl metabolism intersects with oxidative metabolism through homocysteine (HCY). Oxidative metabolism converts HCY to cysteine (and on to glutathione (GSH) [3]), whereas methyl metabolism converts HCY to methionine (Figure 10.1). Sulfhydryl amino acids and their derivatives are used for both methyl group transport and redox reactions. Thus, the forms and availability of sulfhydryl amino acids can have important effects on both methyl and oxidative metabolisms. Depending on dietary intakes of methionine, cysteine, and other metabolites, and on the demands of oxidative and methyl metabolism, the levels of methionine and HCY could be limiting and these, in turn, could affect the amounts of methylation and oxidative protection available to a cell or organism.

There is little a priori reason to think that metabolism or the allocation of nutrients from the diet will contribute to the long-term health of adults or that embryonic and fetal development and maternal metabolism will be geared toward the long-term health of the offspring. Instead, natural selection for reproductive fitness makes animals that are good at early reproduction. Most animals (nearly all individuals of many species) will be killed by predators (hawks, parasites, bacteria, viruses) before they become aged. Animals allocate valuable resources such as nutrients and metabolites for early reproduction and not for the long-term maintenance of the individual (Figure 10.2) [1, 10]. Even maternal metabolism is aimed at providing nutrients to assure that offspring will be efficient early reproducers. When a nutrient is in short supply, it will be allocated to short-term survival and reproduction. Even when a nutrient is plentiful, its main allocation will be toward reproduction.

Homocysteine: A Key Component of Methyl Metabolism

One of the most readily measured indicators of methyl metabolism is blood plasma HCY. High HCY is often an indicator of high SAH levels. SAH is a direct inhibitor of many (possibly all) methyltransferases that use SAM as the methyl donor. HCY and SAH are generally good indicators of methyl deficiency and are often correlated with hypomethylation of DNA and other methylated biomolecules [11–14]. Although the diet is not a significant direct source of HCY, large amounts of HCY are produced via metabolism of dietary methionine. Homocysteine is then recycled to methionine or converted to cysteine by metabolism (Figure 10.1).

In humans, high plasma HCY is one cause of cardiovascular disease [15]. Recent data also implicate HCY in dementia and Alzheimer's disease [16] and other aspects of psychiatric and neurological health [17]. Dietary methyl deficiency is a common cause of high HCY although this can be compounded by genetic susceptibility. Dietary or genetic methyl deficiency has also been linked to a variety of cancers in humans [18]. In rats, the effects of methyl-deficient diets include high plasma HCY and vascular disease [19], cardiac hypertrophy [20], and liver cancer [21]. Studies with rodents and humans show that adult levels of DNA methylation depend on adult nutritional factors such as dietary folate, methionine, and choline and metabolic factors such as HCY and SAH [1, 12, 21–23]. Low levels of folate (and possibly betaine and other nutrients) and high HCY levels during pregnancy

Figure 10.2. Natural selection for reproductive fitness makes animals that are good at early reproduction. Most animals (nearly all individuals of many species) will be killed by predators (hawks, parasites, bacteria, viruses) before they become aged. Animals allocate valuable resources such as nutrients and metabolites for early reproduction and not for the long-term maintenance of the individual.

are causes of neural tube birth defects in humans, although the molecular mechanism is unknown [24, 25].

Nutritional and metabolic effects are important in embryonic and fetal development. Some maternal effects can have lifelong consequences for the offspring and for subsequent generations. The enormous consequences of effects that depend on the short time of pregnancy argue that understanding maternal effects should be a top health priority.

10.3 DNA METHYLATION, EPIGENETICS, AND IMPRINTING

In mammals, as well as higher plants, birds, reptiles, fish, and some fungi, several aspects of genome control are effected by DNA methylation. These include suppression of the expression of intragenomic parasitic sequences [26–28]; the inactivation of X chromosomes [29, 30]; and silencing of some, possibly many, genes including many showing genomic imprinting and epigenetic inheritance [10, 31–34]. At least three Dnmts (Dnmt1, Dnmt3a, Dnmt3b) and at least one methylated DNA binding protein, MeCP2, are necessary for mammalian development [35–37]. At the DNA replication fork, the newly synthesized daughter strand is methylated, duplicating the parental pattern [38–40]. This process is imperfect, however, leading to changes in DNA methylation patterns. Inheritance of parental *germline* DNA methylation patterns by offspring is genomic imprinting [41–43]. Inheritance of *somatic* DNA methylation patterns from parents to offspring is transgenerational epigenetic inheritance [33, 44, 45].

Highly specific histone modifications help determine whether chromatin is in an active or inactive state [9, 46]. These modifications include methylation and acetylation of specific sites on histone tails. Histone acetylation promotes active chromatin. Methylation of some sites promotes active chromatin, while methylation of other sites promotes inactive chromatin. For example, methylation of lysine 9 on histone H3 is strongly associated with gene silencing, whereas acetylation of this lysine is found in transcriptionally active chromatin [46]. Acetyl and methyl groups for these reactions come from metabolism and may therefore be influenced by metabolic state and diet. It has been proposed that gene regulation by methylation of DNA [1] and of histones [18] depends on adequate levels of dietary precursors for methyl metabolism. Interactions between diet, metabolism, and gene regulation may well be carefully tuned, evolved responses to environmental variation.

DNA Methyltransferase Levels: One Determinant of DNA Methylation Levels

There are at least three functional Dnmts in mammals. The best characterized, Dnmt1, usually has mainly maintenance activity (it will copy a DNA methylation pattern from a parental strand to a daughter strand) with a small amount of de novo activity (it will methylate sites not previously methylated) [7, 38, 47]. Many variants of Dnmt1 have been described in relation to development, alternative RNA splicing, and proteolysis of the enzyme molecule [7, 38, 41]. Dnmt1 is also involved in the progress of DNA replication and the cell cycle and often shows an *inverse* correlation with global genomic DNA methylation levels. In most cancers, Dnmt activity (as measured in vitro) is very *high* yet global DNA methylation levels are low when compared to normal cells and tissues. *Dnmt1* transcripts are also high in cancer cells and tumors compared to normal cells and tissues [48]. In normal embryonic and fetal development, Dnmt1 levels are positively correlated with DNA methylation levels [41]. In oocytes and preimplantation development, a transcript variant of Dnmt, called Dnmt1o, along with Dnmt3a, is substantially responsible for DNA methylation [41, 49, 50]. In postimplantation development, Dnmt1, Dnmt3a, and Dnmt3b are required for DNA methylation with Dnmt1 for maintenance, possibly some *de novo* methylation, and Dnmt3a and Dnmt3b for de novo methylation [35, 36, 51].

Mouse models deficient for specific components used in epigenetic regulation show changes in genome stability and cancer incidence. A null allele (knockout) of *Dnmt1* has been established in mice. In homozygotes (*Dnmt1*$^{N/N}$) this allele is lethal at 8–11 days gestation. The heterozygote (*Dnmt1*$^{N/+}$) is viable and has low DNA methylation levels when mice are young [35, 52]. However, *Dnmt1*$^{N/+}$ mice can have improved health prospects. When *Dnmt1*$^{N/+}$ is combined in mice with the multiple intestinal neoplasia gene (*Min*), the intestinal adenoma growth rate and multiplicity are half that of *Min*/+ *Dnmt1*$^{+/+}$ mice [53]. Autoimmunity develops more quickly in *Dnmt1*$^{+/+}$ than in *Dnmt1*$^{N/+}$ mice. *Dnmt1*$^{N/+}$ mice have lower global T cell DNA methylation when young (6 months); however, DNA methylation *increases* with age so that by 18 months of age *Dnmt1*$^{N/+}$ mice have higher T cell global DNA methylation than do *Dnmt1*$^{+/+}$ mice. In contrast, global DNA methylation of T cells decreases with age in *Dnmt1*$^{+/+}$ mice [52].

Dnmt1-deficient mice with a hypomorphic *Dnmt1* allele (chip) and a second, null *Dnmt1* allele (*Dnmt1*$^{chip/N}$ mice) only express about 10% of normal Dnmt1 levels. Mice with these very low levels of Dnmt1 consistently develop thymomas [54]. Mice lacking histone methyltransferase Suv39h (that methylates lysine 9 of histone H3) show greater genomic instability and tumor incidence than control mice [55].

10.4 ENDOGENOUS RETROVIRUSES AND GENOME INTEGRITY

The need to control potentially harmful intragenomic "parasitic" DNA sequences, for example, endogenous retroviruses (ERVs), may have propelled the widespread adoption and expansion of DNA methylation in the evolution of vertebrates and particularly of mammals [26, 27]. Suppression of intragenomic parasites is essential for the development and maintenance of any organism. This is especially true of organisms such as mammals that are complex, highly differentiated, and long-lived. If left uncontrolled, intragenomic parasitic DNA sequences would wreak havoc on our genomes and our health [26]. ERVs are generally methylated [28, 56, 57] starting in early development, where DNA methylation is thought to be critical for suppression of ERV transcription and the integrity of the genome [26–28, 58]. In particular, DNA methylation in long terminal repeats (LTRs) silences expression [10, 58–60]. Hypomethylation and expression of ERVs or other intragenomic parasites may contribute to cancer and aging [57, 61–69].

Establishment of DNA methylation in development and adult DNA methylation maintenance are important factors for determining how long DNA methylation patterns will remain effective. Extensive studies show global loss of DNA methylation, as well as local alterations of DNA methylation, as cancers develop [1, 69–71] and in cell populations in vivo [72] and with variations in cell growth in vitro [73]. Likewise, many studies show similar trends during aging in vitro and in vivo, with both global loss of DNA methylation as well as local alterations of DNA methylation [1, 74, 75]. In mammals, genetic programming is overlaid with epigenetic programming and is likely to include genetic pathways governing longevity. Like genetic programming, some epigenetic programming is passed through the germline (as epigenetic inheritance or genomic imprinting). However, most epigenetic programming is not transmitted via gametes, but instead is established in somatic cells during development [28, 76, 77].

High-density DNA methylation at particular loci is important for maintaining inactivity of the associated DNA (e.g., intracisternal A particle (IAP)-LTR [59]; human immunodeficiency virus (HIV)-LTR [78]). DNA methylation density in mice is considered low compared with that in humans and this may reflect their shorter cellular and organismal life spans [79–81]. It has also been proposed that, within a species, the density of DNA methylation early in life could affect density of DNA methylation remaining later in life, and thus may be a factor in determining the life span of individuals [1]. Improving the DNA methylation in ERVs and other repetitive elements has the potential to improve long-term health.

Some LTRs are independent of other viral sequences in mammalian genomes including those of mice and humans. Solitary LTRs can promote the transcription of nonviral sequences [82]. Like other LTRs, solitary LTRs of mice are methylated to varying degrees [83].

Several examples of LTRs affecting nearby gene expression are discussed in this chapter. It is probable that ectopic and dysregulated gene expression occurs in many individuals due to the juxtaposition of ERVs, solo LTRs, or other mobile sequences. When initial ERV suppression is effective in embryonic and fetal development, ERV expression may nevertheless increase with age as DNA methylation patterns and other epigenetic controls are lost (discussed earlier). In some cases, the initial suppression laid down in embryonic and fetal development may be incomplete and may result in cumulative damage to the offspring from chronic ectopic gene expression.

10.5 EPIGENETICS AND NUTRITION CAN GREATLY MODULATE GENETIC PREDISPOSITIONS

Well-defined genetic diseases of metabolism (e.g., inborn errors of metabolism that may be fatal in childhood) often point the way to discovery of alleles whose effects are less severe but that nevertheless take a heavy toll over time (over decades) or during pregnancy. These may be greatly exacerbated by nutritional deficiencies that would not necessarily affect a normal person or animal. These less severe versions of genetic diseases may be heterozygous (just one copy of a null or defective gene) or homozygous (such as two copies of a gene whose products are functional but have low enzymatic activity due to poor binding of cofactors).

In human homocysteinuria, methyl metabolism is severely compromised, often due to homozygous null mutations in a key enzyme such as cystathionine-beta-synthetase (CBS) or methylenetetrahydrofolate reductase (MTHFR). In spite of this, homocysteinuria can be substantially controlled, and patient health and survival vastly improved, using nutrient balances that allow the genetic and enzymatic deficiencies to be bypassed in alternative pathways. Similarly, alleles with just moderate differences in activity, at least with relation to plasma HCY levels, can be largely or entirely compensated by increasing folate and/or betaine in the diet [84]. These same nutrients are important for maternal nutrition to avoid neural tube birth defects [24, 25] as well as, potentially, for adult nutrition to avoid cardiovascular disease and dementia. Thus, in many cases, genetic "predisposition" to certain diseases can be obviated by a diet specific to genotype—one aim of the field of nutrigenomics.

Much popular culture, as well as many geneticists, promotes the idea that genetics is the major or sole determinant of health and life span. In the science fiction movie, "GATACCA" DNA sequence was used to predict human life span and health prognosis with unrealistic precision. At best, it may be possible to predict human life span by genetics alone with a certainty of ±20 years. Certainly, mice have approximately 2 year life spans compared to human 70 year life spans because of differences between mouse and human genetics (although epigenetics in the two species are also different). However, within a species, epigenetics and other nongenetic influences can have huge effects, probably 50 versus 90 year life spans in humans, and for inbred strains of mice and rats, 18 versus 33 month life spans [10]. Interestingly, many of the variables that would be expected to contribute to such differences have not yet been studied at the epigenetic level.

Some fascinating research on the ability of epigenetics to direct the genetic "blueprint" is done in the area of reproductive cloning. Nuclear transplantation can reprogram terminally differentiated somatic cell nuclei into pluripotent embryonic cells that can develop into adult mammals [85, 86]. This occurs with apparently little or no change in the genetic arrangements (genome structure) between the donor differentiated cell and most of the clone's cells. Instead, most changes are thought to be epigenetic.

Epigenetics can redirect a cell with a normally organized genome, but can epigenetics productively direct differentiation using the extensively rearranged genomes of cancer cells? Jaenisch and co-workers [51] transplanted nuclei of tumor cells into oocytes to show that the epigenetic reprogramming capacity of oocytes could produce pluripotent cells from tumor cells. They made embryonic stem (ES) cells from a blastocyst derived from nuclear transplantation of a RAS-inducible melanoma nucleus. Attempts to make mice derived entirely from these ES cells did not proceed beyond 9.5 days gestation (21 days is term in mice). However, when used to make chimeric mice, these ES cells differentiated into a variety of cell types including melanocytes, lymphocytes, and fibroblasts in the adult mouse [51].

Although these chimeric mice developed melanoma and rhabdomyosarcoma by just a few months of age, this experiment demonstrates the extensive contribution to development and differentiation made by epigenetics. It is remarkable that these cells had the extensive genetic rearrangements typical of cancer (and characteristic of their cancer lineage) and yet could be epigenetically reprogrammed to contribute to normal mouse development.

Yellow agouti mice provide a further example of epigenetics modulating genetics. The epigenetic phenotype in a genetically homogeneous (inbred) background affects the long-term health characteristics and, in some cases, the 2 year survival of mice. Epigenetic silencing (in this case through DNA methylation and likely other concomitant mechanisms) renders a deleterious allele nearly harmless.

10.6 YELLOW MOUSE MODELS OF EPIGENETIC REGULATION

Animals in which variation is expressed in the coat color are easily identified and categorized. Several natural mutations in mouse coat color due to variations at the *agouti* locus have been identified. Most of these involve transposable element insertion in the locus. In some instances, epigenetic modification of a LTR regulates *agouti* expression and coat color. *Agouti* alleles that alter coat colors are very useful tools in epigenetic research.

The wild-type mouse *agouti* allele A^w determines pigmentation in hair follicle melanocytes by regulating the alternative production of black (eumelanin) and yellow (phaeomelanin) pigment in individual hair follicles. This cyclic expression results in the "agouti" coat pattern [87].

The "viable yellow" allele, A^{vy}, is dominant to the wild-type A^w allele and induces synthesis of phaeomelanin, instead of eumelanin, leading to a yellow coat. This A^{vy} mutation arose over 40 years ago from the spontaneous natural insertion of a single IAP transposon within the *agouti* gene [88]. In yellow mice, the *agouti* gene is transcribed continuously in essentially all tissues studied [89]. This ectopic expression of *agouti* in A^{vy} is in contrast to expression of A^w, which only occurs in the hair follicles.

The non-*agouti* allele, *a*, induces virtually no phaeomelanin synthesis [90]. Because the *a* allele produces neither yellow, agouti, nor Y0 (i.e., pseudoagouti) phenotypes, it is used as the second allele in many studies. Mice of the *a/a* genotype have black coats.

These alleles exist in numerous strain backgrounds including C57BL6, VY, and YS. Strain VY (VY/WffC3Hf/Nctr-A^{vy}) was derived from strains C57BL/6 and C3H and need *not* necessarily carry the A^{vy} allele; for example, *a/a* VY strain mice do not carry the A^{vy} allele. Likewise, the A^{vy} allele can be carried by many other strains of mice. In addition to viable yellow, A^{vy}, two other yellow alleles, IAP yellow (A^{iapy}), and hypervariable yellow (A^{hvy}), can be used in epigenetic studies.

Epigenetic suppression of the A^{vy} or A^{iapy} alleles produces *agouti* gene expression similar to that of the wild-type A^w allele and yields *pseudoagouti* mice (Y0), with agouti coats (Figure 10.3). Without DNA methylation, this IAP LTR drives *agouti* gene expression and induces the yellow phenotype. The majority of A^{vy}/a and A^{iapy}/a mice constitute a continuous spectrum of variegated patterns of agouti areas (mottling) on yellow backgrounds. The degree of mottling defines their Y0–Y5 phenotypes. A Y5 "clear yellow" mouse (Figure 10.3B) is not mottled and is at one extreme of this spectrum, whereas a Y0 mouse (Figure 10.3A and 10.3C) has an agouti coat and occupies the other extreme of the spectrum. In the A^{vy}, A^{iapy}, and A^{hy} alleles the degree of agouti mottling (or its A^{hy} equivalent) and the degree of agouti IAP methylation are correlated [10, 33, 91, 92]. For the A^{vy} allele the degree of agouti mottling and the level of methylation specifically within the *agouti* IAP LTR are very highly correlated ($r = 0.98$, $P < 0.03$ [10]).

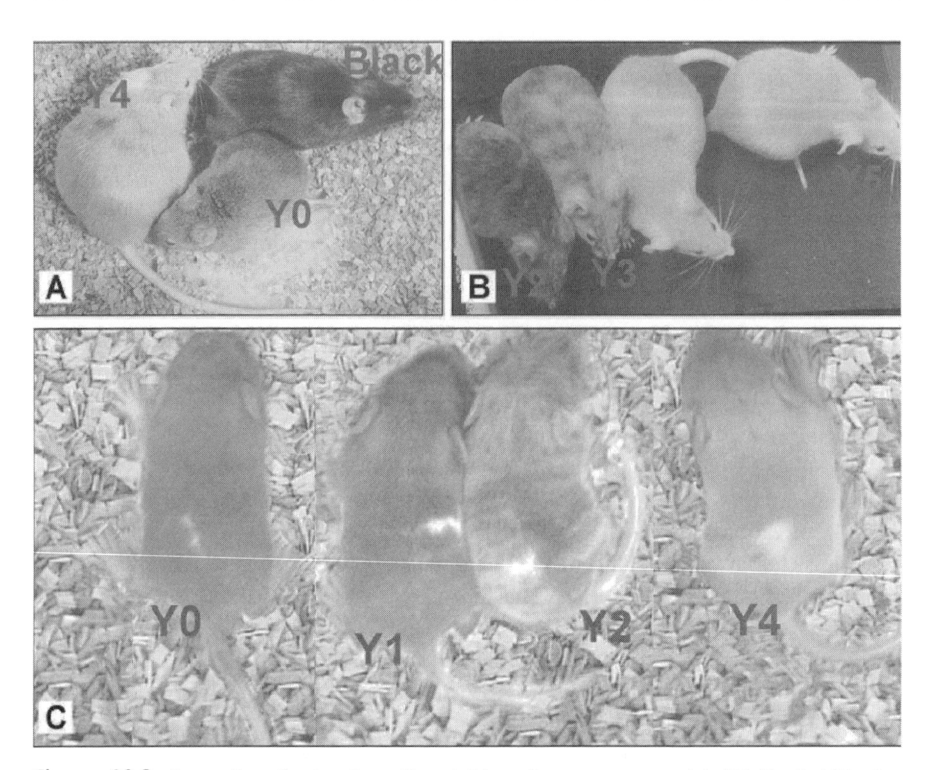

Figure 10.3. Examples of mice from the viable yellow mouse model. (A) Strain VY mice showing *agouti* allele phenotypes. A black "normal" mouse (upper right) has the *a/a* genotype and virtually no *agouti* expression. A pseudoagouti (i.e., Y0) mouse (lower right) with the A^{vy}/a genotype has a "normal" agouti coat pattern. A slightly mottled yellow (i.e., Y4) mouse (left) with A^{vy}/a genotype has ectopic overexpression of *agouti*. (B) Strain VY A^{vy}/a mice, from left to right, a heavily mottled (Y2) mouse, a mottled (Y3) mouse, a slightly mottled (Y4) mouse, and a clear yellow (Y5) mouse. These four mice are genetically identical. Coat color patterns are due to the degree of A^{vy} expression. (C) Strain YS A^{vy}/a mice, from left to right, a pseudoagouti (Y0) mouse, an "almost pseudoagouti" (Y1) mouse, a heavily mottled (Y2) mouse, and a slightly mottled (Y4) mouse. Note the characteristic faint yellow stripes on the right rear quadrant of the Y1 mouse. The white spot on each mouse is due to the recessive spotting allele (*s/s*). These four mice are genetically identical and their coat color patterns are due to the degree of *agouti* gene expression from the A^{vy} allele. (See insert for color representation.)

The *a* and A^{vy} alleles allow the level and pattern of *agouti* expression to be determined from each animal's coat color pattern (Figure 10.3). Coat color is visually apparent just 7 days after birth [89, 93]. Thus, at 7 days of age predictions can be made about the long-term health of mice by merely examining their coat color.

Agouti Overexpression Causes Obesity, Diabetes, and Other Effects

The intermediate steps between ectopic *agouti* overexpression and many gross biological endpoints have been studied in some detail [94]. In hair follicles, alpha-melanocyte stimulating hormone (α-MSH) binds melanocortin receptor 1 (MC1-R) causing increased intracellular cAMP levels, activation of tyrosinase, and subsequent eumelanin synthesis. Agouti protein antagonizes binding of α-MSH to MC1-R, decreasing eumelanin synthesis and resulting in the default synthesis of yellow pigment (pheomelanin). In ectopic expression of *agouti*, melanocortin receptors are affected, but the endpoint of the signaling affects different events in different cell types. For example, in adipocytes, *agouti* causes elevated levels of some signal transduction and activators of transcription (STAT1 and STAT3) as well as a peroxisome proliferator activated receptor γ (PPARγ) protein. STAT1, STAT3, and PPARγ are associated with adipocyte differentiation and may regulate gene expression associated with adipocyte phenotype [94].

Mice with ectopic *agouti* overexpression due to an active A^{vy} allele show increased metabolic efficiency (they convert food calories to fat stores more efficiently) compared to mice with the *a* agouti "null" allele. Caloric restriction results in approximately equal metabolic efficiency for mice with these two alleles [95]. Mice differing in these two alleles and calorie intakes also show different patterns of hepatic gene expression including expression differences in genes likely important in diabetes [96].

Ectopic *agouti* expression leads to increased somatic growth (lean mass and longer bones), obesity (hyperphagia and greater adipose mass), and type 2 diabetes (including insulin resistance, hyperglycemia, pancreatic islet hyperplasia, and hyperinsulinemia [89, 97–99]).

Agouti Overexpression Increases Susceptibility to Cancer

The A^{vy} allele causes increased susceptibility to hyperplasia and adult cancer in many tissues including the bladder [100], liver [101–104], lung [104], mammary gland [101, 105, 106], and skin [107]. However, A^{vy} may not promote all cancers. Becker [108] reported that A^{vy} did not enhance chemical induction of thymic lymphomas.

Agouti Overexpression Results in Lower 2 Year Survival

The mortality at 24 months of age is twice as high for yellow A^{vy}/a mice (Y2–Y5) as it is for pseudoagouti A^{vy}/a (Y0) or black *a/a* mice. Yellow mice had a mortality of 50% at 24 months compared to 24% for Y0 mice and 23% for black mice [93, 98].

Overexpression of Agouti in Transgenes Causes Obesity, Diabetes, and Cancer

Experiments in which *agouti* has been overexpressed in transgenes with housekeeping promoters produce mice that are yellow, obese, hyperinsulinemic, and diabetic [109]. Some similar, localized effects have been produced in transgenic mice intended to mimic the human distribution of normal *agouti* expression (mainly in adipose tissue). In these experiments, *agouti* expression driven by adipocyte-specific promoters produced a heightened sensitivity to insulin, mild obesity, and elevated expression of STAT1, STAT3, and PPARγ in adipocytes [94]. Transgenes in which liver-specific promoters drive *agouti* expression make mice more susceptible to liver cancer without causing obesity [110]. This and previous experiments indicate that overexpression of *agouti* acts independently of obesity to promote cell proliferation in vivo [104, 110].

Maternal Diet, Epigenetics, and Genetics Affect Offspring Epigenetics

Using just two *agouti* alleles (A^{vy} and *a* or A^{iapy} and *a*) allows experimenters to distinguish the coat color genetics and epigenetics of mice by visual observation. The offspring of a cross between A^{vy}/a and *a*/*a* mice are of at least three basic types: (1) genetically normal (*a*/*a*), (2) normal through epigenetic suppression of the A^{vy} mutant allele (A^{vy}/a; Y0 epigenetic phenotype), and (3) phenotypically abnormal through expression of A^{vy} (A^{vy}/a; Y1–Y5 epigenetic phenotypes). The Y0 phenotype mice have extensive epigenetic suppression of the A^{vy} allele and do not become obese or diabetic, and are more resistant to liver tumorigenesis than mice with an unsuppressed A^{vy} allele. Interestingly, the full range of epigenetic phenotypes can be found within a single litter. Maternal effects can readily be studied using the range of phenotypes in this system.

Maternal Epigenetics Partially Determines Offspring Epigenetics

All epigenetic phenotypes of A^{vy}/a mice as well as normal black mice (*a*/*a*) produce viable, fertile offspring and can contribute genetically and epigenetically to each new generation. Remarkably, phenotype is *partially passed to the next generation*, by maternal epigenetic inheritance. For example, Y0 dams are more likely to produce Y0 offspring than are Y2–Y5 dams [93, 111] (Table 10.1). Furthermore, Y0 grandmothers are more likely to produce Y0 grandchildren (through Y0 daughters) than are Y2–Y5 grandmothers through Y0 daughters [33] (Table 10.1). Paternal epigenetic inheritance is *not* observed. These observations demonstrate transgenerational epigenetic inheritance and genomic imprinting at A^{vy}. Maternal epigenetic inheritance suggests that maternal effects at A^{vy} may be heritable to subsequent generations and have multigenerational effects.

TABLE 10.1. Maternal Transgenerational Epigenetic Inheritance in A^{vy}/a Mice

Maternal Epigenetic Phenotype[a]	Percent Y0[b]	N	P	Reference
Strain VY				
Y2–Y5 (mottled yellow)	1.0	1706		[111]
Y0 (pseudoagouti)	5.3	2296	<0.001	[111]
Y2–Y5 (mottled yellow)	1.0	3015		[93]
Y0 (pseudoagouti)	6.3	2323	<0.0001	[93]
Strain YS				
Y2–Y5 (mottled yellow)	0.3	607		[111]
Y0 (pseudoagouti)	1.2	329	NS	[111]
Y2–Y5 (mottled yellow)	0.2	2579		[93]
Y0 (pseudoagouti)	2.3	216	<0.0001	[93]
Strain C57BL6				
Y5 (clear yellow)	0	135		[33]
Y2–Y4 (mottled yellow)	9	45		[33]
Y0 (pseudoagouti)	20	112	<0.0001	[33]
Y0 for two generations	33	95	<0.003	[33]

[a] Maternal epigenetic phenotype affects the epigenetic phenotype of offspring.
[b] Pseudoagouti (Y0) dams produce a significantly higher proportion of Y0 offspring than do dams of other phenotypes. All matings are with *a/a* black sires.

Maternal Diet Affects the Epigenetics of Offspring

Methyl metabolism and DNA methylation are dependent on numerous dietary components, many of which are essential. These components include betaine, choline, folic acid, methionine, vitamin B_{12}, and zinc (Figure 10.1) [1, 10, 93]. In maternal diet studies, dams were fed before and during pregnancy with control diet or with one of two levels of methyl-supplemented diets (called MS and 3SZM, providing substantially increased amounts of cofactors and methyl donors for methyl metabolism) (Table 10.2). Offspring were evaluated for the degree of agouti in their coats (Y0–Y5).

Two levels of supplemented diets were effective for altering offspring coat color when tested in the VY strain of A^{vy}/a mice. Phenotypes with more agouti coats increase in proportion of the population as increasing levels of methyl supplement are added to the maternal diet. The highest level of supplementation (3SZM) was effective on two different strains (VY and YS) of A^{vy}/a mice. Maternal 3SZM diet produced a strong effect on offspring phenotype. The proportion of mice with high mottling (majority agouti coat) increased from 43% for mice fed control diet to 66% for mice on 3SZM ($P < 0.001$). Figure 10.4 shows an example of one experiment. Methyl supplement increased agouti pigmentation in the predicted direction and shifted the distribution of epigenetic phenotype (Figure 10.4). A new phenotype, Y1, was found *only* in litters from dams fed the 3SZM diet (Figure 10.3). These mice have a few thin yellow lines or tiny yellow spots, mainly in the rump area, on a Y0 background and were found in ~13% ($N = 10$) of strain VY offspring [93]. Thus, the 3SZM diet greatly increased the proportion of mice that have almost (Y1) or entirely (Y0) agouti coat color patterns, from 19% to 32% (Figure 10.4). These Y1 mice, unique to the 3SZM diet, have a high degree of DNA methylation on their *agouti* proximal LTR commensurate with their high degree of agouti coat color [10]. Despite very high supplement levels (Table 10.2), no diet exerted any detectable adverse effects on litter size, neonatal mortality, health, and so on [93].

An additional sequence, near *agouti* but not known to affect *agouti* expression, has been shown to be more methylated in response to the MS maternal diet (versus control diet) in the VY mouse strain [112]. Although this sequence is not known to

TABLE 10.2. Composition of MS and 3SZM Supplements[a]

MS Diet Supplement	3SZM Diet Supplement
5 g Choline	15 g Choline
5 g Betaine	15 g Betaine
5 mg Folic acid	15 mg Folic acid
0.5 mg Vitamin B_{12}	1.5 mg Vitamin B_{12}
	7.5 g L-methionine
	150 mg Zinc

[a] The above are added to NIH-31 diet to give 1000 g of the respective final diet. The final total amounts in these diets are substantial increases over the amounts in the base NIH-31 diet [10, 93].

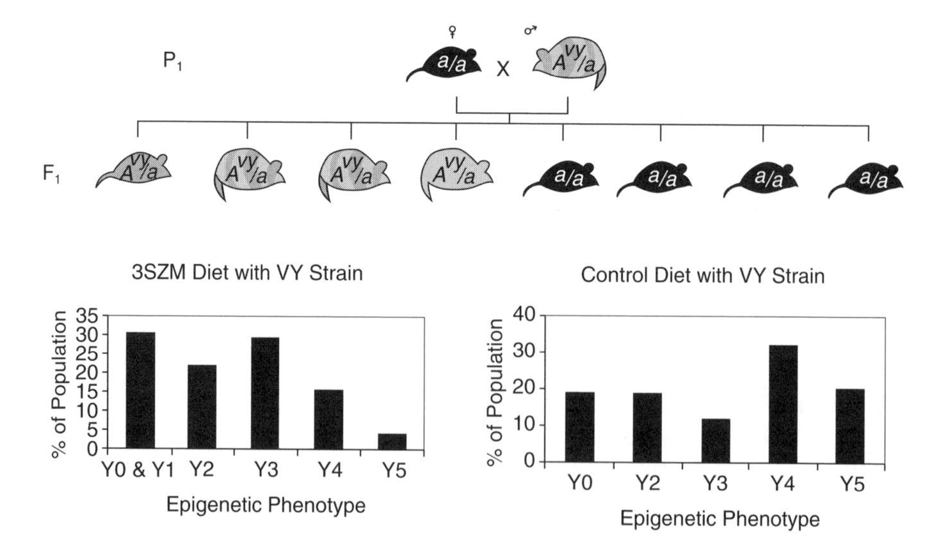

Figure 10.4. Mice mated *a/a* dam × A^{vy}/a sire while the dam is on control or 3SZM diet produce A^{vy}/a and *a/a* offspring. A^{vy}/a offspring have different distributions of epigenetic phenotypes in the population due to maternal diet. (See insert for color representation.)

have a function in *agouti* expression, it is interesting that DNA methylation extends well beyond the *agouti* proximal LTR in silenced alleles.

Embryonic and Fetal Dnmt1 Levels Affect Epigenetics

Experiments that change the levels of specific gene expression at different times in embryonic and fetal development help define the roles of these genes and of developmental timing in epigenetic silencing. Jaenisch and co-workers [113] manipulated levels of *Dnmt1* transcription during embryonic and fetal development of mice carrying the A^{iapy} allele of *agouti*.

The *Dnmt1* gene was modified to prevent production of Dnmt1o (the oocyte alternative transcript of *Dnmt1*), which is normally active in the zygote and cleavage embryo. After this manipulation, some very low levels of some Dnmt1 transcripts remained but, nevertheless, this reduction in Dnmt1o caused the degree of agouti mottling to decline from a control ($Dnmt1^{2lox/1lox}$) level of 70% to 45% ($Dnmt1^{2lox/1lox}$; Msx2cre-oocyte specific CreLox recombination). This shows that some of the maintenance or establishment of epigenetic marks at *agouti* in A^{iapy} depend on Dnmt1o in early development (the zygote and cleavage embryo).

In postimplantation development, reduction of Dnmt1 levels to about 10% of normal caused the degree of agouti mottling to decline from a control ($Dnmt1^{+/+}$) level of at least 60% to less than 30% ($Dnmt1^{chip/-}$). This shows that some of the maintenance or establishment of epigenetic marks at *agouti* in A^{iapy} also depend on *Dnmt1* postimplantation (implantation to neonate). As in other studies, the degree of agouti coat and the degree of DNA methylation in and around the *agouti* IAP LTR were correlated.

These experiments show that the continued inheritance (maintenance) and de novo establishment of epigenetic silencing of A^{iapy} occurs during both preimplantation and postimplantation development and that it has a substantial requirement for Dnmt1. As Jaenisch and co-workers [113] note, there are some mice with a full agouti coat even in populations with the lowest Dnmt1 levels. Although this great variation in epigenetic silencing in the face of very low Dnmt1 levels could be considered stochastic, it may instead be due to the variable availability of other components such as SAM (Dnmt1 substrate) or its cognate reaction product and Dnmt1 inhibitor, SAH.

Other studies show that maternal strain (genotype) has significant effects on *agouti* expression in A^{vy} offspring [10, 93, 111]. Other sequences show epigenetic regulation dependent on maternal genotype [114, 115], suggesting that regulation of A^{vy} is modulated by strain-specific modifiers. This suggests candidates for strain-specific modifiers such as *Dnmts*, genes involved in methyl metabolism, histone methyl- and acetyltransferases, and histone deacetylases as well as other genes less directly affecting metabolic and epigenetic pathways.

Maternal diet, maternal epigenetics, and Dnmt1 levels affect the epigenetic phenotype although these effects have not been parsed to determine if their contributions to epigenetic silencing have proportional contributions to long-term health. It seems probable that they do because maternal epigenetics and diet (standard control diets) and Dnmt1 levels (wild type) presumably also contribute to the "normal" distribution of epigenetic phenotypes.

10.7 A VARIETY OF MATERNAL EFFECTS SEEN IN MICE

Although the A^{vy} and A^{iapy} alleles are probably the most extensively studied epigenetically inherited alleles in mouse, some alleles not associated with coat color in mouse are affected by IAP LTR methylation.

The genetics of the *Axin-fused* ($Axin^{Fu}$) allele (sometimes called fused or *Fu*) in mice was studied extensively by Belyaev and co-workers [116]. $Axin^{Fu}$ can pass between phenotypically active and phenotypically inactive states. Between mouse generations, the state of $Axin^{Fu}$ can also be inherited. This work [116] clearly shows epigenetic inheritance. $Axin^{Fu}$ is an allele of the *axin* developmental gene with an antisense insertion of an IAP retroposon [117]. This mutation and regulation by DNA methylation is much like that of the A^{vy} allele. Unlike the A^{vy} allele, whose epigenetic inheritance is entirely maternal, the inheritance of the $Axin^{Fu}$ allele occurs both maternally and paternally [34].

Sugino and co-workers [118] describe an IAP element in the protocadherin alpha gene cluster in five laboratory mouse strains. They report transcriptional suppression of portions of this gene downstream of a methylated LTR but transcriptional activation by an unmethylated LTR.

Other maternal effects in mouse, for which mechanisms are less well understood, nevertheless affect health for one or more generations. Some antioxidants fed to pregnant mice will affect the life span of their offspring [119]; however, the mechanisms are unknown. Mice exposed prenatally to diethylstilbestrol (DES) can transmit a carcinogenic influence to the next generation (DES-lineage mice) when mated to control mice. This effect persists one generation further (DES-lineage-2 mice) when mating DES-lineage female mice to control males [120].

As discussed earlier with yellow agouti mice, diabetes can be transgenerationally and epigenetically inherited. Some studies have shown that diabetes induced with drugs or glucose loading can induce diabetes in mice and rats even though these strains are not considered to have a genetic susceptibility to diabetes. Shebata and Yasuda [121] describe how the descendants of streptozotocin-induced diabetic mice were diabetic and that this effect extended over several generations. Glucose tolerance was impaired in these mice, especially after the F6 generation. Similar experiments have been done to induce multigenerational diabetes in rats and it is reasonable to consider that similar processes may cause inheritance of diabetes in humans.

10.8 RAT MODELS OF MATERNAL EFFECTS LEADING TO DIABETES

Some of the most important and provocative maternal effects are those that result in multigenerational changes in health. Numerous reports describe how diabetes induced in one generation of rats results in inheritance of diabetes by several subsequent generations of rats.

More than fifty years ago, maternal diabetes in rats was shown to cause hyperglycemia in the offspring (see [122]). Numerous studies have since shown that diabetes in female rats well prior to pregnancy results in diabetes in the offspring, even when the offspring are never directly treated.

Spergel et al. [123] used a single treatment with the drug alloxan to weanling rats to induce latent diabetes that progresses to fasting hyperglycemia by the seventh generation. Initial descendants of these alloxan-treated animals have high blood insulin (hyperinsulinemia), which progresses to abnormally low blood insulin in later generations. Similarly, Van Assche and Aerts [124] reported that certain hallmarks of diabetes (pancreatic islet hyperlasia and beta cell degranulation) were found in the fetuses of third generation rats (F2) from mothers (second generation, F1) born to grandmothers (first generation, P1) who had been made diabetic with streptozotocin. These hallmarks were not found in F2 control fetuses from mothers (F1) born to normal, untreated, grandmothers (P1).

Most effects observed in the above studies are maternal. In the work by Van Assche and Aerts [124], the origin of the father did not matter, even when the father was the offspring of a diabetic mother. Van Assche and Aerts concluded that overstimulation of the fetal endocrine pancreas results in long-term effects to at least the third generation. Other groups have reported similar results where diabetes induced in one generation of rats is passed to subsequent generations [122, 125].

Although the above studies use drugs to induce diabetes, transgenerational diabetes can also be induced nutritionally. Gauguier et al. [126, 127] continuously infused pregnant rats with glucose during the last week of pregnancy (third trimester) to induce mild hyperglycemia. The adult offspring were compared by a number of measures with adult offspring from control dams (who were infused with a glucose-free solution). Compared to controls, young adult offspring (F1) from hyperglycemic mothers had mild glucose intolerance and impaired insulin secretion. With age, these rats developed basal hyperglycemia and severe glucose intolerance. F2 newborns of these F1 hyperglycemic dams were also hyperglycemic, hyperinsulinemic, and macrosomic (showed fetal overgrowth). F2 adults developed basal hyperglycemia and defective glucose tolerance and insulin secretion. These results show that maternal glucose intake in pregnancy can produce a heritable diabetic state in the offspring.

Aerts and Van Assche [128] studied inheritance of induced *gestational* diabetes, that is, diabetes in the mother that occurs mainly during pregnancy. Aerts and Van Assche [128] produced mild diabetes in rats by treating them with streptozotocin. Two generations later, rats have mild diabetic symptoms during pregnancy (increased nonfasting blood glucose and no adaptation of pancreatic beta cells to pregnancy). Effects extend to at least the third generation. Van Assche and Aerts [124] later showed that these effects are mainly maternally transmitted. Gauguier et al. [129] also found higher maternal than paternal inheritance of diabetes in rats.

Although this multigenerational inheritance and progression of diabetes in rats is not defined at the level of gene-specific expression or epigenetic modification, it is nevertheless of great interest because of its potential direct relevance to the current rise in childhood and adult diabetes in the United States. Massively parallel "omic" methods [46, 96, 130, 131] should be useful on models such as these where the molecular mechanisms are undetermined.

Early Events in Development of Diabetes

Numerous studies show that either fetal undergrowth or fetal overgrowth can have later adverse effects on offspring. In the intrauterine growth retardation (IUGR) rat model, fetal growth is retarded by uteroplacental circulatory insufficiency caused by bilateral uterine artery ligation of the pregnant dam. Juvenile offspring show insulin resistance and later, as adults, suffer from type 2 (non-insulin-dependent) diabetes mellitus (hyperglycemia and hyperinsulinemia), hypertension, and hyperlipidemia.

In the IUGR model, there are changes in global DNA methylation and methyl metabolism in liver and increased apoptosis and altered p53 gene methylation in the kidneys of full-term newborn rats [14, 132]. Specifically, global DNA methylation decreased and histone acetylation increased. Levels of Dnmt1, CBS, and methionine adenosyl transferase gene expression decreased. The levels of SAH, adenosine, cysteine, methionine, and HCY were all significantly changed. In particular, SAH and adenosine levels increased and HCY increased more than threefold. In contrast, SAM levels were unchanged. This balance of metabolites would probably inhibit methyltransferase activities including those of Dnmts. It may be that prediabetic metabolic conditions include those that adversely affect methyl metabolism and methyltransferase reactions. However, circulatory insufficiency is so broad an effect that numerous factors from limited nutrients to hypoxia may contribute.

Figure 10.5 is a summary composite of some of the main, long-term maternal effects seen in offspring in the above studies.

10.9 MATERNAL EFFECTS ON MEMORY AND AGING

While many long-term effects of maternal metabolism and diets affect diabetes or cancer susceptibility, some others affect long-term memory and mental function. Choline is an essential nutrient important in the mobilization of fats and in the fluidity of cellular membranes (e.g., as phosphatidylcholine). Choline is also a precursor of the neurotransmitter acetylcholine.

When pregnant rats were fed a normal, adequate diet plus a choline supplement, their offspring had better memory than did offspring of pregnant control rats fed only a normal, adequate diet. The supplement was given from middle to late pregnancy (embryonic days 11 to 17). Otherwise, mothers and offspring received control diet. This data (Figure 10.6) shows that the memory of young rats is better (fewer memory errors) due to maternal choline supplements. Remarkably, unlike control rats (and most humans), the memory of rats from supplemented dams did not decline with age (their number of memory errors stayed the same as when they were young) [133–136].

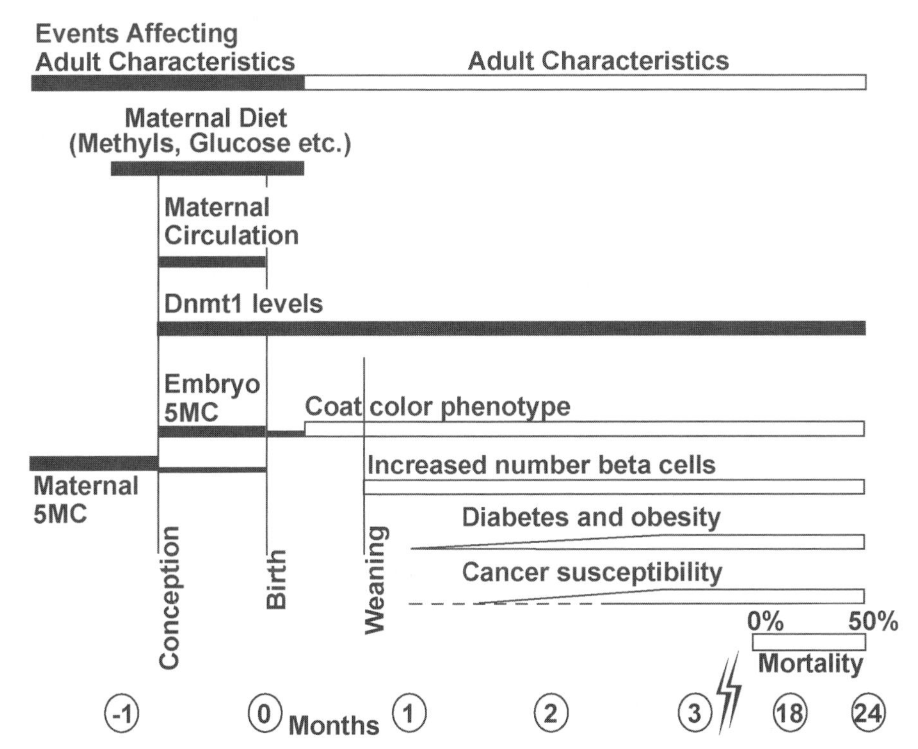

Figure 10.5. Time line of maternal effects and the phenotypic consequences. Prenatal status such as maternal nutrition and maternal epigenetics affect DNA methylation (5MC) of the embryo/fetus/neonate between conception (–3 weeks) and one week (+1 week) after birth. These very early events can have lifelong consequences including obesity, diabetes, and increased mortality. Additional factors involved in epigenetics, such as histone methylation and histone acetylation and the respective enzyme levels, may have similar effects as those shown for 5MC and Dnmt1. Note the break in time scale between 3 and 18 months.

Meck and Williams [137] describe an organizational change in brain function, possibly at the cholinergic synapse, which they call "metabolic imprinting." Further experiments may tell whether metabolic imprinting involves long-term gene expression changes as seen in genomic imprinting and other epigenetic modification. Metabolic imprinting almost certainly involves at least transient effects on gene expression that then lead to long-lasting changes in the number or arrangement of certain cell types.

10.10 EPIGENETIC EFFECTS IN FOXES

Most studies of transgenerational epigenetic inheritance and epigenetic pedigrees have been done in rodents. Without evidence outside of rodents, it might be thought

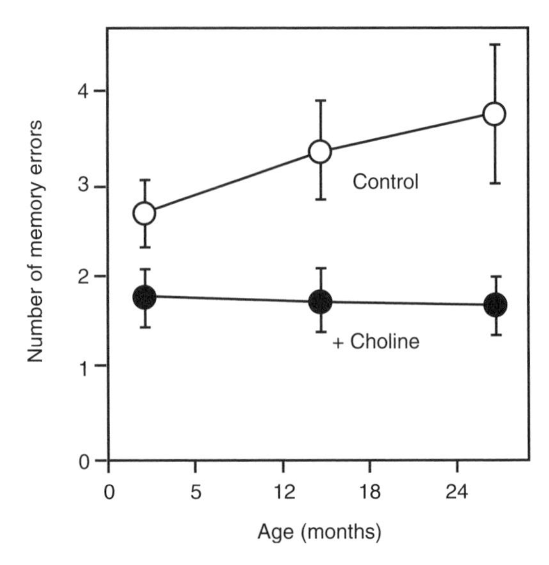

Figure 10.6. Rat dams given a choline supplement from gestational day 11 to 17 produce offspring with better lifelong memory (fewer memory errors) than do dams given a normal, adequate, control diet only. (Redrawn from [136]).

that transgenerational epigenetic inheritance is largely restricted to rodents. However, transgenerational inheritance of apparently epigenetic phenotypes also occurs in foxes.

Silver-black foxes are used for their fur and have been repeatedly domesticated. Their coat colors are of great interest to fox breeders and to some geneticists. In domestication, foxes are selected for friendliness with humans and are otherwise selected for behavior resembling that of domestic dogs. From 1971 to 1979 Belyaev and co-workers [138, 139] frequently observed that domesticated foxes had white spots or "stars" on the tops of their heads between the ears as well as modified ear and tail carriage. This coat color phenotype arose de novo in 48 fox families, much too frequently and too independently to be caused by mutation. The ancestors of these de novo "star" foxes did not have the star phenotype.

Systematic breeding, pedigree mapping, and classical genetic analysis were used to ascribe alleleism to *star*, that is, to determine that there is a dominant allele, *S*, and a recessive allele, *s*. Homozygotes for inactive star (*ss*) are silver black with no spotting and have the visual appearance of wild-type, undomesticated foxes. Heterozygotes (*Ss*) typically have a cluster or spot of white hair between the ears with occasional spotting of the lower jaw, breast, and belly. Homozygotes for active star (*SS*) have a blaze between the ears that spreads along the nose, sometimes making a white face, white collar, or white belt. Often the chest, belly, navel, feet, legs, and/or tail are white. These *SS* foxes always have variable eye color.

Foxes with active *star* alleles (*S*) are more likely to produce offspring with an active *star* allele (*S*) than are animals with an inactive *star* allele (*s*). An *Ss* × *Ss* cross, produces fewer *SS* (phenotype) offspring and more *Ss* (phenotype) offspring

TABLE 10.3. Penetrance of Star Phenotype in Various Crosses of Foxes[a]

Crosses		*SS*	*Ss*	*ss*	Penetrance
Ss × *Ss*	Observed	**65**	261	120	**58%** (homozygotes)
Ss × *Ss*	Expected	111.5	223	111.5	
SS × *ss*	Observed	0	52	**4**	93% (heterozygotes)
SS × *ss*	Expected	0	56	0	
SS × *Ss*	Observed	107	102	**10**	98% (homozygotes)
SS × *Ss*	Expected	109.5	109.5	0	
SS × *SS*	Observed	21	0	0	100% (homozygotes)
SS × *SS*	Expected	21	0	0	

[a] Numbers in bold are significantly different from the expected or, in the cases of *ss* offspring, represent animals of a phenotype unexpected in the cross. These crosses are the combined data from F × M + M × F crosses. Data from [139].

than expected (Table 10.3). Specifically, the proportion of *SS* phenotype offspring was only 58% of expected (*SS* observed $n = 65$ versus *SS* expected $n = 111.5$; 65 not equal to 111.5; $P < 0.001$).

Another unexpected observation was that some *ss* foxes are produced in *SS* × *Ss* and *SS* × *ss* crosses. The partial penetrance/epigenetic nature of *S* expression is demonstrated by these *ss* animals that are not "expected" from normal Mendelian segregation of *S* and *s* alleles. Star behaves as an autosomal, monogenic locus. However, *S* is not fully dominant and its expression fails to penetrate 100%.

Although Belyaev and co-workers do not use the term epigenetics, star expression appears to be inherited epigenetically in domesticated foxes. In some cases there are differences in expression of star depending on the sex of the parent from whom star is received—thus *star* is also imprinted. The inheritance patterns of *star* (*S*) in foxes are in many ways reminiscent of viable yellow (A^{vy}) inheritance in mice (variable penetrance, transgenerational epigenetic inheritance, imprinting). Also, as with A^{vy} in mice, expression of star in foxes varies within litters. There are presumably control mechanisms (molecular basis unknown) suppressing the activity of the dominant *star* allele *S*. It is tempting to speculate that genes for hair pigments or pigment precursor metabolism (e.g., tyrosine metabolism) may be affected by a nearby or interspersed LTR of an ERV or other interspersed repeat. Of considerable scientific and no doubt of some commercial interest is the nature of the parental effects that accompany domestication that lead to changes in offspring and the domesticated fox population. The differences could include diet, physical activity, and social interaction to name a few.

10.11 EPIGENETIC EFFECTS RELATED TO REPRODUCTION IN HUMANS

Although maternal effects and epigenetic inheritance are not as well defined in humans as in rodents, some studies would indicate that many of the same phenomena might occur.

Maintenance of epigenetic states beyond their normal developmental time (e.g., in embryonic and fetal development) and into childhood or even adulthood may occur in some childhood cancers and other diseases. For example, gains or losses of methylation patterns in developmental genes in childhood cancer [140, 141] may represent inappropriate inheritance or loss of a methylation pattern from fetal development. Bousquet et al. [142] have proposed that allergic asthma may be due to inheritance of fetal or neonatal DNA epigenetic patterns and thus persistent fetal gene expression in childhood. They cited genes for IgE and airway remodeling as candidates. The inappropriate maintenance or loss of epigenetic regulation from the embryo or fetus to the child or adult may well be a maternal effect.

Petronis and co-workers [143] have used epigenetic strategies in conjunction with genetics to understand "polygenic" traits that may instead be partially or entirely epigenetic. In a study of schizophrenia, they cloned genes associated with hypomethylated interspersed *Alu* elements as a strategy for identifying aberrantly regulated genes [144].

Harder and co-workers [145] studied family histories of diabetes in women with gestational diabetes. The mothers of these women were significantly more likely to have type 2, non-insulin-dependent diabetes (NIDDM) than were the fathers. Also, grandmothers were more likely to have NIDDM than grandfathers. Maternal transgenerational inheritance of NIDDM due to gestational diabetes was suggested.

Transgenerational inheritance of abnormally methylated genes may occur in humans. Suter et al. [146] identified variable hypermethylation of the DNA mismatch repair gene *MLH1* throughout the bodies of two individuals. In one case, this hypermethylation was found in sperm, suggesting the possibility of transgenerational inheritance.

Tissue culture conditions and other manipulations used in human-assisted reproduction may affect the epigenetics of children [147]. Other manipulations early in development, such as nuclear transplantation, can cause transgenerational epigenetic changes in mice [45]. Substantial data from animal reproduction and cloning experiments [148–150] aimed at improving the epigenetic characteristics of cloned animals may provide technologies that minimize epigenetic change under tissue culture conditions.

Human diets vary greatly in nutrient content including nutrients for methyl metabolism. Two examples are shown in Table 10.4. A dietary guide is shown in Figure 10.7 as a way to provide substantial micronutrients without excessive calories.

10.12 NUTRIENTS AND COMPOUNDS THAT MAY AFFECT EARLY DEVELOPMENT AND EPIGENETICS

In addition to nutrients involved directly in methyl metabolism such as betaine, folate, and methionine, a number of compounds that affect methyl metabolism, Dnmts, or DNA methylation in adult animals have the potential to also cause maternal effects on early development and offspring epigenetics.

TABLE 10.4. Selected Nutrient Levels in Two Human Diets[a] (Based on 2000 calories)

Nutrient (mg)	Hamburger, Fries, and Cola	Salmon, Broccoli, Spinach, and Wheat Germ
Folate	0.26	2.1
Vitamin B_{12}	0.003	0.029
Zinc	8	42
Choline	170	880
Betaine	90	5300
Methionine	820	5100

[a] Variations in the human diet for nutrients important in methyl metabolism. Based on 2000 calorie diets made of equal portions (about 300 g each) of USDA #21107, #21138, and #14400 (hamburger, fries, and cola) and on equal portions (about 400 g each) of USDA #15086, #11091, #11457 (salmon, broccoli, spinach) plus about 200 g of USDA #08084 (wheat germ). Derived from [5] and the USDA National Nutrient Database for Standard Reference, Release 16-1 (http://www.nal.usda.gov/fnic/foodcomp).

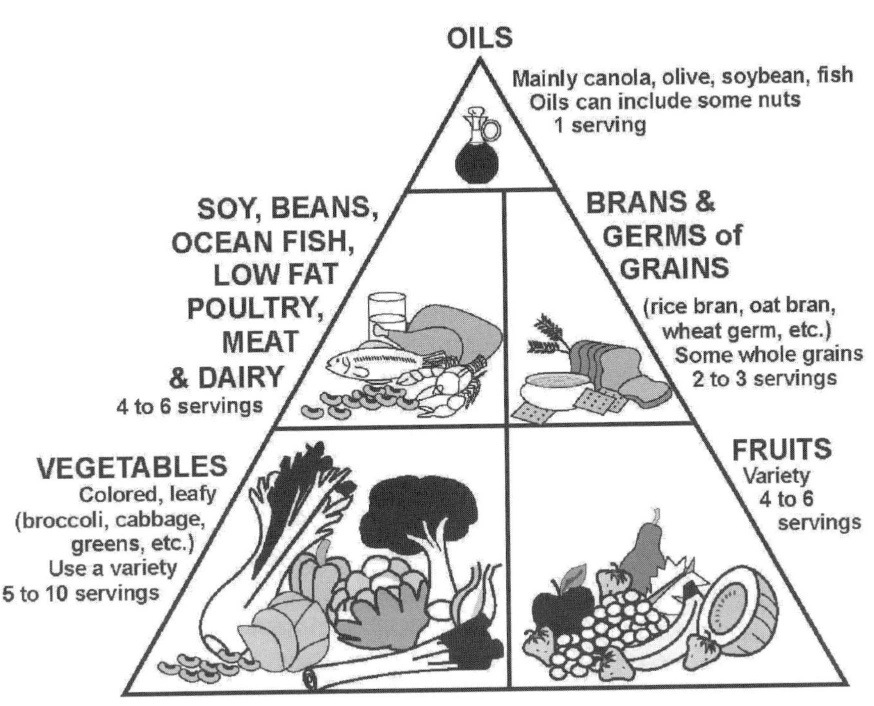

Figure 10.7. A food guide pyramid designed to provide substantial micronutrients while greatly moderating caloric intake.

A number of nutrients, food components, and xenobiotics interact with methyl metabolism and may in turn affect methylation of various molecules including DNA. Methylation is one means of modifying some nutrients (niacin, quercetin) and endogenous compounds (histamine, dopamine) for excretion [151–154]. For example, niacin is metabolized by methylation requiring SAM as a methyl donor [153]. Basu et al. [155] gave rats nutritionally adequate diets with niacin at either 30 mg/kg of diet (control) or at 400 or 1000 mg/kg of diet. Supplementation with niacin at 400 or 1000 mg/kg for 3 months resulted in a significant increase in plasma total HCY levels. This is an almost predictable result of providing a chronic overabundance of a methyltransferase substrate. In other cases, food components can reduce absorption of specific nutrients. For example, phytates (e.g., from grains) can substantially reduce the absorption of zinc [156, 157].

Rowling et al. [158] showed that, in rats, vitamin A and all-*trans*-retinoic acid induced hepatic glycine *N*-methyltransferase and that all-*trans*-retinoic acid induced liver DNA hypomethylation. All-*trans*-retinoic acid is the active vitamin A metabolite in most tissues. It binds and activates the retinoic acid receptor that forms part of a nuclear transcription factor complex [159]. This and related work indicate an important role for vitamin A in regulating methyl metabolism and possibly DNA methylation [160, 161].

Hakkak et al. [162–164] studied rats fed for up to two generations, including during pregnancy, on purified diets made with various protein sources (casein, or whey or soy protein isolate with compensating levels of essential amino acids). Rats fed soy or whey had fewer chemically induced colon and breast tumors than did rats fed casein-based diets. Soy protein is particularly complex [165] and no specific component of these proteins was identified as the chemopreventive agent. Just one soy component, genistein, has been shown to affect DNA methylation of a few DNA sequences in adult mice [166].

In recent screens for mechanisms of chemoprevention, several compounds known to prevent colon cancer, and several known not to prevent colon cancer, were tested for their ability to increase global DNA methylation in colon tumors of rats. Known chemopreventives, calcium chloride, alpha-difluoromethylornithine (DFMO), piroxicam, and sulindac, increased DNA methylation to levels near that of normal colon mucosa. In contrast, several compounds that do not prevent colon cancer, low dose aspirin, 2-carboxyphenyl retinamide, quercetin, 9-*cis*-retinoic acid, and rutin, did not increase global DNA methylation [167]. Aging and cancer in adult mammals are complex, with the genome becoming globally hypomethylated but specific genes, such as tumor suppressor genes, becoming hypermethylated.

Some green tea components are methylated in vivo. Interestingly, the major polyphenol from green tea, (–)-epigallocatechin-3-gallate or EGCG, is both a target of in vivo methylation [168] and an inhibitor of nuclear Dnmt activity. EGCG treatment of some cancer cell lines resulted in hypomethylation and transcriptional activation of previously hypermethylated genes [169].

The desired effect of nutrients or drugs on DNA methylation depends on the genes in question and, of course, on the target (normal cells or cancer cells). Likewise, use of compounds for maternal effects on offspring will be complex.

Normal but denser methylation patterns in young animals have been proposed to be beneficial [1] and thus presumably treatments could be aimed in that direction. However, the long-term benefits of normal but denser methylation patterns remain to be tested. The challenge will be to maintain epigenetic patterns in adulthood with greater fidelity using maternal and adult treatments.

10.13 CONCLUSION

Epigenetics can suppress the expression of harmful alleles and, in at least some cases, can assure the approximately normal development of cells with extensive genetic rearrangements. Epigenetic changes do not change the primary DNA sequence and are reversible. Numerous studies have shown that epigenetics can be changed by maternal treatments although the extent of change at the molecular level is largely unknown. That is, broad effects over the genome, or over the metabolome, remain to be determined. Precise molecular tools are needed to target these epigenetic changes to specific loci or classes of sequences.

The field of nutritional genomics or nutrigenomics aims to tailor diets to genotype and epigenotype in order to prevent disease and optimize health. This includes dietary modulation of gene expression as well as use of diet to compensate for downstream gene products such as enzymes and metabolites. Examples already exist of diet affecting gene expression and of diet compensating for specific alleles. However, the huge potential of these approaches to prevent and treat disease has barely been tapped.

The *agouti* locus in mouse serves as a well-characterized and easily scored model of diet affecting gene expression (through epigenetic correction of gene dysregulation). It is likely that demethylation induces dysregulation of many genes due to the thousands of ERVs and retroposons scattered throughout mammalian genomes. Development of diets to maximize epigenetic silencing of ERVs and retroposons clearly fits in the field of nutrigenomics as one approach that may help prevent a variety of diseases.

Currently, diet is used to regulate specific metabolites such as glucose, cholesterol, or homocysteine. Through advances in nutrigenomics, diets can be developed to regulate a broad range of metabolites. These genotype-specific diets, utilizing data on hundreds or thousands of allelic variations, as well as hundreds or thousands of metabolites, should lead to comprehensive disease prevention and treatment.

Do the diets of individual children and adults fuel the current epidemic of diabetes in the United States or do maternal nutrition and cumulative, multigenerational effects contribute? Maternal effects that lead to persistent increases in glucose or HCY levels in the offspring are likely to have long-term health consequences for the offspring. Some of these effects (especially high glucose) may be passed to future generations. It will be important to identify the probably numerous conditions in early development that cause long-term health problems and multigenerational effects. Also needed are markers for these early events. Transcriptomics, epigenomics, metabolomics, and other omic technologies will aid the discovery of markers and mechanisms. Well-designed maternal nutrition has great potential to direct epigenetics and development in order to stop the passage of diabetes between generations, to improve memory and cognitive function, and to suppress harmful alleles.

ACKNOWLEDGMENTS

I thank Dr. George L. Wolff for discussion and comments on the manuscript and for some of the mouse photographs, Dr. Jim Kaput for comments on the manuscript, Dr. Noel W. Solomons for discussion on the availability of zinc from various foods, and Kimberly Cooney for drawing and organizing most of the figures. This work was supported by grant P01AG20641 from the NIA/NIH.

REFERENCES

1. C. A. Cooney (1993). Are somatic cells inherently deficient in methylation metabolism? A proposed mechanism for DNA methylation loss, senescence and aging, *Growth Dev. Aging* **57**:261–273.
2. N. S. Millian and T. A. Garrow (1998). Human betaine-homocysteine methyltransferase is a zinc metalloenzyme, *Arch. Biochem. Biophys.* **356**:93–98.
3. E. Mosharov, M. R. Cranford, and R. Banerjee (2000). The quantitatively important relationship between homocysteine metabolism and glutathione synthesis by the transsulfuration pathway and its regulation by redox changes, *Biochemistry* **39**:13005–13011.
4. E. O. Uthus and H. M. Brown-Borg (2003). Altered methionine metabolism in long living Ames dwarf mice, *Exp. Gerontol.* **38**:491–498.
5. S. H. Zeisel, M. H. Mar, J. C. Howe, and J. M. Holden (2003). Concentrations of choline-containing compounds and betaine in common foods, *J. Nutr.* **133**:1302–1307.
6. R. Adams and R. Burdon (1985). *Molecular Biology of DNA Methylation.* Springer-Verlag, New York.
7. T. H. Bestor (1992). Activation of mammalian DNA methyltransferase by cleavage of a Zn binding regulatory domain, *EMBO J.* **11**:2611–2617.
8. L. S. Chuang, H. H. Ng, J. N. Chia, and B. F. Li (1996). Characterisation of independent DNA and multiple Zn-binding domains at the N terminus of human DNA-(cytosine-5) methyltransferase: Modulating the property of a DNA-binding domain by contiguous Zn-binding motifs, *J. Mol. Biol.* **257**:935–948.
9. S. A. Jacobs, J. M. Harp, S. Devarakonda, Y. Kim, F. Rastinejad, and S. Khorasanizadeh (2002). The active site of the SET domain is constructed on a knot, *Nat. Struct. Biol.* **9**:833–838.
10. C. A. Cooney, A. A. Dave, and G. L. Wolff (2002). Maternal methyl supplements in mice affect epigenetic variation and DNA methylation of offspring, *J. Nutr.* **132**:2393S–2400S.
11. N. Detich, S. Hamm, G. Just, J. D. Knox, and M. Szyf (2003). The methyl donor *S*-adenosylmethionine inhibits active demethylation of DNA: A candidate novel mecha-

nism for the pharmacological effects of *S*-adenosylmethionine, *J. Biol. Chem.* **278**:20812–20820.

12. R. A. Jacob, D. M. Gretz, P. C. Taylor, S. J. James, I. P. Pogribny (1998). Moderate folate depletion increases plasma homocysteine and decreases lymphocyte DNA methylation in postmenopausal women, *J. Nutr.* **128**:1204–1212.
13. S. J. James, S. Melnyk, M. Pogribna, I. P. Pogribny, and M. A. Caudill (2002). Elevation in *S*-adenosylhomocysteine and DNA hypomethylation: Potential epigenetic mechanism for homocysteine-related pathology, *J. Nutr.* **132**:2361S–2366S.
14. N. K. MacLennan, S. J. James, S. Melnyk, A. Piroozi, S. Jernigan (2004). Uteroplacental insufficiency alters DNA methylation, one-carbon metabolism, and histone acetylation in IUGR rats, *Physiol. Genomics* **18**:43–50.
15. D. S. Wald, M. Law, and J. K. Morris (2002). Homocysteine and cardiovascular disease: Evidence on causality from a meta-analysis, *BMJ* **325**:1202.
16. S. Seshadri, A. Beiser, J. Selhub, P. F. Jacques, I. H. Rosenberg (2002). Plasma homocysteine as a risk factor for dementia and Alzheimer's disease, *N. Engl. J. Med.* **346**:476–483.
17. M. P. Mattson and T. B. Shea (2003). Folate and homocysteine metabolism in neural plasticity and neurodegenerative disorders, *Trends Neurosci.* **26**:137–146.
18. S. Huang (2002). Histone methyltransferases, diet nutrients and tumour suppressors, *Nat. Rev. Cancer* **2**:469–476.
19. F. N. Southern, N. Cruz, L. M. Fink, C. A. Cooney, G. W. Barone, J. F. Eidt, and M. M. Moursi (1998). Hyperhomocysteinemia increases intimal hyperplasia in a rat carotid endarterectomy model, *J. Vasc. Surg.* **28**:909–918.
20. J. Joseph, L. Joseph, N. S. Shekhawat, S. Devi, J. Wang, et al. (2003). Hyperhomocysteinemia leads to pathological ventricular hypertrophy in normotensive rats, *Am. J. Physiol. Heart Circ. Physiol.* **285**:H679–H686.
21. I. P. Pogribny, A. G. Basnakian, B. J. Miller, N. G. Lopatina, L. A. Poirier, and S. J. James (1995). Breaks in genomic DNA and within the p53 gene are associated with hypomethylation in livers of folate/methyl-deficient rats, *Cancer Res.* **55**:1894–1901.
22. L. Tremolizzo, G. Carboni, W. B. Ruzicka, C. P. Mitchell, I. Sugaya, et al. (2002). An epigenetic mouse model for molecular and behavioral neuropathologies related to schizophrenia vulnerability, *Proc. Natl. Acad. Sci. U.S.A.* **99**:17095–17100.
23. M. J. Wilson, N. Shivapurkar, and L. A. Poirier (1984). Hypomethylation of hepatic nuclear DNA in rats fed with a carcinogenic methyl-deficient diet, *Biochem. J.* **218**:987–990.
24. P. M. Ueland and S. E. Vollset (2004). Homocysteine and folate in pregnancy, *Clin. Chem.* **50**:1293–1295.
25. G. M. Shaw, S. L. Carmichael, W. Yang, S. Selvin, and D. M. Schaffer (2004). Periconceptional dietary intake of choline and betaine and neural tube defects in offspring, *Am. J. Epidemiol.* **160**:102–109.
26. T. H. Bestor (1998). The host defence function of genomic methylation patterns, *Novartis Found. Symp.* **214**:187–195; discussion 195–189, 228–132.
27. M. A. Matzke, M. F. Mette, W. Aufsatz, J. Jakowitsch, and A. J. Matzke (1999). Host defenses to parasitic sequences and the evolution of epigenetic control mechanisms, *Genetica* **107**:271–287.

28. N. Lane, W. Dean, S. Erhardt, P. Hajkova, A. Surani, J. Walter, and W. Reik (2003). Resistance of IAPs to methylation reprogramming may provide a mechanism for epigenetic inheritance in the mouse, *Genesis* **35**:88–93.
29. A. D. Riggs and G. P. Pfeifer (1992). X-chromosome inactivation and cell memory, *Trends Genet.* **8**:169–174.
30. T. Sado, M. H. Fenner, S. S. Tan, P. Tam, T. Shioda, and E. Li (2000). X inactivation in the mouse embryo deficient for Dnmt1: Distinct effect of hypomethylation on imprinted and random X inactivation, *Dev. Biol.* **225**:294–303.
31. T. L. Davis, K. D. Tremblay, and M. S. Bartolomei (1998). Imprinted expression and methylation of the mouse H19 gene are conserved in extraembryonic lineages, *Dev. Genet.* **23**:111–118.
32. L. Jackson-Grusby, C. Beard, R. Possemato, M. Tudor, D. Fambrough, et al. (2001). Loss of genomic methylation causes p53-dependent apoptosis and epigenetic deregulation, *Nat. Genet.* **27**:31–39.
33. H. D. Morgan, H. G. Sutherland, D. I. Martin, and E. Whitelaw (1999). Epigenetic inheritance at the agouti locus in the mouse, *Nat. Genet.* **23**:314–318.
34. V. K. Rakyan, S. Chong, M. E. Champ, P. C. Cuthbert, H. D. Morgan, K. V. Luu, and E. Whitelaw (2003). Transgenerational inheritance of epigenetic states at the murine Axin(Fu) allele occurs after maternal and paternal transmission, *Proc. Natl. Acad. Sci. U.S.A.* **100**:2538–2543.
35. E. Li, T. H. Bestor, and R. Jaenisch (1992). Targeted mutation of the DNA methyltransferase gene results in embryonic lethality, *Cell* **69**:915–926.
36. M. Okano, D. W. Bell, D. A. Haber, and E. Li (1999). DNA methyltransferases Dnmt3a and Dnmt3b are essential for de novo methylation and mammalian development, *Cell* **99**:247–257.
37. P. Tate, W. Skarnes, and A. Bird (1996). The methyl-CpG binding protein MeCP2 is essential for embryonic development in the mouse, *Nat. Genet.* **12**:205–208.
38. T. H. Bestor (2000). The DNA methyltransferases of mammals, *Hum. Mol. Genet.* **9**:2395–2402.
39. R. Holliday and J. E. Pugh (1975). DNA modification mechanisms and gene activity during development, *Science* **187**:226–232.
40. A. D. Riggs (1975). X inactivation, differentiation, and DNA methylation, *Cytogenet. Cell Genet.* **14**:9–25.
41. C. Y. Howell, T. H. Bestor, F. Ding, K. E. Latham, C. Mertineit, J. M. Trasler, and J. R. Chaillet (2001). Genomic imprinting disrupted by a maternal effect mutation in the Dnmt1 gene, *Cell* **104**:829–838.
42. J. R. Mann, P. E. Szabo, M. R. Reed, and J. Singer-Sam (2000). Methylated DNA sequences in genomic imprinting, *Crit. Rev. Eukaryot. Gene Expr.* **10**:241–257.
43. W. Reik, A. Collick, M. L. Norris, S. C. Barton, and M. A. Surani (1987). Genomic imprinting determines methylation of parental alleles in transgenic mice, *Nature* **328**:248–251.
44. M. Hadchouel, H. Farza, D. Simon, P. Tiollais, and C. Pourcel (1987). Maternal inhibition of hepatitis B surface antigen gene expression in transgenic mice correlates with de novo methylation, *Nature* **329**:454–456.
45. I. Roemer, W. Reik, W. Dean, and J. Klose (1997). Epigenetic inheritance in the mouse, *Curr. Biol.* **7**:277–280.

46. Y. Kondo, L. Shen, P. S. Yan, T. H. Huang, and J. P. Issa (2004). Chromatin immunoprecipitation microarrays for identification of genes silenced by histone H3 lysine 9 methylation, *Proc. Natl. Acad. Sci. U.S.A.* **101**:7398–7403.

47. M. Fatemi, A. Hermann, S. Pradhan, and A. Jeltsch (2001). The activity of the murine DNA methyltransferase Dnmt1 is controlled by interaction of the catalytic domain with the N-terminal part of the enzyme leading to an allosteric activation of the enzyme after binding to methylated DNA, *J. Mol. Biol.* **309**:1189–1199.

48. K. D. Robertson, E. Uzvolgyi, G. Liang, C. Talmadge, J. Sumegi, F. A. Gonzales, and P. A. Jones (1999). The human DNA methyltransferases (DNMTs) 1, 3a and 3b: Coordinate mRNA expression in normal tissues and overexpression in tumors, *Nucleic Acids Res.* **27**:2291–2298.

49. M. Kaneda, M. Okano, K. Hata, T. Sado, N. Tsujimoto, E. Li, and H. Sasaki (2004). Essential role for de novo DNA methyltransferase Dnmt3a in paternal and maternal imprinting, *Nature* **429**:900–903.

50. S. Ratnam, C. Mertineit, F. Ding, C. Y. Howell, H. J. Clarke, T. H. Bestor, J. R. Chaillet, and J. M. Trasler (2002). Dynamics of Dnmt1 methyltransferase expression and intracellular localization during oogenesis and preimplantation development, *Dev. Biol.* **245**:304–314.

51. K. Hochedlinger, R. Blelloch, C. Brennan, Y. Yamada, M. Kim, L. Chin, and R. Jaenisch (2004). Reprogramming of a melanoma genome by nuclear transplantation, *Genes Dev.* **18**:1875–1885.

52. R. Yung, D. Ray, J. K. Eisenbraun, C. Deng, J. Attwood, et al. (2001). Unexpected effects of a heterozygous dnmt1 null mutation on age-dependent DNA hypomethylation and autoimmunity, *J. Gerontol. A Biol. Sci. Med. Sci.* **56**:B268–B276.

53. R. T. Cormier and W. F. Dove (2000). Dnmt1 N/+ reduces the net growth rate and multiplicity of intestinal adenomas in C57BL/6-multiple intestinal neoplasia (Min)/+ mice independently of p53 but demonstrates strong synergy with the modifier of Min 1(AKR) resistance allele, *Cancer Res.* **60**:3965–3970.

54. F. Gaudet, J. G. Hodgson, A. Eden, L. Jackson-Grusby, J. Dausman, J. W. Gray, H. Leonhardt, and R. Jaenisch (2003). Induction of tumors in mice by genomic hypomethylation, *Science* **300**:489–492.

55. A. H. Peters, D. O'Carroll, H. Scherthan, K. Mechtler, S. Sauer, et al. (2001). Loss of the Suv39h histone methyltransferases impairs mammalian heterochromatin and genome stability, *Cell* **107**:323–337.

56. K. K. Lueders (1995). Multilocus genomic mapping with intracisternal A-particle proviral oligonucleotide probes hybridized to mouse DNA in dried agarose gels, *Electrophoresis* **16**:179–185.

57. K. K. Lueders and E. L. Kuff (1995). Interacisternal A-particle (IAP) genes show similar patterns of hypomethylation in established and primary mouse plasmacytomas, *Curr. Top. Microbiol. Immunol.* **194**:405–414.

58. C. P. Walsh, J. R. Chaillet, and T. H. Bestor (1998). Transcription of IAP endogenous retroviruses is constrained by cytosine methylation, *Nat. Genet.* **20**:116–117.

59. A. Feenstra, J. Fewell, K. Lueders, and E. Kuff (1986). In vitro methylation inhibits the promotor activity of a cloned intracisternal A-particle LTR, *Nucleic Acids Res.* **14**:4343–4352.

60. L. Wang, P. B. Robbins, D. A. Carbonaro, and D. B. Kohn (1998). High-resolution analysis of cytosine methylation in the 5 long terminal repeat of retroviral vectors, *Hum. Gene Ther.* **9**:2321–2330.

61. A. Dupressoir, A. Puech, and T. Heidmann (1995). IAP retrotransposons in the mouse liver as reporters of ageing, *Biochim. Biophys. Acta* **1264**:397–402.

62. L. L. Hsieh, E. Wainfan, S. Hoshina, M. Dizik, and I. B. Weinstein (1989). Altered expression of retrovirus-like sequences and cellular oncogenes in mice fed methyl-deficient diets, *Cancer Res.* **49**:3795–3799.

63. L. L. Mays-Hoopes, A. Brown, and R. C. Huang (1983). Methylation and rearrangement of mouse intracisternal a particle genes in development, aging, and myeloma, *Mol. Cell. Biol.* **3**:1371–1380.

64. A. Puech, A. Dupressoir, M. P. Loireau, M. G. Mattei, and T. Heidmann (1997). Characterization of two age-induced intracisternal A-particle-related transcripts in the mouse liver. Transcriptional read-through into an open reading frame with similarities to the yeast ccr4 transcription factor, *J. Biol. Chem.* **272**:5995–6003.

65. D. Takai, Y. Yagi, N. Habib, T. Sugimura, and T. Ushijima (2000). Hypomethylation of LINE1 retrotransposon in human hepatocellular carcinomas, but not in surrounding liver cirrhosis, *Jpn. J. Clin. Oncol.* **30**:306–309.

66. J. M. Yi, H. M. Kim, and H. S. Kim (2004). Expression of the human endogenous retrovirus HERV-W family in various human tissues and cancer cells, *J. Gen. Virol.* **85**:1203–1210.

67. S. Patzke, M. Lindeskog, E. Munthe, and H. C. Aasheim (2002). Characterization of a novel human endogenous retrovirus, HERV-H/F, expressed in human leukemia cell lines, *Virology* **303**:164–173.

68. A. R. Florl, R. Lower, B. J. Schmitz-Drager, and W. A. Schulz (1999). DNA methylation and expression of LINE-1 and HERV-K provirus sequences in urothelial and renal cell carcinomas, *Br. J. Cancer* **80**:1312–1321.

69. A. R. Florl, C. Steinhoff, M. Muller, H. H. Seifert, C. Hader, R. Engers, R. Ackermann, and W. A. Schulz (2004). Coordinate hypermethylation at specific genes in prostate carcinoma precedes LINE-1 hypomethylation, *Br. J. Cancer* **91**:985–994.

70. M. A. Gama-Sosa, V. A. Slagel, R. W. Trewyn, R. Oxenhandler, K. C. Kuo, C. W. Gehrke, and M. Ehrlich (1983). The 5-methylcytosine content of DNA from human tumors, *Nucleic Acids Res.* **11**:6883–6894.

71. J. G. Herman and S. B. Baylin (2003). Gene silencing in cancer in association with promoter hypermethylation, *N. Engl. J. Med.* **349**:2042–2054.

72. X. Zhu, C. Deng, R. Kuick, R. Yung, B. Lamb, J. V. Neel, B. Richardson, and S. Hanash (1999). Analysis of human peripheral blood T cells and single-cell-derived T cell clones uncovers extensive clonal CpG island methylation heterogeneity throughout the genome, *Proc. Natl. Acad. Sci. U.S.A.* **96**:8058–8063.

73. R. O. Pieper, K. A. Lester, and C. P. Fanton (1999). Confluence-induced alterations in CpG island methylation in cultured normal human fibroblasts, *Nucleic Acids Res.* **27**:3229–3235.

74. J. P. Issa (2000). CpG-island methylation in aging and cancer, *Curr. Top. Microbiol. Immunol.* **249**:101–118.

75. R. P. Singhal, L. L. Mays-Hoopes, and G. L. Eichhorn (1987). DNA methylation in aging of mice, *Mech. Ageing Dev.* **41**:199–210.

76. R. Jaenisch and A. Bird (2003). Epigenetic regulation of gene expression: How the genome integrates intrinsic and environmental signals, *Nat. Genet.* **33**(Suppl): 245–254.
77. E. Li (2002). Chromatin modification and epigenetic reprogramming in mammalian development, *Nat. Rev. Genet.* **3**:662–673.
78. K. A. Gutekunst, F. Kashanchi, J. N. Brady, and D. P. Bednarik (1993). Transcription of the HIV-1 LTR is regulated by the density of DNA CpG methylation, *J. Acquir. Immune Defic. Syndr.* **6**:541–549.
79. M. Gardiner-Garden and M. Frommer (1987). CpG islands in vertebrate genomes, *J. Mol. Biol.* **196**:261–282.
80. R. Holliday (1989). X-chromosome reactivation and ageing, *Nature* **337**:311.
81. D. Toniolo, M. Filippi, R. Dono, T. Lettieri, and G. Martini (1991). The CpG island in the 5′ region of the G6PD gene of man and mouse, *Gene* **102**:197–203.
82. A. Di Cristofano, M. Strazullo, L. Longo, and G. La Mantia (1995). Characterization and genomic mapping of the ZNF80 locus: Expression of this zinc-finger gene is driven by a solitary LTR of ERV9 endogenous retroviral family, *Nucleic Acids Res.* **23**:2823–2830.
83. N. B. Kuemmerle, L. Y. Ch'ang, C. K. Koh, L. R. Boone, and W. K. Yang (1987). Characterization of two solitary long terminal repeats of murine leukemia virus type that are conserved in the chromosome of laboratory inbred mouse strains, *Virology* **160**:379–388.
84. S. A. Craig (2004). Betaine in human nutrition, *Am. J. Clin. Nutr.* **80**:539–549.
85. I. Wilmut, A. E. Schnieke, J. McWhir, A. J. Kind, and K. H. Campbell (1997). Viable offspring derived from fetal and adult mammalian cells, *Nature* **385**:810–813.
86. T. Wakayama, A. C. Perry, M. Zuccotti, K. R. Johnson, and R. Yanagimachi (1998). Full-term development of mice from enucleated oocytes injected with cumulus cell nuclei, *Nature* **394**:369–374.
87. G. L. Wolff (2003). Regulation of yellow pigment formation in mice: A historical perspective, *Pigment Cell Res.* **16**:2–15.
88. D. M. Duhl, H. Vrieling, K. A. Miller, G. L. Wolff, and G. S. Barsh (1994). Neomorphic agouti mutations in obese yellow mice, *Nat. Genet.* **8**:59–65.
89. T. T. Yen, A. M. Gill, L. G. Frigeri, G. S. Barsh, and G. L. Wolff (1994). Obesity, diabetes, and neoplasia in yellow A(vy)/– mice: Ectopic expression of the agouti gene, *FASEB J.* **8**:479–488.
90. S. J. Bultman, M. L. Klebig, E. J. Michaud, H. O. Sweet, M. T. Davisson, and R. P. Woychik (1994). Molecular analysis of reverse mutations from nonagouti (a) to black-and-tan (a(t)) and white-bellied agouti (Aw) reveals alternative forms of agouti transcripts, *Genes Dev.* **8**:481–490.
91. E. J. Michaud, M. J. van Vugt, S. J. Bultman, H. O. Sweet, M. T. Davisson, and R. P. Woychik (1994). Differential expression of a new dominant agouti allele (Aiapy) is correlated with methylation state and is influenced by parental lineage, *Genes Dev.* **8**:1463–1472.
92. A. C. Argeson, K. K. Nelson, and L. D. Siracusa (1996). Molecular basis of the pleiotropic phenotype of mice carrying the hypervariable yellow (Ahvy) mutation at the agouti locus, *Genetics* **142**:557–567.

93. G. L. Wolff, R. L. Kodell, S. R. Moore, and C. A. Cooney (1998). Maternal epigenetics and methyl supplements affect agouti gene expression in Avy/a mice, *FASEB J.* **12**:949–957.

94. R. L. Mynatt and J. M. Stephens (2001). Agouti regulates adipocyte transcription factors, *Am. J. Physiol. Cell Physiol.* **280**:C954–C961.

95. G. L. Wolff, R. L. Kodell, J. A. Kaput, and W. J. Visek (1999). Caloric restriction abolishes enhanced metabolic efficiency induced by ectopic agouti protein in yellow mice, *Proc. Soc. Exp. Biol. Med.* **221**:99–104.

96. J. Kaput, K. G. Klein, E. J. Reyes, W. A. Kibbe, C. A. Cooney, B. Jovanovic, W. J. Visek, and G. L. Wolff (2004). Identification of genes contributing to the obese yellow Avy phenotype: Caloric restriction, genotype, diet × genotype interactions, *Physiol. Genomics* **18**:316–324.

97. G. L. Wolff, D. W. Roberts, and D. B. Galbraith (1986). Prenatal determination of obesity, tumor susceptibility, and coat color pattern in viable yellow (Avy/a) mice. The yellow mouse syndrome, *J. Hered.* **77**:151–158.

98. G. L. Wolff, D. W. Roberts, and K. G. Mountjoy (1999). Physiological consequences of ectopic agouti gene expression: The yellow obese mouse syndrome, *Physiol. Genomics* **1**:151–163.

99. E. J. Michaud, R. L. Mynatt, R. J. Miltenberger, M. L. Klebig, J. E. Wilkinson, M. B. Zemel, W. O. Wilkison, and R. P. Woychik (1997). Role of the agouti gene in obesity, *J. Endocrinol.* **155**:207–209.

100. G. L. Wolff, D. W. Gaylor, C. H. Frith, and R. L. Suber (1983). Controlled genetic variation in a subchronic toxicity assay: Susceptibility to induction of bladder hyperplasia in mice by 2-acetylaminofluorene, *J. Toxicol. Environ. Health* **12**:255–265.

101. W. E. Heston and G. Vlahakis (1968). C3H-Avy—a high hepatoma and high mammary tumor strain of mice, *J. Natl. Cancer Inst.* **40**:1161–1166.

102. G. L. Wolff, R. L. Morrissey, and J. J. Chen (1986). Susceptible and resistant subgroups in genetically identical populations: Response of mouse liver neoplasia and body weight to phenobarbital, *Carcinogenesis* **7**:1935–1937.

103. G. L. Wolff, R. L. Morrissey, and J. J. Chen (1986). Amplified response to phenobarbital promotion of hepatotumorigenesis in obese yellow Avy/A (C3H × VY) F-1 hybrid mice, *Carcinogenesis* **7**:1895–1898.

104. G. L. Wolff, D. W. Roberts, R. L. Morrissey, D. L. Greenman, R. R. Allen, et al. (1987). Tumorigenic responses to lindane in mice: Potentiation by a dominant mutation, *Carcinogenesis* **8**:1889–1897.

105. G. L. Wolff, D. Medina, and R. L. Umholtz (1979). Manifestation of hyperplastic alveolar nodules and mammary tumors in "viable yellow" and non-yellow mice, *J. Natl. Cancer Inst.* **63**:781–785.

106. G. L. Wolff, R. L. Kodell, A. M. Cameron, and D. Medina (1982). Accelerated appearance of chemically induced mammary carcinomas in obese yellow (Avy/A) (BALB/c × VY) F1 hybrid mice, *J. Toxicol. Environ. Health* **10**:131–142.

107. L. A. Hansen, D. E. Malarkey, J. E. Wilkinson, M. Rosenberg, R. E. Woychik, and R. W. Tennant (1998). Effect of the viable-yellow (A(vy)) agouti allele on skin tumorigenesis and humoral hypercalcemia in v-Ha-ras transgenic TG × AC mice, *Carcinogenesis* **19**:1837–1845.

108. F. F. Becker (1988). Failure of the viable yellow (Avy) and lethal yellow (Ay) genes to enhance chemical induction of thymic lymphomas, *Carcinogenesis* **9**:1673–1675.

109. M. L. Klebig, J. E. Wilkinson, J. G. Geisler, and R. P. Woychik (1995). Ectopic expression of the agouti gene in transgenic mice causes obesity, features of type II diabetes, and yellow fur, *Proc. Natl. Acad. Sci. U.S.A.* **92**:4728–4732.

110. A. I. Kuklin, R. L. Mynatt, M. L. Klebig, L. L. Kiefer, W. O. Wilkison, R. P. Woychik, and E. J. Michaud (2004). Liver-specific expression of the agouti gene in transgenic mice promotes liver carcinogenesis in the absence of obesity and diabetes, *Mol. Cancer* **3**:17.

111. G. L. Wolff (1978). Influence of maternal phenotype on metabolic differentiation of agouti locus mutants in the mouse, *Genetics* **88**:529–539.

112. R. A. Waterland and R. L. Jirtle (2003). Transposable elements: Targets for early nutritional effects on epigenetic gene regulation, *Mol. Cell. Biol.* **23**:5293–5300.

113. F. Gaudet, W. M. Rideout, 3rd, A. Meissner, J. Dausman, H. Leonhardt, and R. Jaenisch (2004). Dnmt1 expression in pre- and postimplantation embryogenesis and the maintenance of IAP silencing, *Mol. Cell. Biol.* **24**:1640–1648.

114. N. D. Allen, M. L. Norris, and M. A. Surani (1990). Epigenetic control of transgene expression and imprinting by genotype-specific modifiers, *Cell* **61**:853–861.

115. P. Engler, D. Haasch, C. A. Pinkert, L. Doglio, M. Glymour, R. Brinster, and U. Storb (1991). A strain-specific modifier on mouse chromosome 4 controls the methylation of independent transgene loci, *Cell* **65**:939–947.

116. D. K. Belyaev, A. O. Ruvinsky, and P. M. Borodin (1981). Inheritance of alternative states of the fused gene in mice, *J. Hered.* **72**:107–112.

117. T. J. Vasicek, L. Zeng, X. J. Guan, T. Zhang, F. Costantini, and S. M. Tilghman (1997). Two dominant mutations in the mouse fused gene are the result of transposon insertions, *Genetics* **147**:777–786.

118. H. Sugino, T. Toyama, Y. Taguchi, S. Esumi, M. Miyazaki, and T. Yagi (2004). Negative and positive effects of an IAP-LTR on nearby Pcdaalpha gene expression in the central nervous system and neuroblastoma cell lines, *Gene* **337**:91–103.

119. D. Harman (1998). Extending functional life span, *Exp. Gerontol.* **33**:95–112.

120. B. E. Walker and M. I. Haven (1997). Intensity of multigenerational carcinogenesis from diethylstilbestrol in mice, *Carcinogenesis* **18**:791–793.

121. M. Shibata and B. Yasuda (1980). New experimental congenital diabetic mice (N.S.Y. mice), *Tohoku J. Exp. Med.* **130**:139–142.

122. G. Dorner, A. Plagemann, J. Ruckert, F. Gotz, W. Rohde, et al. (1988). Teratogenetic maternofoetal transmission and prevention of diabetes susceptibility, *Exp. Clin. Endocrinol.* **91**:247–258.

123. G. Spergel, F. Khan, and M. G. Goldner (1975). Emergence of overt diabetes in offspring of rats with induced latent diabetes, *Metabolism* **24**:1311–1319.

124. F. A. van Assche and L. Aerts (1985). Long-term effect of diabetes and pregnancy in the rat, *Diabetes* **34**(Suppl 2):116–118.

125. V. G. Baranov, I. M. Sokoloverova, A. M. Sitnikova, and R. F. Onegova (1988). Development of diabetes mellitus in the progeny of 6 generations of female rats with alloxan diabetes, *Biull. Eksp. Biol. Med.* **105**:13–15.

126. D. Gauguier, M. T. Bihoreau, A. Ktorza, M. F. Berthault, and L. Picon (1990). Inheritance of diabetes mellitus as consequence of gestational hyperglycemia in rats, *Diabetes* **39**:734–739.

127. D. Gauguier, M. T. Bihoreau, L. Picon, and A. Ktorza (1991). Insulin secretion in adult rats after intrauterine exposure to mild hyperglycemia during late gestation, *Diabetes* **40**(Suppl 2):109–114.

128. L. Aerts and F. A. Van Assche (1979). Is gestational diabetes an acquired condition? *J. Dev. Physiol.* **1**:219–225.

129. D. Gauguier, I. Nelson, C. Bernard, V. Parent, C. Marsac, D. Cohen, and P. Froguel (1994). Higher maternal than paternal inheritance of diabetes in GK rats, *Diabetes* **43**:220–224.

130. H. M. Abdolmaleky, C. L. Smith, S. V. Faraone, R. Shafa, W. Stone, S. J. Glatt, and M. T. Tsuang (2004). Methylomics in psychiatry: Modulation of gene–environment interactions may be through DNA methylation, *Am. J. Med. Genet.* **127B**: 51–59.

131. S. Maier and A. Olek (2002). Diabetes: A candidate disease for efficient DNA methylation profiling, *J. Nutr.* **132**:2440S–2443S.

132. T. D. Pham, N. K. MacLennan, C. T. Chiu, G. S. Laksana, J. L. Hsu, and R. H. Lane (2003). Uteroplacental insufficiency increases apoptosis and alters p53 gene methylation in the full-term IUGR rat kidney, *Am. J. Physiol. Regul. Integr. Comp. Physiol.* **285**: R962–R970.

133. W. H. Meck, R. A. Smith, and C. L. Williams (1989). Organizational changes in cholinergic activity and enhanced visuospatial memory as a function of choline administered prenatally or postnatally or both, *Behav. Neurosci.* **103**:1234–1241.

134. W. H. Meck and C. L. Williams (1997). Perinatal choline supplementation increases the threshold for chunking in spatial memory, *Neuroreport* **8**:3053–3059.

135. W. H. Meck and C. L. Williams (1997). Simultaneous temporal processing is sensitive to prenatal choline availability in mature and aged rats, *Neuroreport* **8**:3045–3051.

136. J. K. Blusztajn (1998). Choline, a vital amine, *Science* **281**:794–795.

137. W. H. Meck and C. L. Williams (2003). Metabolic imprinting of choline by its availability during gestation: Implications for memory and attentional processing across the lifespan, *Neurosci. Biobehav. Rev.* **27**:385–399.

138. D. K. Belyaev (1979). The Wilhelmine E. Key 1978 invitational lecture. Destabilizing selection as a factor in domestication, *J. Hered.* **70**:301–308.

139. D. K. Belyaev, A. O. Ruvinsky, and L. N. Trut (1981). Inherited activation–inactivation of the star gene in foxes: Its bearing on the problem of domestication, *J. Hered.* **72**:267–274.

140. B. Chen, P. Dias, J. J. Jenkins, 3rd, V. H. Savell, and D. M. Parham (1998). Methylation alterations of the MyoD1 upstream region are predictive of subclassification of human rhabdomyosarcomas, *Am. J. Pathol.* **152**:1071–1079.

141. R. T. Kurmasheva, C. A. Peterson, D. M. Parham, B. Chen, R. E. McDonald, and C. A. Cooney (2005). Upstream CpG island methylation of the PAX3 gene in human rhabdomyosarcomas, *Pediatr. Blood Cancer.* **44**:328–337.

142. J. Bousquet, W. Jacot, H. Yssel, A. M. Vignola, and M. Humbert (2004). Epigenetic inheritance of fetal genes in allergic asthma, *Allergy* **59**:138–147.

143. A. Petronis, V. Popendikyte, P. Kan, and T. Sasaki (2002). Major psychosis and chromosome 22: Genetics meets epigenetics, *CNS Spectr.* **7**:209–214.

144. P. X. Kan, V. Popendikyte, Z. A. Kaminsky, R. H. Yolken, and A. Petronis (2004). Epigenetic studies of genomic retroelements in major psychosis, *Schizophr. Res.* **67**:95–106.

145. T. Harder, K. Franke, R. Kohlhoff, and A. Plagemann (2001). Maternal and paternal family history of diabetes in women with gestational diabetes or insulin-dependent diabetes mellitus type I, *Gynecol. Obstet. Invest.* **51**:160–164.

146. C. M. Suter, D. I. Martin, and R. L. Ward (2004). Germline epimutation of MLH1 in individuals with multiple cancers, *Nat. Genet.* **36**:497–501.

147. M. De Rycke, H. Van de Velde, K. Sermon, W. Lissens, A. De Vos, et al. (2001). Preimplantation genetic diagnosis for sickle-cell anemia and for beta-thalassemia, *Prenat. Diagn.* **21**:214–222.

148. L. E. Young and N. Beaujean (2004). DNA methylation in the preimplantation embryo: the differing stories of the mouse and sheep, *Anim. Reprod. Sci.* **82-83**: 61–78.

149. L. E. Young, A. E. Schnieke, K. J. McCreath, S. Wieckowski, G. Konfortova, et al. (2003). Conservation of IGF2-H19 and IGF2R imprinting in sheep: Effects of somatic cell nuclear transfer, *Mech. Dev.* **120**:1433–1442.

150. K. D. Sinclair, L. E. Young, I. Wilmut, and T. G. McEvoy (2000). In-utero overgrowth in ruminants following embryo culture: Lessons from mice and a warning to men, *Hum. Reprod.* **15**(Suppl 5):68–86.

151. C. De Santi, A. Pietrabissa, F. Mosca, and G. M. Pacifici (2002). Methylation of quercetin and fisetin, flavonoids widely distributed in edible vegetables, fruits and wine, by human liver, *Int. J. Clin. Pharmacol. Ther.* **40**:207–212.

152. Y. P. Pang, X. E. Zheng, and R. M. Weinshilboum (2001). Theoretical 3D model of histamine *N*-methyltransferase: insights into the effects of a genetic polymorphism on enzymatic activity and thermal stability, *Biochem. Biophys. Res. Commun.* **287**:204–208.

153. K. Shibata and H. Matsuo (1989). Correlation between niacin equivalent intake and urinary excretion of its metabolites, *N'*-methylnicotinamide, *N'*-methyl-2-pyridone-5-carboxamide, and *N'*-methyl-4-pyridone-3-carboxamide, in humans consuming a self-selected food, *Am. J. Clin. Nutr.* **50**:114–119.

154. B. T. Zhu (2004). CNS dopamine oxidation and catechol-*O*-methyltransferase: importance in the etiology, pharmacotherapy, and dietary prevention of Parkinson's disease, *Int. J. Mol. Med.* **13**:343–353.

155. T. K. Basu, N. Makhani, and G. Sedgwick (2002). Niacin (nicotinic acid) in non-physiological doses causes hyperhomocysteineaemia in Sprague-Dawley rats, *Br. J. Nutr.* **87**:115–119.

156. B. Lonnerdal (2000). Dietary factors influencing zinc absorption, *J. Nutr.* **130**: 1378S–1383S.

157. S. L. Fitzgerald, R. S. Gibson, J. Quan de Serrano, L. Portocarrero, A. Vasquez, et al. (1993). Trace element intakes and dietary phytate/Zn and Ca × phytate/Zn millimolar ratios of periurban Guatemalan women during the third trimester of pregnancy, *Am. J. Clin. Nutr.* **57**:195–201.

158. M. J. Rowling, M. H. McMullen, and K. L. Schalinske (2002). Vitamin A and its derivatives induce hepatic glycine *N*-methyltransferase and hypomethylation of DNA in rats, *J. Nutr.* **132**:365–369.

159. J. L. Napoli (1999). Interactions of retinoid binding proteins and enzymes in retinoid metabolism, *Biochim. Biophys. Acta* **1440**:139–162.

160. M. J. Rowling and K. L. Schalinske (2003). Retinoic acid and glucocorticoid treatment induce hepatic glycine *N*-methyltransferase and lower plasma homocysteine concentrations in rats and rat hepatoma cells, *J. Nutr.* **133**:3392–3398.

161. M. K. Ozias and K. L. Schalinske (2003). All-*trans*-retinoic acid rapidly induces glycine *N*-methyltransferase in a dose-dependent manner and reduces circulating methionine and homocysteine levels in rats, *J. Nutr.* **133**:4090–4094.

162. R. Hakkak, S. Korourian, S. R. Shelnutt, S. Lensing, M. J. Ronis, and T. M. Badger (2000). Diets containing whey proteins or soy protein isolate protect against 7,12-dime thylbenz(*a*)anthracene-induced mammary tumors in female rats, *Cancer Epidemiol. Biomarkers Prev.* **9**:113–117.

163. R. Hakkak, S. Korourian, M. J. Ronis, J. M. Johnston, and T. M. Badger (2001). Dietary whey protein protects against azoxymethane-induced colon tumors in male rats, *Cancer Epidemiol. Biomarkers Prev.* **10**:555–558.

164. R. Hakkak, S. Korourian, M. J. Ronis, J. M. Johnston, and T. M. Badger (2001). Soy protein isolate consumption protects against azoxymethane-induced colon tumors in male rats, *Cancer Lett.* **166**:27–32.

165. N. Fang, S. Yu, and T. M. Badger (2004). Comprehensive phytochemical profile of soy protein isolate, *J. Agric. Food Chem.* **52**:4012–4020.

166. J. K. Day, A. M. Bauer, C. DesBordes, Y. Zhuang, B. E. Kim, et al. (2002). Genistein alters methylation patterns in mice, *J. Nutr.* **132**:2419S–2423S.

167. L. Tao, W. Wang, P. M. Kramer, R. A. Lubet, V. E. Steele, and M. A. Pereira (2004). Modulation of DNA hypomethylation as a surrogate endpoint biomarker for chemoprevention of colon cancer, *Mol. Carcinog.* **39**:79–84.

168. K. Okushio, M. Suzuki, N. Matsumoto, F. Nanjo, and Y. Hara (1999). Methylation of tea catechins by rat liver homogenates, *Biosci. Biotechnol. Biochem.* **63**:430–432.

169. M. Z. Fang, Y. Wang, N. Ai, Z. Hou, Y. Sun, H. Lu, W. Welsh, and C. S. Yang (2003). Tea polyphenol (–)-epigallocatechin-3-gallate inhibits DNA methyltransferase and reactivates methylation-silenced genes in cancer cell lines, *Cancer Res.* **63**:7563–7570.

11

NUTRIENT–GENE INTERACTIONS INVOLVING SOY PEPTIDE AND CHEMOPREVENTIVE GENES IN PROSTATE EPITHELIAL CELLS

Mark Jesus M. Magbanua,[1] Kevin Dawson,[1,2,5]
Liping Huang,[3,5] Wasyl Malyj,[1,2,5] Jeff Gregg,[4,5]
Alfredo Galvez,[1,2,5] and Raymond L. Rodriguez[1,2,5]

[1] *Section of Molecular and Cellular Biology, University of California–Davis, Davis, California*
[2] *Bioinformatics Shared Resources Core, University of California–Davis, Davis, California*
[3] *Western Human Nutrition Research Center, University of California–Davis, Davis, California*
[4] *Molecular Pathology Core, University of California–Davis Medical Center, Davis, California*
[5] *Center of Excellence in Nutritional Genomics, University of California–Davis, Davis, California*

11.1 INTRODUCTION

Epidemiological studies repeatedly show associations between food intake and the incidence and severity of chronic diseases [1, 2]. The idea, however, that foods contain bioactive molecules that can affect physiology and gene expression is a relatively new one [3]. Dietary molecules are now known to affect gene expression

Nutritional Genomics: Discovering the Path to Personalized Nutrition
Edited by Jim Kaput and Raymond L. Rodriguez

directly or indirectly after modification by primary or secondary metabolism. Some components of foods can act (1) as ligands for transcription factors [4, 5], (2) as positive or negative activators of signal transduction pathways [6, 7] or (3) as ligands for nuclear receptors [8]. Genistein, vitamin A, and hyperforin are just a few well-documented examples of dietary chemicals that can bind directly to nuclear receptors and influence gene expression (see Chapter 7 in this volume).

The role of dietary factors in the etiology of different types of cancer is also well documented [9]. For example, catechins such as epigallocatechin-3-gallate (EGCG) found in green tea block signaling pathways to cell proliferation by binding to membrane receptors and inhibiting tyrosine kinases (see Chapter 8 in this volume). Diets high in EGCG are associated with reduced risk of proliferative heart disease and various types of cancers.

Diets rich in soy are also associated with lower cancer mortality rates, particularly for cancers of the colon, breast, and prostate [10–12]. Isoflavones and the Bowman–Birk protease inhibitor (BBI) are some of the components in soybeans believed to be responsible for suppressing carcinogenesis [13]. In the case of BBI, extensive studies at the University of Pennsylvania have clearly shown that this 8 kDa inhibitor of chymotrypsin and trypsin prevents chemical carcinogen and X-ray induced tumor formation in mammalian cells and in laboratory animals. Purified BBI, or its soy concentrate called BBIC, has comparable suppressive effects on the carcinogenic process in a variety of in vitro and in vivo systems. BBI (or BBIC) suppresses carcinogenesis in mice, rats, and hamsters and in several organ systems and tissue types. It also suppresses carcinogenesis in cells of hematopoietic origin and in cells of connective and epithelial tissue origin [14, 15]. Carcinogenesis suppression by BBI or BBIC can be achieved through different routes of administration, including the diet. To date, however, BBI's mechanism of action has not been elucidated. Recently, lunasin, a small peptide also found in soybean seeds, has shown promise as a chemopreventive agent [16].

11.2 LUNASIN STRUCTURE AND FUNCTION

Lunasin is a 43 amino acid small subunit of a soybean 2S albumin [17]. The carboxyl end of lunasin contains a chromatin-binding domain, a cell adhesion motif Arg-Gly-Asp (RGD) followed by eight Asp residues [17, 18]. As shown in Figure 11.1, the chromatin-binding domain consists of a 10 amino acid helical region homologous to a short conserved region found in other chromatin-binding proteins [19]. Studies in animals and mammalian cells have shown that lunasin may play a role in cell cycle control. For example, transfection of the lunasin gene into mammalian cells results in mitotic arrest and subsequent cell death [18]. In vitro studies [16] demonstrated that lunasin is capable of preferentially binding to deacetylated histones. Figure 11.2 shows lunasin's preference for deacetylated histone H4 over acetylated H4. In this experiment, truncation of lunasin's polyaspartyl tail (trLunasin-del) reduced binding to deacetylated H4 while substitution of the RGD motif (Lunasin-GRG) did not [16]. Furthermore, when chemically synthesized lunasin was added

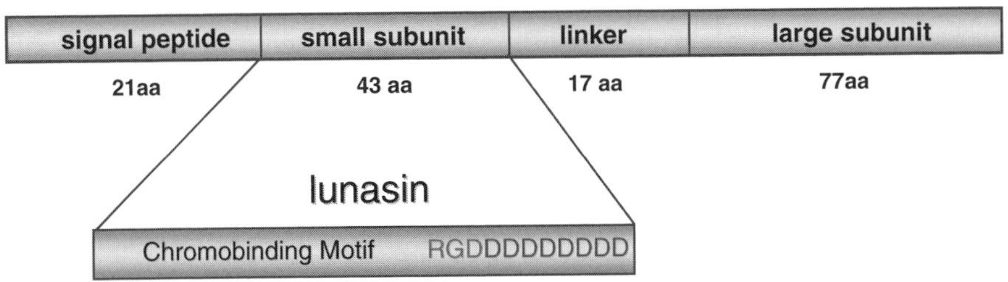

Figure 11.1. Lunasin is encoded as part of the cDNA, *Gm2S-1*, which also encodes the large subunit of a 2S albumin seed protein in soybean [17]. aa = amino acids, RGD = arginine–glycine–aspartic acid, D = aspartic acid.

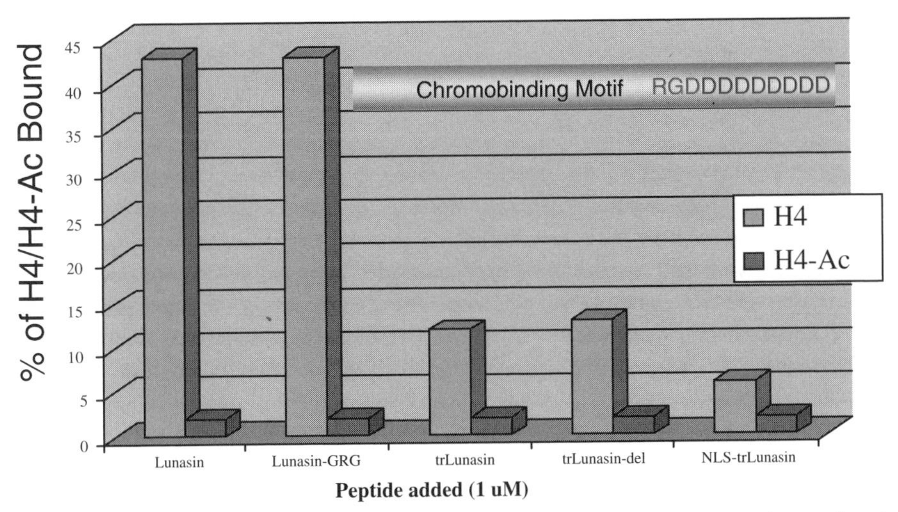

Figure 11.2. Immunobinding assays show lunasin preferentially binds to deacetylated histone H4. GRG = substitution of RGD motif; trLunasin = deletion of the polyaspartyl tail of lunasin; trLunasin-del = deletion of both the polyaspartyl tail and RDG; and NLS-trLunasin = deletion of the polyaspartyl tail and substituion of the RGD with a nuclear localization sequence (NLS). (Reprinted with permission from [16].)

exogenously to mammalian cells, it is observed to colocalize with hypoacetylated chromatin and prevent histone H3 and H4 acetylation in vivo in the presence of a histone deacetylase inhibitor [16]. Recently, lunasin was isolated from barley and was reported to possess the same biological activity ascribed to chemically synthesized lunasin [20].

While transfection of lunasin leads to cell death (see Figure 11.3), exogenously added lunasin peptide has been shown to be chemopreventive [16]. For example, significant suppression of chemical carcinogen DMBA-induced foci formation in C3H/10T1/2 mouse embryo fibroblast cells was observed when lunasin was added

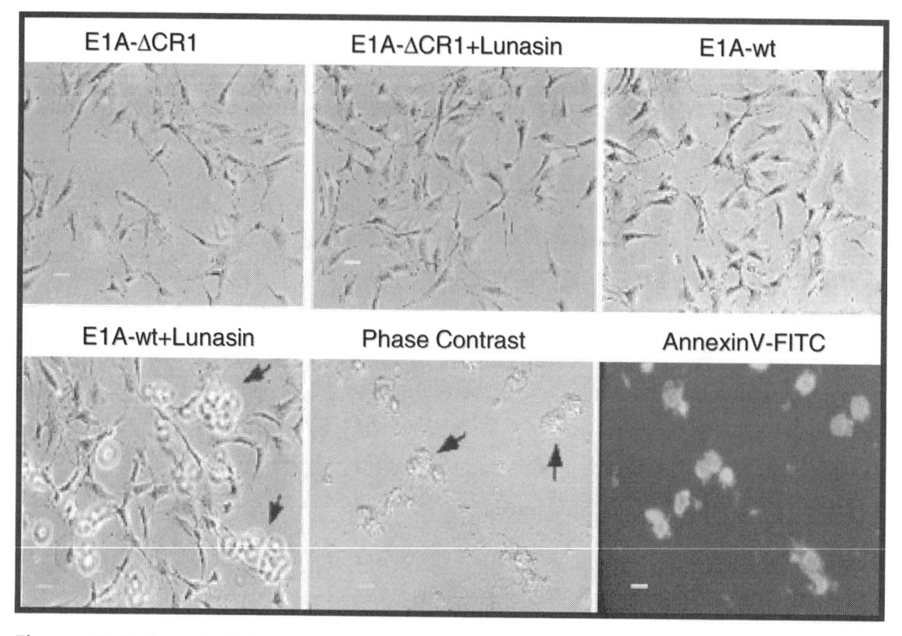

Figure 11.3. Lunasin induces apoptosis in normal C3H cells transfected with the E1A viral oncogene. C3H (murine embryonic fibroblast) cells were released from confluency and then treated with and without 2 mmolar lunasin for 20 hours, before transfection with gene constructs containing E1Awt and E1A-deletionCR1. Phase contrast images of the cells were taken 20 hours after transfection. Arrows indicate nonadherent and apoptotic cells in lunasin-treated and E1Awt-transfected cells. Annexing V-FITC (Annexin 5) stained cells visualized by fluorescent microscopy. (Reprinted with permission from [16].)

exogenously at nanomolar concentrations (Figure 11.4). In addition, topical application of lunasin peptide inhibited skin tumorigenesis in female SENCAR mice [16]. Furthermore, when C3H/10T1/2 and MCF-7 human breast cancer cells were treated with lunasin in the presence of the histone deacetylase inhibitor, sodium butyrate, a 10- to 95-fold reduction in acetylation of core histones H3 and H4 was observed [16]. The genome-wide reduction in core histone acetylation suggests an epigenetic mechanism of action for lunasin, which can influence expression of genes required for carcinogenesis. Finally, lunasin suppresses foci formation in E1A-transfected mouse fibroblast NIH 3T3 cells [21]. E1A is a viral oncoprotein that inactivates the Rb (retinoblastoma) tumor suppressor [22].

11.3 LUNASIN TREATMENT OF PROSTATE CANCER AND GENE EXPRESSION PROFILING

Prostate cancer is the most common nondermatological carcinoma in the United States with an estimated 220,900 new cases and 28,900 deaths in 2003 [23]. This type

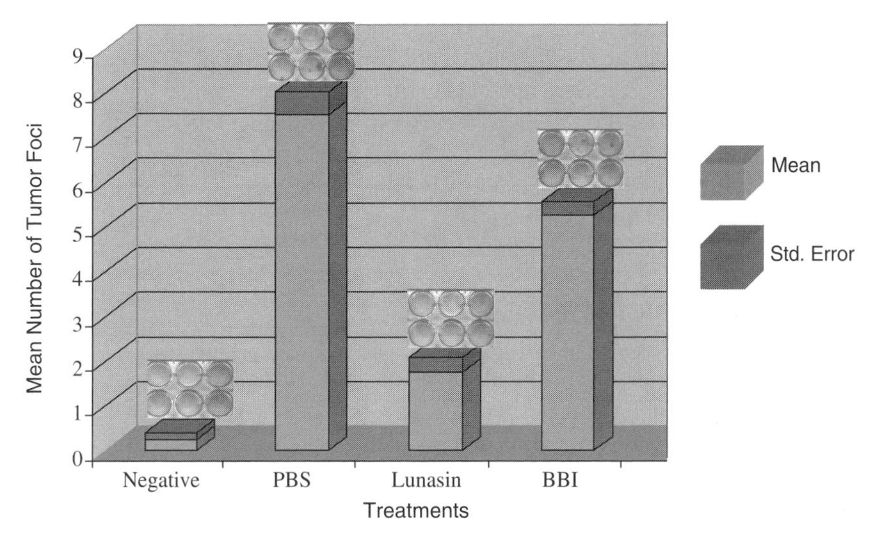

Figure 11.4. Lunasin and Bowman–Birk inhibitor (BBI) were added separately to C3H10T1/2 cells after a 4 hour treatment with 1.5 mg of DMBA. Cells were exposed to the carcinogen for 20 hours, washed with phosphate buffered saline (PBS), and fresh medium was added. For the duration of the experiment, lunasin was added (at 125 nM) to fresh medium up to the indicated time point. After 6 weeks, wells in each plate were washed with 0.9% NaCl, fixed with methanol, stained with Giemsa, and scored for transformed foci. Treatments were replicated four times in 6-well plates and twice in 24-well plates, a total of 24–48 wells/treatment. Transformation assays were conducted in duplicates at different cell passages and well sizes. Plating efficiency of each treatment was determined with a Coulter cell counter. Negative = negative untreated control; PBS = carcinogen treatment without lunasin or BBI.

of cancer is the second leading cause of death among American men [23, 24]. The effects of anticancer agents on gene expression profiles of prostate cell lines using cDNA microarray analysis have been reported [25–27]. Microarray analysis has been a useful tool for simultaneously determining changes in gene expression levels of tens of thousands of genes. Microarray analysis was recently used to determine the changes in gene expression profiles of nontumorigenic (RWPE-1) and tumorigenic (RWPE-2) prostate epithelial cell lines after 24 hour exposure to synthetic lunasin. The results of these microarray experiments are described below and they substantiate the chemopreventive property of lunasin and help elucidate its mechanism of action. Details of these experiments are reported elsewhere.

11.4 LUNASIN-INDUCED GENE EXPRESSION PROFILES

The gene expression profiles of non-tumorigenic (RWPE-1) and tumorigenic (RWPE-2) cells treated with lunasin were assessed using microarray analysis. The results of this analysis indicated that of the 14,500 genes interrogated, 123 genes had a greater than twofold change in expression in the cells exposed to 2 μM lunasin for 24 hours. Of these genes, 121 genes were upregulated in RWPE-1 cells while only two genes were upregulated in RWPE-2 cells. No genes were downregulated in either nontumorigenic or tumorigenic epithelial cells treated with 2 μM lunasin. Genes upregulated in RWPE-1 cells include those involved in tumor suppression, apoptosis, and the control of cell division (Table 11.1).

11.5 GENES FOR APOPTOSIS

Transfection of lunasin into mammalian cells results in arrest of mitosis leading to apoptosis [18]. The antimitotic effect of lunasin is believed to be due to the binding of its polyaspartyl carboxyl end to regions of hypoacetylated chromatin, like that found in kinetochores in centromeres [18]. Apoptosis is triggered when the kinetochore complex does not form properly and the microtubules fail to attach to the centromeres, leading to mitotic arrest and eventually cell death. Our microarray results show that exogenous addition of lunasin upregulates genes that are known to play a direct or indirect role in the induction of apoptosis (Figure 11.5). These genes include thrombospondin 1 (THBS1), BCL2/adenovirus E1B 19 kDa interacting protein 3 (BNIP3), and protein kinase C-like 2 (PRKCL2). THBS1 (also called TSP-1) is a member of a family of extracellular proteins that participate in cell-to-cell and cell-to-matrix communication [32]. The role of THBS1 in the process of

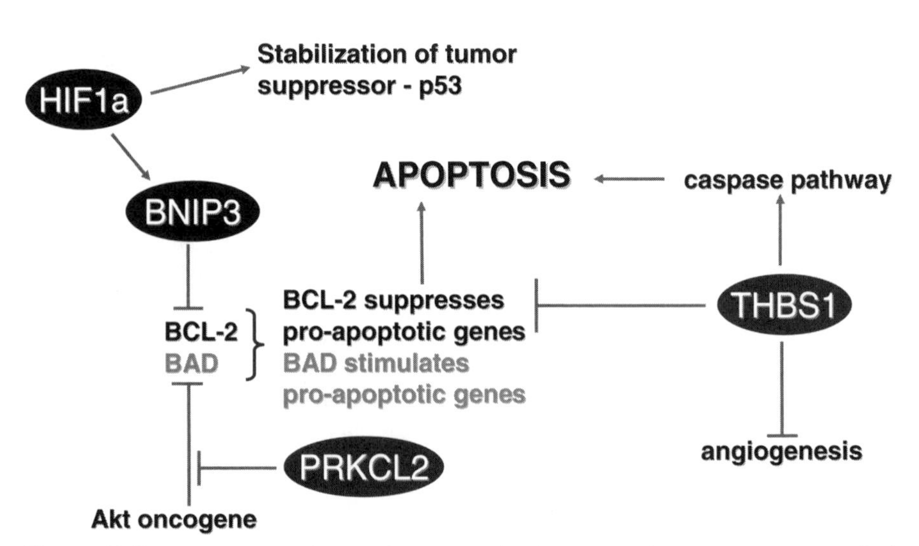

Figure 11.5. Lunasin upregulation of genes involved in apoptosis. Apoptosis is controlled by genes that suppress (*bcl-2*) or promote (*bad* and the caspases) the apoptotic process. In addition to promoting apoptosis, lunasin upregulates genes (black ellipses) that can also stabilize tumor suppressors like p53 and suppress genes involved in angiogenesis.

TABLE 11.1. Genes Upregulated by Lunasin in Nontumorigenic Cell Line (RWPE-1) and Tumorigenic Cell Line (RWPE-2)

	Affymetrix Probesets	Gene Identifiers	Gene Name	Official Gene Symbol
			A. Genes Upregulated in Nontumorigenic Prostate Epithelial Cell Line (RWPE-1)	
1	200914_x_at	3895	Kinectin 1 (kinesin receptor)	*KTN1*
2	201730_s_at	7175	Translocated promoter region (to activated MET oncogene)	*TPR*
3	200915_x_at	3895	Kinectin 1 (kinesin receptor)	*KTN1*
4	213229_at	23405	Dicer1, Dcr-1 homologue (*Drosophila*)	*DICER1*
5	200050_at	7705	Zinc finger protein 146	*ZNF146*
6	201667_at	2697	Gap junction protein, alpha 1, 43 kDa (connexin 43)	*GJA1*
7	201699_at	5706	Proteasome (prosome, macropain) 26S subunit, ATPase, 6	*PSMC6*
8	204455_at	667	Bullous pemphigoid antigen 1, 230/240 kDa	*BPAG1*
9	208896_at	8886	DEAD (Asp-Glu-Ala-Asp) box polypeptide 18	*DDX18*
10	200603_at	5573	Protein kinase, camp-dependent, regulatory, type I, alpha (tissue specific extinguisher 1)	*PRKAR1A*
11	212648_at	54505	DEAH (Asp-Glu-Ala-His) box polypeptide 29	*DHX29*
12	216268_s_at	182	Jagged 1 (Alagille syndrome)	*JAG1*
13	218181_s_at	54912	Hypothetical protein FLJ20373	
14	219918_s_at	259266	Asp (abnormal spindle)-like, microcephaly associated (*Drosophila*)	*ASPM*
15	203789_s_at	10512	Sema domain, immunoglobulin domain (Ig), short basic domain, secreted, (semaphorin) 3C	*SEMA3C*
16	217975_at	51186	pp21 homologue	
17	204240_s_at	10592	SMC2 structural maintenance of chromosomes 2-like 1 (yeast)	*SMC2L1*
18	202169_s_at	60496	Aminoadipate-semialdehyde dehydrogenase-phosphopantetheinyl transferase	*AASDHPPT*
19	214709_s_at	3895	Kinectin 1 (kinesin receptor)	*KTN1*
20	221802_s_at	57698	KIAA1598 protein	
21	201713_s_at	5903	RAN binding protein 2	*RANBP2*
22	212629_s_at	5586	Protein kinase C-like 2	*PRKCL2*
23	220085_at	3070	Helicase, lymphoid-specific	*HELLS*
24	213168_at	6670	Sp3 transcription factor	*SP3*

TABLE 11.1. (*Continued*)

	Affymetrix Probesets	Gene Identifiers	Gene Name	Official Gene Symbol
25	212888_at	23405	Dicer1, Dcr-1 homolog (Drosophila)	*DICER1*
26	212245_at	90411	Multiple coagulation factor deficiency 2	*MCFD2*
27	213294_at		Hypothetical protein FLJ38348	
28	208808_s_at	3148	High-mobility group box 2	*HMGB2*
29	202704_at	10140	Transducer of ERBB2, 1	*TOB1*
30	212582_at	114882	Oxysterol binding protein-like 8	*OSBPL8*
31	202381_at	8754	A disintegrin and metalloproteinase domain 9 (meltrin gamma) hypothetical	*ADAM9*
32	212640_at		protein LOC201562	
33	211953_s_at	3843	Karyopherin (importin) beta 3	*KPNB3*
34	205596_s_at	64750	E3 ubiquitin ligase SMURF2	*SMURF2*
35	211929_at	10151	Heterogeneous nuclear ribonucleoprotein A3	*HNRPA3*
36	201889_at	10447	Family with sequence similarity 3, member C	*FAM3C*
37	203820_s_at	10643	IGF-II mRNA-binding protein 3	
38	208731_at	5862	RAB2, member RAS oncogene family	*RAB2*
39	211967_at	114908	Pro-oncosis receptor inducing membrane injury gene	*PORIMIN*
40	217945_at	53339	BTB (POZ) domain containing 1	*BTBD1*
41	211257_x_at	27332	NP220 nuclear protein	
42	212250_at		LYRIC/3D3	
43	207941_s_at	9584	RNA-binding region (RNP1, RRM) containing 2	*RNPC2*
44	200605_s_at	5573	Protein kinase, cAMP-dependent, regulatory, type I, alpha (tissue specific extinguisher 1)	*PRKAR1A*
45	201486_at	5955	Reticulocalbin 2, EF-hand calcium binding domain	*RCN2*
46	217941_s_at	55914	erbb2 interacting protein	*ERBB2IP*
47	203755_at	701	BUB1 budding uninhibited by benzimidazoles 1 homolog beta (yeast)	*BUB1B*
48	201242_s_at	481	ATPase, Na+/K+ transporting, beta 1 polypeptide	*ATP1B1*
49	201737_s_at	10299	Similar to *S. Cerevisiae* SSM4	
50	221773_at	2004	ELK3, ETS-domain protein (SRF accessory protein 2)	*ELK3*
51	217816_s_at	57092	PEST-containing nuclear protein	
52	209187_at	1810	Downregulator of transcription 1, TBP-binding (negative cofactor 2)	*DR1*
53	202760_s_at	11217	A kinase (PRKA) anchor protein 2	*AKAP2*

54	212352_s_at	10972	Transmembrane trafficking protein	
55	213285_at	NA	Hypothetical protein LOC161291	
56	212692_s_at	987	LPS-responsive vesicle trafficking, beach and anchor containing	*LRBA*
57	212718_at	27249	Hypothetical protein CL25022	
58	209259_s_at	9126	Chondroitin sulfate proteoglycan 6 (bamacan)	*CSPG6*
59	201595_s_at	55854	Likely orthologue of mouse immediate early response, erythropoietin 4	
60	202551_s_at	51232	Cysteine-rich motor neuron 1	*CRIM1*
61	218067_s_at	55082	Hypothetical protein FLJ10154	
62	211945_s_at	3688	Integrin, beta 1 (fibronectin receptor, beta polypeptide, antigen CD29 includes MDF2, MSK12)	*ITGB1*
63	209025_s_at	10492	Synaptotagmin binding, cytoplasmic RNA interacting protein	*SYNCRIP*
64	202599_s_at	8204	Nuclear receptor interacting protein 1	*NRIP1*
65	201991_s_at	3799	Kinesin family member 5B	*KIF5B*
66	202413_s_at	7398	Ubiquitin specific protease 1	*USP1*
67	209422_at	51230	Chromosome 20 open reading frame 104	*C20orf104*
68	201132_at	3188	Heterogeneous nuclear ribonucleoprotein H2 (H′)	*HNRPH2*
69	214055_x_at	23215	HBxAg transactivated protein 2	
70	203743_s_at	6996	Thymine-DNA glycosylase	*TDG*
71	211969_at	3320	Heat shock 90 kDa protein 1, alpha	*HSPCA*
72	204059_s_at	4199	Malic enzyme 1, NADP(+)-dependent, cytosolic	*ME1*
73	221505_at	81611	Acidic (leucine-rich) nuclear phosphoprotein 32 family, member E	*ANP32E*
74	212408_at	26092	Lamina-associated polypeptide 1B	
75	212615_at	80205	Hypothetical protein FLJ12178	
76	218577_at	55631	Hypothetical protein FLJ20331	
77	201735_s_at	1182	Chloride channel 3	*CLCN3*
78	200995_at	10527	Importin 7	*IPO7*
79	217758_s_at	56889	SM-11044 binding protein	
80	201505_at	3912	Laminin, beta 1	*LAMB1*
81	201690_s_at	7163	Tumor protein D52	*TPD52*
82	203362_s_at	4085	MAD2 mitotic arrest deficient-like 1 (yeast)	*MAD2L1*
83	202277_at	10558	Serine palmitoyltransferase, long chain base subunit 1	*SPTLC1*
84	208783_s_at	4179	Membrane cofactor protein (CD46, trophoblast-lymphocyte cross-reactive antigen)	*MCP*
85	217523_at		Data not available	

TABLE 11.1. (*Continued*)

	Affymetrix Probesets	Gene Identifiers	Gene Name	Official Gene Symbol
86	204822_at	7272	TTK protein kinase	*TTK*
87	218542_at	55165	Chromosome 10 open reading frame 3	*C10orf3*
88	201939_at	10769	Serum-inducible kinase	
89	204976_s_at	9949	Alport syndrome, mental retardation, midface hypoplasia and elliptocytosis chromosomal region, gene 1	*AMMECR1*
90	202234_s_at	6566	Solute carrier family 16 (monocarboxylic acid transporters), member 1	*SLC16A1*
91	204258_at	1105	Chromodomain helicase DNA binding protein 1	*CHD1*
92	201862_s_at	9208	Leucine-rich repeat (in FLII) interacting protein 1	*LRRFIP1*
93	213070_at	6165	Ribosomal protein L35a	*RPL35A*
94	204058_at	4199	Malic enzyme 1, NADP(+)-dependent, cytosolic chloride	*ME1*
95	201734_at		channel 3	*CLCN3*
96	209272_at	4664	NGFI-A binding protein 1 (EGR1 binding protein 1)	*NAB1*
97	201745_at	5756	PTK9 protein tyrosine kinase 9	*PTK9*
98	213729_at	55660	Formin binding protein 3	*FNBP3*
99	209115_at	9039	Ubiquitin-activating enzyme E1C (UBA3 homolog, yeast)	*UBE1C*
100	201849_at	664	BCL2/adenovirus E1B 19 kDa interacting protein 3	*BNIP3*
101	212455_at	91746	Splicing factor YT521-B	
102	201689_s_at	7163	Tumor protein D52	*TPD52*
103	201110_s_at	7057	Thrombospondin 1	*THBS1*
104	212149_at	23167	KIAA0143 protein	
105	212192_at	115207	Potassium channel tetramerization domain containing 12	*KCTD12*
106	209476_at	81542	Thioredoxin domain containing	*TXNDC*
107	214363_s_at	9782	Matrin 3	*MATR3*
108	201567_s_at	2803	Golgi autoantigen, golgin subfamily a, 4	*GOLGA4*
109	200626_s_at	9782	Matrin 3	*MATR3*
110	203804_s_at	10414	Acid-inducible phosphoprotein	
111	200977_s_at	8887	Tax1 (human T-cell leukemia virus type I) binding protein 1	*TAX1BP1*
112	212893_at	26009	Zinc finger, ZZ domain containing 3	*ZZZ3*

113	201435_s_at	1977	Eukaryotic translation initiation factor 4E	*EIF4E*
114	215548_s_at	23256	Sec1 family domain containing 1	*SCFD1*
115	212248_at		LYRIC/3D3	
116	201398_s_at	23471	Translocation associated membrane protein 1	*TRAM1*
117	201304_at	4698	NADH dehydrogenase (ubiquinone) 1 alpha subcomplex, 5, 13 kDa	*NDUFA5*
118	203987_at	8323	Frizzled homologue 6 (*Drosophila*)	*FZD6*
119	208925_at	56650	Chromosome 3 open reading frame 4	*C3orf4*
120	204094_s_at	9819	KIAA0669 gene product	
121	200989_at	3091	Hypoxia-inducible factor 1, alpha subunit (basic helix–loop–helix transcription factor)	*HIF1A*
			B. Genes Upregulated in Tumorigenic Prostate Epithelial Cell Line (RWPE-2)	
1	203325_s_at	1289	Collagen, type V, alpha 1	*COL5A1*
2	201551_s_at	3916	Lysosomal-associated membrane protein 1	*LAMP1*

The genereal conditions for microarray analysis are as follows. Prostate epithelial cells, RWPE-1 and RWPE-2, were maintained in the keratinocyte–serum free medium with 5 ng/mL rEGF and 0.05 mg/mL bovine pituitary extract (Invitrogen). The RWPE-1 cell line was immortalized from histological normal prostate epithelial cells. The RWPE-2 cell line was derived from RWPE-1 by transforming the cells with Ki-ras. RWPE-1 and RWPE-2 share the same genotype, differing only in their tumorigenic status [28, 29]. Cells were grown to 70% confluence, after which cells were harvested and transferred to a 150 mm^2 plate at a cell density of 1×10^7 cells/mL. The cells were incubated with synthetic lunasin peptide (American Peptide Company, Sunnyvale, California) at a final concentration of 2 µM. The cells were incubated for 24 hours after lunasin treatment for 24 hours. RWPE-1 and RWPE-2 cells that were not treated with lunasin served as controls. After lunasin treatment, cells were washed twice with 1X PBS and total RNA was extracted. The cDNA used in microarray analysis was synthesized from 10 µg of total RNA using the SuperScript Choice system (Invitrogen). The cDNA was then transcribed in vitro in the presence of biotin-labeled nucleotide triphosphates using T7 RNA polymerase after phenol-chloroform extraction and ethanol precipitation. cRNA was purified using the RNeasy mini kit (Qiagen) and fragmented at 94 °C for 30 minutes in the buffer containing 0.2 M Tris-acetate (pH 8.1), 0.5 M potassium acetate, and 0.15 M magnesium acetate. Fragmented cRNA was hybridized overnight at 45 °C to the human genome U133A GeneChip (Affymetrix) representing approximately 14,500 genes. Hybridization was then detected using a confocal laser scanner (Affymetrix). The gene expression levels of samples were normalized and analyzed using Robust Multichip Average analysis [30]. Average-linkage hierarchical clustering of data was applied using the Gene Cluster 3.0 [31] and the results were displayed with Java TreeView 1.0.1 [31]. The cDNA used for quantitative PCR was synthesized from 3 µg of total RNA using the SuperScript First-Strand Synthesis for RT-PCR kit (Invitrogen) and the cDNA was diluted fourfold and 2 µL of cDNA was added to the quantitative PCR reaction using FAM-labeled TaqMan probes purchased from Applied Biosystems. The quantitative PCR reactions were performed in triplicates using a PRISM® ABI 7900HT Sequence Detection System (Applied Biosystems). The expression of the β-actin gene (BACT) was used for normalization. Changes in expression were calculated using relative quantification as follows: $\Delta\Delta Ct = \Delta Ct q - \Delta Ct cb$, where Ct is the cycle number at which amplification rises above the background threshold, ΔCt is the change in Ct between two test samples, q is the target gene, and cb is the calibrator gene, BACT. Gene expression is then calculated as $2^{-\Delta\Delta Ct}$ (Applied Biosystems). EASE Version 2.0 was used for annotating the genes listed above.

apoptosis or programmed cell death in cancer cells has recently been reviewed [33]. THBS1 induces apoptosis by activating the caspase cell death pathway [34]. In addition, THBS1 has also been shown to have a potent antiangiogenic activity [34] and the induction of apoptosis by THBS1 is associated with decreased expression of the antiapoptotic gene, *bcl-2* [34]. We find it intriguing that lunasin also upregulates the expression of BNIP3, a gene that inhibits the antiapoptotic activities of BCL-2. BNIP3, formerly known as Nip3, is a mitochondrial protein that activates apoptosis and overcomes BCL-2 activity [35–37]. A more recent study shows that BNIP3 mediates a necrosis-like cell death independent of apoptotic events, such as release of cytochrome *c*, caspase activation, and nuclear translocation of apoptosis-inducing factor [38]. Evidence indicates that BNIP3 causes cell death through opening of the mitochondrial permeability transition pores, resulting in mitochondrial dysfunction and plasma membrane damage [38].

Another gene upregulated by lunasin is the hypoxia-inducible factor-1, alpha subunit (HIF-1A), a basic helix–loop–helix transcription factor that regulates the expression of BNIP3 [39]. The BNIP3 promoter contains a functional HIF-1A responsive element and is potently activated by both hypoxia and forced expression of HIF-1A [40]. Studies have also shown that HIF-1A binds and stabilizes p53, a tumor suppressor [41, 42]. On the other hand, PRKCL2, also termed PRK2, promotes apoptosis by inhibiting the antiapoptotic activities of the oncogene Akt [43]. Akt exerts its antiapoptotic effects by inactivating BAD, a proapoptotic BCL-2 family protein, by phosphorylation [44]. However, a PRKCL2 C-terminal fragment generated during the early stages of apoptosis binds Akt, resulting in the inhibition of the Akt-mediated phosphorylation of BAD, thereby allowing apoptosis to occur [43]. Based on our microarray data, we propose that lunasin primes the precancerous cell for apoptosis by upregulating proapoptotic genes like *bad* and inhibiting the antiapoptotic activities of Akt and some members of the *bcl-2* gene family. We also suggest that lunasin may be indirectly involved in stabilization of tumor suppressor p53 via its interaction with HIF-1A.

11.6 GENES INVOLVED IN SUPPRESSION OF CELL PROLIFERATION

Lunasin's inhibitory effects on carcinogenesis can be explained by the upregulation of genes that play a role in tumor suppression or antiproliferation (Figure 11.6). For example, lunasin upregulates a tumor suppressor gene encoding the cyclic AMP-dependent protein kinase A type I-α regulatory subunit, PRKAR1A [45]. Mutations in the PRKAR1A result in the Carney complex (CNC), a multiple neoplasia syndrome that is associated with thyroid tumorigenesis [45, 46]. It is proposed that PRKAR1A mutant cells have deregulated control of gene expression, which results in the activation of cAMP signaling pathways and abnormal growth and proliferation [46]. Another gene upregulated by lunasin, BTB (POZ) containing domain 1 (ABTB1 or BPOZ), is thought to be one of the mediators of the growth-suppressive signaling pathway of the tumor suppressor PTEN [47]. Inactivation of the PTEN gene is

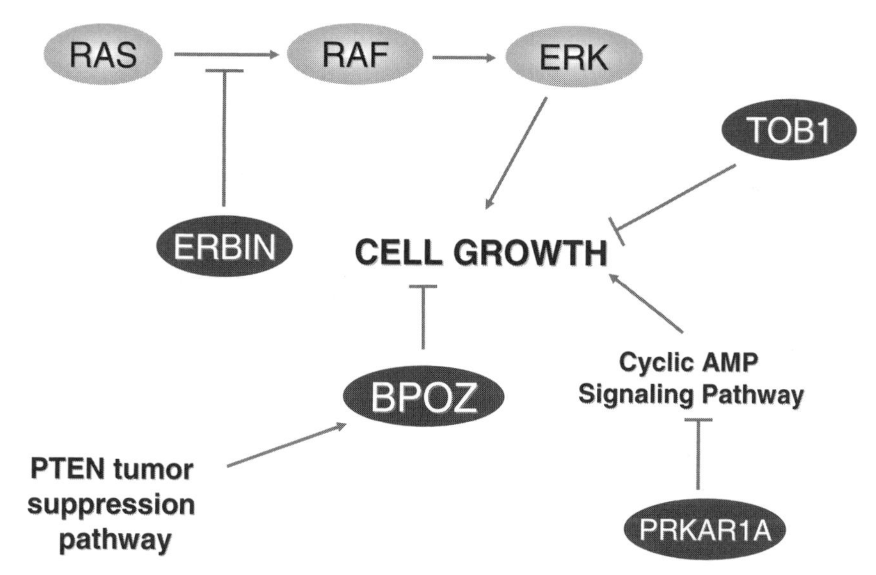

Figure 11.6. Lunsin upregulates genes (black ellipses) involved in tumor suppression and suppression of cell growth.

extremely common in human cancer, including cancer of the prostate [48]. Overexpression of BPOZ inhibits cell cycle progression and suppresses growth of cancer cells while the transfection of BPOZ antisense accelerates cell growth [47]. Another antiproliferative gene upregulated by lunasin is Tob (also referred to as Tob1). Tob is a member of the antiproliferative BTG/Tob family [49] and mice that are Tob deficient are prone to spontaneous formation of tumors [50].

The gene *erbb2* interacting protein (ERBIN), a novel suppressor of Ras signaling, is upregulated by lunasin. ERBIN, a leucine-rich repeat-containing protein, interacts with Ras and interferes with the interaction between Ras and Raf, resulting in the negative regulation of Ras-mediated activation of extracellular signal regulated kinases (Erk) [51]. The Ras oncogene is one of the most common mutations occurring in about 30% of human cancers [52]. Mutations that cause constitutive activation of Ras result in a continuous signal that tell the cells to grow regardless of whether or not receptors on the cell surface are activated by growth factors [53].

11.7 MITOTIC CHECKPOINT GENES

Other genes upregulated by lunasin include the mitotic checkpoint genes like budding uninhibited benzimidazoles 1 homologue beta (BUB1B or BubR1), TTK protein kinase (a homologue of yeast MPS1) and mitotic arrest deficient 2-like 1 (MAD2L1). Mitotic spindle checkpoint proteins monitor proper microtubule attachment to chromosomes prior to progression through mitosis, allowing correct segregation of chromosomes into progeny cells [54]. TTK is a protein kinase that phosphorylates

MAD1p—a process essential for the activation of the mitotic checkpoint [55]. In yeast, MPS1 is required early in the spindle assembly checkpoint [56]. It is considered to be a limiting step in checkpoint activation, since it can activate the pathway when overexpressed. Overexpression of MPS1 is able to delay cell cycle progression into anaphase in a manner similar to checkpoint activation by spindle damage [56]. MPS1 is also required for the essential process of spindle pole body duplication [57]. BubR1 is a protein kinase required for checkpoint control. Evidence shows that inactivation of BubR1 by microinjection of specific antibodies abolishes the checkpoint control [58]. Another study revealed that endogenous BubR1 protein levels are reduced in some breast cancer cell lines. Furthermore, the breast cancer-specific gene 1 (BCSG1), coding for an oncogenic protein, directly interacts with BubR1, resulting in the degradation of BubR1 through the proteosome machinery [59]. It is speculated that BCSG1-induced reduction of the BubR1 protein allows breast cancer to progress at least in part by compromising the mitotic checkpoint control through the inactivation of BubR1 [59]. Another checkpoint gene MAD2L1 was reported to have reduced expression in a human breast cancer cell line exhibiting chromosome instability and aneuploidy [60]. Some breast cancer cell lines carry a mutation in the MAD2L1 gene that creates a truncated protein product; however, the specific role of MAD2L1 in breast cancer is still under investigation [61]. We propose that lunasin upregulates these mitotic checkpoint genes to allow a heightened level of molecular surveillance to prevent premature cell division, chromosome instability, and aneuploidy.

11.8 GENES INVOLVED IN PROTEIN DEGRADATION

Our microarray results also show that lunasin upregulates several genes involved in protein degradation and turnover via the ubiquitin pathway. These genes include the proteosome 26S subunit ATPase 6 (PSMC6); E3 ubiquitin ligase (SMURF2); ubiquitin specific protease 1 (USP1); and ubiquitin-activating enzyme E1C (UBE1C). It is possible that lunasin upregulates these genes to mediate the degradation of proteins that are required for the onset of cell transformation and foci formation.

11.9 CONNEXIN 43 GENE FOR THE GAP JUNCTION PROTEIN

Adjacent cells communicate with each other through gap junctional channels that allow the passage of small molecules [62]. This process is referred to as "gap junctional intercellular communication" (GJC) and is blocked in many cancer cells, including malignant human prostate cells [63]. Gap junctional channels are composed of proteins called connexins [64]. Lunasin upregulates the expression of a gap junction protein called connexin 43, which has been shown to have a tumor suppressive role. Decreased expression and impaired post-translational modification of connexin 43 were observed in several prostate tumor cell lines but not in normal cells, suggesting that the loss of junctional communication is a critical step in the progression to human prostate cancer [63]. Studies have shown that the viral oncogene Src disrupts cell growth regulation by adding a phosphate group to a tyrosine residue in connexin 43, thereby blocking gap junction communication [65]. Transfection of a functional connexin 43 gene into tumorigenic mouse cells results in the restoration of GJC, normal growth regulation, and cell-to-cell communication, as well as suppression of tumorigenesis [66].

11.10 TARGET VERIFICATION USING RT-PCR

Changes in mRNA levels detected by microarray analysis were confirmed using real-time PCR (RT-PCR) analysis of four genes: thrombospondin 1 (THBS1); protein kinase, cAMP-dependent, regulatory, type 1 alpha (PRKAR1A); transducer of ERBB2, 1 (TOB1); and hypoxia-inducible factor 1, alpha subunit (HIF-1A). These genes were selected because they represent genes involved in apoptosis and cell proliferation pathways. The results of the RT-PCR analysis for these selected genes were consistent with the microarray data (data not shown). These results support our interpretation of the microarray data, that lunasin upregulates the expression of genes of normal prostate epithelial cells but not in established malignant prostate epithelial cells.

11.11 CONCLUSION

The microarray data presented here reveal that lunasin upregulates genes involved in mitotic checkpoint, tumor suppressor genes, and genes that promote apoptosis. This suggests that lunasin acts, either directly or indirectly, as a transcriptional activator of genes that protect normal cells from transformation. These findings are in contrast to previous studies suggesting that lunasin prevents normal cell transformation into tumors by preventing the acetylation of hypoacetylated histones H3 and H4 [16]. It is believed that blocking the acetylation of these histones results in chromatin condensation and transcriptional inactivity of oncogenes. In a recent study, however, lunasin-treated NIH 3T3 cells transfected with the E1A oncogene showed a fivefold increase in p21/WAF1/Cip1 protein levels [21]. The protein p21/WAF1/Cip1 is a potent and universal inhibitor of cyclin-dependent kinases, which are major control points of cell cycle progression [67]. To investigate the effect of lunasin on global gene expression, microarray analysis was performed on nontumorigenic (RWPE-1) and tumorigenic (RWPE-2) prostate epithelial cells. Our microarray results did not show upregulation of p21/WAF1/Cip1 within 24 hours. Western blot analysis has shown that p21/WAF1/Cip1 protein levels increase fivefold in NIH 3T3 cells (pretreated with lunasin for 24 hours) 8 days after E1A transfection [21]. However, the gene SP3, a transcriptional activator of p21/WAF1/Cip1 [68], was upregulated by lunasin at 24 hours, which can explain the later increase in expression of p21/WAF1/Cip1 in the NIH 3T3 cells.

In addition, our microarray results help explain the 70% reduction of foci formation observed when C3H/T101/2 cells pretreated with lunasin are exposed to the chemical carcinogens 7,12-dimethylbenz[*a*]-anthracene (DMBA) and 3-methylcholanthrene (MCA) [16]. A single 24 hour exposure of these cells to 125 nM lunasin was sufficient in suppressing foci formation in chemical carcinogenesis assays that lasted for 6 weeks [16]. We speculate that the 24 hour pretreatment of C3H/T101/2 cells with lunasin upregulates expression of chemopreventive genes that protect cells from transformation induced by DMBA and MCA.

It is also interesting to note that lunasin has little effect on the gene expression profiles of RWPE-2, a tumorigenic line of prostate epithelial cells. This finding agrees with previous experiments, which showed that lunasin has no cytotoxic effect on transformed cells [21]. The transformation of RWPE-2 with *KI-RAS* oncogene, however, does result in changes in growth properties, increase in genetic instability, and morphological alterations [69]. We hypothesize that the morphological changes in RWPE-2 cells, in effect, resulted in its insensitivity to lunasin treatment, probably because these tumorigenic cells neither bind nor internalize lunasin as efficiently as RWPE-1 cells. It is reasonable to assume that the presence or absence of integrin receptors on the cell membrane will determine whether or not a small peptide like lunasin is internalized. For example, a human breast carcinoma cell line (MDA-MB-435) lacks $\alpha_V\beta_3$ and $\alpha_V\beta_5$ integrins and, therefore, cannot bind a chemotherapeutic proapoptotic peptide containing an Asn-Gly-Arg (NGR) cell adhesion motif. On the other hand, a Kaposi sarcoma-derived cell line can bind and incorporate the same peptide very efficiently [70, 71]. In addition, the specific anticancer activity of another chemotherapeutic proapoptotic peptide containing the RGD motif has been attributed to the fact that it binds to human α_V integrins, which are known to be selectively expressed in human tumor blood vessels [72].

Aside from lunasin, several other components of soybean have been proposed to act as anticancer agents [13]. For example, the Bowman–Birk inhibitor (BBI), a soybean-derived serine protease inhibitor, has been shown to posses anticarcinogenic activity in both in vitro and in vivo systems [73]. Unlike lunasin, which is not cytotoxic to normal and cancer cells [21], BBI and BBI concentrate (BBIC), a soybean concentrate enriched in BBI, inhibited the growth, invasion, and clonogenic survival of prostate cancer cell lines [15]. The precise mechanism(s) for the suppressive effects of BBI and BBIC is currently under investigation. Another possible chemopreventive agent from soy is genistein, a major isoflavone in soybeans (see Chapter 14 in this volume). Genistein has been found to inhibit carcinogenesis both in vitro and in vivo [74] and it is known to inhibit the activation of the nuclear transcription factor, NF-κB, and the Akt signaling pathway, both of which are known to maintain the balance between cell survival and apoptosis [75, 76]. Evidence shows that genistein induces apoptosis by upregulating BAX, a protein that antagonizes the antiapoptotic function of Bcl-2 [75]. A microarray analysis of the gene expression profiles of PC3 prostate cancer cells treated with genistein showed downregulation of 774 genes and the upregulation of 58 genes [77]. Similar to lunasin's effect, genistein altered the expression of genes that are mainly involved in the regulation of cell growth, cell cycle, apoptosis, cell signal transduction, angiogenesis, tumor cell invasion, and metastasis [77, 78]. In contrast to BBI and genistein, lunasin has little or no inhibitory effect on transformed prostate cancer cells, which is confirmed by our microarray data showing that lunasin upregulates the expression of only two genes (out of 14,500 arrayed genes) in tumorigenic prostate cancer cells. These genes are collagen, type V, alpha 1 and lysosomal-associated membrane protein 1.

Other short peptides have also been reported to prevent cancer. Ellerby and colleagues [70] designed 21–26 amino acid peptides that were selectively toxic to angiogenic endothelial cells and showed anticancer activity in mice. Interestingly, there are two striking structural similarities between lunasin and these peptides. Like lunasin, these peptides contain a cell adhesion motif (RGD or NGR) in their "homing domain" and a short helical region in their proapoptotic domain. However, unlike lunasin, these peptides do not possess a negatively charged tail. These targeted proapoptotic peptides may represent a new class of anticancer therapeutics [70].

This study is the first to use gene expression profiling to investigate the potential chemopreventive properties of a peptide derived from soybean. Based on the data obtained from our microarray analysis, we propose that lunasin can act as a transcriptional activator to upregulate genes necessary for protecting the cells from transformation events induced by either chemical carcinogens, oncogenes, or mutated tumor suppressor genes. Lunasin primes the cell for apoptosis and upregulates the expression of some tumor suppressor genes and genes involved in the activation of mitotic checkpoint. Furthermore, lunasin's selective effect on human gene expression supports an extensive body of epidemiological evidence linking high soybean intake and reduced risks of certain types of cancer such as breast, prostate, and colon [10–12].

The apparent action of lunasin as a transcriptional activator and/or an inhibitor of histone acetylation still needs further study. It is possible that lunasin affects the transcriptional activity of chemopreventive genes indirectly by altering intracellular NAD homeostasis and energy balance [79]. NADH reoxidation is associated with mitochondrial electron transport activity and NAD is a cofactor for histone deacetylases involved in chromatin remodeling [80, 81]. The short- and long-term consequences of chromatin remodeling (e.g., histone acetylation and deacetylation) on gene regulation is well documented [82]. That lunasin upregulates the expression of genes that can influence NAD homeostasis may provide the link between lunasin's chemopreventive properties and its previously described inhibition of histone acetylation.

Additional research on lunasin will help elucidate how it affects the regulation of chemopreventive genes at the transcriptional, post-transcriptional, post-translational, and genome levels. Some of the questions that still need to be addressed include the following:

1. What is the precise mechanism of action that gives lunasin its chemopreventive properties?
2. Does lunasin work independently, or in combination with other proteins in the cell, to upregulate gene expression?

3. Is lunasin modified (e.g., phosphorylated) in mammalian cells and is this modification required for its chemopreventive properties?
4. Why are prostate tumor cells refractory to lunasin treatment?
5. Are there genetic variants (e.g., SNPs) in human genes that can affect the way lunasin interacts with the components of the cell?

A better understanding of how lunasin interacts with the mammalian cell, and its many components, should lead to more effective chemopreventive strategies based on dietary interventions.

ACKNOWLEDGMENTS

The work was supported in part by a grant from the National Institutes of Health, P60MD00222. We would like to thank Benito de Lumen for his helpful advice and FilGen Bioscience Inc. for the generous gift of chemically synthesized lunasin peptide. This chapter is dedicated to Professor Bruce N. Ames for his pioneering scientific achievements and leadership in promoting human health through a better understanding of gene–environment interactions.

REFERENCES

1. D. Jenkins, C. Kendall, and T. Ransom (1998). Dietary fiber, the evolution of the human diet and coronary heart disease, *Nutr. Res.* **18**:633–652.
2. W. Willett (2002). Isocaloric diets are of primary interest in experimental and epidemiological studies, *Int. J. Epidemiol.* **31**:694–695.
3. J. Kaput and R. L. Rodriguez (2004). Nutritional genomics: The next frontier in the postgenomic era, *Physiol. Genomics* **16**:166–177.
4. M. N. Jacobs and D. F. Lewis (2002). Steroid hormone receptors and dietary ligands: A selected review, *Proc. Nutr. Soc.* **61**:105–122.
5. M. J. Dauncey, P. White, K. A. Burton, and M. Katsumata (2001). Nutrition–hormone receptor–gene interactions: Implications for development and disease, *Proc. Nutr. Soc.* **60**:63–72.
6. M. Masuda, M. Suzui, and I. Weinstein (2001). Effects of epigallocatechin-3-gallate on growth, epidermal growth factor receptor signaling pathways, gene expression, and chemosensitivity in human head and neck squamous cell carcinoma cell lines, *Clin. Cancer Res.* **7**:4220–4229.
7. S. Pianetti, S. Guo, K. Kavanagh, and G. Sonenshein (2002). Green tea polyphenol epigallocatechin-3 gallate inhibits Her-2/neu signaling, proliferation, and transformed phenotype of breast cancer cells, *Cancer Res.* **62**:652–655.
8. S. Kliewer, H. Xu, M. Lambert, and T. Willson (2001). Peroxisome proliferator-activated receptors: From genes to physiology, *Recent Prog. Horm. Res.* **56**:239–263.
9. P. Greenwald, C. K. Clifford, and J. A. Milner (2001). Diet and cancer prevention. *Eur. J. Cancer* **37**:948–965.

10. M. J. Messina, V. Persky, K. D. R. Setchell, and S. Barnes (1994). Soy intake and cancer risk: A review of the *in vitro* and *in vivo* data, *Nutr. Cancer* **21**:113–131.
11. M. J. Messina and S. Barnes (1991). The role of soy products in reducing risk of cancer, *J. Natl. Cancer Inst.* **83**:541–546.
12. K. D. R. Setchell and A. Cassidy (1999). Dietary isoflavones: Biological effects and relevance to human health, *J. Nutr.* **129**:758S–767S.
13. A. R. Kennedy (1995). The evidence for soybean products as cancer preventive agents, *J. Nutr.* **125**:733S–743S.
14. A. R. Kennedy (1998). The Bowman–Birk inhibitor from soybean as a anticancinogenic agent, *Am. J. Clin. Nutr.* **68**(Suppl):14012S–14065S.
15. A. R. Kennedy and X. S. Wan (2002). Effects of the Bowman–Birk inhibitor on growth, invasion, and clonogenic survival of human prostate epithelial cells and prostate cancer cells, *Prostate* **50**:125–133.
16. A. F. Galvez, N. Chen, J. Macasieb, and B. O. de Lumen (2001). Chemopreventive property of a soybean peptide (lunasin) that binds to deacetylated histones and inhibits acetylation, *Cancer Res.* **61**:7473–7478.
17. A. F. Galvez and B. O. de Lumen (1997). A novel methionine-rich protein from soybean cotyledon: Molecular characterization of cDNA (Accession No. AF005030) (PGR97-103), *Plant Physiol.* **114**:567.
18. A. F. Galvez and B. O. de Lumen (1999). A soybean cDNA encoding a chromatin-binding peptide inhibits mitosis of mammalian cells, *Nat. Biotechnol.* **17**:495–500.
19. R. Aasland and A. F. Stewart (1995). The chromo shadow domain, a second chromo domain in heterochromatin-binding protein 1, HP1, *Nucleic Acids Res.* **23**:3168–3173.
20. H. J. Jeong, Y. Lam, and B. O. de Lumen (2002). Barley lunasin suppresses ras-induced colony formation and inhibits core histone acetylation in mammalian cells, *J. Agric. Food Chem.* **50**:5903–5908.
21. Y. Lam, A. Galvez, and B. O. de Lumen (2003). Lunasin suppresses E1A-mediated transformation of mammalian cells but does not inhibit growth of immortalized and established cancer cell lines, *Nutr. Cancer* **47**:88–94.
22. J. R. Nevins (1992). E2F: a link between the Rb tumor suppressor protein and viral oncoproteins, *Science* **258**:424–429.
23. American Cancer Society (2003). *ACS Cancer Facts & Figures*, p. 4. ACS, Altanta, GA.
24. G. P. Haas and W. A. Sakr (1997). Epidemiology of prostate cancer. *CA Cancer J. Clin.* **47**:273–287.
25. Y. Li, X. Li, and F. H. Sarkar (2003). Gene expression profiles of I3C and DIM treated PC3 prostate cancer cells by cDNA microarray analysis. *Proc. Am. Assoc. Cancer Res. Annu. Meeting* **44**:1311.
26. K. Kudoh, M. Ramanna, R. Ravatn, A. G. Elkahloun, M. L. Bittner, et al. (2000). Monitoring the expression profiles of doxorubicin-induced and doxorubicin-resistant cancer cells by cDNA microarray, *Cancer Res.* **60**:4161–4166.
27. H. Zembutsu, Y. Ohnishi, Y. Daigo, T. Katagiri, T. Kikuchi, et al. (2003). Gene-expression profiles of human tumor xenografts in nude mice treated orally with the EGFR tyrosine kinase inhibitor ZD1839, *Int. J. Oncol.* **23**:29–39.
28. D. Bello, M. M. Webber, H. K. Kleinman, D. D. Wartinger, and J. S. Rhim (1997). Androgen responsive adult human prostatic epithelial cell lines immortalized by human papillomavirus 18, *Carcinogenesis* **18**:1215–1223.

29. M. M. Webber, D. Bello, H. K. Kleinman, and M. P. Hoffman (1997). Acinar differentiation by non-malignant immortalized human prostatic epithelial cells and its loss by malignant cells, *Carcinogenesis* **18**:1225–1231.

30. R. A. Irizarry, B. M. Bolstad, F. Collin, L. M. Cope, B. Hobbs, and T. P. Speed (2003). Summaries of Affymetrix GeneChip probe level data, *Nucleic Acids Res.* **31**:e15.

31. M. B. Eisen, P. T. Spellman, P. O. Brown, and D. Botstein (1998). Cluster analysis and display of genome-wide expression patterns, *Proc. Natl. Acad. Sci. U.S.A.* **95**: 14863–14868.

32. P. Bornstein (1995). Diversity of function is inherent in matricellular proteins: An appraisal of thrombospondin 1. *J. Cell Biol.* **130**:503–506.

33. P. Friedl, P. Vischer, and M. A. Freyberg (2002). The role of thrombospondin-1 in apoptosis, *Cell. Mol. Life Sci.* **59**:1347–1357.

34. J. E. Nor, R. S. Mitra, M. M. Sutorik, D. J. Mooney, V. P. Castle, and P. J. Polverini (2000). Thrombospondin-1 induces endothelial cell apoptosis and inhibits angiogenesis by activating the caspase death pathway, *J. Vasc. Res.* **37**:209–218.

35. G. Chen, R. Ray, D. Dubik, L. Shi, J. Cizeau, et al. (1997). The E1B 19K/Bcl-2-binding protein Nip3 is a dimeric mitochondrial protein that activates apoptosis, *J. Exp. Med.* **186**:1975–1983.

36. R. Ray, G. Chen, V. C. Vande, J. Cizeau, J. H. Park, and A. H. Greenberg (2000). BNIP3 heterodimerizes with Bcl-2/Bcl-XL and induces cell death independent of a Bcl-2 homology 3 (BH3) domain at both mitochondrial and nonmitochondrial sites, *J. Biol. Chem.* **275**:1439–1448.

37. G. Chen, J. Cizeau, V. C. Vande, J. H. Park, G. Bozek, J. Bolton L. Shi, D. Bubik, and A. Greenberg (1999). Nix and Nip3 form a subfamily of pro-apoptotic mitochondrial proteins, *J. Biol. Chem.* **274**:7–10.

38. C. V. Velde, J. Cizeau, D. Dubik, J. Alimonti, T. Brown, S. Israels, R. Hakem, and A. H. Greenberg (2000). BNIP3 and genetic control of necrosis-like cell death through the mitochondrial permeability transition pore, *Mol. Cell. Biol.* **20**:5454–5468.

39. G. L. Wang, B.-H. Jiang, E. A. Rue, and G. L. Semenza (1995). Hypoxia-inducible factor 1 is a basic-helix-loop-helix-PAS heterodimer regulated by cellular O[2] tension, *Proc. Natl. Acad. Sci. U.S.A.* **92**:5510–5514.

40. R. K. Bruick (2000). Expression of the gene encoding the proapoptotic Nip3 protein is induced by hypoxia, *Proc. Natl. Acad. Sci. U.S.A.* **97**:9082–9087.

41. D. Chen, M. Li, J. Luo, and W. Gu (2003). Direct interactions between HIF-1alpha and Mdm2 modulate p53 function, *J. Biol. Chem.* **278**:13595–13598.

42. W. G. An, M. Kanekal, M. C. Simon, E. Maltepe, M. V. Blagosklonny, and L. M. Neckers (1998). Stabilization of wild-type p53 by hypoxia-inducible factor 1alpha, *Nature* **392**:405–408.

43. H. Koh, K. Lee, D. Kim, S. Kim, J. W. Kim, and J. Chung (2000). Inhibition of Akt and its anti-apoptotic activities by tumor necrosis factor-induced protein kinase C-related kinase 2 (PRK2) cleavage, *J. Biol. Chem.* **275**:34451–34458.

44. A. Khwaja (1999). Apoptosis: Akt is more than just a Bad kinase, *Nat. Rev. Mol. Cell Biol.* **401**:33–34.

45. F. Sandrini, L. Matyakhina, N. J. Sarlis, L. S. Kirschner, C. Farmakidis, O. Gimm, and C. A. Stratakis (2002). Regulatory subunit type I-alpha of protein kinase A (PRKAR1A):

A tumor-suppressor gene for sporadic thyroid cancer, *Genes Chromosom. Cancer* **35**:182–192.

46. S. G. Stergiopoulos and C. A. Stratakis (2003). Human tumors associated with Carney complex and germline PRKAR1A mutations: A protein kinase A disease! *FEBS Lett.* **546**:59–64.
47. M. Unoki and Y. Nakamura (2001). Growth-suppressive effects of BPOZ and EGR2, two genes involved in the PTEN signaling pathway, *Oncogene* **20**:4457–4465.
48. L. C. Trotman, M. Niki, Z. A. Dotan, J. A. Koutcher, A. D. Cristofano, et al. (2003). Pten dose dictates cancer progression in the prostate, *PLoS Biol.* **1**:e59.
49. H. Sasajima, K. Nakagawa, and H. Yokosawa (2002). Antiproliferative proteins of the BTG/Tob family are degraded by the ubiquitin-proteasome system, *Eur. J. Biochem.* **269**: 3596–3604.
50. Y. Yoshida, T. Nakamura, M. Komoda, H. Satoh, T. Suzuki, et al. (2003). Mice lacking a transcriptional corepressor Tob are predisposed to cancer, *Genes Dev.* **17**:1201–1206.
51. Y. Z. Huang, M. Zang, W. C. Xiong, Z. Luo, and L. Mei (2003). Erbin suppresses the MAP kinase pathway, *J. Biol. Chem.* **278**:1108–1114.
52. A. M. Duursma and R. Agami (2003). Ras interference as cancer therapy, *Semin. Cancer Biol.* **13**:267–273.
53. M. Macaluso, G. Russo, C. Cinti, V. Bazan, N. Gebbia, and A. Russo (2002). Ras family genes: An interesting link between cell cycle and cancer, *J. Cell. Physiol.* **192**: 125–130.
54. C. Lengauer, K. W. Kinzler, and B. Vogelstein (1998). Genetic instabilities in human cancers, *Nature* **396**:643–649.
55. K. A. Farr and M. A. Hoyt (1998). Bub1p kinase activates the *Saccharomyces cerevisiae* spindle assembly checkpoint, *Mol. Cell. Biol.* **18**:2738–2747.
56. K. G. Hardwick, E. Weiss, F. C. Luca, M. Winey, and A. W. Murray (1996). Activation of the budding yeast spindle assembly checkpoint without mitotic spindle disruption, *Science* **273**:953–956.
57. M. Winey, L. Goetsch, P. Baum, and B. Byers (1991). Mps1 and Mps2 novel yeast genes defining distinct steps of spindle pole body duplication, *J. Cell Biol.* **114**:745–754.
58. G. Chan, S. A. Jablonski, V. Sudakin, J. C. Hittle, and T. J. Yen (1999). Human BUBR1 is a mitotic checkpoint kinase that monitors CENP-E functions at kinetochores and binds the cyclosome/APC, *J. Cell Biol.* **146**:941–954.
59. A. Gupta, S. Inaba, O. K. Wong, G. Fang, and J. Liu (2003). Breast cancer-specific gene 1 interacts with the mitotic checkpoint kinase BubR1, *Oncogene* **22**:7593–7599.
60. Y. Li and R. Benezra (1996). Identification of a human mitotic checkpoint gene: hsMAD2, *Science* **274**:246–248.
61. M. J. Percy, K. A. Myrie, C. K. Neeley, J. N. Azim, S. P. Ethier, and E. M. Petty (2000). Expression and mutational analyses of the human MAD2L1 gene in breast cancer cells, *Genes Chromosom. Cancer* **29**:356–362.
62. W. R. Loewenstein and B. Rose (1992). The cell–cell channel in the control of growth, *Semin. Cell Biol.* **3**:58–79.
63. M. Z. Hossain, A. B. Jagdale, P. Ao, C. Leciel, R.-P. Huang, and A. L. Boynton (1999). Impaired expression and posttranslational processing of connexin 43 and downregulation of gap junctional communication in neoplastic human prostate cells, *Prostate* **38**:55–59.

64. R. Bruzzone, T. W. White, and D. L. Paul (1996). Connections with connexins: The molecular basis of direct intercellular signaling, *Eur. J. Biochem.* **238**:1–27.

65. M. Y. Kanemitsu, L. W. M. Loo, S. Simon, A. F. Lau, and W. Eckhart (1997). Tyrosine phosphorylation of connexin 43 by v-Src is mediated by SH2 and SH3 domain interactions, *J. Biol. Chem.* **272**:22824–22831.

66. B. Rose, P. P. Mehta, and W. R. Loewenstein (1993). Gap-junction protein gene suppresses tumorigenicity, *Carcinogenesis* **14**:1073–1075.

67. O. Coqueret (2003). New roles for p21 and p27 cell-cycle inhibitors: A function for each cell compartment? *Trends Cell Biol.* **13**:65–70.

68. Y. Sowa, T. Orita, S. Minamikawa-Hiranabe, T. Mizuno, H. Nomura, and T. Sakai (1999). Sp3, but not Sp1, mediates the transcriptional activation of the p21/WAF1/Cip1 gene promoter by histone deacetylase inhibitor, *Cancer Res.* **59**:4266–4270.

69. J. S. Rhim, M. M. Webber, D. Bello, M. S. Lee, P. Arnstein, L.-S. Chen, and G. Jay (1994). Stepwise immortalization and transformation of adult human prostate epithelial cells by a combination of HPV-18 and v-Ki-ras, *Proc. Natl. Acad. Sci. U.S.A.* **91**: 11874–11878.

70. H. M. Ellerby, W. Arap, L. M. Ellerby, R. Kain, R. Andrusiak, et al. (1999). Anti-cancer activity of targeted pro-apoptotic peptides, *Nat. Med.* **5**:1032–1038.

71. W. Arap, R. Pasqualini, and E. Ruoslahti (1998). Cancer treatment by targeted drug delivery to tumor vasculature in a mouse model, *Science* **279**:377–380.

72. R. Max, R. R. C. M. Gerritsen, P. T. G. A. Nooijen, S. L. Goodman, A. Sutter, U. Keilholz, D. J. Ruiter, and R. M. W. De Waal (1997). Immunohistochemical analysis of integrin alpha-v-beta-3 expression on tumor-associated vessels of human carcinomas, *Int. J. Cancer* **71**:320–324.

73. A. R. Kennedy (1998). Chemopreventive agents: Protease inhibitors, *Pharmacol. Ther.* **78**:167–209.

74. S. Barnes (1995). Effect of genistein on in vitro and in vivo models of cancer, *J. Nutr.* **125**:777S–783S.

75. F. H. Sarkar and Y. Li (2002). Mechanisms of cancer chemoprevention by soy isoflavone genistein, *Cancer Metastasis Rev.* **21**:265–280.

76. J. N. Davis, O. Kucuk, and F. H. Sarkar (1999). Genistein inhibits NF-kappaB activation in prostate cancer cells, *Nutr. Cancer* **35**:167–174.

77. Y. Li and F. H. Sarkar (2002). Gene expression profiles of genistein-treated PC3 prostate cancer cells, *J. Nutr.* **132**:3623–3631.

78. Y. Li and F. H. Sarkar (2002). Down-regulation of invasion and angiogenesis-related genes identified by cDNA microarray analysis of PC3 prostate cancer cells treated with genistein, *Cancer Lett.* **186**:157–164.

79. S. J. Lin and L. Guarente (2003). Nicotinamide adenine dinucleotide, a metabolic regulator of transcription, longevity and disease, *Curr. Opin. Cell Biol.* **15**:241–246.

80. D. Moazed (2001). Enzymatic activities of Sir2 and chromatin silencing, *Curr. Opin. Cell Biol.* **13**:232–238.

81. S. M. Gasser and M. M. Cockell (2001). The molecular biology of the SIR proteins, *Gene* **279**:1–16.

82. A. Eberharter and P. B. Becker (2002). Histone acetylation: a switch between repressive and permissive chromatin. Second in review series on chromatin dynamics, *EMBO Rep* **3**:224–229.

12

ENZYMES LOSE BINDING AFFINITY (INCREASED K_m) FOR COENZYMES AND SUBSTRATES WITH AGE: A STRATEGY FOR REMEDIATION

Bruce N. Ames, Jung H. Suh, and Jiankang Liu

Nutrition and Metabolism Center, Children's Hospital Oakland Research Institute, Oakland, California and University of California–Berkeley, Berkeley, California

12.1 INTRODUCTION

In this chapter we discuss (1) our previous review on the remediation by high B vitamin intake in human mutant enzymes with poor binding affinity for coenzymes; (2) the effect of the increased oxidants produced with age in deforming proteins in mitochondria and possible remediation by B vitamins and substrates; and (3) the effect of age in deforming nonmitochondrial proteins and possible remediation.

12.2 REMEDIATION BY HIGH B VITAMIN INTAKE OF VARIANT ENZYMES WITH POOR BINDING AFFINITY (K_m) FOR COENZYMES

Gene mutations commonly result in the corresponding enzyme becoming deformed and having a decreased binding affinity (increased K_m) for a coenzyme, resulting

Nutritional Genomics: Discovering the Path to Personalized Nutrition
Edited by Jim Kaput and Raymond L. Rodriguez

in a lower rate of reaction [1]. The K_m is a measure of the binding affinity of an enzyme for its ligand (substrate or coenzyme) and is defined as the concentration of ligand required to fill half of the available binding sites. About 50 human genetic diseases due to defective enzymes can be remedied or ameliorated by the administration of high doses of the vitamin component of the corresponding coenzyme [1], which at least partially restores enzymatic activity. The therapeutic vitamin regimens increase intracellular coenzyme concentrations, thereby permitting binding to the defective enzyme and alleviating the primary defect. In some cases these enzymes have been shown to have an increased K_m. Consistent with the explanation, it has been shown that high doses of each B vitamin, often 100 times the RDA, can markedly increase the level of the corresponding coenzyme by an order of magnitude or more [1]. The B vitamins discussed in the review [1] are pyridoxine, thiamine, riboflavin, niacin, biotin, cobalamin, folic acid, and pantothenic acid. From what is known of enzyme structure, it seems plausible that, in addition to direct changes in the amino acids at the coenzyme binding site, mutations can commonly affect the conformation of the protein, thus causing an indirect change in the binding site. The proportion of mutations in a disease gene that is responsive to high concentrations of a vitamin or substrate may be as much as one-third [1, 2], though determining the true percentage from the literature is difficult because exact response rates in patients are not always reported and much of the literature deals only with individual case reports.

The 2002 review also discussed five genes with single-nucleotide polymorphisms, in which the variant protein amino acid reduces coenzyme binding and enzymatic activity, and thus may be remediable by raising cellular concentrations of the cofactor through high-dose vitamin therapy. It is likely that many more will be discovered as polymorphisms continue to be characterized from defective proteins involved in disease.

Since only a tiny fraction of genetic defects and polymorphisms has been tested for remediation, it seems clear that extending these studies should yield some of the first hints of the coming nutritional genomic revolution. Of the 50 remediable genetic diseases, 14 are in enzymes located in the mitochondria. This suggests that coenzyme levels are raised in mitochondria as well as the cytosol by feeding high levels of a B vitamin [1]. It also raises the issue of whether some of the mitochondrial decay of aging may be remediable.

12.3 DEFORMATION OF PROTEINS IN MITOCHONDRIA WITH AGING

Mechanisms of Protein Oxidation

Oxidants can cause deformation resulting in inactivation of enzymes. Mechanisms of protein deformation include (1) direct protein oxidation, (2) adduction of aldehydes from lipid peroxidation, and, (3) in the case of membrane proteins, decreases in fluidity of oxidized membranes.

Direct Protein Oxidation. Free amino acids can be reversibly or irreversibly modified or destroyed by oxidation [3–6]. Oxidized protein accumulates with age in both rodents and humans due to an increase in oxidant leakage from mitochondria with age [7, 8] that is not compensated by cell defense systems [9, 10]. Oxidation of aliphatic amino acids causes deamination and yields mainly carboxylic acids and aldehydes, while oxidation of aromatic amino acids targets the indole and aromatic rings. Mild oxidation in vitro, which mimics the situation in vivo, can modify a number of different amino acid residues in proteins without cleavage of peptide bonds [11, 12]. Thus, the effect of amino acid modifications on a protein may vary from a slight structure modification to extensive denaturation accompanied by fragmentation. The most significant modifications are those of histidine, arginine, lysine, proline, methionine, and cysteine. Enzyme inactivation can be inhibited by antioxidants, metal chelators, or high concentrations of substrates [13].

Adduction of Aldehydes from Lipid Peroxidation. Lipid peroxidation increases with age [14], and the released aldehydes bind to protein. Aldehyde products derived from lipid peroxidation of membranes react with amino and sulfhydryl groups [15], thus altering proteins [16, 17]. Malondialdehyde (MDA) and 4-hydroxy-2-nonenal (4-HNE) are two of the many known active aldehydes formed from lipid peroxidation that contribute significantly to enzyme inactivation. For example, in vitro experiments showed that MDA and 4-HNE cause loss of activity and decrease in binding affinity (increased K_m) to substrates of carnitine acetyltransferase and pyruvate dehydrogenase [14]. The enzyme dysfunction induced by lipid peroxidation products such as MDA and 4-HNE may be a common mechanism of age-associated dysfunction of enzymes with amino and sulfhydryl groups at or near their active sites. Reactive aldehydic products generated from lipid peroxidation might be a major cause of mitochondrial dysfunction during aging. The main pathways of 4-HNE elimination in mitochondria are through aldehyde dehydrogenase (ALDH) catalyzed oxidation and the glutathione transferase (GST) catalyzed conjugation of HNE. Age reduces both ALDH and GST activities in both mice and rats; mitochondrial 4-HNE oxidation by ALDH declines at 18 and 24 months of age, and the glutathione conjugation of 4-HNE reduces at 24 months of age [18]. These findings are consistent with the earlier proposal that indicates an age-associated decrease in mitochondrial detoxification as a major underlying process for MDA and lipofuscin accumulation in older animals [18].

Decrease in Fluidity of Membranes. Mitochondrial membrane fluidity declines with age [19]. Lipid peroxidation causes changes in membrane morphology [20] and all membrane organelles are susceptible to it. Lipid peroxidation produces a number of lipid soluble reactive aldehydes, such as 4-HNE and MDA, which increase in mitochondria with age. Correlated with the increased lipid oxidation, the osmotic stability of the mitochondria declines with age during the normal course of aging, but can be prevented by caloric restriction [21]. The preincubation of rat hepatocyte mitochondria with either 4-HNE or MDA significantly reduced membrane fluidity [20], showing that products of lipid peroxidation can affect mitochondrial membrane integrity.

Mitochondrial Decay with Age

Mitochondria provide energy for basic metabolic processes, detoxify oxygen, provide essential metabolism such as heme biosynthesis, and produce oxidants as inevitable by-products. During aging, oxidation deforms many proteins in mitochondria, thereby decreasing their affinity for their substrates or coenzymes [14]. Mitochondria are particularly vulnerable to this damage as they are the sites of oxidant leakage and are the most complex organelle in the cell. Cells defend themselves from damaged mitochondria leaking oxidants in part by lysosomes degrading defective mitochondria, though this defense system also declines markedly in efficiency with age [22]. Mitochondrial decay with age impairs cellular metabolism and leads to cellular decline. Mitochondrial membrane potential, respiratory control ratios, and cellular oxygen consumption decline with age, and oxidant production increases [9, 23, 24]. Oxidant damage to DNA, RNA [43], proteins [13], and lipid membranes in mitochondria are most likely involved in this decay, resulting in oxidative damage that may compromise the ability to meet cellular energy demands.

Age-Related Changes in the Activities of Mitochondrial Krebs Cycle Enzymes

The Krebs cycle plays a central role in intermediary metabolism and is the major site of oxidation of carbon chains from carbohydrates, fatty acids, and amino acids to carbon dioxide and water. NADH and $FADH_2$ produced from the Krebs cycle serve as substrates of oxidative phosphorylation. In addition to its role in energy metabolism, intermediates produced during the Krebs cycle are essential components required for the synthesis of amino acids and porphyrins. Three major Krebs cycle enzymes are discussed: these are aconitase, pyruvate dehydrogenase complex (PDC), and α-ketoglutarate dehydrogenase (KGDC).

Aconitase, in particular, is exquisitely sensitive to oxidation and subsequent inactivation [25]. In houseflies, mitochondrial aconitase is subjected to oxidative damage with age that is paralleled by a decrease in catalytic activity [25]. The inactivation of aconitase slows down the Krebs cycle, which leads to reductive stress. This buildup of NADH and other electron donors can contribute to an increase in oxidants as well as an increase in the oxidative modification of other proteins.

Pyruvate dehydrogenase complex (PDC) is a multimeric complex that utilizes lipoic acid, thiamine pyrophosphate, CoA, NAD, and FAD cofactors to convert pyruvate to acetyl CoA [26]. PDC is an essential enzyme for carbohydrate metabolism. The activity of PDC is inhibited upon phosphorylation by pyruvate kinase (PK) and activated by the influx of calcium that activates the phosphopyruvate dehydrogenase phosphatase [26]. In rodents and primates, the expression of PK decreases significantly with age, leading to increased PDC activity [27, 28]. Interestingly, it was found that the catalytic efficiency of myocardial PDC increased rather than decreased with age, owing to increased overall V_{max} and lower K_m for pyruvate [28, 29]. The higher catalytic efficiency of PDC and decreased expression of PK in the aging cardiac tissue may represent an adaptation to the age-related decrease in fatty acid utilization [28].

α-Ketoglutarate dehydrogenase (KGDH) complex shares structural similarities and cofactors with PDC but is not regulated by phosphorylation [30]. The incubation of an isolated mitochondrial preparation with 4-hydroxy 2-nonenal (4-HNE) decreases the mitochondrial respiratory rate; KGDH is especially sensitive to 4-HNE adduction [16, 31]. In the rat heart, the steady-state levels of 4-HNE-adducted KGDH increase as a function of age but no significant loss in its activity was observed [31]. Interestingly, under the stressful conditions of ischemia and reperfusion, the myocardial KGDH activity in older rats declines at a higher rate than in younger rats [32]. A decreased KGDH activity was also found in patients suffering from Alzheimer's and Parkinson's diseases [33, 34]. These data suggest that the KGDH may be more vulnerable to oxidative inactivation under pathophysiological conditions than other mitochondrial proteins [16, 32, 33].

In light of the thiamine responsiveness of some people with various KGDH mutations [1], it would be of interest to know if high thiamine, in conjunction with the other KGDH cofactors, could reduce oxidative stress and thus treat some forms of KGDH-associated aging. It is noteworthy that despite the fact that most PDC and KGDH complexes are fully saturated with thiamine, exogenous thiamine beyond their saturating point can enhance catalytic efficiency of these enzymes by lowering the K_m for pyruvate, NAD, and CoA. The ability of thiamine to improve catalytic efficiency of PDC and/or KGDH may also be helpful for intervention in patients with Alzheimer's and Parkinson's diseases [35] with decreased activity of KGDH.

Age-Related Increase in K_m of Mitochondrial Enzymes Required for Fatty Acid Metabolism

In highly energetic tissues, such as the heart, the mitochondrial β-oxidation of fatty acids is the primary source of energy production [36]. Aging is associated with a decreased myocardial ability to utilize fat for energy production [36]. The mechanism(s) leading to the age-related loss in fatty acid utilization is likely to involve multiple factors including the changes in mitochondrial lipid content, changes in lipid membrane composition, and changes in membrane transport proteins.

Carnitine acetyltransferase 1 (CAT 1) catalyzes the initial step of fatty acid import into the mitochondrial matrix. CAT 1 is associated with the outer membrane of the mitchondria. CAT 1 is composed of two main domains: the N terminal domain is important for binding malonyl CoA while the C terminal domain is essential for CoA and carnitine binding [37, 38]. Structural studies with carnitine palmitoyltransferase (CPT 1) isolated from rat liver reveal that the two domains are connected by a short loop that traverses through the outer membrane [37, 38]. Within the outer membrane, CPT 1 is especially enriched at the contact sites that occur between the outer and inner membrane of mitochondria [39]. Because of the integral connection of this protein to the outer membrane, CAT 1 is exquisitely sensitive to changes in membrane composition and fluidity [40]. Treating carnitine palmitoyltransferase 1 (CPT 1) liposomes with benzyl alcohol, an agent that increases membrane fluidity, ameliorated the inhibitory effects of malonyl CoA [40]. Moreover, the kinetic characteristics of CPT 1 markedly change depending on whether CPT 1 is located within the outer membrane or at the contact sites, suggesting that lipid environment is a critical regulator of CPT 1 activity [39]. Based on the known properties of CPT 1, we asked whether the kinetic parameters of CAT 1 become altered in the aging rat brain [14]. Our results showed a significant loss in cerebral CAT activity in mitochondria from old rat brain in comparison to young [14]. A kinetic analysis revealed that the affinity of this enzyme for both CoA and acetyl carnitine decreased with age [14] (Figure 12.1). As shown in Figure 12.1, the double reciprocal plots of reaction velocity against varying substrate concentrations of acetyl carnitine and CoA show

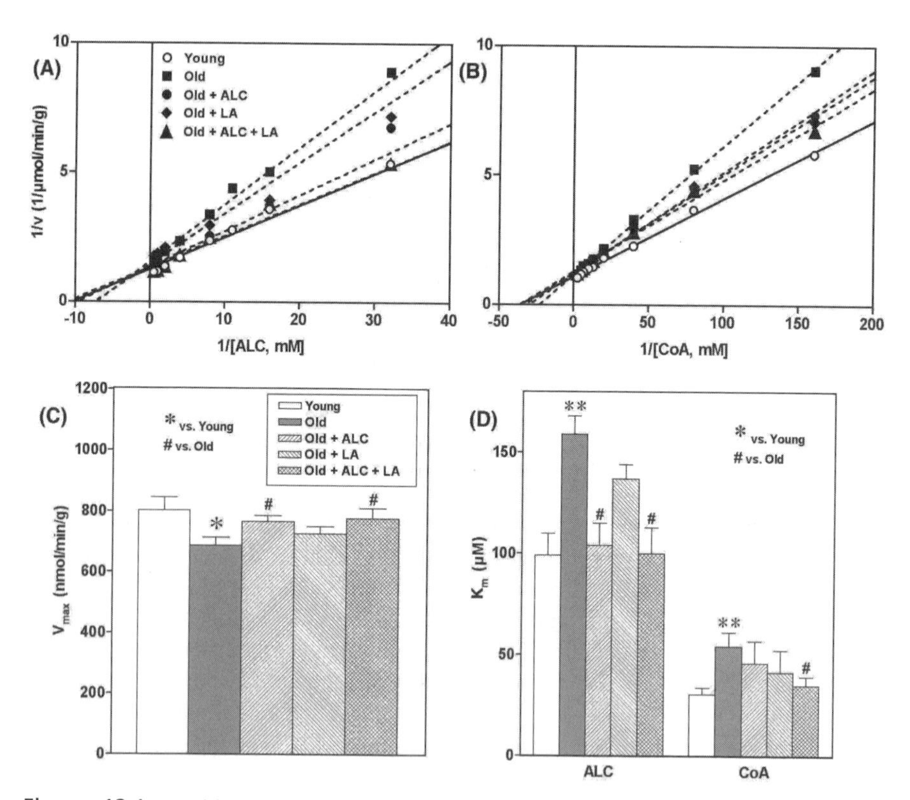

Figure 12.1. Double reciprocal plots of reaction velocity versus substrate ALC (A) or CoA (B) concentrations in rat brain. (C) V_{max}. (D) Apparent K_m for ALC and CoA. All values are mean ± SE of 10 animals for young and old groups, 5 for the LA group, and 6 for the ALC and ALC plus LA groups. Significant difference was calculated by using Student's t test between young and old groups ($^*P = 0.05$, $^{**}P = 0.01$) and by using one-way ANOVA with Dunnett's multiple comparison test between old and other treated groups ($^{\#}P = 0.05$) [15].

a linear response. If a K_m change was introduced due to a direct conformational change on the protein itself, the substrate binding curve should reveal at least two different sets of proteins that are either modified or unmodified. The linearity of response suggests a strong homogeneous effect of aging on CAT substrate binding affinity. One potential interpretation of these data is that the homogeneous increase in the cerebral CAT K_m is due to age-related alterations in membrane lipid composition (such as the decrease in cardiolipin) and/or fluidity. A similar increase in the K_m for these substrates was also observed in the young brain upon oxidation by transition metals and reactive aldehydes [14].

R-α-lipoic acid (LA) is a coenzyme for pyruvate dehydrogenase and KGDH. Aside from its role in metabolism, dihydrolipoic acid (DHLA) is a potent antioxidant, which inhibits metal-dependent lipid peroxidation and recycles other low molecular weight antioxidants, such as vitamins C and E, and raises intracellular levels of glutathione (GSH) [41]. We have found previously that a dietary intervention with a combination of LA and acetyl carnitine (ALC) significantly improved the age-related loss in cognition and ambulatory activities in rats [24, 42]. To discern whether the combined dietary interventions with LA and ALC would remediate the observed age-related decrease in CAT activity, young and old Fischer 344 rats were fed a diet with or without 0.2% LA and 0.5% ALC [14]. Feeding the substrate ALC together with LA restored the velocity of the reaction, the K_m value of carnitine acetyltransferase for ALC and CoA, and mitochondrial function [14].

Age-Related Increase in K_m for Mitochondrial Membrane Enzyme Complexes

Decreased capacity to produce ATP and increased oxidant production are two properties of aging mitochondria supported by multiple lines of direct and indirect observations. First, the analysis of gene expression profiles in mice showed significant age-associated declines in the mRNA levels of mitochondrially encoded subunits of complex I, III, IV, and V in old compared to young mice [43]. Second, in addition to reduced gene expression, the levels of 4-HNE and carbonylated mitochondrial proteins increase in aging tissues [44, 45]. Lastly, the activities of NADH–cytochrome *c* reductase (complexes I–III) exhibit approximately 30% decline in the aging rat liver and the brain when compared to young [45–47]. Similar declines in complex IV activity were also reported [44, 45].

While the gross age-related declines in respiratory complex activities have been reported extensively, much less attention has been given to the extent of age-related changes in their substrate binding affinities. One study [48] examined the substrate binding affinities of complexes I, III, and IV in mitochondria isolated from the gastrocnemius muscles of young and old mice. A kinetic analysis of complex III revealed a significant 29% age-associated increase in the K_m for ubiquinol-2 [48]. These findings are congruent with more recent work [49] that reported a defect in the ubiquinol binding site of cytochrome *b* in complex III in the interfibrillary mitochondria isolated from old rats [49]. The resulting defect in ubiquinol binding affinity is likely to increase superoxide production at this site. The specific post-translational modification responsible for the K_m increase is currently unknown. The affinity of complex IV for reduced cytochrome *c* also exhibits a similar age-related loss but again the exact mechanisms for the increased K_m is incompletely understood [48]. A decreased electron flux due in part to the reduced affinity of complex IV for cytochrome *c* would result in an enhanced rate of oxidant production at complex III.

The above studies support the hypothesis that the substrate binding affinities of the major complexes decline with age and may contribute to increased mitochondrial oxidant production. Age-related changes in mitochondrial membrane fluidity discussed above may provide an explanation for the changes in K_m for multiple respiratory complexes.

There is an age-associated loss of membrane cardiolipin, which may be a driving force behind the increase in K_m for various substrates in the aging rat mitochondria [50, 51] (Figure 12.2). A loss of cardiolipin significantly decreases complex IV activity in isolated mitochondria [51]. Cardiolipin is an anionic phospholipid composed of two phosphatidyl residues that are linked by a glycerol molecule [52]. In mammalian cells, cardiolipin is found almost exclusively in mitochondrial membranes [52]. Cardiolipin deficiency causes a marked change in fluidity and compromises the osmotic stability of the mitochondrial membrane [53]. Decreases in membrane fluidity can also lead to deformation of membrane proteins, which may increase K_m. It has been shown that cardiolipin binding is essential for the optimal enzymatic activities of complexes III, IV, and V [52]. Moreover, cardiolipin is required from the membrane attachment of cytochrome *c* [54]. Structural analysis of the *Saccharomyces cerevisiae* mitochondria revealed that the respiratory com-

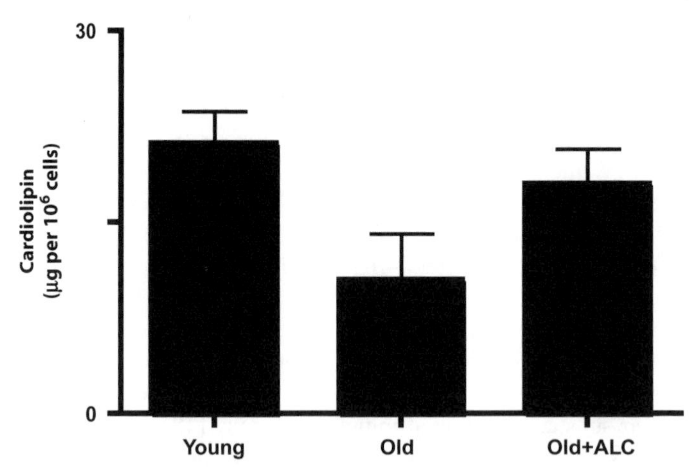

Figure 12.2. Cardiolipin, a key phospholipid necessary for mitochondrial substrate transport, was extracted from hepatocytes and the levels were assessed by UV spectrophotometric detection following HPLC separation. Results show that mitochondria in cells from old rats ($n = 5$) have significantly lower cardiolipin when compared with cells isolated from young rats ($n = 4$). Cardiolipin levels in ALC supplemented old rats ($n = 6$) were normalized to levels seen in young unsupplemented rats [51].

plexes III and IV are assembled into a large supercomplex that is comprised of a dimeric complex III and one or possibly two complex IV units [55]. From the theoretical standpoint, the formation of supracomplex may provide an advantage by closely coupling the electron transfer process such that the overall efficiency of electron transfer is greatly enhanced and the rate of oxidant leakage is minimized. Thus, the age-related loss in cardiolipin may cause increases in K_m for various substrates of respiratory complexes.

Our own experiments with high-dose LA and ALC supplementation suggest that age-related increases in K_m can be partially remediated by high-dose supplementation with the two mitochondrial metabolites [14]. An acute interperitoneal injection with ALC (300 mg/kg body weight) for 3 hours significantly improved the myocardial pyruvate uptake and its dependent oxygen consumption rates in aging rats [56]. Pyruvate carrier (PC) is a transporter that resides within the inner mitochondrial membrane and is sensitive to changes to cardiolipin loss [56]. The binding affinity of PC for pyruvate markedly declines with age and appeared to be tightly correlated with cardiolipin loss [56]. Remarkably, only a short duration (3 hours) of high-dose carnitine administration was sufficient to normalize the age-related loss in mitochondrial carnitine contents [56]. These studies provide a strong rationale for remediation of at least some of the biochemical defects associated with mitochondria.

Delaying the Age-Related Decay of Mitochondrial Enzymes with Age

About one-fourth of genetic diseases remediable by high doses of particular vitamins involve defects in mitochondrial enzymes [1]. We suggest that some genetically normal enzymes are inactivated by oxidation during aging and that some of this lost activity, in so far as it affects the K_m, may be partially recoverable by treatment with high doses of the same vitamins shown to remediate the genetic diseases. Since toxic metabolites accumulate in many genetic diseases, these metabolites could be measured in elderly people with the symptoms of the particular genetic diseases, and, if high, the utility of the vitamin to reduce the levels of toxic metabolite, and thus to ameliorate the disease, could be tested. Some cases have been reported where aged individuals who developed symptoms of a genetic disease responded to treatment with high doses of the vitamin. It is uncertain, however, whether the symptoms were due to aging, or whether they simply represent delayed-onset forms of the genetic disease.

12.4 NONMITOCHONDRIAL ENZYMES THAT ARE DEFORMED WITH AGE

Microsomal Proteins That Increase K_m with Age

The biosynthesis of long chain polyunsaturated fatty acids (PUFAs) such as arachidonic acid (AA) and docosahexaenoic (DHA) from linoleic acid (18 : 2 n6) is mainly regulated by the activity of delta-6-desaturase (D6D) [57, 58]. Biosynthesis of PUFAs is essential for a variety of physiological functions, including membrane synthesis, eicosanoid signaling, and cognitive functions [57, 58]. Alteration in the biosynthesis of PUFAs is hypothesized to play a role in the pathogenesis of age-associated pathologies such as diabetes, insulin resistance, and hypertension. The age-dependent decrease in capacity to metabolize n-6 PUFA may negatively influence mitochondrial function by increasing n-6 PUFA levels in the mitochondrial membrane [59].

D6D catalyzes the initial rate-limiting conversion of linoleic acid to its product, γ-linoleic acid (18:3 n6; GLA) [57, 58]. Studies show that aging is associated with a significant decline in D6D activity [60]. This loss in activity is accompanied by a significant loss in the D6D catalytic efficiency; the V_{max} of D6D decreases, and K_m for delta-6 desaturation of linoleic acid increases with age in rats [61]. A dietary GLA supplementation to aging rats normalizes the adverse impact of D6D activity loss on microsomal membrane viscosity and unsaturated index [60].

Cytosolic Proteins That Increase K_m with Age

Glutathione (γ-glutamyl-cyteinyl-glycine; GSH) plays a central role in maintaining the intracellular thiolredox environment in mammalian cells [62]. The synthesis of GSH requires the concerted actions of two ATP requiring enzymes, γ-glutamylcysteine ligase (GCL) and GSH synthetase [62]. GCL catalyzes the formation of γ-glutamylcysteine, which is the rate-limiting step in GSH de novo synthesis [62]. The intracellular concentration of cysteine is kept low and is maintained at a concentration that is below the K_m for GCL [63]. Thus, any increase in intracellular cysteine uptake can rapidly increase the intracellular concentration of GSH. A large body of evidence suggests that intracellular GSH synthesis capacity declines with age [63–67]. In the liver, it was found that the basal expression of both the catalytic and regulatory subunits of GCL decrease significantly with age [65]. The paradoxical decrease in GSH synthesis enzymes in aging cells that experience higher levels of oxidative stress has been determined to be in part due to a lower basal and inducible activation of NF-E2 related factor-2 (Nrf2), an essential transcriptional regulator for GCL [65]. Aside from the transcriptional dysregulation that occurs in aging tissues, the kinetic efficiency of GCL also markedly declines with age. Kinetic analysis of GCL in brains of old rats displayed a significant increase in the apparent K_m for cysteine versus young rats (84.3 ± 25.4 versus 179.0 ± 49.0; young and old, respectively), resulting in a 40% loss in apparent catalytic turnover of the enzyme [65]. Thus, the age-related decline in total GSH appears to be mediated, in part, by a general decrement in the catalytic efficiency of GCL. Treating old rats with LA (40 mg/kg body weight; by i.p.) markedly increased tissue cysteine levels by 54% 12 hours post-treatment, followed by restoration of cerebral GSH levels [65]. Moreover, LA improved the age-related changes in the tissue GSH/GSSG ratios in both heart and brain [65].

The age-associated changes in the substrate binding affinity have also been reported in an insect model of aging. GCL has significantly higher affinities for its substrates in the young counterparts. In addition, young but not old flies exhibited substrate-dependent inhibition of GCL activity at high cysteine concentrations (>5 mM), indicating a loss of metabolic regulation. The age-associated differences suggest that de novo synthesis of glutathione would be relatively less efficient in old houseflies [68].

Considering how critical GSH is in maintaining the cytosolic redox environment, the age-associated loss in GSH synthesis is likely to disrupt the activities of proteins that rely on the redox status of critical cysteine residues. Reversible modification of free protein cysteine sulfhydryl moieties is now a well-established means by which cells detect and adapt to increased oxidative stress [62].

Glyceraldehye-3-phosphate dehydrogenase (GAPDH) is a key cytosolic glycolytic enzyme that catalyzes the conversion of glyceraldehyde-3-phosphate to 1,3-bisphosphoglycerate. During the course of its catalysis, GAPDH generates NADH from NAD^+. GAPDH contains a critical cysteine residue (cys-149) in the NAD binding site that is subject to oxidation [69, 70]. In vitro, oxidation of that cysteine residue significantly decreases the NAD binding affinity [69, 70]. In aging rat muscles, the NAD binding affinity of GAPDH decreases significantly [71].

In addition to its metabolic role, other functions of GAPDH have recently been described [72]. GAPDH serves as a kinase, a tubulin and actin binding protein, and facilitates membrane fusion [72]. In growing cells, GAPDH can also act as a uracil-DNA-glycosylase and a helicase and participate in the transcription regulation [73]. Interestingly, the induction of cellular apoptosis by various agents causes a rapid nuclear translocation of GAPDH [74]. The proapoptotic activities of GAPDH may be related in part to its ability to bind to nucleic acids. Interestingly, the oxidation of an active site cysteine residue in GAPDH heightens its affinity for nucleic acids [75]. The binding of NAD to GAPDH has been noted to decrease its DNA and RNA binding affinity [75]. Maintaining higher levels of NAD by increasing niacin intake may protect against oxidative inactivation of GAPDH and decrease the proapoptotic activities.

A number of enzymes critical for neurotransmitter synthesis also exhibit changes in enzymatic efficiency with age. Choline acetyltransferase is responsible for biosynthesis of acetylcholine. The activity of choline acetyltransferase decreases with age in humans, rats, and also mice [76]. A kinetic study demonstrated that the activity of choline acetyltransferase decreases to about 60% and K_m for both substrates (choline and acetyl CoA) increases up to sixfold from the age of 3 weeks to 87 weeks in all brain regions in rats [77]. These results could suggest that feeding high doses of choline or pantothenates, which raise CoA levels, would retard the age-associated decay of choline acetyltransferase.

Tryptophan hydroxylase performs the first and rate-limiting step in the conversion of tryptophan to serotonin in the brain. After hydroxylation of tryptophan by tryptophan hydroxylase, 5-hydroxytryptophan is decarboxylated to form 5-hydroxytryptamine, or serotonin [78]. A significant elevation in K_m has been found in both midbrain and pons from old (24 months) rats as compared to mature (12 months) rats. In the midbrain, K_m was elevated sevenfold from 22.5 µM in mature rats to 141.1 µM in old rats ($p < 0.01$). In the pons, K_m was elevated from 38.2 to

126.0 ($p < 0.01$). There was also a threefold decline in clearance formation (V_{max}/K_m) of 5-hydroxytryptophan in midbrain ($p < 0.01$), medulla ($p < 0.01$), and pons ($p < 0.05$) [78]. A further study by the same group demonstrated that sulfhydryl oxidation of tryptophan hydroxylase plays a role in the age-associated dysfunction of this enzyme [78]. It would be of interest to know if the aged rat brain displays similar kinetic properties for the vitamin cofactor, tetrahydrobiopterin (BH_4). If tryptophan hydroxylase did have a decreased affinity for BH_4 cofactor as well, there might be basis for designing a rational vitamin regimen to overcome the binding defect [78].

12.5 CONCLUSION

Enzymes are commonly deformed with age due to direct and indirect affects of oxidation, particularly in the mitochondria, the source and target of oxidants. Deformation of an enzyme commonly decreases binding affinity (increased K_m) for its coenzyme or substrate. Enzyme substrates and vitamin precursors of coenzymes can be elevated by feeding and may enhance the activity of a deformed enzyme. B vitamins, for example, are quite nontoxic and associated coenzymes can be raised tenfold or more with high-dose supplementation. This raises the question of whether the RDA for B vitamins should be reexamined in the old, as the levels set are based on experiments in the young. It also raises the question of whether many metabolites, as well as vitamins, might be fed to improve functioning of enzymes in the old. The remediation of deformed enzymes, whether due to mutation or aging, is a field that shows promise and may be an inexpensive way to improve health.

ACKNOWLEDGMENTS

Supported from the Ellison Medical Foundation SS-0422-99, the National Foundation for Cancer Research Grant M2661, the National Center for Minority Health and Health Disparities Grant P60 MD00222, the National Center for Complementary and Alternative Medicine R21 AT001918, and Research Scientist Award K05 AT001323, the National Institute of Aging Grant AG023265-01, and the Bruce and Giovanna Ames Foundation.

REFERENCES

1. B. N. Ames, I. Elson-Schwab, and E. A. Silver (2002). High-dose vitamin therapy stimulates variant enzymes with decreased coenzyme binding affinity (increased k(m)): Relevance to genetic disease and polymorphisms, *Am. J. Clin. Nutr.* **75**(4):616–658.
2. S. H. Mudd, F. Skovby, H. L. Levy, K. D. Pettigrew, B. Wilcken, R. E. Pyeritz, G. Andria, G. H. Boers, I. L. Bromberg, R. Cerone, et al. (1985). The natural history of homocystinuria due to cystathionine beta-synthase deficiency, *Am. J. Hum. Genet.* **37**(1): 1–31.

3. S. Fu, L. A. Hick, M. M. Sheil, and R. T. Dean (1995). Structural identification of valine hydroperoxides and hydroxides on radical-damaged amino acid, peptide, and protein molecules, *Free Radic. Biol. Med.* **19**(3):281–292.
4. J. M. Gutteridge (1981). Thiobarbituric acid-reactivity following iron-dependent free-radical damage to amino acids and carbohydrates, *FEBS Lett.* **128**(2):343–346.
5. B. Morin, W. A. Bubb, M. J. Davies, R. T. Dean, and S. Fu (1998). 3-hydroxylysine, a potential marker for studying radical-induced protein oxidation, *Chem. Res. Toxicol.* **11**(11):1265–1273.
6. E. Bourdon and D. Blache (2001). The importance of proteins in defense against oxidation, *Antioxid. Redox Signal.* **3**(2):293–311.
7. T. M. Hagen, D. L. Yowe, J. C. Bartholomew, C. M. Wehr, K. L. Do, J. Y. Park, and B. N. Ames (1997). Mitochondrial decay in hepatocytes from old rats: Membrane potential declines, heterogeneity and oxidants increase, *Proc. Natl. Acad. Sci. U.S.A.* **94**(7): 3064–3069.
8. E. R. Stadtman (2001). Protein oxidation in aging and age-related diseases, *Ann. NY Acad. Sci.* **928**:22–38.
9. M. K. Shigenaga, T. M. Hagen, and B. N. Ames (1994). Oxidative damage and mitochondrial decay in aging, *Proc. Natl. Acad. Sci. U.S.A.* **91**(23):10771–10778.
10. K. B. Beckman and B. N. Ames (1998). The free radical theory of aging matures, *Physiol. Rev.* **78**(2):547–581.
11. E. R. Stadtman, P. E. Starke-Reed, C. N. Oliver, J. M. Carney, and R. A. Floyd (1992). Protein modification in aging, *Exs.* **62**:64–72.
12. E. R. Stadtman and R. L. Levine (2000). Protein oxidation, *Ann. NY Acad. Sci.* **899**:191–208.
13. L. Fucci, C. N. Oliver, M. J. Coon, and E. R. Stadtman (1983). Inactivation of key metabolic enzymes by mixed-function oxidation reactions: Possible implication in protein turnover and ageing, *Proc. Natl. Acad. Sci. U.S.A.* **80**(6):1521–1525.
14. J. Liu, E. Head, A. M. Gharib, W. Yuan, R. T. Ingersoll, T. M. Hagen, C. W. Cotman, and B. N. Ames (2002). Memory loss in old rats is associated with brain mitochondrial decay and rna/DNA oxidation: Partial reversal by feeding acetyl-l-carnitine and/or r-alpha-lipoic acid, *Proc. Natl. Acad. Sci. U.S.A.* **99**(4):2356–2361.
15. H. Esterbauer, R. J. Schaur, and H. Zollner (1991). Chemistry and biochemistry of 4-hydroxynonenal, malonaldehyde and related aldehydes, *Free Radic. Biol. Med.* **11**(1):81–128.
16. K. M. Humphries and L. I. Szweda (1998). Selective inactivation of alpha-ketoglutarate dehydrogenase and pyruvate dehydrogenase: Reaction of lipoic acid with 4-hydroxy-2-nonenal, *Biochemistry* **37**(45):15835–15841.
17. K. M. Humphries, Y. Yoo, and L. I. Szweda (1998). Inhibition of nadh-linked mitochondrial respiration by 4-hydroxy-2-nonenal, *Biochemistry* **37**(2):552–557.
18. J. J. Chen and B. P. Yu (1996). Detoxification of reactive aldehydes in mitochondria: Effects of age and dietary restriction, *Aging (Milano)* **8**(5):334–340.
19. J. J. Chen and B. P. Yu (1994). Alterations in mitochondrial membrane fluidity by lipid peroxidation products, *Free Radic. Biol. Med.* **17**(5):411–418.
20. Q. Chen and B. N. Ames (1994). Senescence-like growth arrest induced by hydrogen peroxide in human diploid fibroblast f65 cells, *Proc. Natl. Acad. Sci. U.S.A.* **91**(10):4130–4134.

21. J. H. Choi and B. P. Yu (1995). Brain synaptosomal aging: Free radicals and membrane fluidity, *Free Radic. Biol. Med.* **18**(2):133–139.
22. A. Terman and U. T. Brunk (2004). Aging as a catabolic malfunction, *Int. J. Biochem. Cell Biol.* **36**(12):2365–2375.
23. D. Harman (1972). The biologic clock: The mitochondria? *J. Am. Geriatr. Soc.* **20**(4):145–147.
24. T. M. Hagen, J. Liu, J. Lykkesfeldt, C. M. Wehr, R. T. Ingersoll, V. Vinarsky, J. C. Bartholomew, and B. N. Ames (2002). Feeding acetyl-l-carnitine and lipoic acid to old rats significantly improves metabolic function while decreasing oxidative stress, *Proc. Natl. Acad. Sci. U.S.A.* **99**(4):1870–1875.
25. L. J. Yan, R. L. Levine, and R. S. Sohal (1997). Oxidative damage during aging targets mitochondrial aconitase, *Proc. Natl. Acad. Sci. U.S.A.* **94**(21):11168–11172.
26. O. H. Wieland (1983). The mammalian pyruvate dehydrogenase complex: Structure and regulation, *Rev. Physiol. Biochem. Pharmacol.* **96**:123–170.
27. K. Hagopian, J. J. Ramsey, and R. Weindruch (2003). Influence of age and caloric restriction on liver glycolytic enzyme activities and metabolite concentrations in mice, *Exp. Gerontol.* **38**(3):253–266.
28. R. Moreau, S. H. Heath, C. E. Doneanu, R. A. Harris, and T. M. Hagen (2004). Age-related compensatory activation of pyruvate dehydrogenase complex in rat heart, *Biochem. Biophys. Res. Commun.* **325**(1):48–58.
29. L. Yan, H. Ge, H. Li, S. C. Lieber, F. Natividad, R. R. Resuello, S. J. Kim, S. Akeju, A. Sun, K. Loo, A. P. Peppas, F. Rossi, E. D. Lewandowski, A. P. Thomas, S. F. Vatner, and D. E. Vatner (2004). Gender-specific proteomic alterations in glycolytic and mitochondrial pathways in aging monkey hearts, *J. Mol. Cell Cardiol.* **37**(5):921–929.
30. K. F. Sheu and J. P. Blass (1999). The alpha-ketoglutarate dehydrogenase complex, *Ann. NY Acad. Sci.* **893**:61–78.
31. R. Moreau, S. H. Heath, C. E. Doneanu, J. G. Lindsay, and T. M. Hagen (2003). Age-related increase in 4-hydroxynonenal adduction to rat heart alpha-ketoglutarate dehydrogenase does not cause loss of its catalytic activity, *Antioxid. Redox Signal.* **5**(5): 517–527.
32. D. T. Lucas and L. I. Szweda (1999). Declines in mitochondrial respiration during cardiac reperfusion: Age-dependent inactivation of alpha-ketoglutarate dehydrogenase, *Proc. Natl. Acad. Sci. U.S.A.* **96**(12):6689–6693.
33. G. E. Gibson, K. F. Sheu, J. P. Blass, A. Baker, K. C. Carlson, B. Harding, and P. Perrino (1988). Reduced activities of thiamine-dependent enzymes in the brains and peripheral tissues of patients with alzheimer's disease, *Arch. Neurol.* **45**(8):836–840.
34. R. F. Butterworth and A. M. Besnard (1990). Thiamine-dependent enzyme changes in temporal cortex of patients with alzheimer's disease, *Metab. Brain Dis.* **5**(4):179–184.
35. G. E. Gibson, L. C. Park, K. F. Sheu, J. P. Blass, and N. Y. Calingasan (2000). The alpha-ketoglutarate dehydrogenase complex in neurodegeneration, *Neurochem. Int.* **36**(2): 97–112.
36. A. M. Kates, P. Herrero, C. Dence, P. Soto, M. Srinivasan, D. G. Delano, A. Ehsani, and R. J. Gropler (2003). Impact of aging on substrate metabolism by the human heart, *J. Am. Coll. Cardiol.* **41**(2):293–299.
37. F. Fraser, C. G. Corstorphine, and V. A. Zammit (1997). Topology of carnitine palmitoyl-transferase i in the mitochondrial outer membrane, *Biochem. J.* **323**(Pt 3):711–718.

38. F. Fraser, C. G. Corstorphine, and V. A. Zammit (1996). Evidence that both the acyl-coa and malonyl-coa binding sites of mitochondrial overt carnitine palmitoyltransferase (cpt i) are exposed on the cytosolic face of the outer membrane, *Biochem. Soc. Trans.* **24**(2):184S.
39. F. Fraser, R. Padovese, and V. A. Zammit (2001). Distinct kinetics of carnitine palmitoyl-transferase i in contact sites and outer membranes of rat liver mitochondria, *J. Biol. Chem.* **276**(23):20182–20185.
40. J. D. McGarry and N. F. Brown (2000). Reconstitution of purified, active and malonyl-coa-sensitive rat liver carnitine palmitoyltransferase i: Relationship between membrane environment and malonyl-coa sensitivity, *Biochem. J.* **349**(Pt 1):179–187.
41. L. Packer, S. Roy, and C. K. Sen (1997). Alpha-lipoic acid: A metabolic antioxidant and potential redox modulator of transcription, *Adv. Pharmacol.* **38**:79–101.
42. J. Liu, D. W. Killilea, and B. N. Ames (2002). Age-associated mitochondrial oxidative decay: Improvement of carnitine acetyltransferase substrate-binding affinity and activity in brain by feeding old rats acetyl-l-carnitine and/or r-alpha-lipoic acid, *Proc. Natl. Acad. Sci. U.S.A.* **99**(4):1876–1881.
43. M. Manczak, Y. Jung, B. S. Park, D. Partovi, and P. H. Reddy (2005). Time-course of mitochondrial gene expressions in mice brains: Implications for mitochondrial dysfunction, oxidative damage, and cytochrome c in aging, *J. Neurochem.* **92**(3):494–504.
44. J. H. Suh, S. H. Heath, and T. M. Hagen (2003). Two subpopulations of mitochondria in the aging rat heart display heterogenous levels of oxidative stress, *Free Radic. Biol. Med.* **35**(9):1064–1072.
45. A. Navarro and A. Boveris (2004). Rat brain and liver mitochondria develop oxidative stress and lose enzymatic activities on aging, *Am. J. Physiol. Regul. Integr. Comp. Physiol.* **287**(5):R1244–1249.
46. V. G. Desai, R. Weindruch, R. W. Hart, and R. J. Feuers (1996). Influences of age and dietary restriction on gastrocnemius electron transport system activities in mice, *Arch. Biochem. Biophys.* **333**(1):145–151.
47. G. Paradies, F. M. Ruggiero, G. Petrosillo, M. N. Gadaleta, and E. Quagliariello (1994). The effect of aging and acetyl-l-carnitine on the function and on the lipid composition of rat heart mitochondria, *Ann. NY Acad. Sci.* **717**:233–243.
48. R. J. Feuers (1998). The effects of dietary restriction on mitochondrial dysfunction in aging, *Ann. NY Acad. Sci.* **854**:192–201.
49. S. Moghaddas, C. L. Hoppel, and E. J. Lesnefsky (2003) Aging defect at the qo site of complex iii augments oxyradical production in rat heart interfibrillar mitochondria, *Arch. Biochem. Biophys.* **414**(1):59–66.
50. T. M. Hagen, R. T. Ingersoll, C. M. Wehr, J. Lykkesfeldt, V. Vinarsky, J. C. Bartholomew, M. H. Song, and B. N. Ames (1998). Acetyl-l-carnitine fed to old rats partially restores mitochondrial function and ambulatory activity, *Proc. Natl. Acad. Sci. U.S.A.* **95**(16): 9562–9566.
51. G. Paradies, F. M. Ruggiero, G. Petrosillo, M. N. Gadaleta, and E. Quagliariello (1994). Effect of aging and acetyl-l-carnitine on the activity of cytochrome oxidase and adenine nucleotide translocase in rat heart mitochondria, *FEBS Lett.* **350**(2–3):213–215.
52. K. Pfeiffer, V. Gohil, R. A. Stuart, C. Hunte, U. Brandt, M. L. Greenberg, and H. Schagger (2003). Cardiolipin stabilizes respiratory chain supercomplexes, *J. Biol. Chem.* **278**(52):52873–52880.

53. V. Koshkin and M. L. Greenberg (2002). Cardiolipin prevents rate-dependent uncoupling and provides osmotic stability in yeast mitochondria, *Biochem. J.* **364**(Pt 1):317–322.

54. M. Schlame, D. Rua, and M. L. Greenberg (2000). The biosynthesis and functional role of cardiolipin, *Prog. Lipid Res.* **39**(3):257–288.

55. H. Schagger and K. Pfeiffer (2000). Supercomplexes in the respiratory chains of yeast and mammalian mitochondria, *EMBO. J.* **19**(8):1777–1783.

56. G. Paradies, G. Petrosillo, M. N. Gadaleta, and F. M. Ruggiero (1999). The effect of aging and acetyl-l-carnitine on the pyruvate transport and oxidation in rat heart mitochondria, *FEBS Lett.* **454**(3):207–209.

57. M. T. Nakamura and T. Y. Nara (2004). Structure, function, and dietary regulation of delta6, delta5, and delta9 desaturases, *Annu. Rev. Nutr.* **24**:345–376.

58. M. T. Nakamura and T. Y. Nara (2003). Essential fatty acid synthesis and its regulation in mammals, *Prostaglandins Leukot. Essent. Fatty Acids.* **68**(2):145–150.

59. S. Pepe, N. Tsuchiya, E. G. Lakatta, and R. G. Hansford (1999). Pufa and aging modulate cardiac mitochondrial membrane lipid composition and ca2+ activation of pdh, *Am. J. Physiol.* **276**(1 Pt 2):H149–158.

60. P. L. Biagi, A. Bordoni, S. Hrelia, M. Celadon, and D. F. Horrobin (1991). Gamma-linolenic acid dietary supplementation can reverse the aging influence on rat liver microsome delta 6-desaturase activity, *Biochim. Biophys. Acta.* **1083**(2):187–192.

61. S. Hrelia, A. Bordoni, P. Motta, M. Celadon, and P. L. Biagi (1991). Kinetic analysis of delta-6-desaturation in liver microsomes: Influence of gamma-linoleic acid dietary supplementation to young and old rats, *Prostaglandins Leukot. Essent. Fatty Acids.* **44**(3):191–194.

62. F. Q. Schafer and G. R. Buettner (2001). Redox environment of the cell as viewed through the redox state of the glutathione disulfide/glutathione couple, *Free Radic. Biol. Med.* **30**(11):1191–1212.

63. J. H. Suh, H. Wang, R. M. Liu, J. Liu, and T. M. Hagen (2004). (r)-alpha-lipoic acid reverses the age-related loss in gsh redox status in post-mitotic tissues: Evidence for increased cysteine requirement for gsh synthesis, *Arch. Biochem. Biophys.* **423**(1): 126–135.

64. T. M. Hagen, V. Vinarsky, C. M. Wehr, and B. N. Ames (2000). (r)-alpha-lipoic acid reverses the age-associated increase in susceptibility of hepatocytes to tert-butylhydroperoxide both in vitro and in vivo, *Antioxid. Redox Signal.* **2**(3):473–483.

65. J. H. Suh, S.V. Shenvi, B. M. Dixon, H. Liu, A. K. Jaiswal, R. M. Liu, and T. M. Hagen (2004). Decline in transcriptional activity of nrf2 causes age-related loss of glutathione synthesis, which is reversible with lipoic acid, *Proc. Natl. Acad. Sci. U.S.A.* **101**(10): 3381–3386.

66. R. M. Liu (2002). Down-regulation of gamma-glutamylcysteine synthetase regulatory subunit gene expression in rat brain tissue during aging, *J. Neurosci Res.* **68**(3): 344–351.

67. R. Liu and J. Choi (2000). Age-associated decline in gamma-glutamylcysteine synthetase gene expression in rats, *Free Radic. Biol. Med.* **28**(4):566–574.

68. D. Toroser and R. S. Sohal (2005). Kinetic characteristics of native gamma-glutamylcysteine ligase in the aging housefly, musca domestica l, *Biochem. Biophys. Res. Commun.* **326**(3):586–593.

69. A. Gafni and N. Noy (1984). Age-related effects in enzyme catalysis, *Mol. Cell Biochem.* **59**(1–2):113–129.

70. A. Gafni (1984). Age-related effects on subunit interactions in rat muscle glyceraldehyde-3-phosphate dehydrogenase, *Curr. Top. Cell Regul.* **24**:273–285.

71. V. Dulic and A. Gafni (1987). Mechanism of aging of rat muscle glyceraldehyde-3-phosphate dehydrogenase studied by selective enzyme-oxidation, *Mech. Ageing Dev.* **40**(3):289–306.

72. J. L. Mazzola and M. A. Sirover (2002). Alteration of intracellular structure and function of glyceraldehyde-3-phosphate dehydrogenase: A common phenotype of neurodegenerative disorders? *Neurotoxicology* **23**(4–5):603–609.

73. W. G. Tatton, R. M. Chalmers-Redman, M. Elstner, W. Leesch, F. B. Jagodzinski, D. P. Stupak, M. M. Sugrue, and N. A. Tatton (2000). Glyceraldehyde-3-phosphate dehydrogenase in neurodegeneration and apoptosis signaling, *J. Neural Transm.* Suppl(60): 77–100.

74. M. D. Berry and A. A. Boulton (2000). Glyceraldehyde-3-phosphate dehydrogenase and apoptosis, *J. Neurosci. Res.* **60**(2):150–154.

75. E. I. Arutyunova, P. V. Danshina, L. V. Domnina, A. P. Pleten, and V. I. Muronetz (2003). Oxidation of glyceraldehyde-3-phosphate dehydrogenase enhances its binding to nucleic acids, *Biochem. Biophys. Res. Commun.* **307**(3):547–552.

76. S. N. Pradhan (1980). Central neurotransmitters and aging, *Life Sci.* **26**(20):1643–1656.

77. C. Mohan and E. Radha (1978). Age dependent kinetic changes in the activities of central cholinergic enzymes, *Exp. Gerontol.* **13**(5):349–356.

78. A. M. Hussain and A. K. Mitra (2004). Effect of reactive oxygen species on the metabolism of tryptophan in rat brain: Influence of age, *Mol. Cell Biochem.* **258**(1–2): 145–153.

13

DIETARY AND GENETIC EFFECTS ON ATHEROGENIC DYSLIPIDEMIA

Ronald M. Krauss[1,2,3] and Patty W. Siri[1]

[1]*Children's Hospital Oakland Research Institute, Oakland, California*
[2]*Department of Genome Science, Lawrence Berkeley National Laboratory*
[3]*Department of Nutritional Sciences, University of California–Berkeley, Berkeley, California*

13.1 INTRODUCTION

Atherogenic dyslipidemia is a major feature of a number of metabolic conditions associated with increased risk for coronary artery disease, including overweight and/or obesity, insulin resistance and the metabolic syndrome, and Type 2 diabetes. High triglyceride concentrations, low HDL cholesterol levels, and an increase in small LDL particles are the hallmark characteristics of atherogenic dyslipidemia. The phenotype can be influenced by both genetic and environmental determinants. Importantly, the ability of environmental factors to promote the dyslipidemic profile may be contingent on the genetic susceptibility of an individual. This chapter reviews the evidence for functionally important genetic variants that have been shown to affect atherogenic dyslipidemia, in part via their role in the determination of responsiveness to dietary variables. The identification of such gene–diet interactions may help to better define subgroups of persons who may benefit from specific dietary regimens.

Nutritional Genomics: Discovering the Path to Personalized Nutrition
Edited by Jim Kaput and Raymond L. Rodriguez

13.2 LDL REPRESENTS A HETEROGENEOUS POPULATION OF PARTICLES

Major LDL Subclasses

Subpopulations of LDL are differentiated based on particle density, size, charge, and chemical composition. At least seven distinct subspecies of LDL have been identified by nondenaturing gradient gel electrophoresis, which separates particles based on variations in size and shape [1]. These subspecies can be further classified, based on density, into four subclasses that range in size from the largest and most buoyant (i.e., LDL1) to the smallest and most dense (i.e., LDL4). These LDL subclasses are distinguished by variations in lipid and carbohydrate composition as well as by the conformation of the major structural protein (apo B) on the particle. Of clinical relevance, increased concentrations of LDL3 and LDL4, which together comprise small, dense LDL, have been associated with increased coronary heart disease risk [2].

Plasma triglyceride and VLDL levels have been shown to be strongly associated with decreasing size and increasing density of LDL particles (Figure 13.1). Furthermore, increased VLDL and increased small, dense LDL have been shown to be inversely associated with plasma levels of HDL, particularly the HDL2 subclass [5, 6]. While the metabolic basis for these relationships has not yet been completely delineated, there is evidence that a number of mechanisms may be relevant. The increased secretion of triglyceride [7, 8] and apo B [9, 10] has been documented in patients with dyslipidemic profiles [11]. Impaired clearance of lipid-rich particles may also contribute to the profile of increased triglyceride, decreased HDL, and increased small, dense LDL. In this regard, deficient LPL activity [12] and increased HL activity [12, 13] have been associated with dyslipidemic profiles.

Importantly, insulin resistance represents an underlying abnormality closely tied to the pathophysiology of atherogenic dyslipidemia. The dyslipidemia associated with insulin resistance is characterized by high triglyceride concentrations and low HDL cholesterol, features that have, in turn, been linked metabolically to smaller LDL diameter, or a "pattern B" phenotype. Nonetheless, insulin resistance and its associated hypertriglyceridemia can explain only a proportion of the variation in LDL particle size [14]. Indeed, the phenotype of smaller LDL diameter can occur independently of the metabolic syndrome as observed in a recently completed weight loss study by our group [15]. Of 178 moderately overweight participants enrolled, 84 individuals (or 47% of the population) were classified as pattern B, but only 16% of the population could be defined as having the metabolic syndrome by ATP III criteria. Importantly, among these persons 75% exhibited the pattern B phenotype.

Hypertension has also often been associated with insulin resistance and its associated dyslipidemia; pattern B patients have been shown to have an increased risk for high blood pressure [14, 16]. Type 2 diabetics have been shown to exhibit a predominance of small dense LDL particles, an association that likely reflects the insulin resistance present in these patients [17, 18]. Moreover, in well-

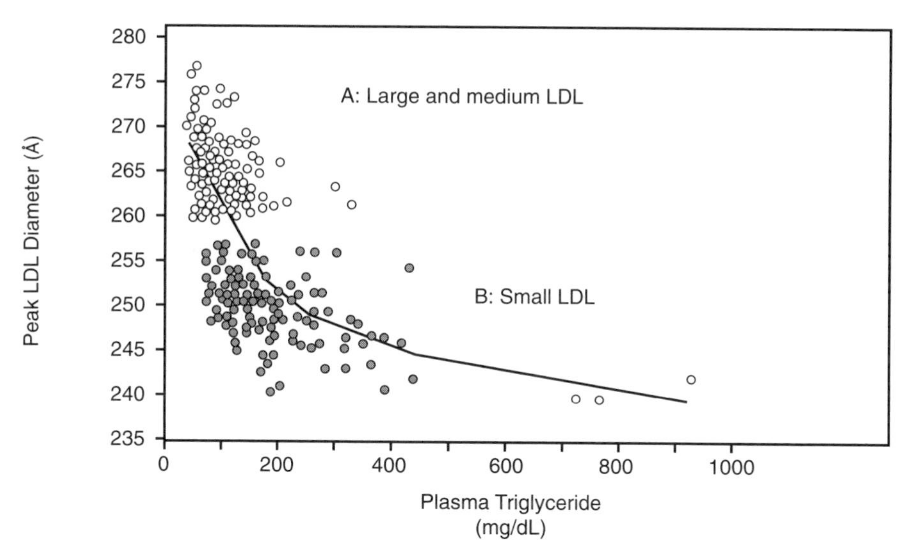

Figure 13.1. Peak LDL diameters vary inversely with plasma triglyceride levels. LDL size determination by nondenaturing gradient gel electrophoresis [1] in a group of 422 subjects from the Berkeley Lipid Study Population [3, 4] has allowed for the identification of two groups of subjects, designated pattern A (predominance of large and medium LDL1 and LDL2) and shown with open circles and pattern B (predominance of small LDL3 and LDL4) and shown with closed circles.

controlled Type 1 diabetics, relative increases in LDL3 and IDL have been reported [19].

Model for Origins of LDL Subclasses

A proposed model for the metabolic origins of the four major subclasses of LDL has been previously described [20]. Briefly, the model outlines a scheme in which the production of heterogeneous LDL particles originates in the parallel operation of several independent pathways. Under conditions of low hepatic triglyceride production, secreted apo B-containing lipoproteins are relatively triglyceride depleted and give rise to the larger and more buoyant LDL1 and LDL2. With higher hepatic triglyceride secretion, VLDL particles are larger and more triglyceride enriched. Following lipolysis by lipoprotein lipase (LPL) and remodeling via cholesterol ester transfer protein (CETP)-mediated triglyceride enrichment and hepatic lipase (HL)-catalyzed lipolysis, these particles form small, dense LDL particles; with yet higher triglyceride loading, the liver secretes very large VLDL particles that normally yield only small amounts of LDL since their lipolytic remnant products may be cleared directly from plasma [21, 22]. The further processing of the remnants remaining in plasma by LPL and HL yields LDL4, the smallest and most dense particles among the LDL subclasses.

13.3 LDL SUBCLASSES INFLUENCED BY GENES AND THE ENVIRONMENT

Pattern B Defined

Multiple genetic and environmental factors can influence whether an individual has a predominance of large (LDL1 and LDL2) or small (LDL3 or LDL4) LDL particles. Indeed, individuals cluster into two distinct categories based on peak LDL particle size, designated pattern A (large LDL) and pattern B (small LDL). Peak particle size has been shown to be well-correlated with plasma triglyceride levels (Figure 13.1).

The prevalence of pattern B in adult men is approximately 30–35% [23]. Lower prevalence rates are found in premenopausal women and young men [24]. A potential role for estrogen is implicated when one considers the increased prevalence (15–25%) of pattern B observed in postmenopausal women compared to premenopausal women [16, 25]. However, estrogen treatment has been shown to reduce LDL size in postmenopausal women [26]. Factors other than estrogen deficiency must therefore be considered as contributing to pattern B in women.

Genetic Studies of Phenotype B

Data from complex segregation analyses suggest an autosomal dominant or codominant model for the inheritance of phenotype B with varying additive and polygenic effects. Two candidate gene linkage studies have found evidence for linkage of this phenotype with a site in the vicinity of the LDL receptor gene locus on chromosome 19p [27, 28]. Linkages to other genetic loci were found in the latter study, specifically, the apoA-I/C-III/A-IV gene cluster on chromosome 11, the *CETP* locus on chromosome 16, and the manganese superoxide dismutase gene on chromosome 6. Subsequent studies have confirmed the linkage of peak LDL size to the *CETP* gene [29].

In a dyzygotic twin study in women, the *APOB* gene was linked to peak LDL size, plasma triglycerides, HDL cholesterol, and apo B levels [30]. The HL gene locus was found to contribute to LDL size in patients with familial combined hyperlipidemia [31], although a polymorphism in the promoter previously associated with HDL cholesterol levels [32] was not linked to LDL size. While polymorphisms in the *APOE* gene were found to be associated with LDL size in the San Antonio Heart Study [33], several other studies were unable to demonstrate linkage [30, 34].

Despite evidence for major gene determinants of LDL phenotype B in family studies, twin studies have indicated that the heritability of LDL particle size ranges only from 30% to 50% [35]. These relatively low heritability estimates emphasize the importance of nongenetic and environmental influences on the expression of phenotype B. As noted above, the presence of insulin resistance or its metabolic determinants can lead to reduced LDL particle size. In this regard, the presence of abdominal or visceral obesity may be particularly important [36, 37]. Acquired alterations in triglyceride metabolism may also induce pattern B, as shown in the case of the hypertriglyceridemia of AIDS [38].

Dietary Effects on LDL Subclasses

The effects of diet on the determination of LDL subclass phenotype have been extensively studied by our group. In a randomized crossover design study with 6 week long treatment regimens of either a high-fat (46% fat, 34% carbohydrate) or low-fat (24% fat, 56% carbohydrate) diet, reductions in LDL cholesterol concomitant with increases in plasma triglyceride were observed on the low-fat diet [39, 40]. When switched to the high-fat diet, pattern B patients exhibited a twofold greater reduction in LDL cholesterol compared to pattern A patients. Most of the reduction was attributable to a decrease in mid-sized LDL2.

Notably, a significant percentage (41%) of persons who were pattern A on the high-fat diet converted to the pattern B phenotype on the low-fat, high-carbohydrate diet. In this group, in contrast to those who remained pattern B on the low-fat diet, plasma levels of apo B were not reduced (Figure 13.2). Since apo B level is a measure of the number of LDL particles, decreased plasma LDL cholesterol without a reduction in apo B indicates that the reduction of LDL by a low-fat diet in phenotype A subjects is due primarily to a change to smaller LDL with reduced cholesterol content, rather than reduced particle number [23].

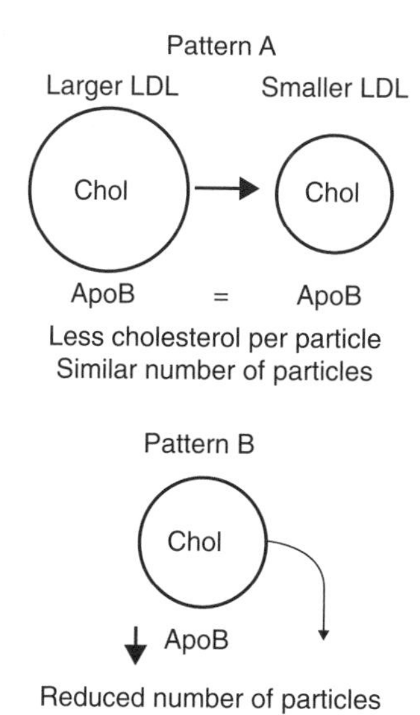

Figure 13.2. Low-fat, high-carbohydrate diets induce different lipoprotein responses in individuals with different LDL phenotypes. While both pattern A and pattern B individuals respond to low-fat diets with reductions in LDL cholesterol, the mechanism by which this reduction occurs differs. Pattern A individuals show reduced cholesterol per apo B particle without a significant change in the number of particles. In contrast, pattern B individuals show a reduction in the number of particles. (Reprinted with permission from the *Annual Review of Nutrition*, Volume 21; copyright © 2001 by Annual Reviews. www.annualreviews.org.)

Additional studies have shown that with further reduction of dietary fat and substitution of carbohydrate in isocaloric diets, there is a progressive increase in the percentage of men who convert from phenotype A to B such that the majority of men express phenotype B with very-low-fat diets (10–20% of calories) [41]. Recently, we have shown that LDL peak diameter and expression of phenotype B are significantly reduced in overweight men by lowering dietary carbohydrate from 54% to 39% and substituting protein without changing fat content. We have shown that this effect is due primarily to changes in carbohydrate intake [15]. This indicates that carbohydrate is a major dietary determinant of LDL size profiles, a conclusion consistent with longstanding evidence for the triglyceride-raising effects of increased carbohydrate intake [42]. A reduced fat (20%) and higher carbohydrate (65%) diet was also shown to promote the expression of phenotype B in premenopausal women, although the overall prevalence was much lower than in men [43]. While these results have been based on short-term studies, it is likely that they reflect the effects of longer-term dietary patterns, since the prevalence for phenotype B in free-living men and women on their usual diets [24] is similar to that which would be predicted from the relationship between dietary carbohydrate and phenotype B in our studies [41].

Genetic Influences on LDL Subclass Responses to Dietary Change

The evidence for genetic determinants of phenotype B led to the hypothesis that the differential LDL response to low-fat, high-carbohydrate diets in phenotype B versus A subjects is also genetically influenced. This hypothesis was supported by the study cited above of dietary effects on LDL subclasses in premenopausal women [43]. Consistent with the low prevalence of phenotype B in premenopausal women, only 3 of 72 expressed phenotype B on their usual diets (mean 35% fat). However, the presence of a gene predisposing to phenotype B was inferred on the basis of at least one parent with this phenotype in 29 of the women. Following consumption of a 20% fat diet for 8 weeks, reductions in LDL cholesterol were significantly related to the increasing number of pattern B parents (two versus one versus none, $p = 0.005$) and could not be explained by diet adherence or baseline characteristics including initial lipoprotein profile. Among LDL subclasses, parental phenotype B was most strongly related to reductions in LDL1 ($p = 0.05$) and LDL2 ($p = 0.02$), as well as with reduction in IDL ($p = 0.008$) and increase in triglyceride ($p = 0.02$). These results suggest that genetic predisposition to phenotype B contributes to the altered lipoprotein responses that are characteristic of this phenotype.

In the study described above, six of the phenotype A women converted to phenotype B on the lower fat diet, a number too small to permit testing of the hypothesis that parental phenotype was associated with susceptibility to this conversion. There was, however, a trend for a greater reduction in peak LDL diameter as a function of the number of phenotype B parents ($p = 0.09$). Therefore, a more extreme short-term dietary challenge (10% fat, 75% carbohydrate for 10 days) was devised to elicit expression of phenotype B in 50 male and female offspring of parental pairs classified according to LDL phenotype [44]. This study demonstrated that conversion from LDL phenotype A to B occurred in 6 of the 50 subjects, all of whom were offspring of two phenotype B parents ($p < 0.01$). Moreover, despite similar LDL peak particle size at baseline for the 19 offspring of two phenotype A parents and the 21 offspring of two phenotype B parents, the latter showed a significant reduction of LDL size with the dietary challenge, as opposed to no change in the former ($p < 0.01$ between groups). These studies therefore point to a heritable basis for the effect of a low-fat, high-carbohydrate diet on LDL size phenotypes.

A genetic basis for heritable effects on response of LDL subclasses to dietary fat and carbohydrate is supported by studies of associations of candidate gene polymorphisms with dietary effects. We have shown that the relatively common E4 polymorphism of the *APOE* gene is associated with reduction in large LDL1 particles with a low-fat diet, but not with changes in levels of small LDL [45]. In additional preliminary studies, we have found that common polymorphisms in other candidate genes related to LDL size phenotypes, including *CETP*, *LDLR*, and the recently identified *APO5*, are specifically associated with low-fat diet-induced reductions in mid-sized LDL2, a dietary response that is characteristic for LDL phenotype B as described above [46]. Furthermore, the *CETP* and *LDLR* polymorphisms are associated with significantly greater diet-induced conversion from phenotype A to B [46]. While the molecular and functional bases for these genotype associations have not been established, they support a genetic contribution to susceptibility to diet-induced changes in LDL subclasses.

13.4 CONCLUSION

Heritable factors contribute to differences in the LDL response to low-fat diets. Analysis of the changes in specific LDL subclasses is important in determining whether reductions in LDL cholesterol are more or less likely to be beneficial for reducing cardiovascular disease risk. Individuals with higher risk genetically influenced LDL phenotype B show enhanced LDL response to low-fat diets, with the major change occurring in less atherogenic, medium-sized LDL2. In contrast, the low-fat diet-induced reductions in LDL cholesterol observed in persons with phenotype A are largely due to changes in composition—specifically, decreases in LDL cholesterol—without a change in particle number that result in a shift in distribution toward a predominance of smaller and denser LDL particles.

The identification of specific genes underlying these differing LDL subclass responses to dietary fat and carbohydrate may provide a basis for preventing and managing atherogenic dyslipidemia in at-risk population subgroups. This may, in turn, provide a model system for assessing the value of genetic information in determining appropriate diets for subgroups of individuals with other diet-influenced conditions, including obesity, insulin resistance, and hypertension.

REFERENCES

1. R. M. Krauss and P. J. Blanche (1992). Detection and quantitation of LDL subfractions, *Curr. Opin. Lipid.* 377.

2. R. M. Krauss and P. W. Siri (2004). Metabolic abnormalities: Triglyceride and low-density lipoprotein, *Endocrinol. Metab. Clin North Am.* **33**:405.
3. L. A. Pennacchio, M. Olivier, J. A. Hubacek, J. C. Cohen, D. R. Cox, J. C. Fruchart, R. M. Krauss, and E. M. Rubin (2001). An apolipoprotein influencing triglycerides in humans and mice revealed by comparative sequencing, *Science* **294**:169.
4. L. A. Pennacchio, M. Olivier, J. A. Hubacek, R. M. Krauss, E. M. Rubin, and J. C. Cohen (2002). Two independent apolipoprotein A5 haplotypes influence human plasma triglyceride levels, *Hum. Mol. Genet.* **11**:3031.
5. R. M. Krauss, F. T. Lindgren, and R. M. Ray (1980). Interrelationships among subgroups of serum lipoproteins in normal human subjects, *Clin. Chim. Acta.* **104**:275.
6. R. M. Krauss, P. T. Williams, F. T. Lindgren, and P. D. Wood (1988). Coordinate changes in levels of human serum low and high density lipoprotein subclasses in healthy men, *Arteriosclerosis* **8**:155.
7. E. A. Nikkila (1973). Plasma triglyceride transport kinetics in diabetes mellitus, *Metabolism* **22**:1.
8. M. R. Taskinen, W. F. Beltz, I. Harper, R. M. Fields, G. Schonfeld, S. M. Grundy, and B. V. Howard (1986). Effects of NIDDM on very-low-density lipoprotein triglyceride and apolipoprotein B metabolism. Studies before and after sulfonylurea therapy, *Diabetes* **35**:1268.
9. M. H. Cummings, G. F. Watts, A. M. Umpleby, T. R. Hennessy, R. Naoumova, B. M. Slavin, G. R. Thompson, and P. H. Sonksen (1995). Increased hepatic secretion of very-low-density lipoprotein apolipoprotein B-100 in NIDDM, *Diabetologia* **38**: 959.
10. L. Duvillard, F. Pont, E. Florentin, C. Galland-Jos, P. Gambert, and B. Verges (2000). Metabolic abnormalities of apolipoprotein B-containing lipoproteins in non-insulin-dependent diabetes: A stable isotope kinetic study, *Eur. J. Clin. Invest.* **30**:685.
11. J. B. Marsh (2003). Lipoprotein metabolism in obesity and diabetes: Insights from stable isotope kinetic studies in humans, *Nutr. Rev.* **61**:363.
12. H. Campos, D. M. Dreon, and R. M. Krauss (1995). Associations of hepatic and lipoprotein lipase activities with changes in dietary composition and low density lipoprotein subclasses, *J. Lipid Res.* **36**:462.
13. J. H. Auwerx, C. A. Marzetta, J. E. Hokanson, and J. D. Brunzell (1989). Large buoyant LDL-like particles in hepatic lipase deficiency, *Arteriosclerosis* **9**:319.
14. G. M. Reaven, Y. D. Chen, J. Jeppesen, P. Maheux, and R. M. Krauss (1993). Insulin resistance and hyperinsulinemia in individuals with small, dense low density lipoprotein particles, *J. Clin. Invest.* **92**:141.
15. R. M. Krauss, P. J. Blanche, R. S. Rawlings, H. S. Fernstrom, and P. T. Williams (2006). Separate effects of reduced carbohydrate and weight loss on atherogenic dyslipidemia, *Amer. J. Clin. Nutr.* in press.
16. J. V. Selby, M. A. Austin, B. Newman, D. Zhang, C. P. Quesenberry, Jr., E. J. Mayer, and R. M. Krauss (1993). LDL subclass phenotypes and the insulin resistance syndrome in women, *Circulation* **88**:381.
17. H. A. Barakat, V. D. McLendon, R. Marks, W. Pories, J. Heath, and J. W. Carpenter (1992). Influence of morbid obesity and non-insulin-dependent diabetes mellitus on high-density lipoprotein composition and subpopulation distribution, *Metabolism* **41**:37.

18. K. R. Feingold, C. Grunfeld, M. Pang, W. Doerrler, and R. M. Krauss (1992). LDL subclass phenotypes and triglyceride metabolism in non-insulin-dependent diabetes, *Arterioscler. Thromb.* **12**:1496.
19. E. Manzato, A. Zambon, S. Zambon, R. Nosadini, A. Doria, R. Marin, and G. Crepaldi (1993). Lipoprotein compositional abnormalities in type 1 (insulin-dependent) diabetic patients, *Acta. Diabetol.* **30**:11.
20. K. K. Berneis and R. M. Krauss (2002). Metabolic origins and clinical significance of LDL heterogeneity, *J. Lipid. Res.* **43**:1363.
21. C. J. Packard, A. Munro, A. R. Lorimer, A. M. Gotto, and J. Shepherd (1984). Metabolism of apolipoprotein B in large triglyceride-rich very low density lipoproteins of normal and hypertriglyceridemic subjects, *J. Clin. Invest.* **74**:2178.
22. A. F. Stalenhoef, M. J. Malloy, J. P. Kane, and R. J. Havel (1984). Metabolism of apolipoproteins B-48 and B-100 of triglyceride-rich lipoproteins in normal and lipoprotein lipase-deficient humans, *Proc. Natl. Acad. Sci. U.S.A.* **81**:1839.
23. R. M. Krauss (2001). Dietary and genetic effects on low-density lipoprotein heterogeneity, *Annu. Rev. Nutr.* **21**:283.
24. M. A. Austin, M. C. King, K. M. Vranizan, and R. M. Krauss (1990). Atherogenic lipoprotein phenotype. A proposed genetic marker for coronary heart disease risk, *Circulation* **82**:495.
25. H. Campos, E. Blijlevens, J. R. McNamara, J. M. Ordovas, B. M. Posner, P. W. Wilson, W. P. Castelli, and E. J. Schaefer (1992). LDL particle size distribution. Results from the Framingham Offspring Study, *Arterioscler. Thromb.* **12**:1410.
26. H. Campos, F. M. Sacks, B. W. Walsh, I. Schiff, M. A. O'Hanesian, and R. M. Krauss (1993). Differential effects of estrogen on low-density lipoprotein subclasses in healthy postmenopausal women, *Metabolism* **42**:1153.
27. P. M. Nishina, J. P. Johnson, J. K. Naggert, and R. M. Krauss (1992). Linkage of atherogenic lipoprotein phenotype to the low density lipoprotein receptor locus on the short arm of chromosome 19, *Proc. Natl. Acad. Sci. U.S.A.* **89**:708.
28. J. I. Rotter, X. Bu, R. M. Cantor, C. H. Warden, J. Brown, R. J. Gray, et al. (1996). Multilocus genetic determinants of LDL particle size in coronary artery disease families, *Am. J. Hum. Genet.* **58**:585.
29. P. J. Talmud, K. L. Edwards, C. M. Turner, B. Newman, J. M. Palmen, S. E. Humphries, and M. A. Austin (2000). Linkage of the cholesteryl ester transfer protein (CETP) gene to LDL particle size: Use of a novel tetranucleotide repeat within the CETP promoter, *Circulation* **101**:2461.
30. M. A. Austin, P. J. Talmud, L. A. Luong, L. Haddad, I. N. Day, et al. (1998). Candidate-gene studies of the atherogenic lipoprotein phenotype: A sib-pair linkage analysis of DZ women twins, *Am. J. Hum. Genet.* **62**:406.
31. H. Allayee, K. M. Dominguez, B. E. Aouizerat, R. M. Krauss, J. I. Rotter, et al. (2000). Contribution of the hepatic lipase gene to the atherogenic lipoprotein phenotype in familial combined hyperlipidemia, *J. Lipid Res.* **41**:245.
32. S. S. Deeb, A. Zambon, M. C. Carr, A. F. Ayyobi, and J. D. Brunzell (2003). Hepatic lipase and dyslipidemia: Interactions among genetic variants, obesity, gender, and diet, *J. Lipid Res.* **44**:1279.
33. S. M. Haffner, M. P. Stern, H. Miettinen, D. Robbins, and B. V. Howard (1996). Apolipoprotein E polymorphism and LDL size in a biethnic population, *Arterioscler. Thromb. Vasc. Biol.* **16**:1184.

34. S. P. Zhao, M. H. Verhoeven, J. Vink, L. Hollaar, A. van der Laarse, P. de Knijff, and F. M. van't Hooft (1993). Relationship between apolipoprotein E and low density lipoprotein particle size, *Atherosclerosis* **102**:147.

35. M. A. Austin (1992). Genetic epidemiology of low-density lipoprotein subclass phenotypes, *Ann. Med.* **24**:477.

36. R. B. Terry, P. D. Wood, W. L. Haskell, M. L. Stefanick, and R. M. Krauss (1989). Regional adiposity patterns in relation to lipids, lipoprotein cholesterol, and lipoprotein subfraction mass in men, *J. Clin. Endocrinol. Metab.* **68**:191.

37. D. J. Nieves, M. Cnop, B. Retzlaff, C. E. Walden, J. D. Brunzell, R. H. Knopp, and S. E. Kahn (2003). The atherogenic lipoprotein profile associated with obesity and insulin resistance is largely attributable to intra-abdominal fat, *Diabetes* **52**:172.

38. K. R. Feingold, R. M. Krauss, M. Pang, W. Doerrler, P. Jensen, and C. Grunfeld (1993). The hypertriglyceridemia of acquired immunodeficiency syndrome is associated with an increased prevalence of low density lipoprotein subclass pattern B, *J. Clin. Endocrinol. Metab.* **76**:1423.

39. D. M. Dreon, H. A. Fernstrom, B. Miller, and R. M. Krauss (1994). Low-density lipoprotein subclass patterns and lipoprotein response to a reduced-fat diet in men, *FASEB J.* **8**:121.

40. R. M. Krauss and D. M. Dreon (1995). Low-density-lipoprotein subclasses and response to a low-fat diet in healthy men, *Am. J. Clin. Nutr.* **62**:478S.

41. D. M. Dreon, H. A. Fernstrom, P. T. Williams, and R. M. Krauss (1999). A very low-fat diet is not associated with improved lipoprotein profiles in men with a predominance of large, low-density lipoproteins, *Am. J. Clin. Nutr.* **69**:411.

42. E. J. Parks and M. K. Hellerstein (2000). Carbohydrate-induced hypertriacylglycerolemia: historical perspective and review of biological mechanisms, *Am. J. Clin. Nutr.* **71**:412.

43. D. M. Dreon, H. A. Fernstrom, P. T. Williams, and R. M. Krauss (1997). LDL subclass patterns and lipoprotein response to a low-fat, high-carbohydrate diet in women, *Arterioscler. Thromb. Vasc. Biol.* **17**:707.

44. D. M. Dreon, H. A. Fernstrom, P. T. Williams, and R. M. Krauss (2000). Reduced LDL particle size in children consuming a very-low-fat diet is related to parental LDL-subclass patterns, *Am. J. Clin. Nutr.* **71**:1611.

45. D. M. Dreon, H. A. Fernstrom, B. Miller, and R. M. Krauss (1995). Apolipoprotein E isoform phenotype and LDL subclass response to a reduced-fat diet, *Arterioscler. Thromb. Vasc. Biol.* **15**:105.

46. R. M. Krauss (2005). Dietary and genetic probes of atherogenic dyslipidemia, *Arterioscler. Thromb. Vasc. Biol.* in press.

14

GENISTEIN AND POLYPHENOLS IN THE STUDY OF CANCER PREVENTION: CHEMISTRY, BIOLOGY, STATISTICS, AND EXPERIMENTAL DESIGN

Stephen Barnes,[2,3,4] David B. Allison,[1,3,4] Grier P. Page,[1,3,4] Mark Carpenter,[4,5] Gary L. Gadbury,[4,6] Sreelatha Meleth,[3,4] Pamela Horn-Ross,[4,7] Helen Kim,[2,3,4] and Coral A. Lamartinere[2,3,4]

[1] *Department of Biostatistics, Section on Statistical Genetics, University of Alabama-Birmingham, Birmingham, Alabama*
[2] *Department of Pharmacology and Toxicology, University of Alabama-Birmingham, Birmingham, Alabama*
[3] *Comprehensive Cancer Center, University of Alabama-Birmingham, Birmingham, Alabama*
[4] *Center for Nutrient-Gene Interaction, University of Alabama-Birmingham, Birmingham, Alabama*
[5] *Department of Mathematics and Statistics, Auburn University, Auburn, Alabama*
[6] *Department of Mathematics and Statistics, University of Missouri, Rolla, Missouri*
[7] *Northern California Cancer Center, Fremont, California*

Nutritional Genomics: Discovering the Path to Personalized Nutrition
Edited by Jim Kaput and Raymond L. Rodriguez

14.1 INTRODUCTION

The past 25 years have seen a vast increase in human knowledge, the rise of both personal and large-scale computing, rapid developments in molecular biology, and the resultant sequencing of human and other genomes. At the same time an important aspect of cancer has become more apparent—that diet can alter the risk of this collection of related diseases. In this chapter, we show that all of these advances, disparate as they may at first seem, can be connected and in doing so may advance knowledge still further. Together they offer the potential for remarkable new insights into how the conditions that give rise to cancer can be altered. However, without careful consideration of experimental design and subsequent sample and data analysis, the promise offered by these approaches will not be achieved. We provide a road map to this path in order to facilitate research.

14.2 DIET AND CANCER

Epidemiologists found from worldwide studies that there are significant differences in the rates of cancers from country-to-country [1]. While this could be due to racial/ethnic differences or different methods of medical care in each country, careful study suggested that in many cases these are minor contributors to cancer risk. A seminal analysis by Doll [2] found that diet may account for 35–70% of the risk of cancer. Studies of immigrant populations further supported the diet–cancer risk hypothesis [3–5]. Thus, the hunt began to discover which components in the diet were responsible for altering cancer risk. Initially, differences in the macrocomponents of the diet were investigated—too much red meat [6] and fat [7] and not enough complex carbohydrate [8] were each intimated as increasing the risk of cancer. In this view of the problem, countries with high cancer rates had diets that *caused* cancer.

Other groups pointed out that the differences in cancer risk could also result from dietary factors that are *preventive* in nature [9, 10]. Interestingly, this was hard for many biomedical scientists and clinicians to accept since they were used to pharmaceuticals, not the diet, preventing and curing disease. Nonetheless, the race was on to identify the dietary factors that prevent cancer. One of the most interesting factors studied is the polyphenols.

14.3 CHEMISTRY OF THE POLYPHENOLS

Polyphenols are a large group of compounds found in plants [11]. They consist of several closely related classes and include bioflavonoids as well as coumestanes and stilbenes (Figure 14.1). The bioflavonoids are further divided into flavonoids and isoflavonoids dependent on the position of the phenyl substituent in the benzopyran ring (Figure 14.1). Genistein, an isoflavonoid (Figure 14.1A), and apigenin, a flavonoid (Figure 14.1C), are positional isomers. Most of the polyphenols in plants are

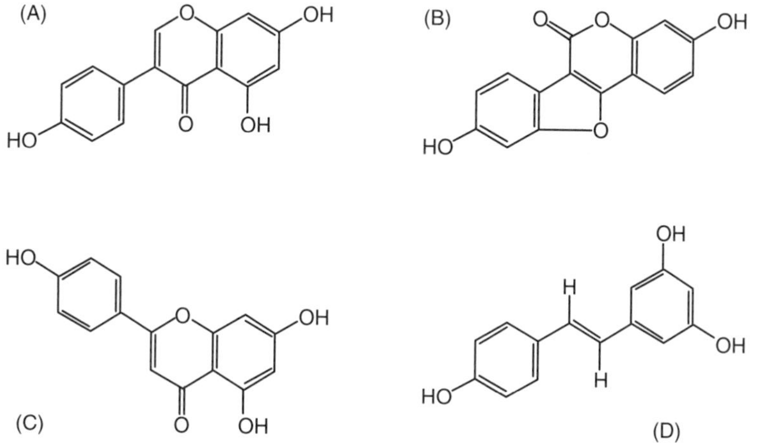

Figure 14.1. Chemical structures of polyphenols found in fruits and vegetables. The polyphenols are (A) genistein, an isoflavone, (B) coumestrol, a coumestane, (C) apigenin, a flavonoid, and (D) resveratrol, a stilbene. Note that genistein and apigenin are positional isomers—the phenyl ring is in the 2 position for apigenin and in the 3 position for genistein.

O-glycosides, although there are a few examples of *C*-linked glycosides (Figure 14.2) [11, 12]. The sugar moiety is usually glucose, but arabinose and rhamnose glycosides also occur [13, 14]. In some cases the sugar may also be a disaccharide [11, 15]. The oxidation state of the heterocyclic ring in flavonoids leads to other subclassifications—thus there are flavones, flavonols, flavanones, flavanols, and flavans (Figure 14.3). Some of these are not present in the diet, but instead are formed by metabolism following consumption of the polyphenol-containing food. The bioflavonoids are also susceptible to forming condensed or oligomeric forms such as the proanthocyanins and the tannins. These are found in many types of berries and in grapes and may have strong antioxidant properties, at least as measured in vitro [16].

14.4 UPTAKE, DISTRIBUTION, METABOLISM, AND EXCRETION OF POLYPHENOLS

Uptake of polyphenol aglycones (without their glycosyl groups) into cells is generally more rapid than their glycosylated counterparts. However, when polyphenol-containing foods are eaten, hydrolysis of the glycosides by intestinal hydrolases (lactose phlorizin hydrolase, LPH) leads to release of the polyphenol aglycones [17]. This facilitates their absorption by passive mechanisms, the extent of uptake being dependent on the hydrophobicity of the polyphenol aglycones. Genistein is quite hydrophobic and greater than 80% of an oral dose is taken up from rat small intestine

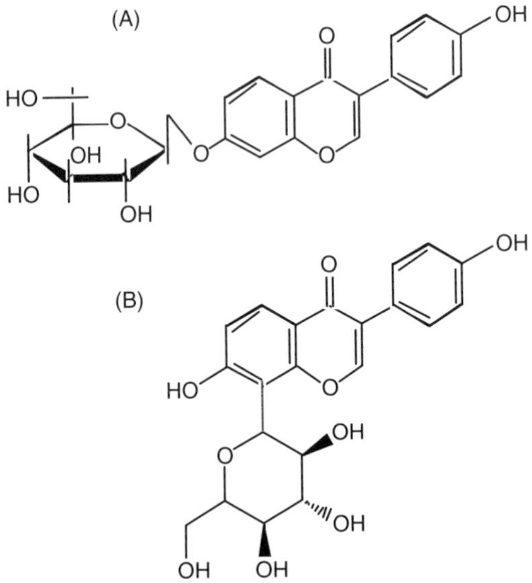

Figure 14.2. Chemical structures of isoflavone *O*- and *C*-glycosides. Most isoflavones are found as their *O*-glycosides (A) in plants. *C*-glycosides (B) of isoflavones are rare.

and is excreted into bile [18]. In contrast, polyphenols such as quercetin with many hydroxyl groups are quite hydrophilic and are poorly absorbed [19]. Polyphenols in which the sugar moiety is esterified are resistant to β-glucosidases and their hydrolysis may be delayed until they encounter significant concentrations of intestinal bacteria, usually in the colon [20]. Bioflavonoid C-glycosides (Figure 14.2B) are an exception to this: they are resistant to hydrolase activity. Puerarin, the *C*-glycoside of daidzein, is taken up quickly and rapidly from the small intestine, presumably via the intestinal sodium-dependent glucose transporter [21].

Most polyphenols are immediately converted to β-glucuronide conjugates in the small intestinal wall [18, 22]. Those that aren't undergo a similar reaction in the liver, as well as other phase I and phase II reactions [23, 24]. Thus, only a small proportion of the absorbed polyphenols circulate in the bioactive form. While this may seem hopeless from the standpoint of inducing biological effects, it is no different from the situation for the physiologic steroids. Following the oral ingestion of polyphenols, there is, nonetheless, a period (1–2 hours) where the aglycones are present in the blood [25]. Certain reports have indicated that in microcompartments such as prostatic fluid [26] or the aqueous humor of the eye [27], polyphenols either accumulate or are selectively absorbed as their aglycones.

Many polyphenols undergo an enterohepatic circulation [18, 28]: after their entry into the blood compartment, they are taken up by the liver and excreted into bile. This returns them to the small intestine. Although polyphenol glucuronide and sulfate conjugates are poor substrates for intestinal absorption, they are readily hydrolyzed by intestinal bacteria, thereby releasing their aglycones for reabsorption

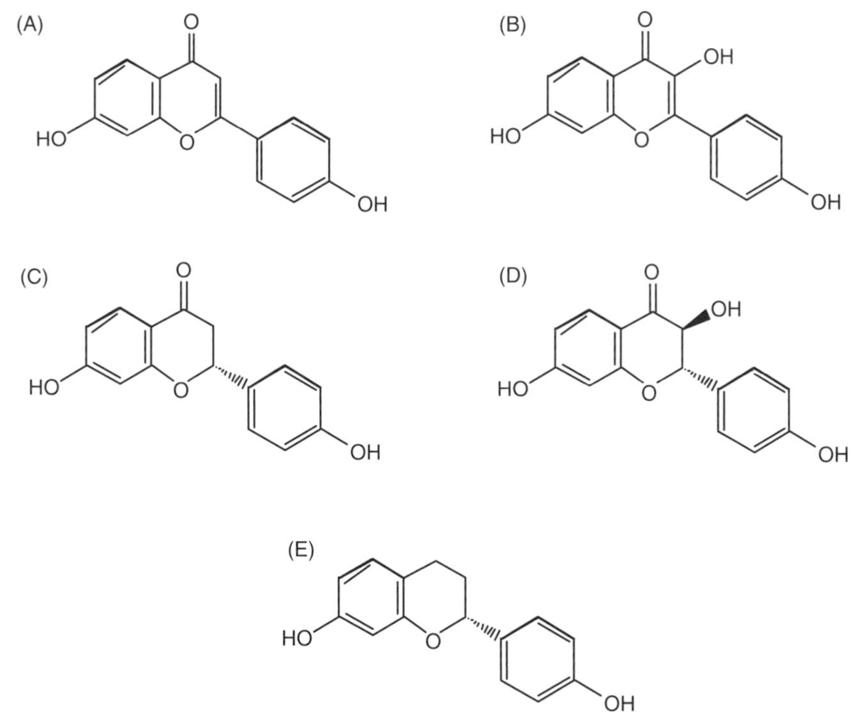

Figure 14.3. Oxidation states of the benzopyran ring in bioflavonoids. There are many chemical variants of the heterocyclic ring—(A) flavonoid, (B) flavonol, (C) flavanone, (D) flavanol, and (E) flavan.

[18, 28]. This process occurs more distally than for the plant glycosides and therefore ensures a more extensive contact with colonic bacteria. Colonic bacterial metabolism causes reduction of the heterocyclic ring of the bioflavonoids and may also lead to cleavage, resulting in the formation of phenolic acids. These too may undergo reabsorption [29] and lead to an ever more complex array of metabolites to consider.

14.5 POLYPHENOLS AND CANCER PREVENTION

Polyphenols from numerous foods have been investigated for cancer prevention. Barnes et al. [30] demonstrated that rats, on soy-based diets from young adulthood on, developed a lower number of mammary tumors induced by *N*-methyl-*N*-nitrosourea (MNU). This confirmed an earlier report by Troll et al. [31], who had used whole soybeans to lower the incidence of mammary tumors in a radiation-induced model of breast cancer. Then, Lamartiniere et al. [32–35] showed that neonatal and prepubertal exposures to genistein suppressed 7,12-dimethylbenz[*a*]anthracene

(DMBA)-induced mammary cancer. Hilakivi-Clarke et al. [36] and Badger et al. [37] confirmed that prepubertal and neonatal exposures to genistein reduce mammary tumorigenesis. It was also demonstrated that neonatal and prepubertal genistein action in the undifferentiated mammary gland increased cell proliferation in terminal ductal structures of the rat mammary gland to enhance gland differentiation [33–35]. The result of these biological actions was to yield a gland that is ultimately less proliferative and less susceptible for chemical carcinogenesis in mature animals.

In the prostate, genistein has been reported to inhibit the growth of human and rat prostate cancer cell lines [38–40]. Soy has been found to have a protective effect against prostatic dysplasia [41] and to inhibit the growth of transplantable human prostate carcinomas and tumor angiogenesis in mice [42]. High isoflavone-supplemented soy diet and the addition of genistein to a phytoestrogen-free diet fed to Lobund–Wistar rats reduced the incidence of MNU-induced prostate-related cancer [43, 44]. Also, genistein in the diet reduced the incidence of poorly differentiated spontaneously developing prostatic adenocarcinomas in the transgenic TRAMP mouse model [35, 45].

14.6 MECHANISMS OF ACTION OF POLYPHENOLS

Polyphenols have many reported mechanisms of action. Isoflavonoids and their physiological metabolites bind with high affinity to the estrogen receptor [46]. The discovery in 1996 of a second estrogen receptor [47], ERβ, provided a tissue target whose affinity for isoflavonoids matched their physiological concentrations (nM) [48]. In addition, genistein and several other bioflavonoids are inhibitors of protein tyrosine kinases, although such inhibition only occurs at micromolar concentrations [49, 50]. Other processes that might be regulated by polyphenols include prevention of oxidation [51]. Genistein is a good inhibitor of tumor necrosis factor-α activation [52], particularly of neutrophil rolling and adherence to endothelial cells, although this is only observed under conditions of flow [53]. Belenky et al. [54] used an affinity approach to isolate proteins with high affinity for genistein from human breast cancer MCF-7 cells. Curiously, the amino acid sequence of the protein (named DING) that was isolated is not represented in public DNA sequence databases [54, 55].

14.7 IMPORTANCE OF TIMING OF EXPOSURE TO POLYPHENOLS

It has long been known that the perinatal period of life is vulnerable to hormonally active chemicals. As an example of this, rats injected prenatally with diethylstilbestrol enhanced DMBA-induced mammary cancer development [56]. However, rats injected neonatally with diethylstilbestrol suppressed DMBA-induced mammary cancer [57]. Lamartiniere [58] hypothesized that selective manipulation of the prenatal and neonatal periods by hormonally active chemicals could result in determining susceptibility for disease later in life. Indeed, it had been previously shown that a combination of 17β-estradiol and a progestin administered prepubertally lowered the number of mammary tumors in rats treated with MNU [59]. This has been also demonstrated with genistein, an estrogenically active phytochemical. Rat dams were fed AIN-76A diet with and without the addition of 250 mg genistein/kg diet at different time periods. Their female offspring were treated intragastrically with DMBA on day 50. Rats treated with DMBA who had been exposed to control diet (AIN-76A) only from birth until the end of the experiment had the highest number of tumors (8.9 tumors/rat). DMBA-treated rats exposed to genistein prenatally only developed 8.8 tumors/rat, demonstrating that prenatal genistein in the diet neither increased mammary carcinogenesis nor conferred protection against DMBA-induced mammary cancer [35, 58]. In contrast, Hilakivi-Clarke et al. [60] reported that injecting pregnant rats with genistein resulted in increased susceptibility of the offspring for mammary cancer. However, this appears to be due to different routes of administration and bioavailability in the two study designs. Circulating blood genistein concentrations from 21-day fetal, 7-day neonatal, and 21-day prepubertal rats exposed to 250 mg genistein/kg AIN-76A diet in their mother's diet were determined to be 43, 726, and 1810 nM, respectively [33]. This demonstrates excellent genistein bioavailability during postnatal life but poor bioavailability prenatally. Also, it was shown that approximately 46% of circulating total genistein is free genistein 24 hours after injection of rats [61]. This is in contrast to less than 2% being free (aglycone) genistein from dietary administration [34]. Hence, it was concluded that route of administration and timing of exposure determines the metabolism, bioavailability, biological action, and toxicity of genistein.

Rats fed AIN-76A diet containing genistein starting at day 100 (50 days after DMBA exposure and the approximate time of first tumor development) averaged 8.2 tumors/rat, similar to rats on the control AIN-76A diet. This demonstrates the importance of genistein action during active mammary gland development/maturation. On the other hand, offspring exposed to genistein via the mother's milk and directly via the diet from days 1 to 21 postpartum and then treated with DMBA developed significantly fewer tumors/rat (4.3) than animals not provided genistein in the diet (8.9 tumors/rat) [35, 58]. Interestingly, DMBA-treated rats exposed from days 1 to 21 postpartum to genistein and then again from days 100 onward developed the fewest tumors/rat (2.8). The latter result demonstrates that genistein fed to adult rats previously exposed neonatally and prepubertally to genistein provided these animals with additional protection against mammary cancer. In this case, genistein fed during the neonatal and prepubertal periods programmed future genistein response against mammary cancer susceptibility in adults [35, 58]. It is hypothesized that programming in the mammary gland with genistein determines the "blueprint" of how the mammary tissue will respond to future similar stimuli in adults. Programming is the term used to describe lifelong changes in the function that follow a particular event in an earlier period of the life span [62, 63].

These data also suggest that even when mammary cancer has been initiated, genistein exposure in adulthood does not promote cancer in intact normal rats. This is to be contrasted to the reports of genistein promoting mammary cancer in athymic (immunocompromised), ovariectomized mice subcutaneously implanted with MCF-7 breast cancer cells [64]. Finally, to put this in perspective to the human situation, in a case–control epidemiological study using the Shanghai Cancer Registry, an inverse relationship (50%) between adolescent (13–15 years) soyfood intake and breast cancer incidence later in life has been shown [65], thus validating laboratory data with epidemiology reports. This result was also confirmed in Asians living in California [66]. This latter study more importantly showed that intake of soyfood only in adult life was not associated with a reduction in breast cancer. In contrast, intake of soyfood (1 serving per week or more) during adolescence, but not during adult life, nonetheless lowered breast cancer incidence.

14.8 ASSESSING EVENTS LEADING TO CANCER: LOW-DIMENSIONAL APPROACHES

During much of the latter part of the 20th century, identification of key steps that could account for cancer or regulate its spread has been largely monodimensional research. A particular process (expression of a gene or amount or activity of a protein, usually one in which an investigator had previous experience or was the most familiar, or is currently notorious) was the focus of attention. Most experiments were confined to the effect of a single substance in rodent or cell models, in some cases involving multiple concentrations, on this specific protein or gene target. The analysis of the resulting data is familiar to investigators, typically consisting of Student's t test, analysis of variance, or a general linear models procedure. Investigators tested null hypotheses and understood the importance of $p < 0.05$. If they worked closely with a statistician before starting the experiment, they assessed the biological and measurement variance of the system they were studying, calculated the number of samples needed to obtain sufficient power and precision to test and estimate the effect of their test compound, and would have been confident that the results of the experiment were not unduly affected by a poor experimental design.

14.9 STATISTICAL CONSEQUENCES OF HIGH-DIMENSIONAL APPROACHES

Changes occurred when high-throughput methods were developed in the 1990s to more rapidly *discover* compounds that would interact with selected protein targets. However, these types of experiments created the need for a more careful analysis of the data since multiple effects on the target of choice were being tested. To do so required the full range of modern computing that had developed in the 20 year period from 1980 to 2000. Indeed, powerful computers were now to be found on just about everyone's desk, providing not only software for the statistical algorithms, but much more temptingly browsers connecting information centers across the Internet.

When analyzing data from two treatment groups, it is generally considered that when there is only a 1 in 20 chance (or less) of the observed data confirming the null hypothesis, the null hypothesis should be rejected. In modern *high-dimensional biology,* by increasing the number of potential variables, the probability of erroneously concluding that at least one variable is affected by the treatment increases, *ceteris paribus*, directly with the number of variables tested. If all of the variables were independent and all were truly unaffected by the treatment (two admittedly implausible assumptions), then the probability of making one or more such erroneous conclusion is $1 - (1 - \alpha)^k$, where α is the per-test Type 1 error rate (e.g., 0.05) and k is the number of tests performed. Similarly, the expected number of Type 1 errors is αk. Thus, if 20 independent variables are studied and if all 20 null hypotheses were true, we nonetheless expect that, on average, for one of them we will obtain data that cause the null hypothesis to be rejected at the conventional 0.05 α level. The traditional solution to account for the multiple testing is to apply more stringent statistical criteria, such as the Bonferroni correction [67], that hold the probability of rejecting a true null hypothesis (i.e., of making one or more type 1 errors) *over the entire collection of tests performed* to $\geqslant \alpha$ (typically 0.05). With the Bonferroni correction, the per-test alpha level is set at α_{fw}/k, where α_{fw} is the "family-wise" alpha one desires (e.g., 0.05) and k is the number of tests conducted. Thus, to reject the null hypothesis when conducting five (5) tests, a p-value $\leqslant 0.01$ is required—and for 1000 variables, the required p-value is reduced to $\geqslant 0.00005$. If all other factors are held constant, the effect of reducing the possibility of a Type 1 error, however, is to increase the probability of a Type 2 error (failing to reject a null hypothesis that is false)—namely, reducing the power to detect a true difference. Increasing sample size is one way to compensate for reduced power, but the dramatic reductions in power wrought by adjustments such as the Bonferroni demand economically impractical increases in sample sizes to maintain power. Moreover, most biologists doing discovery-based research seem uninterested in maintaining their α_{fw} below some very low level such as 0.05 and instead are quite happy to accept that some errors will occur provided that, among those "discoveries" they make, a sufficiently large portion can be expected to be true discoveries. In other words, they wish to control their "false discovery rates" (FDRs) rather than their family-wise type error rates (see [68]).

14.10 HIGH-DIMENSIONAL SYSTEMS AND THE IMPORTANCE OF THE FALSE DISCOVERY RATE

Benjamini and Hochberg [69] first coined the phrase *false discovery rate*. Because of the potentially tens of thousands of comparisons in a given proteomic/genomic study, the recent microarray literature is replete with references to the FDR in significance testing specifically designed for high-dimensional biology [70–73]. A false discovery refers to the act of incorrectly concluding that one or more groups are truly different from one another with respect to a specific protein or gene. For a particular experiment, the FDR is the expected or estimated proportion of false discoveries out of the total number of significantly different proteins/genes. For example, if an experiment yielded 100 significant genes out of a total of 1000 on a gene-chip and an estimated FDR of 3%, then the interpretation of this FDR is that it is estimated that only 3 out of the 100 are false discoveries. The complementary interpretation is that it is estimated that 97 out of the 100 (97%) are true discoveries. Note that a Bonferroni adjusted alpha, $\alpha_{fw} = 0.05/1000 = 0.00005$, would lead to much different conclusions resulting in several potentially significant genes/proteins being ignored arbitrarily. Therefore, with a reliably estimated FDR, the investigator is armed with information to assist in deciding if further research and resources on the 100 genes/proteins is worthwhile. Certainly a large FDR of, say, 50%, in this scenario would lead the investigator to a different decision with respect to allocation of resources than if the FDR were 5%. Defining the "best" FDR method and how various methods perform under various circumstances is an active area of investigation [74].

14.11 DNA MICROARRAY ANALYSIS: HIGH-DIMENSIONAL RESEARCH INTO GENE EXPRESSION

The introduction of DNA microarray analysis has produced considerable excitement in the scientific community. It came largely as a result of the sequencing of the human genome and other experimentally important genomes. The genes from these genomes are printed onto glass and nylon slides as short (20–25 bp) and long (35–70 bp) representative oligonucleotides, or as complete or near-complete cDNAs. mRNAs are recovered from cells or tissues and converted to their cDNA equivalents. These in turn are converted to cRNAs, in doing so incorporating either a radiolabel or a fluorescent label for probing the arrays. An advantage of the latter is that different fluorescent labels can be used for the control and treatment groups. The fluorescent labeling allows the two (or more) samples to be mixed prior to the hybridization step, resulting in their joint analysis on the same microarray chip.

For DNA microarray analysis, whereas the number of external factors (experimental levels of a modifying agent) remains only one or two, the number of variable targets (the genes) is increased substantially. The latest human DNA microarrays have over 400,000 features, representing over 20,000 human genes. The Affymetrix DNA chip has not only oligonucleotides that are exact matches for a particular gene sequence, but also related mismatch oligonucleotides. This approach improves the confidence that the observed signal for the exact match probe is real. Nonetheless, the data offer a substantial challenge to be analyzed statistically because of both the large number of genes to be examined and the few replicates that are typically analyzed. The typical analysis strategy for microarray data is shown in Figure 14.4.

14.12 PROTEOMICS ANALYSIS: AN EVEN BIGGER CHALLENGE

Analysis of the proteome of a cell is a massive undertaking. Proteins are not only found in forms that correspond to the open reading frame of genes; they also result

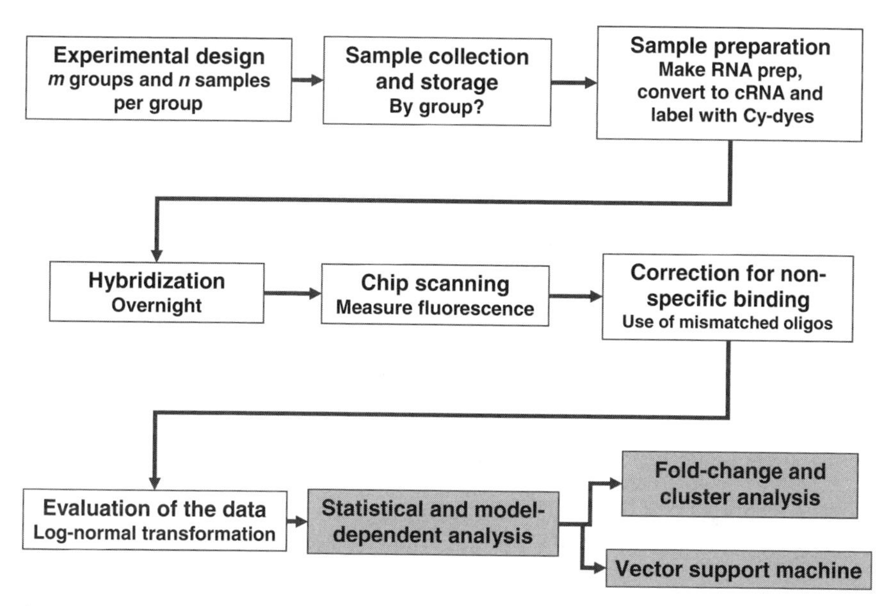

Figure 14.4. Steps in an experiment in DNA microarray analysis. This is the sequence of events used in most current methods reported in the literature.

from differential RNA splicing and extensive post-translational modifications. A single cell type may have 200,000 or more different protein forms with a range of expression over nine orders of magnitude. Analysis of the proteome by two-dimensional isoelectric focusing/sodium dodecylsulfate–polyacrylamide gel electrophoresis (2D-IEF/SDS-PAGE) has similar issues to DNA microarrays, although the number of detectable protein spots is much lower (depending on the gel size, 400–1500). However, the exact location of the same protein spot from gel to gel varies a lot and this introduces error in identifying individual proteins across gels. As with DNA microarray analysis, proteins can be labeled with fluorescent Cy dyes [75] and this overcomes some of the localization problems on the 2D gel. For comparison of two samples, the control and treated sample are labeled separately with fluorescent Cy3 and Cy5, respectively. The two samples are then mixed and run on the same 2D gel. An individual protein in the control group labeled with Cy3 migrates to the same position as the identical protein in the treated group labeled with Cy5 (Figure 14.5). The ratio of Cy3 to Cy5 fluorescence for each spot is used to calculate changes in protein abundance.

Other methods of proteomics analysis are being used: two-dimensional liquid chromatography of proteins [76] offers a better approach since it allows separation of hydrophobic as well as hydrophilic proteins and permits a greater protein loading and hence the detection of lower abundance proteins. A second method is based on two-dimensional liquid chromatography of peptides derived by protease treatment of whole proteomes [77]. This method with the acronym MUDPIT results in 20–50

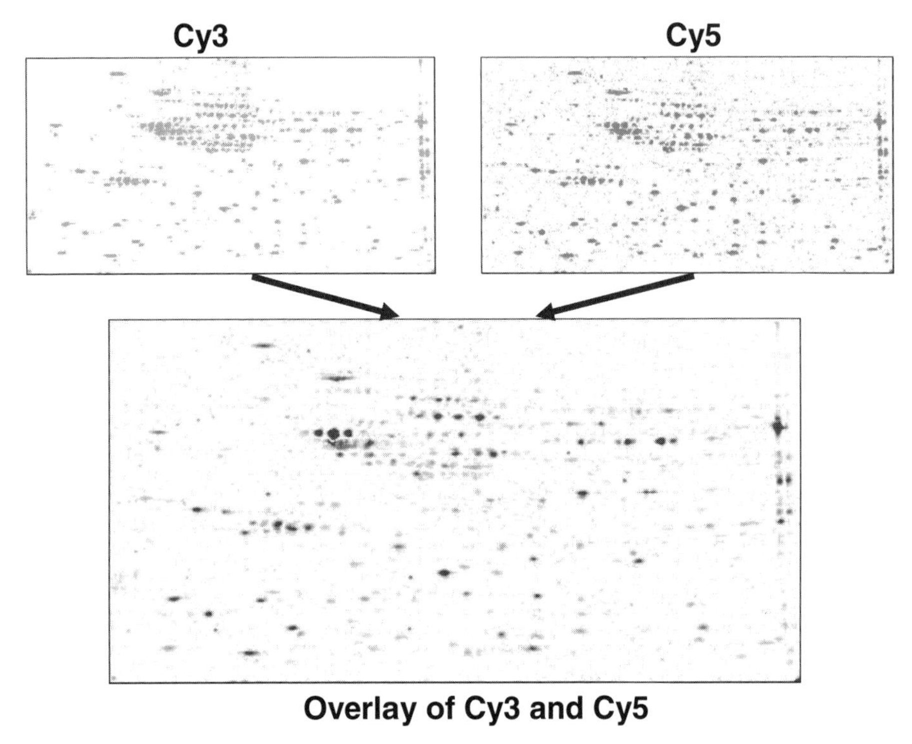

Figure 14.5. Two dimensional (2D) electrophoresis analysis using a DIGE approach. Proteins from the control mammary gland were labeled with Cy3 and those from the mammary gland of a treated rat were labeled with Cy5. The two samples were mixed and subjected to 2D electrophoresis. The gels were scanned for their fluorescence at two sets of excitation and emission wavelengths. Cy3 labeled proteins are shown in light gray and Cy5 proteins in red. The lower figure is an overlay of the images. In this image, light gray spots are proteins only observed in the control sample, whereas black spots are proteins largely confined to the treated sample. Dark gray spots are proteins expressed in both control and treated samples. Imaging of the gels were done at BioRad Laboratories.

times the number of observables—substantially increasing the statistical challenges. It should be noted that much of the data are correlated, which poses both additional challenges and opportunities. There are separate concerns in this method about the statistical quality of the identification of the peptides. It is more of a discovery tool and is not well suited to quantitative analysis. To address the latter issue, isotopically labeled (d_9 and $^{13}C_8$) reagents have been used for the treated samples [78]—the control group is reacted with nonisotopically enriched reagent. Since once hydrolyzed by proteases the resulting peptides co-chromatograph, the ratio of the intensities of the two peptides is thereby a quantitative measure of differences/similarities. The principle of co-chromatography is true for $^{13}C_n$-labeled reagents, but generally not so for d_n-labeled reagents.

14.13 STATISTICAL PROBLEMS WITH FOLD-CHANGE IN DNA MICROARRAY AND PROTEOMICS ANALYSES

Analysis of the effects of agents on systems where the number (400–20,000) of variables (genes, proteins) is large is compromised when the number of cases or replicates is small (1–20). In this situation, the ability to detect real differences is minimal and the sampling variability of statistical estimates is quite large. As a result, what might appear to be a "significant" change in the value of a variable (e.g., if it increased by a factor of 1 to 2.5, or decreased from 1.0 to 0.4, compared to control) may as likely to be a false positive as a true finding. In addition, the ratio methodology used in Cy-dye based or isotope methods to calculate fold-change has statistical problems [79]. It is unlikely that the significance of the ratio is independent of the intensities of the observed values [80]. The statistical criterion for a significant change will depend on the magnitude of the measured signal, in addition to the biological variability of the gene or protein being examined.

By analogy to DNA microarray analysis, proteomics investigators will benefit from consulting with statisticians who are now analyzing the two or three channels of Cy-dye information from DNA microarray analysis separately rather than as a ratio [81]. In general, statisticians no longer deem analyses based only on fold-change acceptable and journals considering manuscripts containing such data suggest further analysis. This involves assessment of variance or nonparametric, rank-based procedures, as well as independent analyses (RT-PCR or Northern blotting for mRNA analysis and Western blotting for proteins). Nonetheless, there is considerable debate as to what should constitute validation of microarray and proteomics data [82, 83].

14.14 DESIGN IN EXPERIMENTS INVOLVING DNA MICROARRAY AND PROTEOMICS ANALYSIS

Nonetheless, given that gene and protein expression really does change in response to anticancer therapeutics or cancer prevention agents, how can we best design a high-dimensional experiment to produce meaningful data? As in any experiment, it takes careful control and appreciation of the sources of error. Put very simply, the total sum of squares (SST) for a set of observations is equal to the sums of squares due to real effects (SSR) plus the sums of squares due to error (SSE). In this context, by "errors" we refer not to "mistakes" but to individual observations' deviations from expected mean levels and such deviations are contributed to by both true within-individual biological deviations as well as measurement errors. Clearly, the larger the ratio of SSR to SSE, the more statistically significant differences will be found. In an ideal world, the SSE should be minimized—however, not all of the SSE can be reduced. Analytical variation can certainly be lowered by better procedures, but only so far. Even those who only carry out experiments in a one-dimensional world will know that day-to-day and operator-to-operator variations are substantial. In a high-dimensional analysis, a crucial point is to ensure that error is carefully and evenly partitioned among all the factors being measured. This can be facilitated by the inclusion from the outset of experienced statisticians. The remainder of this chapter is devoted to optimization of the design of experiments involving polyphenols and prevention of cancer.

14.15 THE DESIGN

A crucial point is to ensure that there is a question or objective or hypothesis that is driving the experiment. It is not sufficient to say that "the data will speak to us." If there is a clear question, then an experimental design can be optimized to answer that question or achieve that goal.

Choice of Model and Agent

First of all, for the experiments on the role of polyphenols in prevention of cancer, it is necessary to consider whether to carry out the research in a clinical trial, in a suitable animal model, or in a cell culture system. The cell culture system appears to provide the greatest amount of control. The cell type can be defined and so can the agent to be studied. But is this relevant to prevention of cancer? Is the agent in the form that reaches the cell that is or is destined to be a cancer cell? This can be corrected by finding out what the metabolic form(s) is (are). Of course, trying to answer questions about cancer prevention is not going to be helped by using transformed cancer cell lines that stably grow in cell culture. A better approach is to use cell lines that are not transformed but can be grown in culture. In studying breast cancer prevention, human mammary MCF10A cells are one option [84]. Another is the use of ultrathin (2–3 cell thicknesses) sections of the rodent mammary tissue [85]. The advantage of this preparation is that all cell types in the mammary tissue are represented as well as the matrix that surrounds them. The disadvantage is that the observed signals will come from various cells, and associations that might be otherwise inferred from correlations in expression may have no biological meaning. To overcome this disadvantage, laser capture microdissection has been used to collect a single cell layer and presumably one type of cell [86]. However, this method only isolates 1000–10,000 cells; while this is enough for DNA microarray analysis (because of the amplification capability of this technique), it is difficult to carry out proteomic identifications. Even for housekeeping proteins expressed at 10^6 copies per cell, to have 100 fmol for analysis requires a minimum of 600 cells.

Animal models allow for the issues of agent uptake, metabolism, and tissue distribution. However, these may not be the same for the animals when compared to humans. For instance, most animals accumulate the daidzein metabolite equol to micromolar levels when fed soy diets [87]—in contrast, only one-third of humans make equol at all and in most the blood equol level is less than 500 nM [88, 89]. Then there is the question of what polyphenol to test—a single agent (e.g., genistein), a mixture (polyphenol extract), or a polyphenol-containing food. Using a food preparation requires consideration of the potential effects of the rest of the food matrix. This may also be an issue regarding the overall diet used for animal experiments. In a recent report, genistein and grape seed extract were not chemopreventive in the semipurified AIN-76A diet, but were when added to a Teklad laboratory chow diet [90].

DNA microarray analysis and proteomics can be carried out on the whole mammary tumor or cells from laser capture microdissection with the advantages and limitations of the latter noted above. The mammary tumors induced by the carcinogens MNU and DMBA are estrogen dependent and may not represent carcinogenesis that occur in humans. Other animal models based on the overexpression of specific oncogenes are available [91, 92].

Clinical tumor samples can be obtained at the time of surgical resection. As for the animal tumors, analyses can be carried out on the whole tumor or on cells from laser capture microdissection. A considerable effort is being put into examining changes in the serum proteome; however, it is like looking for needles in a haystack because of the vast amounts of the normal serum proteins (albumin, γ-globulins, etc.). The latter can be depleted by an affinity column containing antibodies to these proteins [93], but this step may also remove minor proteins of interest. Because of the filtering mechanism provided by kidney glomeruli that retains the larger proteins, urine may be more useful in identifying cancer-specific proteins and peptides. Other useful physiological fluids are nipple aspirate fluid (for breast cancer) [94] and prostatic fluid (for prostate cancer).

Randomization

The purpose of randomization in experimental design is to convert the variance introduced by the "nuisance" factors, such as technician, day, and gel lot, in the experimental design to random error. This will allow the technical variation to be distinguished from the differences in variance introduced by the factors of interest in the study. In order to do this, the statistician has to develop a clear understanding of the primary objectives of the experiment, the procedures that are involved in the conduct of the experiment, the measurement techniques involved, and the nuisance factors involved, which are controllable (such as which technician does the work) and which may be uncontrollable (such as the humidity in the lab). Working with the biologist, the statistician can then help decide which factors' cases (i.e., subjects) can be randomized, and which factors need to be controlled, measured, and distinguished from the effects of interest. One example of the introduction of a systematic bias is the time an animal is euthanized. Also, since the analysis of a large experiment may take 8–12 different 2D gel setups, it is important that samples from the control and experimental groups are evenly and randomly distributed across the analysis days. Furthermore, because fluorescence depends on the absolute intensity of the incident light source rather than being a ratio as in measurements of absorbance, even randomization of image analysis within a day should be considered.

Choices in the Analytical Method

It is increasingly appreciated that the choice of analysis methods for gene expression and proteomic data is enormously influential in the outcome that is measured. For microarray analysis, in most cases the sequences placed on the array were chosen by commercial companies such as Affymetrix. Their rationale for selecting the target oligonucleotide to represent a gene is a compromise between complete specificity and a consistency in optimum hybridization temperature, and therefore may not be ideal for each gene. To compensate for this problem, investigators have selected longer oligonucleotides (~35–60 bp) for each gene and have made customized arrays. In another approach, whole cDNAs for the gene of interest have been spotted onto arrays. These allow for much higher hybridization and washing temperatures but run the risk that cross-hybridization of genes can occur. There's no guarantee that each microarray method will identify the same set of changes. Recently, Churchill's group [95] compared these three methods. While there was a good correlation between data obtained using the Affymetrix chip and spotted long oligo arrays, the data from the spotted cDNA arrays were not well related to those from the oligo-based arrays, even though they had a higher correlation among biological replicates. These results raise a degree of uncertainty about any set of microarray results, irrespective of the platform used.

Whatever microarray procedure is used, from the point of view of controlling error, ideally only one batch of arrays and the same batch of reagents creating the probes for the arrays is used (or if more arrays are needed then batch and reagents supply should be orthogonalized with respect to treatment variables). In addition, each aspect of the hybridization procedure should be standardized. It's noteworthy that rarely are the incubations carried out until a true equilibrium binding of the labeled cRNA species has occurred; this raises the possibility that even when all other aspects of the microarray analysis are the same, investigators may get different results. Lowly expressed RNA species will come into equilibrium more slowly than highly expressed ones (Figure 14.6). Even similarly expressed RNA species may hybridize to their specific DNA spot on the array at different rates.

A challenge in the analysis of the proteome is effective coverage. In 2D gel analysis, the numbers of resolvable protein spots from a complex proteome are usually in the range from 500 to 1500. This may represent only 1–5% of the total proteome. Essentially, only the high abundance proteins can be detected. A solution to this is to use a preliminary fractionation procedure based on physical or chemical properties of the proteins (or a combination of both). The disadvantage of doing so is that any fractionation method can introduce bias if it does not fractionate members of a class of protein equally efficiently. A good example of this is the use of an immunopurification approach with an antiphosphotyrosine antibody. Each commercial form of this antibody binds to a different subgroup of the total tyrosine-phosphorylated proteins. Although the popular MUDPIT technique that is based on automated LC-MS can give rise to recognition of 50,000 peptides per sample, when all the data are boiled down, and weak, poorly defined spectra and redundant spectra (the same peptide appearing in multiple LC-MS fractions and multiple peptides from

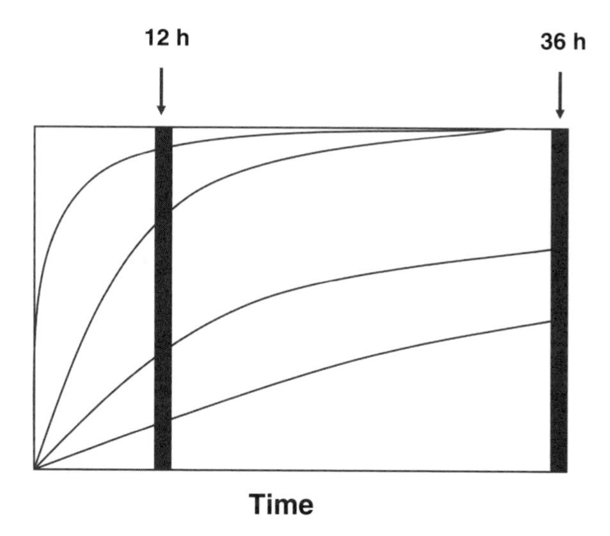

Figure 14.6. Effects of sampling times for oligonucleotides with differences in rates of equilibration during hybridization. The rates of hybridization are dependent on the nucleotide sequence. The relative binding of each oligonucleotide in these simulated binding curves is different at 12 hours as opposed to 36 hours.

one protein) are eliminated, the total number of identified proteins are not particularly different from those identified using the 2D gel approach [96]. Perhaps not surprisingly, the two groups of proteins are largely nonoverlapping [97], suggesting that each technique is differentially sampling the overall proteome. However, since little or no information is available on what would be observed if the whole procedure were repeated, one cannot be sure that some of the difference in the proteins that are identified is in fact due to reproducibility problems. The marked difference in the proteins that are identified from a proteome is also true of 2D-LC separation of proteins. This technique does a better job of identifying hydrophobic proteins and proteins with alkaline isoelectric points than 2D gels or MUDPIT [98].

Randomization in the Order of Analysis

Many systematic effects can be introduced by failing to randomize the order of sample analysis. Not all samples from an experiment can be analyzed on the same day. A plan for establishing the design for an experiment based on DNA microarray analysis is shown in Figure 14.7.

For 2D gels, once run and stained with Sypro Ruby, they have to be scanned for their fluorescence. Since each gel takes 20–30 minutes for scanning to be complete, the overall scanning process may take several days. Again, this is a situation where randomization of scanning order is crucial (Figure 14.8). Similar issues apply to DNA microarray analysis.

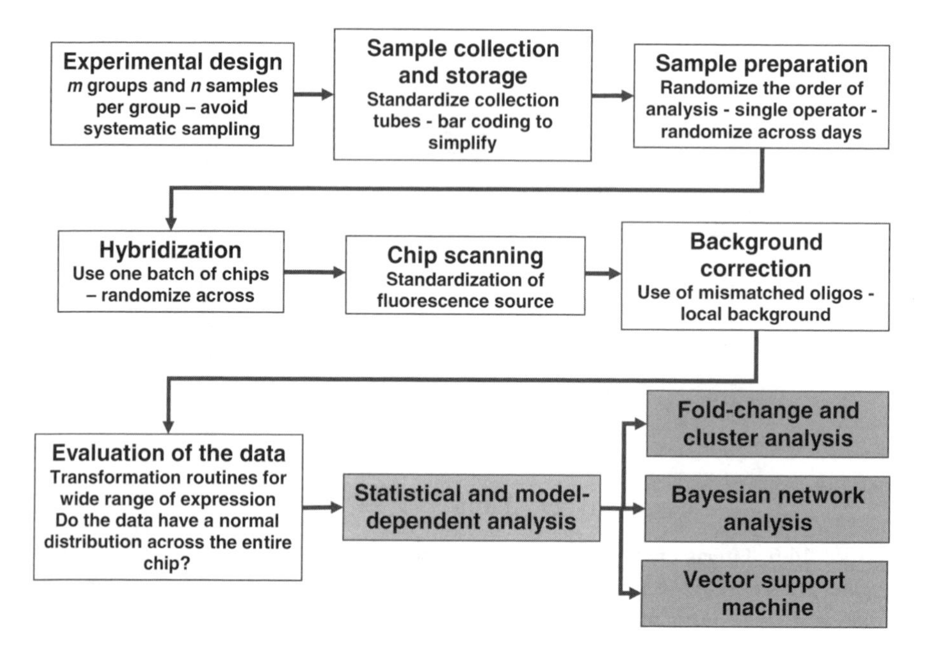

Figure 14.7. Optimizing the design of an experiment based on DNA microarray analysis. This is the suggested protocol for this type of analysis.

14.16 ROLE OF THE COMPUTER IN HIGH-DIMENSIONAL ANALYSIS

High-dimensional research creates another dimension for computing—keeping track of every stage of the analysis using a *laboratory information management system* (LIMS). Because a high-dimensional model system verges on being mathematically singular, a small uncontrolled, unrealized source of error can bring about a catastrophic misinterpretation of the data. Bar coding of samples is the only way to keep track of them during the extensive randomization required during collection, storage, processing, and analysis. A single proteomic experiment can generate huge amounts of data; as a result, sophisticated databases are required for the accurate storage and dissemination of proteomics. While some databases have been developed for microarray data (GEO at http://www.ncbi.nlm.nih.gov/geo/), good public databases for public dissemination of the great variety of proteomic data have not yet been developed.

In summary, the high-dimensional research that is in vogue for the study of cancer prevention and many other chronic diseases requires a new level of scientific discipline and application. As happened in the world of physics in 19th and 20th centuries, improvements in analytical methods and experimental design will open

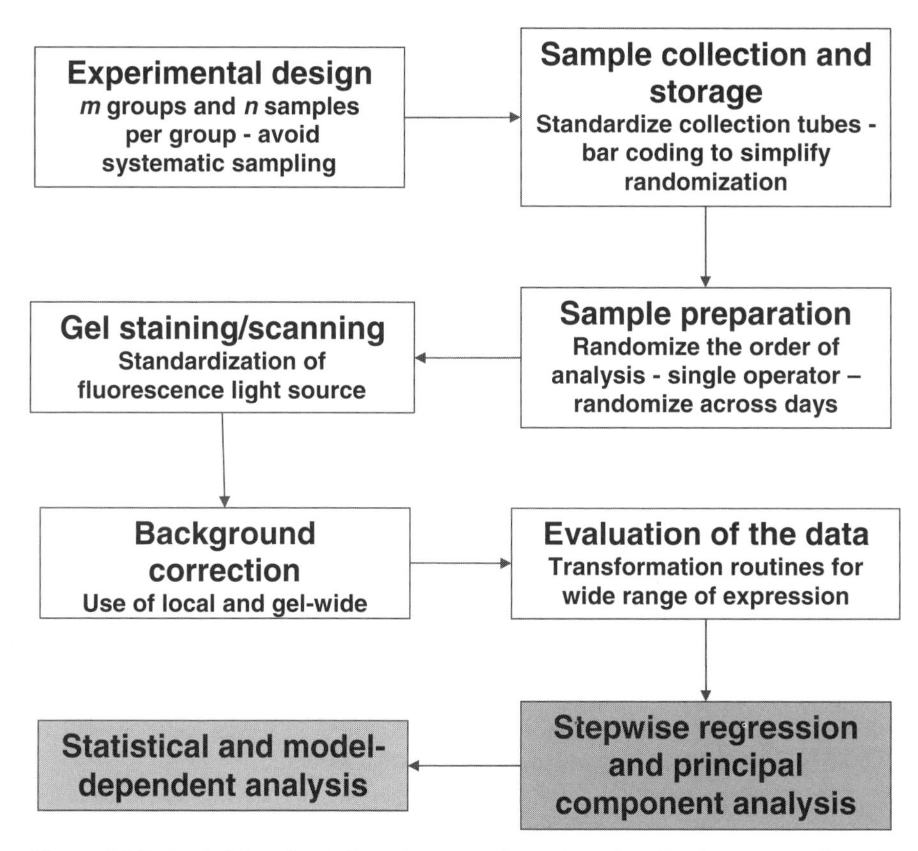

Figure 14.8. Optimizing the design of an experiment based on 2D electrophoresis analysis. This is the suggested protocol for this type of analysis.

up new vistas in prevention research and will require the cooperation of and collaboration with scientists from outside traditional areas of biomedicine. These will include bioinformaticists, engineers, mathematicians, experts in materials science, physicists, and statisticians, as well as those who specialize in theory and practice of systems analysis.

ACKNOWLEDGMENTS

Research on high-dimensional systems is supported by grants to the Center for Nutrient–Gene Interaction at UAB by the National Cancer Institute (U54 CA—100949, PI SB) and to the Section on Statistical Genetics in the UAB Department of Biostatistics by the National Science Foundation (DBI—0217651; PI, DBA).

REFERENCES

1. J. H. Hankin and V. Rawlings (1978). Diet and breast cancer: A review, *Am. J. Clin. Nutr.* **31**:2005–2016.
2. R. Doll (1980). The epidemiology of cancer, *Cancer* **45**:2475–2485.
3. H. Shimizu, R. K. Ross, L. Bernstein, R. Yatani, B. E. Henderson, and T. M. Mack (1991). Cancers of the prostate and breast among Japanese and white immigrants in Los Angeles County, *Br. J. Cancer* **63**:963–966.
4. R. G. Ziegler, R. N. Hoover, M. C. Pike, A. Hildesheim, A. M. Nomura, et al. (1993). Migration patterns and breast cancer risk in Asian-American women, *J. Natl. Cancer Inst.* **85**:1819–1827.
5. L. S. Cook, M. Goldoft, S. M. Schwartz, and N. S. Weiss (1999). Incidence of adenocarcinoma of the prostate in Asian immigrants to the United States and their descendants, *J. Urol.* **161**:152–155.
6. M. S. Sandhu, I. R. White, and K. McPherson (2001). Systematic review of the prospective cohort studies on meat consumption and colorectal cancer risk: A meta-analytical approach, *Cancer Epidemiol. Biomarkers Prev.* **10**:439–446.
7. E. Cho, D. Spiegelman, D. J. Hunter, W. Y. Chen, M. J. Stampfer, G. A. Colditz, and W. C. Willett (2003). Premenopausal fat intake and risk of breast cancer, *J. Natl. Cancer Inst.* **95**:1079–1085.
8. M. J. Hill (1997). Cereals, cereal fibre and colorectal cancer risk: A review of the epidemiological literature, *Eur. J. Cancer Prev.* **6**:219–225.
9. J. H. Weisburger (1991). Nutritional approach to cancer prevention with emphasis on vitamins, antioxidants, and carotenoids, *Am. J. Clin. Nutr.* **53**:226S–237S.
10. J. H. Weisburger (1999). Antimutagens, anticarcinogens, and effective worldwide cancer prevention, *J. Environ. Pathol. Toxicol. Oncol.* **18**:85–93.
11. J. B. Harborne (1993). *The Flavonoids: Advances in Research Since 1986*. CRC Press, Boca Raton, FL.
12. J. K. Prasain, K. Jones, M. Kirk, L. Wilson, M. Smith-Johnson, C. M. Weaver, and S. Barnes (2003). Identification and quantitation of isoflavonoids in Kudzu dietary supplements by HPLC and electrospray ionization tandem mass spectrometry, *J. Agric. Food Chem.* **51**:4213–4218.
13. W. Eloesser and K. Herrmann (1975). Flavonols and flavones of vegetables. V. Flavonols and flavones of root vegetables, *Z. Lebensm. Unters. Forsch.* **159**:265–270.
14. J. Kunzemann and K. Herrmann (1977). Isolation and identification of flavon(ol)-*O*-glycosides in caraway (*Carum carvi* L.), fennel (*Foeniculum vulgare* Mill.), anise (*Pimpinella anisum* L.), and coriander (*Coriandrum sativum* L.), and of flavon-*C*-glycosides in anise. I. Phenolics of spices, *Z. Lebensm. Unters. Forsch.* **164**:194–200.
15. S. Barnes, C.-C. Wang, M. Kirk, M. Smith-Johnson, L. Coward, N. C. Barnes, G. Vance, and B. Boersma (2002). HPLC-mass spectrometry of isoflavonoids in soy and the American groundnut, *Apios americana*. In: *Flavonoids in Cell Function*, Béla S. Buslig and John A. Manthey, eds. Kluwer Academic/Plenum Publishers, New York, pp. 77–88.
16. F. Natella, F. Belelli, V. Gentili, F. Ursini, and C. Scaccini (2002). Grape seed proanthocyanidins prevent plasma postprandial oxidative stress in humans, *J. Agric. Food Chem.* **50**:7720–7725.

17. A. P. Wilkinson, J. M. Gee, M. S. Dupont, P. W. Needs, F. A. Mellon, G. Williamson, and I. T. Johnson (2003). Hydrolysis by lactase phlorizin hydrolase is the first step in the uptake of daidzein glucosides by rat small intestine *in vitro*, *Xenobiotica* **33**:255–264.
18. J. Sfakianos, L. Coward, M. Kirk, and S. Barnes (1997). Intestinal uptake and biliary excretion of the isoflavone genistein in the rat, *J. Nutr.* **127**:1260–1268.
19. C. Manach, A. Scalbert, C. Morand, C. Remesy, and L. Jimenez (2004). Polyphenols: food sources and bioavailability, *Am. J. Clin. Nutr.* **79**:727–747.
20. S. Barnes, J. Sfakianos, L. Coward, and M. Kirk (1996). Soy isoflavonoids and cancer prevention. Underlying biochemical and pharmacological issues, *Adv. Exp. Med. Biol.* **401**:87–100.
21. J. K. Prasain, K. Jones, N. Brissie, D. R. Moore II, J. M. Wyss, and S. Barnes (2004). Identification of puerarin and its metabolites in rats by liquid chromatography–tandem mass spectrometry, *J. Agric. Food Chem.* **52**:3708–3712.
22. M. K. Piskula and J. Terao (1998). Accumulation of (–)-epicatechin metabolites in rat plasma after oral administration and distribution of conjugation enzymes in rat tissues, *J. Nutr.* **128**:1172–1178.
23. A. Scalbert, C. Morand, C. Manach, and C. Remesy (2002). Absorption and metabolism of polyphenols in the gut and impact on health, *Biomed. Pharmacother.* **56**:276–282.
24. P. A. Kroon, M. N. Clifford, A. Crozier, A. J. Day, J. L. Donovan, C. Manach, and G. Williamson (2004). How should we assess the effects of exposure to dietary polyphenols *in vitro*? *Am. J. Clin. Nutr.* **80**:15–21.
25. T. Izumi, M. K. Piskula, S. Osawa, A. Obata, K. Tobe, et al. (2000). Soy isoflavone aglycones are absorbed faster and in higher amounts than their glucosides in humans, *J. Nutr.* **130**:1695–1699.
26. M. S. Morton, P. S. Chan, C. Cheng, N. Blacklock, A. Matos-Ferreira, et al. (1997). Lignans and isoflavonoids in plasma and prostatic fluid in men: Samples from Portugal, Hong Kong, and the United Kingdom, *Prostate* **32**:122–128.
27. S. Barnes, J. K. Prasain, C.-C. Wang, and D. R. Moore, II (2005). Applications of LC-MS in the study of the uptake, distribution, metabolism and excretion of bioactive polyphenols from dietary supplements. *Life Sci.* in press.
28. T. Akao, K. Kawabata, E. Yanagisawa, K. Ishihara, Y. Mizuhara, Y. Wakui, Y. Sakashita, and K. Kobashi (2000). Baicalin, the predominant flavone glucuronide of scutellariae radix, is absorbed from the rat gastrointestinal tract as the aglycone and restored to its original form, *J. Pharm. Pharmacol.* **52**:1563–1568.
29. A. R. Rechner, G. Kuhnle, P. Bremner, G. P. Hubbard, K. P. Moore, and C. A. Rice-Evans (2002). The metabolic fate of dietary polyphenols in humans, *Free Radic. Biol. Med.* **33**:220–235.
30. S. Barnes, C. Grubbs, K. D. R. Setchell, and J. Carlson (1990). Soybeans inhibit mammary tumors in models of breast cancer. In: *Mutagens and Carcinogens in the Diet*, M. Pariza, ed. Wiley-Liss, Hoboken, NJ, pp. 239–253.
31. W. Troll, R. Wiesner, C. J. Shellabarger, S. Holtzman, and J. P. Stone (1980). Soybean diet lowers breast tumor incidence in irradiated rats, *Carcinogenesis* **1**:469–472.
32. C. A. Lamartiniere, J. B. Moore, M. Holland, and S. Barnes (1995). Neonatal genistein chemoprevents mammary cancer, *Proc. Soc. Exp. Biol. Med.* **208**:120–123.
33. C. A. Lamartiniere, J. B. Moore, N. M. Brown, R. Thompson, M. J. Hardin, and S. Barnes (1996). Prepubertal genistein exposure suppresses mammary cancer and enhances gland differentiation in rats, *Carcinogenesis* **17**:1451–1457.

34. W. A. Fritz, L. Coward, J. Wang, and C. A. Lamartiniere (1998). Dietary genistein: Perinatal mammary cancer prevention, bioavailability and toxicity testing in the rat, *Carcinogenesis* **19**:2151–2158.
35. C. A. Lamartiniere, M. S. Cotroneo, W. A. Fritz, J. Wang, R. Mentor-Marcel, and A. Elgavish (2002). Genistein chemoprevention: Timing and mechanisms of action in murine mammary and prostate, *J. Nutr.* **132**:552S–558S.
36. L. Hilakivi-Clarke, I. Onojafe, M. Raygada, E. Cho, T. Skaar, I. Russo, and R. Clarke (1999). Prepubertal exposure to zearalenone or genistein reduces mammary tumorigenesis, *Br. J. Cancer* **80**:1682–1688.
37. T. M. Badger, M. J. Ronis, R. Hakkak, J. C. Rowlands, and S. Korourian (2002). The health consequences of early soy consumption, *J. Nutr.* **132**:559S–565S.
38. T. G. Peterson and S. Barnes (1993). Genistein and biochanin A inhibit the growth of human prostate cancer cells but not epidermal growth factor receptor tyrosine autophosphorylation, *Prostate* **22**:335–345.
39. H. R. Naik, J. E. Lehr, and K. J. Pienta (1994). An *in vitro* and *in vivo* study of antitumor effects of genistein on hormone refractory prostate cancer, *Anticancer Res.* **14**: 2617–2620.
40. B. A. J. Evans, K. Griffiths, and M. S. Morton (1995). Inhibition of 5-alpha reductase in genital skin fibroblasts and prostate tissue by ligands and isoflavonoids, *J. Endocrinol.* **147**:295–302.
41. S. Makela, L. Pylkanen, R. Santti, and H. Adlercreutz (1991). Role of plant estrogens in normal and estrogen-related altered growth of the mouse prostate. In: *Proceedings of the Interdisciplinary Conference on Effects of Food on the Immune and Hormonal Systems*, Zurich, Switzerland, pp. 135–139.
42. J.-R. Zhou, E. T. Gugger, T. Tanaka, Y. Guo, G. L. Blackburn, and S. K. Clinton (1999). Soybean phytochemicals inhibit the growth of transplantable human prostate carcinoma and tumor angiogenesis in mice, *J. Nutr.* **129**:1628–1635.
43. M. Pollard and P. H. Luckert (1997). Influences of isoflavones in soy protein isolates on development of induced prostate-related cancers in L-W rats, *Nutr. Cancer* **28**:41–45.
44. J. Wang, I.-E. Eltoum, and C. A. Lamartiniere (2002). Dietary genistein suppresses chemically-induced prostate cancer in Lobund-Wistar rats, *Cancer Lett.* **186**:11–18.
45. R. Mentor-Marcel, C. A. Lamartiniere, N. Greenberg, and A. Elgavish (2001). Genistein in the diet reduces the incidence of prostate tumors in a transgenic mouse (TRAMP), *Cancer Res.* **61**:6777–6782.
46. K. Verdeal, R. R. Brown, T. Richardson, and D. S. Ryan (1980). Affinity of phytoestrogens for estradiol-binding proteins and effect of coumestrol on growth of 7,12-dimethylbenz[*a*]anthracene-induced rat mammary tumors, *J. Natl. Cancer Inst.* **64**:285–290.
47. G. G. Kuiper, E. Enmark, M. Pelto-Huikko, S. Nilsson, and J. A. Gustafsson (1996). Cloning of a novel receptor expressed in rat prostate and ovary, *Proc. Natl. Acad. Sci. U.S.A.* **93**:5925–5930.
48. G. G. Kuiper, B. Carlsson, K. Grandien, E. Enmark, J. Haggblad, S. Nilsson, and J. A. Gustafsson (1997). Comparison of the ligand binding specificity and transcript tissue distribution of estrogen receptors alpha and beta, *Endocrinology* **138**:863–870.
49. H. Ogawara, T. Akiyama, J. Ishida, S. Watanabe, and S. Suzuki (1986). A specific inhibitor for tyrosine protein kinase from *Pseudomonas*, *J. Antibiot.* (*Tokyo*) **39**:606–608.

50. R. L. Geahlen, N. M Koonchanok, J. L. McLaughlin, and D. E. Pratt (1989). Inhibition of protein-tyrosine kinase activity by flavanoids and related compounds, *J. Nat. Prod.* **52**:982–986.

51. O. J. Park and Y. J. Surh (2004). Chemopreventive potential of epigallocatechin gallate and genistein: Evidence from epidemiological and laboratory studies, *Toxicol. Lett.* **150**:43–56.

52. J. N. Davis, O. Kucuk, and F. H. Sarkar (1999). Genistein inhibits NF-kappa B activation in prostate cancer cells, *Nutr. Cancer* **35**:167–174.

53. B. K. Chacko, R. T. Chandler, A. Mundhekar, H. M. Pruitt, D. F. Kucik, C. G. Kevil, S. Barnes, and R. P. Patel (2005). Revealing anti-inflammatory mechanisms of soy-isoflavones by flow: Modulation of leukocyte–endothelial cell interactions, *Am. J. Physiol.* in press.

54. M. Belenky, H. Kim, J. K. Prasain, and S. Barnes (2003). DING—a protein without a gene, *J. Nutr.* **133**:2497S–2501S.

55. A. Berna, F. Bernier, K. Scott, and B. Stuhlmuller (2002). Ring up the curtain on DING proteins, *FEBS Lett.* **524**:6–10.

56. E. S. Boylan and R. E. Calhoon (1979). Mammary tumorigenesis in the rat following prenatal exposure to diethylstilbestrol and postnatal treatment with 7,12-dimethylbenz(*a*)anthracene, *J. Toxicol. Environ. Health* **5**:1059–1071.

57. C. A. Lamartiniere and M. B. Holland (1992). Neonatal diethylstilbestrol prevents spontaneously developing mammary tumors. In: *Proceedings of First International Symposium on Hormonal Carcinogenesis*, J. J. Li, S. Nandi, and S. A. Li, eds. Springer-Verlag, New York, pp. 305–308.

58. C. A. Lamartiniere (2002). Timing of exposure and mammary cancer risk, *J. Mammary Gland Biol. Neoplasia* **7**:67–76.

59. C. J. Grubbs, D. R. Farnell, D. L. Hill, and K. C. McDonough (1985). Chemoprevention of *N*-nitroso-*N*-methylurea-induced mammary cancers by pretreatment with 17 beta-estradiol and progesterone, *J. Natl. Cancer Inst.* **74**:927–931.

60. L. Hilakivi-Clarke, E. Cho, I. Onojafe, M. Raygada, and R. Clarke (1999). Maternal exposure to genistein during pregnancy increases carcinogen-induced mammary tumorigenesis in female rat offspring, *Oncol. Rep.* **6**:1089–1095.

61. M. S. Cotroneo and C. A. Lamartiniere (2001). Pharmacologic, but not dietary genistein supports endometriosis in a rat model, *Toxicol. Sci.* **61**:68–75.

62. R. A. Gorski (1974). The neuroendocrine regulation of sexual behavior. In: *Advances in Psychobiology*, Vol. 2. Wiley, Hoboken. NJ, pp. 1–58.

63. C. A. Lamartiniere, C. A. Sloop, J. Clark, H. A. Tilson, and G. W. Lucier (1982). Organizational effects of hormones and hormonally-active xenobiotics on postnatal development. In: *12th Conference on Environmental Toxicology*, Dayton, Ohio. U.S. Air Force Publication: AFAMRL-TR-81-149, pp. 96–121.

64. Y. H. Ju, C. D. Allred, K. F. Allred, K. L. Karko, D. R. Doerge, and W. G. Helferich (2001). Physiological concentrations of dietary genistein dose-dependently stimulate growth of estrogen-dependent human breast cancer (MCF-7) tumors implanted in athymic nude mice, *J. Nutr.* **131**:2957–2962.

65. X. O. Shu, F. Jin, Q. Dai, W. Wen, J. D. Potter, et al. (2001). Soyfood intake during adolescence and subsequent risk of breast cancer among Chinese women, *Cancer Epidemiol. Biomarkers Prev.* **10**:483–488.

66. A. H. Wu, P. Wan, J. Hankin, C. C. Tseng, M. C. Yu, and M. C. Pike (2002). Adolescent and adult soy intake and risk of breast cancer in Asian-Americans, *Carcinogenesis* **23**:1491–1496.

67. C. E. Bonferroni (1935). Il calcolo delle assicurazioni su gruppi di teste. In: *Studi in Onore del Professore Salvatore Ortu Carboni*, Rome, Italy, pp. 13–60.

68. A. Reiner, D. Yekutieli, and Y. Benjamini (2003). Identifying differentially expressed genes using false discovery rate controlling procedures, *Bioinformatics* **19**:368–375.

69. Y. Benjamini and Y. Hochberg (1995). Controlling the false discovery rate: A practical and powerful approach to multiple testing, *J. R. Stat. Soc. Ser. B* **85**:289–300.

70. D. B. Allison, G. Gadbury, M. Heo, J. Fernandez, C. K. Lee, T. A. Prolla, and R. Weindruch (2002). A mixture model approach for the analysis of microarray gene expression data, *Comput. Stat. Data Anal.* **39**:1–20.

71. J. G. Liao, Y. Lin, Z. E. Selvanayagam, and W. J. Shih (2004). A mixture model for estimating the local false discovery rate in DNA microarray analysis, *Bioinformatics* **20**:2694–2701.

72. J. D. Storey (2002). A direct approach to false discovery rates, *J. R. Stat. Soc. Ser. B* **64**:479–498.

73. J. D. Storey (2003). The positive false discovery rate: A Bayesian interpretation and the *q*-value, *Ann. Stat.* **31**:2013–2035.

74. S. Pounds and C. Cheng (2004). Improving false discovery rate estimation, *Bioinformatics* **20**:1737–1745.

75. M. Unlu, M. E. Morgan, and J. S. Minden (1997). Difference gel electrophoresis: A single gel method for detecting changes in protein extracts, *Electrophoresis* **18**:2071–2077.

76. D. B. Wall, S. J. Parus, and D. M. Lubman (2002). Three-dimensional protein map according to pI, hydrophobicity and molecular mass, *J. Chromatogr. B Anal. Technol. Biomed. Life Sci.* **774**:53–58.

77. M. P. Washburn, D. Wolters, and J. R. Yates, 3rd (2001). Large-scale analysis of the yeast proteome by multidimensional protein identification technology, *Nat. Biotechnol.* **19**: 242–247.

78. S. P. Gygi, B. Rist, S. A. Gerber, F. Turecek, M. H. Gelb, and R. Aebersold (1999). Quantitative analysis of complex protein mixtures using isotope-coded affinity tags, *Nat. Biotechnol.* **17**:994–999.

79. G. P. Page, J. W. Edwards, S. Barnes, R. Weindruch, and D. B. Allison (2003). A design and statistical perspective on microarray gene expression studies in nutrition: The need for playful creativity and scientific hard-mindedness, *Nutrition* **19**:997–1000.

80. M. A. Newton, C. M. Kendziorski, C. S. Richmond, F. R. Blattner, and K. W. Tsui (2001). On differential variability of expression ratios: Improving statistical inference about gene expression changes from microarray data, *J. Comput. Biol.* **8**:37–52.

81. S. Attoor, E. R. Dougherty, Y. Chen, M. L. Bittner, and J. M Trent (2004). Which is better for cDNA-microarray-based classification: Ratios or direct intensities? *Bioinformatics* **20**:2513–2520.

82. R. F. Chuaqui, R. F. Bonner, C. J. Best, J. W. Gillespie, M. J. Flaig, et al. (2002). Postanalysis follow-up and validation of microarray experiments, *Nat. Genet.* **32** (Suppl.): 509–514.

83. J. C. Rockett and G. M. Hellmann (2004). Confirming microarray data—is it really necessary? *Genomics* **83**:541–549.

84. P. V. Shekhar, M. L. Chen, J. Werdell, G. H. Heppner, F. R. Miller, and J. K. Christman (1998). Transcriptional activation of functional endogenous estrogen receptor gene expression in MCF10AT cells: A model for early breast cancer, *Int. J. Oncol.* **13**:907–915.
85. K. Brendel, R. L. McKee, V. J. Hruby, D. G. Johnson, A. J. Gandolfi, and C. L. Krumdieck (1987). Precision cut tissue slices in culture: A new tool in pharmacology, *Proc. West. Pharmacol. Soc.* **30**:291–293.
86. A. P. Fuller, D. Palmer-Toy, M. G. Erlander, and D. C. Sgroi (2003). Laser capture microdissection and advanced molecular analysis of human breast cancer, *J. Mammary Gland Biol. Neoplasia* **8**:335–345.
87. M. Axelson, J. Sjovall, B. E. Gustafsson, and K. D. R. Setchell (1984). Soya—a dietary source of the non-steroidal oestrogen equol in man and animals, *J. Endocrinol.* **102**:49–56.
88. D. Urban, W. Irwin, M. Kirk, M. A. Markiewicz, R. Myers, et al. (2001). The effect of isolated soy protein on plasma biomarkers in elderly men with elevated serum prostate specific antigen, *J. Urol.* **165**:294–300.
89. M. S. Morton, O. Arisaka, N. Miyake, L. D. Morgan, and B. A. Evans (2002). Phytoestrogen concentrations in serum from Japanese men and women over forty years of age, *J. Nutr.* **132**:3168–3171.
90. H. Kim, P. Hall, M. Smith, M. Kirk, J. K. Prasain, S. Barnes, and C. Grubbs (2004). Chemoprevention by grape seed extract and genistein in carcinogen-induced mammary cancer in rats is diet-dependent, *J. Nutr.* **134**:3445S–3552S.
91. M. N. Gould (1993). The introduction of activated oncogenes to mammary cells in vivo using retroviral vectors: A new model for the chemoprevention of premalignant and malignant lesions of the breast, *J. Cell. Biochem. Suppl.* **17G**:66–72.
92. S. Rossi and M. Loda (2003). The role of the ubiquitination-proteasome pathway in breast cancer: Use of mouse models for analyzing ubiquitination processes, *Breast Cancer Res.* **5**:16–22.
93. N. I. Govorukhina, A. Keizer-Gunnink, A. G. van der Zee, S. de Jong, H. W. de Bruijn, and R. Bischoff (2003). Sample preparation of human serum for the analysis of tumor markers. Comparison of different approaches for albumin and γ-globulin depletion, *J. Chromatogr. A* **1009**:171–178.
94. S. M. Varnum, C. C. Covington, R. L. Woodbury, K. Petritis, L. J. Kangas, et al. (2003). Proteomic characterization of nipple aspirate fluid: Identification of potential biomarkers of breast cancer, *Breast Cancer Res. Treat.* **80**:87–97.
95. Y. Woo, J. Affourtit, S. Daigle, A. Viale, K. Johnson, J. Naggert, and G. A. Churchill (2004). Comparison of cDNA, oligonucleotide, and Affymetrix GeneChip gene expression microarray platforms, *J. Biomol. Techniques* **15**:276–284.
96. L. Breci, L. Bennett, J. Letarte, M. Keeler, E. Hattrup, R. Johnson, and P. A. Haynes (2004). Comprehensive proteomics using gels, isoelectric focusing, chromatographic fractionation and gas phase fractionation. In: *Proceedings of the 52nd Conference of the American Society for Mass Spectrometry*, Nashville, TN.
97. T. Berggren, L. Bonilla, T. Richmond, C. Rozanas, R. Asbury, et al. (2004). Comparison of quantitative proteomics methods on human embryonic stem cells. In: *Proceedings of the 52nd Conference of the American Society for Mass Spectrometry*, Nashville, TN.
98. E. Chemodanova, H. A. Brown, J. E. Van Eyk, and I. Neverova (2004). 2-DE and RP-HPLC of intact proteins are complementary approaches for analysis of swine heart proteome, *Biophys. J.* **86**:625A.

15

SUSCEPTIBILITY TO EXPOSURE TO HETEROCYCLIC AMINES FROM COOKED FOOD: ROLE OF UDP-GLUCURONOSYLTRANSFERASES

Michael A. Malfatti and James S. Felton

Biosciences Program, Lawrence Livermore National Laboratory, Livermore, California

15.1 INTRODUCTION

A number of carcinogenic heterocyclic amines (PhIP, MeIQx, and DiMeIQx) are produced from the condensation of creatinine, hexoses, and amino acids during the cooking of meat [1]. There are many variables that impact the production and subsequent ingestion of these compounds in our diet. Temperature, type of meat product, cooking method, doneness, and other factors affect the quantity of these carcinogens consumed by humans. Estimates of ingestion of these carcinogens are 1–20 ng/kg body weight per day [2]. Human case–control studies that correlate meat consumption from well-done cooking practices with cancer incidence indicate excess tumors for breast, colon, stomach, esophagus, and possibly prostate [3–5].

15.2 GENETIC SUSCEPTIBILITY

Heterocyclic amines (HAs) are activated and detoxified through a number of different pathways. The initial N-oxidation at the exocyclic amino group, present in all HAs, primarily occurs in the liver, and almost exclusively by cytochrome P4501A2

Nutritional Genomics: Discovering the Path to Personalized Nutrition
Edited by Jim Kaput and Raymond L. Rodriguez

[6, 7]. Further activation is possible by conjugating enzymes like *N*-acetyltransferase and sulfotransferase to eventually give a very reactive nitrenium ion that theoretically binds to the guanine of DNA at the C8 position [8]. This resulting adduct then leads to mutations and cancer if not removed by the nucleotide excision repair pathway. If precursors to the reactive intermediates can be removed from the cells by detoxification, then the amount of DNA damage can be minimized. UDP-glucuronosyltransferase appears to fulfill this role. Although genetic variants in any of the activation pathways can have an impact on the total amount of DNA damage, we believe a major influence on the whole pathway and thus susceptibility to cancer is the ability to detoxify. We describe here the basics of this pathway (genes to protein products) and discuss the impact of variation of the numerous UGTs on DNA damage and ultimately cancer.

15.3 UDP-GLUCURONOSYLTRANSFERASE

The elimination of many endogenous and xenobiotic chemicals from the body is highly dependent on UDP-glucuronosyltransferase (UGT)-dependent conjugation with glucuronic acid. Compared to other conjugating enzymatic reactions, UGT conjugation is somewhat unique. The most notable differences are the location of the UGT protein and the atypical gene structure of one of the enzyme subfamilies. Similar to the cytochrome P450s, the UGTs have a wide and varied substrate selectivity and specificity. This trait makes them able to conjugate many types of chemical classes. UGTs are widely distributed among tissues making site-specific metabolism an important factor in UGT-mediated conjugation. Chemical inducers and repressors, as well as genetic variation in some UGT genes, regulate UGT expression. Furthermore, alterations in gene expression, due to genetic polymorphisms, can have profound effects on glucuronidation capacity. The metabolism of several chemical carcinogens, including the cooked-food carcinogen 2-amino-1-methyl-6-phenylimidazo[4,5-*b*]pyridine (PhIP), can be influenced by the differential expression of certain UGT isozymes. This differential expression of UGT isoforms can change the metabolic ratio between bioactivation and detoxification. A change favoring bioactivation, caused by decreased glucuronidation activity, would likely lead to an increase in the susceptibility of potential tumor formation from carcinogen exposure.

The metabolism of certain chemical carcinogens by UGT-mediated glucuronidation is one of the central pathways in maintaining health. The structure, function, and especially regulation of specific UGTs all contribute to how chemical carcinogens can be bioactivated or detoxified. Differential glucuronidation capacity affects the bioactivation of the cooked-food carcinogen PhIP.

15.4 UDP-GLUCURONOSYLTRANSFERASE BIOCHEMISTRY

The *UGTs* are a multigene superfamily of constitutively and inducible membrane-bound enzymes that participate in the biotransformation of many different chemical compounds. Glucuronidation is an especially important pathway for detoxifying reactive intermediates from metabolic reactions, which otherwise can be biotransformed into cytotoxic or carcinogenic species [9]. The UGTs catalyze the conjugation of glucuronic acid to a nucleophilic substrate that increases the polarity of the substrate to facilitate its excretion through the urine or bile [10, 11]. The sugar cosubstrate for the reaction is uridine 5′-diphosphoglucuronic acid (UDPGA), which is generated in the cytosol in a series of reactions starting with glucose. The rate-limiting enzyme in the UDPGA synthetic pathway is UDP-glucose dehydrogenase and is most likely an important factor affecting the rate of glucuronidation [12]. The subcellular localization of the UGTs is within the endoplasmic reticulum in a conformation such that the majority of the protein is luminal. This luminal localization results in the phenomenon of latency, possibly due to the ER membrane acting as a diffusional barrier for substrate and cofactor access, and metabolite and by-product removal [13].

15.5 UDP-GLUCURONOSYLTRANSFERASE GENE STRUCTURE

The human UGT proteins are divided into two gene families, *UGT1* and *UGT2*, based on sequence homologies. These families are further divided into three subfamilies, *UGT1A*, *UGT2A*, and *UGT2B*. The proteins range from 529 to 534 amino acids in length for a molecular weight of 52–57 kDa, with several highly conserved regions that are important for membrane targeting and activity. The carboxyl terminus of all UGTs shares a high degree of similarity, whereas the amino terminal domain is divergent. The *UGT1A* family is located on chromosome 2 (2q37) and is derived from a single gene locus composed of 5 exons (Figure 15.1). The amino half of the gene (280 amino acids) is encoded by one of thirteen exon 1 sequences that produce individual UGT1A proteins [14]. Each unique exon 1 sequence is proximal to its own distinctive promoter. Of the thirteen exon 1 sequences, nine code for functional UGT proteins (UGT1A1, UGT1A3, UGT1A4, UGT1A5, UGT1A6, UGT1A7, UGT1A8, UGT1A9, and UGT1A10) and four represent pseudogenes (UGT1A2p, UGT1A11p, UGT1A12p, and UGT1A13p) [14, 15]. The regulatory sequences flanking each of the exon 1 regions are thought to dictate the individual expression profile of each UGT1A isoform [16]. The carboxyl terminus sequence (245 amino acids) is identical for all UGT1A subfamily members and is comprised of exons 2–5. This gene structure is atypical among enzymes involved in xenobiotic metabolism [12]. It has been proposed that the unique amino terminal half of each UGT protein codes for the specific substrate binding domain while the common carboxyl terminal half codes for the UDPGA cosubstrate binding domain [17, 18]. The existence of the unique substrate binding domains provides for the large substrate specificity and selectivity observed in the UGT1A proteins.

In contrast to the *UGT1* family, the *UGT2* genes are located on chromosome 4 (4q13) and are comprised of eight individual structural genes (six genes for *UGT2B*; two genes for *UGT2A* subfamilies). Like the UGT1A proteins, the UGT2B proteins share a high degree of similarity at the carboxyl terminal region and are divergent

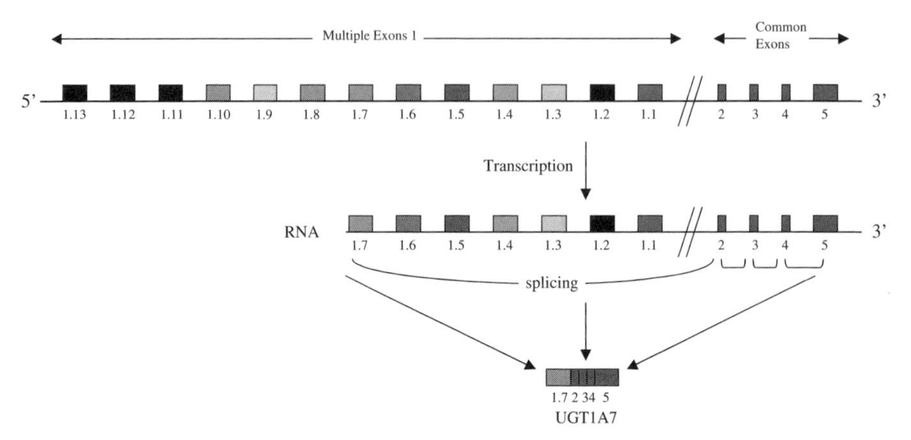

Figure 15.1. Organization of the *UGT1A* gene locus and an example of how different UGT1A RNAs are processed. Exons 2–5 are common to all UGT1A isoforms. Exon 1 contains sequences that code for the divergent portion of each UGT1A protein, represented by exons 1.1–1.13. Transcription is initiated at promoters that flank each of the exon 1 sequences. The 5′ and 3′ consensus-splice sites are recognized by the spliceosome and the intervening sequences are removed. (Adapted from [19].)

at the amino terminal domain. The UGT2B proteins are primarily responsible for steroid metabolism.

15.6 SUBSTRATE SPECIFICITY AND SELECTIVITY

The UGTs have wide and overlapping substrate specificities and selectivities. The substrates include endogenous steroids, hormones, bilirubin, bile acids, dietary constituents, and numerous xenobiotic drugs, environmental pollutants, and carcinogens. Functional groups known to be conjugated include phenols and aliphatic alcohols, carboxylic acids, primary, secondary, and tertiary amines, and nucleophilic carbon atoms. A listing of UGT glucuronidation activity toward specific substrate classes for each UGT isozyme can be found in a review by Tukey and Strassburg [19]. Glucuronidation is generally thought of as a detoxification reaction; however, there are examples where glucuronide conjugation results in increased biological activity. These include the *N-O*-glucuronides of hydroxamic acids and the acyl glucuronides of carboxylic acids. In addition, several drugs are known to produce glucuronide conjugates that have pharmacological or toxic activities higher than their parent compound. Examples of these include glucuronides of morphine, all-*trans*-retinol and all-*trans*-retinoic acid, and natural and synthetic estrogens (reviewed in [12]).

15.7 TISSUE DISTRIBUTION OF UDP-GLUCURONOSYLTRANSFERASE

Expression of UGT proteins has been detected in multiple tissues, with the liver being regarded as the site with the greatest glucuronidation capacity. High levels of activity have also been observed in the kidney and intestine, indicating a significant capacity for extrahepatic glucuronidation. Other tissues expressing UGT activity include lung, olfactory epithelium, ovary, mammary gland, testis, and prostate (reviewed in [19]). The mechanisms that determine the wide UGT tissue distribution and expression have yet to be determined. In addition to having a wide tissue distribution, UGT proteins can be preferentially expressed in different tissues. The *UGT1A* gene locus in the liver codes for *UGT1A1*, *UGT1A3*, *UGT1A4*, *UGT1A6*, and *UGT1A9*. Expression of *UGT1A7*, *UGT1A8*, and *UGT1A10* is found exclusively in extrahepatic tissue, primarily in gastric, colon, and biliary tissue, respectively. Expression of the *UGT1A* locus in the colon is the most diverse with gene transcripts detected for *UGT1A1*, *UGT1A3*, *UGT1A4*, *UGT1A6*, *UGT1A8*, *UGT1A9*, and *UGT1A10*. The differential expression of the *UGT1A* family of enzymes demonstrates the importance of site-specific metabolism and substrate selectivity of the UGT1A proteins. Low expression of a particular UGT in a specific tissue could alter the metabolism of a xenobiotic compound in that tissue, potentially diverting it to pathways that would produce a more biologically active compound that could bind proteins and/or DNA.

15.8 GENE REGULATION

Expression of *UGT* genes can be regulated by chemically mediated induction or repression. Studies have shown that human *UGT1A1*, *UGT1A6*, and *UGT1A9* expression is induced by dioxin via binding of the aryl hydrocarbon receptor (AhR) to a xenobiotic response element [20, 21]. Upon ligand binding, the AhR–ligand complex translocates to the nucleus where it dimerizes with the AhR nuclear translocator (ARNT). This heterodimer complex binds to a dioxin response element, which results in enhanced *UGT* transcription and subsequent expression. Other inducers include antioxidants such as *tert*-butylhydroquinone and quercetin that can induce *UGT1A6* [22]. *UGT1A1* has been shown to be upregulated by 3-methylcholanthrene and oltripaz in human hepatocytes and by flavonoids in human HepG2 cells [23, 24]. Furthermore, the dietary anticarcinogens coumarin, curcumin, α-angelicalactone, fumaric acid, and flavones caused an increase in the glucuronidation of 4-nitrophenol and 4-methylumbelliferone in rat hepatic microsomes [25].

More recently, the human orphan nuclear receptors, human pregnane X receptor (hPXR) and constitutive androstane receptor (CAR), have been implicated in the regulation of *UGT* genes. Studies have shown that phenobarbital and rifampicin, which are ligands for hPXR and CAR, can mediate *UGT* expression. In human HepG2 cells it was shown that CAR binds to a nuclear response element within the

UGT1A1 promoter and enhances *UGT1A1* gene transcription. This enhancement is increased by phenobarbital and decreased by androstenol (an inhibitory ligand for CAR) [26]. Deletion analysis of the *UGT1A1* promoter region resulted in the identification of a regulatory sequence that conferred hPXR regulation of the *UGT1A1* gene [27]. Western blot analysis showed that both hPXR and CAR induce UGT1A1 and UGT1A6 activities. In addition, transgenic mice that possess a constitutively activated form of hPXR demonstrated a significant increase in UGT activity toward steroids, carcinogens, and enhanced bilirubin clearance. Evidence also suggests *UGT1A1* and *UGT1A6* are direct transcriptional targets for hPXR [27].

In addition to chemical induction of *UGT*s, repression of UGT activity has been reported for a number of chemical compounds. The antibiotic novobiocin was shown to inhibit UGT-mediated bilirubin conjugation in rats by disrupting Mg^{2+} complexes. This inhibition was dose dependent and caused hyperbilirubinemia in the animals and was reversible when novobiocin was removed [28]. In other studies, long chain acyl CoAs, which are intermediates in fatty acid metabolism pathways, have been shown to be excellent inhibitors of glucuronidation. Both oleoyl CoA and palmityl CoA caused a dose-dependent inhibition of UGT activity in rat liver microsomes and hepatocytes. This inhibition was noncompetitive and occurred at physiological concentrations of acyl CoAs [29, 30]. This inhibition by fatty acid metabolites illustrates how complex the whole organism detoxification can be for dietary constituents. If this has physiological significance, well-done meats with higher fat content will not only have higher HA content but the detoxification pathways will be inhibited, leading to increased DNA binding and cancer. Furthermore, significant downregulation of hepatic *UGT1A1*, *UGT1A3*, *UGT1A4*, and *UGT1A9* has been reported in malignant hepatocellular carcinoma and its premalignant precursor, hepatic adenoma, but not in benign focal nodular hyperplasia [31]. This finding indicates that downregulation of the *UGT1A* gene is an early event in hepatocarcinogenesis.

15.9 GENETIC VARIATION

UGT activity can also be regulated by genetic variability. Interindividual expression patterns have been reported for all *UGT*s except *UGT1A10*. This variation in expression was observed only in extrahepatic tissue, whereas hepatic expression showed no difference in the expression of *UGT* gene transcripts [32–35]. Polymorphisms in *UGT1A1*, *UGT1A6*, *UGT1A7*, *UGT2B4*, *UGT2B7*, and *UGT2B15* have been reported. These polymorphisms have been implicated as risk factors for certain clinical diseases and cancers [19, 35–40]. The most notable polymorphisms are variants in the *UGT1A1* gene that result in significant downregulation of *UGT1A1* activity. UGT1A1 is involved in the glucuronidation of estradiol, simple and complex phenols, and several chemical carcinogens and is the only UGT isoform known to catalyze the glucuronidation of bilirubin in humans [19, 41, 42]. Bilirubin is a toxic breakdown product of heme that can accumulate in tissues resulting in jaundice if not eliminated through transport via albumin binding or conjugation by UGT1A1. At high serum levels, bilirubin can cross the blood–brain barrier and lead to fatal necrosis of neurons and glial tissue. Therefore, downregulation of *UGT1A1* activity can result in an increase in serum levels of unconjugated bilirubin, which can lead to bilirubin toxicity.

In humans, three forms of inheritable unconjugated hyperbilirubinemia exist. The most serious, although very rare, is Crigler–Najjar syndrome type I, which is transmitted as an autosomal recessive trait in humans and is characterized by an inability to form bilirubin glucuronides. This condition is caused by mutant coding regions in various *UGT1A1* alleles, which results in either a lack of UGT1A1 production or the production of a nonfunctional protein (reviewed in [19]). The onset of Crigler–Najjar syndrome type I results in early childhood death. Crigler–Najjar syndrome type II is less severe than type I and is characterized by having very low UGT1A1 activity (10% of normal activity). This condition can be treated with enzyme induction therapy using phenobarbital. The most common form of hyperbilirubinemia is Gilbert syndrome, which occurs, in 3–10% of the general population. This condition is characterized by chronic, mild hyperbilirubinemia, which is exacerbated by stress, infection, fasting, or physical activity [19]. Gilbert syndrome is characterized by an allelic variant in the *UGT1A1* gene, which contains an additional (TA) dinucleotide repeat in the $A(TA)_nTAA$ box region of the promoter [43, 44]. Functional studies have shown that the reference *UGT1A*1 activity is associated with six TA repeats (*UGT1A1*1*). Increasing the number of TA repeats leads to a decrease in the rate of *UGT1A1* transcription [35]. The most common variant allele contains seven TA repeats (*UGT1A1*28*) and is the polymorphism associated with Gilbert syndrome [16]. This polymorphism results in hepatic bilirubin UGT conjugation being reduced to about 30% of normal [13, 45]. There is evidence to suggest that individuals with Gilbert syndrome may be at greater risk from toxicants and carcinogens that are conjugated by UGT1A1 because their ability to detoxify these compounds would be diminished [35]. For example, a recent study has reported a correlation between the *UGT1A1* ATATAA box polymorphism and a decreased ability to glucuronidate and detoxify benzo[*a*]pyrene-*trans*7*R*,8*R*-dihydrodiol [BAPD(−)] in human liver microsomes [46]. Although we have been primarily discussing HAs, PAHs, such as BaP, are accumulated on the surface of meat when the meat is cooked on a barbeque grill. As the fat drips onto the coals, it is then pyrolized to PAHs. The PAHs are then deposited on the meat from the smoke plume. Clearly, dual exposure to both classes of carcinogens is probably additive with respect to DNA damage, but specific genetic variants as described above can attenuate the exposure. Subjects possessing the *UGT1A1*28* allelic variant (which contains $(TA)_7$ repeats) showed significant decreases in UGT1A1 protein expression and BAPD(−) glucuronidation activity in liver microsomes when compared to subjects having the wild-type UGT1A1(*1/*1) genotype.

Other significant polymorphisms include allelic variation in *UGT1A6*, *UGT1A7*, *UGT2B7*, and *UGT2B15*. *UGT1A6* has been found to be polymorphic with at least four alleles characterized by three single nucleotide polymorphisms (SNPs) in the coding region [19, 47]. UGT1A6 catalyzes the conjugation of simple phenols and planar arylamines, as well as many drugs including antidepressants and β-adrenoceptor blockers. The *UGT1A7* gene that is not expressed in liver is responsible for glucuronidating many different drugs and toxicants, including several carcinogens, and contains nine different variant alleles [16]. One of the alleles (*UGT1A7*3*) is found in 17% of the general population. All the mutations in both the *UGT1A6* and *UGT1A7* genes are associated with lower glucuronidation activity compared to the wild-type genotype [16]. Both UGT2B7 and UGT2B15 catalyze the glucuronidation of steroid hormones, as well as several classes of xenobiotic substrates [11]. Allelic variants have been identified for UGT2B7 and UGT2B15 isozymes resulting in a histidine to tyrosine and an aspartic acid to tyrosine amino acid substitution, respectively [16, 48]. In vitro studies have reported no difference in catalytic activity between the UGT2B7 and UGT2B15 reference alleles and the variant alleles [49, 50]. In vivo studies have yet to be completed.

15.10 UDP-GLUCURONOSYLTRANSFERASE AND CANCER SUSCEPTIBILITY

Polymorphisms in *UGT* genes have also been associated with a potential increase in the susceptibility to certain forms of cancer due to a reduced capacity to detoxify carcinogenic compounds. Several chemical carcinogens have been shown to be substrates for glucuronidation, including many primary amines, several BaPs, and some heterocyclic amines (reviewed in [19]). Studies have shown that a reduced glucuronidation capacity can lead to an increase in bioactivation for these compounds. For example, when *UGT1A*-deficient Gunn rats were exposed to BaP, covalent binding to hepatic DNA and microsomal protein was enhanced, and production of BaP glucuronide conjugates was reduced when compared to rats with normal UGT1A activity [51]. In human tissue, glucuronidating activity toward benzo[*a*]pyrene-*trans*-7*R*,8*R*-dihydrodiol (BAPD(−)) was compared with liver microsomes from humans expressing *UGT1A1*28* (Gilbert syndrome) versus normal liver microsomes. Results showed a significant decreased in BAPD(−) glucuronide conjugate formation in the subjects possessing the *UGT1A1*28* genotype [46]. In another study, an over 200-fold interindividual variability was observed in both the glucuronidation and covalent binding of BaP metabolites in human lymphocytes exposed to BaP [52]. A decrease in BaP glucuronidation activity correlated with both an increase in BaP covalent binding and enhanced BaP cytotoxicity. Reduced UGT activity can also alter the biotransformation of carcinogenic heterocyclic amines that are found in the diet.

15.11 HETEROCYCLIC AMINE CARCINOGENS IN FOOD

Diet has long been associated with cancer etiology [53–55]. This association may be related to such nutritional factors as fat or fiber intake, antioxidant exposure, or exposure to carcinogenic substances present in the diet. Historically, sources of mutagens/carcinogens in food have been derived from pesticides and artificially added chemicals such as food preservatives and coloring agents. In addition to synthetic chemicals, naturally occurring mutagens can be present in certain foods as well. These include pyrrolizidine alkaloids, which occur in many plant species, hydrazines found in mushrooms, alkylating agents found in spice oils, and nitrites that produce nitrosamines from the degradation products of proteins or other food components [56]. A more recently discovered source of food-derived mutagens/carcinogens are those produced during the cooking of food [57]. Carcinogenic compounds derived from cooking food include pyrolysis products, which are formed at temperatures of 300–600°C, and low-temperature (<300°C) thermic mutagens, which are produced in high protein foods derived from muscle. One of the first observations of the carcinogenic potential of cooked foods was made in 1939 by Widmark, who reported malignant tumors in the mammary glands of female mice chronically exposed to extracts of horse muscle cooked at a temperature of 275°C [58]. It was not until the mid- to late 1970s, however, when better analytical techniques were developed, that significant advances were made in the study of dietary thermic mutagens. These studies ultimately led to the discovery of a class of heterocyclic amines (HAs) that are formed during the cooking of foods that are commonly consumed in a typical Western diet [57, 59, 60]. These compounds are part of the amino-imidazoazaarene (AIA) class of HAs due to a common imidazole-ring structure and exocyclic amine group [61] (Figure 15.2). They are formed by the condensation of creatinine with amino acids during the cooking of meat, under normal household cooking conditions. The concentration of HAs found in cooked meats can range from less than 1 part per billion (ppb) to greater than 500 ppb depending on meat type, precursor concentration, and cooking method [1, 2]. In general, frying or grilling at high temperatures for longer time periods will increase the amount of HAs in meats.

These AIA HAs are among the most potent mutagens ever tested in the Ames/*Salmonella* mutation assay [57]. In rodents, these compounds produced tumors in a variety of tissues including the liver, lung, intestine, breast, and prostate [62]. In humans, they have been shown to cause DNA adducts in multiple tissues [63–65]. Epidemiology studies have indicated an increased risk of colon tumors associated with HA exposure from well-done red meat consumption [3, 66].

15.12 CARCINOGENICITY OF PhIP

Of all the HAs currently identified, PhIP is the most mass abundant and has been detected at the highest levels in grilled or fried beef and chicken [2, 67]. PhIP has been shown to be carcinogenic in both mice and rats. Exposure to PhIP produced lymphomas in the mesenteric lymph nodes, mediastinal lymph nodes, and spleen of CDF1 mice exposed to 400 parts per million dietary PhIP for 579 days [68], as well as hepatic adenomas in neonatal mice [69]. In F344 rats, PhIP produced high incidences of colon, mammary, and prostate carcinomas when administered at a concentration of 400 parts per million in the diet for 365 days [4, 5]. PhIP has also been shown to induce DNA strand breaks and sister-chromatid exchanges in Chinese hamster ovary cells [70, 71], and to form DNA adducts in both rodent and human

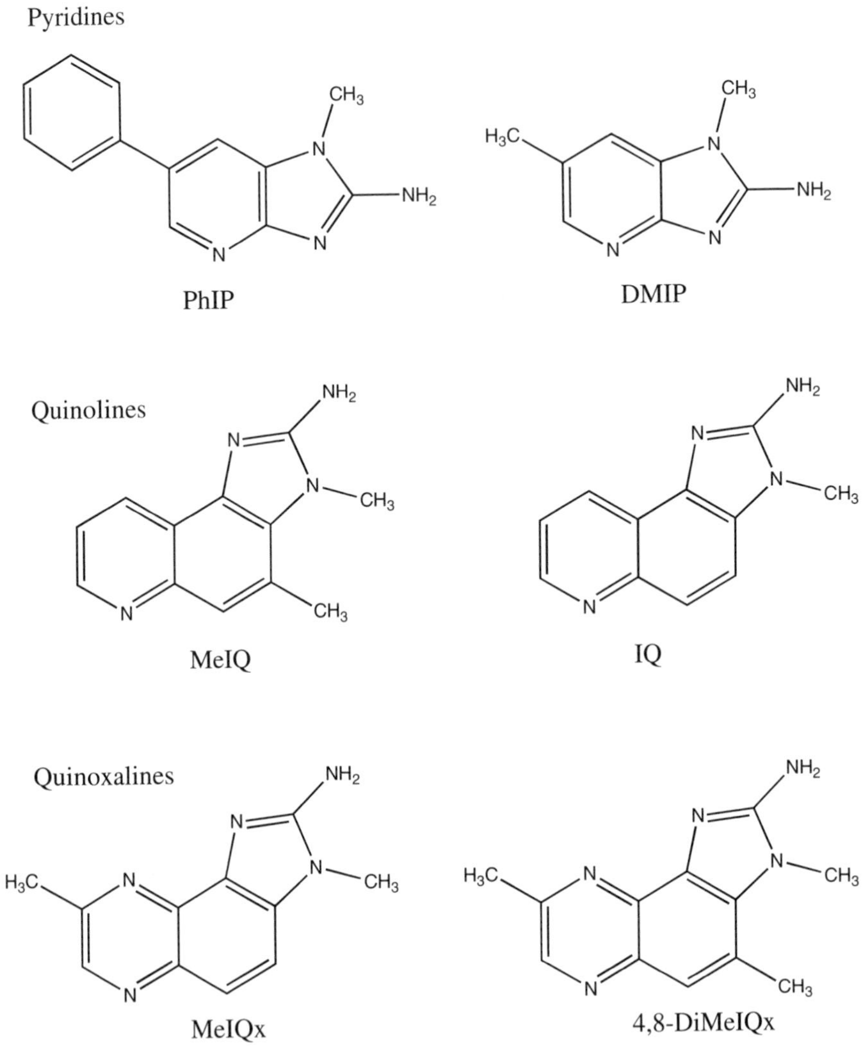

Pyridines

PhIP

DMIP

Quinolines

MeIQ

IQ

Quinoxalines

MeIQx

4,8-DiMeIQx

Figure 15.2. Carcinogenic heterocyclic amines isolated from cooked foods. PhIP, 2-amino-1-methyl-6-phenylimidazo[4,5-*b*]pyridine; DMIP, 2-amino-1,6-dimethylimidazo [4,5-*b*]pyridine; MeIQ, 2-amino-3,5-dimethylimidazo[4,5-*f*]quinoline; IQ, 2-amino-3-methylimidazo[4,5-*f*]quinoline; MeIQx, 2-amino-3,8-dimethylimidazo[4,5-*f*]quinoxaline; 4,8-DiMeIQx, 2-amino-3,4,8-trimethylimidazo[4,5-*f*]quinoxaline.

tissues [63, 72–75]. In addition, intake of PhIP from well-done red meat consumption has been associated with an increased risk for breast cancer in women [76]. These findings, together with the relative abundance of PhIP in cooked foods, indicate that PhIP may pose a significant risk to the development of certain human cancers.

15.13 METABOLISM OF PhIP

The metabolism of PhIP involves both phase I and phase II pathways for bioactivation and/or detoxification (Figure 15.3). PhIP bioactivation is highly dependent on the hepatic cytochrome P4501A2 (*CYP1A2*) mediated N-hydroxylation to the corresponding 2-hydroxyamino-1-methyl-6-phenylimidazo[4,5-*b*]pyridine (*N*-hydroxyl-PhIP) [6, 7]. In extrahepatic tissue, CYP1A1 and CYP1B1 have also been reported to activate PhIP [77]. *N*-Hydroxy-PhIP is subsequently esterified by phase II sulfotransferases and/or acetyltransferases that generate the highly electrophilic *O*-sulfonyl and *O*-acetyl esters, respectively. These esters are capable of heterolytic cleavage to generate the reactive nutrenium ion, which is considered the ultimate carcinogenic species [78]. This reactive nutrenium ion is able to form DNA adducts, mainly at the C8 position of guanine [8, 72, 79]. Other conjugating enzymes have also been shown to further activate *N*-hydroxyl-PhIP and bind DNA but are considered minor contributors to PhIP bioactivation [78, 80]. *N*-Glucuronidation can compete with these activation reactions resulting in the formation of the less reactive *N*-hydroxyl-PhIP-N^2-glucuronide and *N*-hydroxyl-PhIP-*N*3-glucuronide. These compounds can be excreted through the urine or bile or can be transported to extrahepatic tissue, where deconjugation by β-glucuronidase can occur leading to the regeneration of the reactive intermediate *N*-hydroxyl-PhIP [81, 82]. Detoxification of PhIP involves formation of the CYP450-mediated nonreactive 2-amino-1-methyl-6-(4′-hydroxy)phenylimidazo [4,5-*b*]pyridine (4′-hydroxyl-PhIP). This compound can undergo sulfotransferase— and/or glucuronosyltransferase-mediated conjugation, producing more polar unreactive compounds, which are readily excreted [83, 84]. This pathway is more prevalent in rodents than in humans because the rate of 4′-hydroxyl-PhIP formation is much more dependent on CYP1A1 than CYP1A2, and CYP1A1 is not expressed in human liver, the site most responsible for PhIP hydroxylation [77, 85, 86]. PhIP can also form nonreactive direct glucuronides at the N^2 and *N*3 positions [87].

In humans, studies have indicated that CYP1A2 catalyzed *N*-hydroxylation and subsequent UGT-mediated glucuronidation is quantitatively the most important pathway in the metabolism of PhIP. When human volunteers were exposed to PhIP, *N*-hydroxyl-PhIP glucuronide conjugates accounted for approximately 60% of the total PhIP urinary metabolites [88]. *N*-hydroxyl-PhIP-N^2-glucuronide was also the major metabolite in human urine after consumption of a single cooked chicken meal [89]. Further investigations have determined that the *UGT1A* subfamily of *UGT*s plays a major role in the glucuronidation of PhIP [90–92]. Microsomes containing human UGT1A isoforms were able to convert *N*-hydroxyl-PhIP to both *N*-hydroxyl-PhIP-N^2-glucuronide and *N*-hydroxyl-PhIP-*N3*-glucuronide. A more recent study has implicated UGT1A1 as the primary UGT1A isoform responsible for PhIP glucuronidation [93]. Microsomal preparations from *UGT1A1* expressing baculovirus-infected insect cells produced five times more *N*-hydroxyl-PhIP-N^2-glucuronide compared to microsomes containing UGT1A4, the next most active UGT protein.

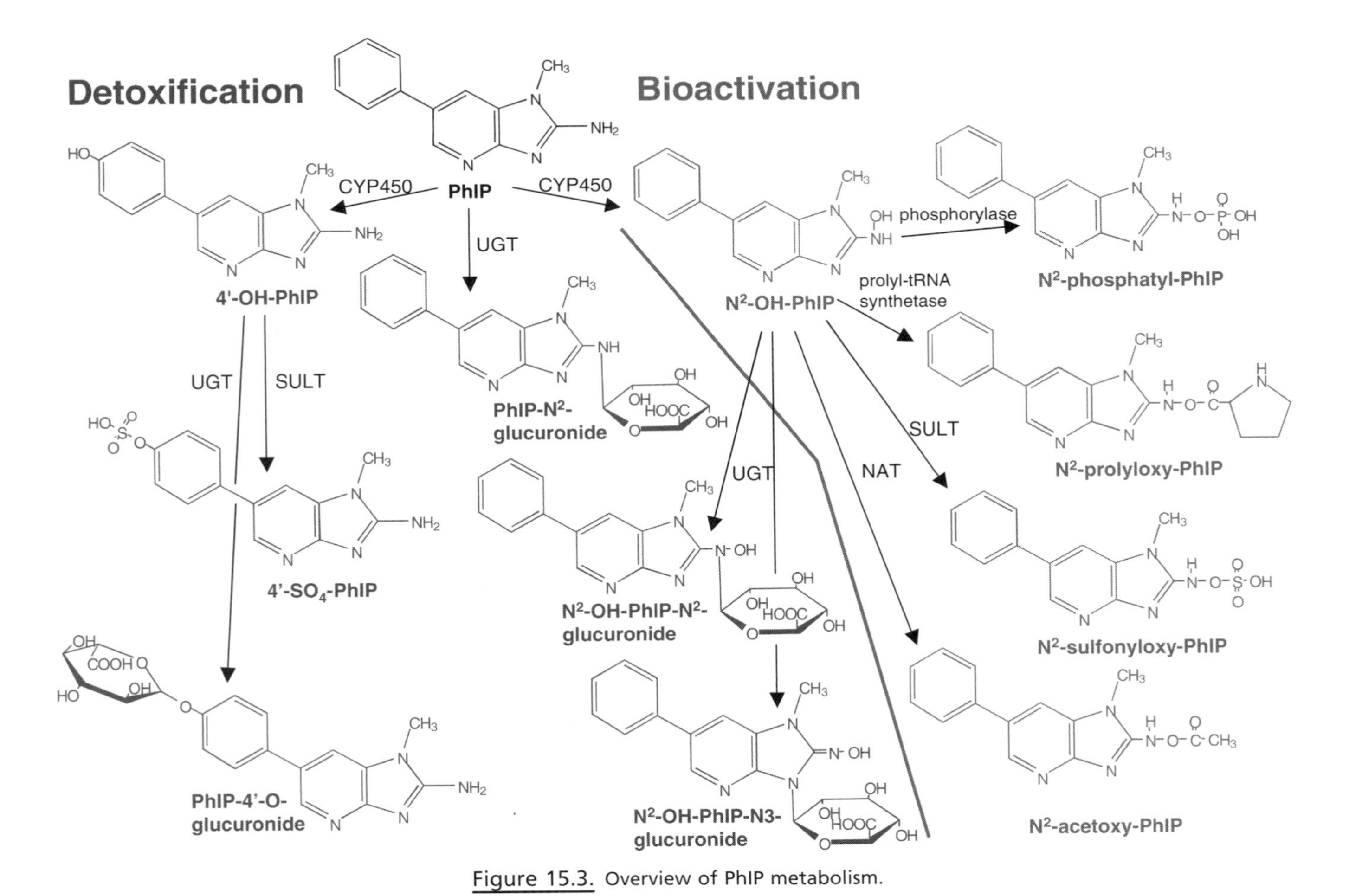

Figure 15.3. Overview of PhIP metabolism.

15.14 UDP-GLUCURONOSYLTRANSFERASE AND PhIP RISK SUSCEPTIBILITY

It is difficult to assess PhIP risk susceptibility in humans due to wide variations in PhIP metabolism. There are numerous factors that can influence metabolism, which are regarded as important determinants of individual susceptibility to the carcinogenic effects of PhIP. These include but are not limited to PhIP dose, variations in diet, species differences in metabolism, and polymorphic distribution of PhIP-metabolizing enzymes. High doses of PhIP, typically used in animal studies, can potentially saturate enzymatic pathways, which can divert PhIP to metabolic pathways that would not usually be used in a low-dose exposure. Species differences in metabolism can also affect risk susceptibility determinations. In humans and dogs, UGT-mediated *N*-hydroxyl-PhIP glucuronidation is a major pathway in PhIP biotransformation, whereas in rodents this is a minor pathway (Figure 15.4). Polymorphic expression of PhIP-metabolizing enzymes can play a large role in determining risk susceptibility. These polymorphisms can arise from both heritable and environmental factors [94]. Humans display a large interindividual variation in the expression of CYP450 and several of the phase II conjugation enzymes (reviewed in [95]). Studies have shown that the polymorphic expression of certain UGTs can lead to differential metabolism of both endogenous and exogenous substrates including PhIP [13]. For example, when *UGT1A*-deficient Gunn rats were dosed orally with PhIP, a decrease in PhIP glucuronide levels in the urine of these rats correlated with an increase in hepatic DNA adducts compared to control rats with normal UGT1A activity [96]. PhIP glucuronides in the bile of *UGT1A*-deficient rats dosed with PhIP intravenously were also reduced compared to control animals [97]. Furthermore, when Chinese hamster ovary cells were transfected with the human *UGT1A1*, gene a significant reduction in PhIP-induced cytotoxicity and mutation induction was observed when compared to control cells that did not contain the *UGT1A1* gene [98]. These results suggest that UGT1A1 plays a major role in the metabolism of PhIP by providing a protective effect against PhIP-induced toxicity and mutation induction, and that variations in UGT protein expression can potentially alter the bioactivation of PhIP.

15.15 CONCLUSION

Since UGT-mediated *N*-hydroxyl-PhIP glucuronidation is such a prominent metabolic step in the biotransformation of PhIP in humans, understanding the mechanistic aspects of PhIP glucuronidation and identifying the specific UGT enzymes involved, and their regulation, are especially important. The failure to conjugate *N*-hydroxyl-PhIP by glucuronidation could lead to further activation by sulfotransferase and/or acetyltransferase, resulting in highly reactive esters that can bind DNA and potentially cause mutations. Furthermore, differential expression of UGT isoforms in specific tissues can change the metabolic ratio between bioactivation and detoxification. A change favoring bioactivation, due to decreased glucuronidation activity,

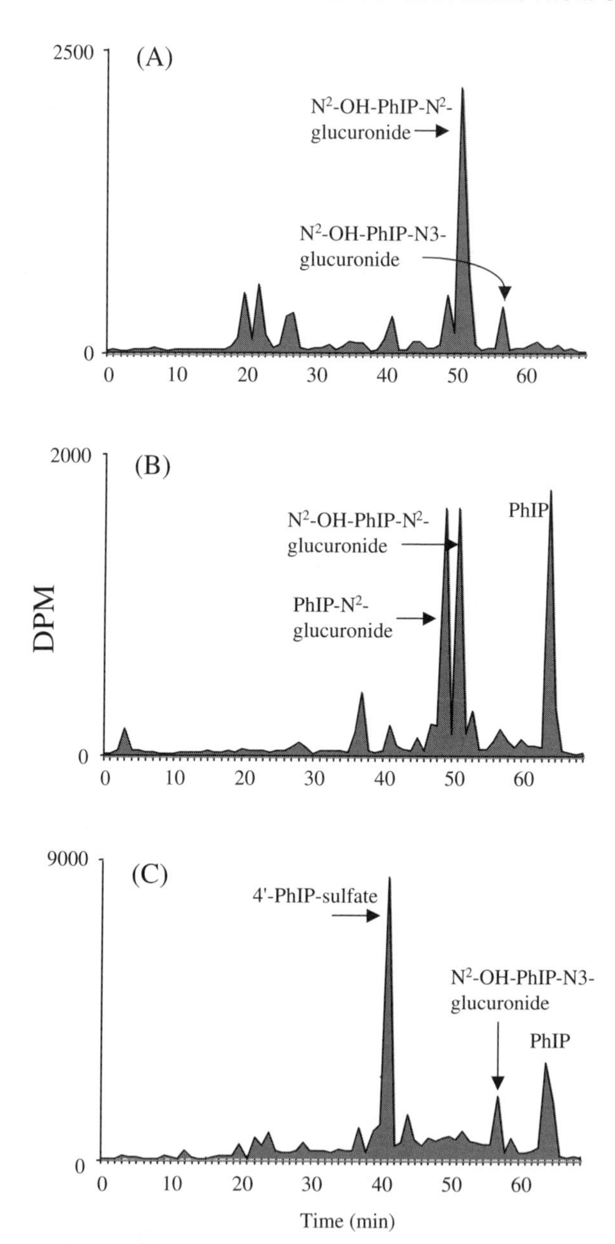

Figure 15.4. Species comparison of urinary PhIP metabolic profiles—(A) human, (B) dog, and (C) mouse.

would likely lead to an increase in the susceptibility of potential tumor formation from PhIP exposure. Knowing the glucuronidation capacity of the specific UGT isozymes involved in *N*-hydroxyl-PhIP glucuronidation, and understanding the role of UGT in PhIP metabolism, will allow for a better understanding of the overall bioactivation/detoxification mechanisms of PhIP. This will help in evaluating the individual susceptibility to the potential cancer risks associated with exposure to PhIP. It is hypothesized that individuals with low levels of specific UGT proteins will have a diminished capacity to detoxify PhIP, making them more susceptible to the deleterious effects from PhIP exposure.

ACKNOWLEDGMENTS

This work was performed under the auspices of the U.S. Department of Energy by the University of California, Lawrence Livermore National Laboratory under contract No. W-7405-Eng-48 and supported by NCI grant CA55861.

REFERENCES

1. J. S. Felton and M. G. Knize (1990). Heterocyclic-amine mutagens/carcinogens in foods. In: *Handbook of Experimental Pharmacology*, C. S. Cooper and P. L. Grover, eds. Springer-Verlag, Berlin, pp. 471–502.
2. R. Sinha, N. Rothman, E. D. Brown, C. P. Salmon, M. G. Knize, et al. (1995). High concentrations of the carcinogen 2-amino-1-methyl-6-phenylimidazo[4,5-*b*]pyridine (PhIP) occur in chicken but are dependent on the cooking method, *Cancer Res.* **55**:4516–4519.
3. R. Sinha and N. Rothman (1999). Role of well-done, grilled red meat, heterocyclic amines (HCAs) in the etiology of human cancer, *Cancer Lett.* **143**:189–194.
4. N. Ito, R. Hasegawa, M. Sano, S. Tamano, H. Esumi, S. Takayama, and T. Sugimura (1991). A new colon and mammary carcinogen in cooked food, 2-amino-1-methyl-6-phenylimidazo[4,5-*b*]pyridine (PhIP), *Carcinogenesis* **12**:1503–1506.
5. T. Shirai, M. Sano, S. Tamano, S. Takahashi, M. Hirose, et al. (1997). The prostate: A target for carcinogenicity of 2-amino-1-methyl-6-phenylimidazo[4,5-*b*]pyridine (PhIP) derived from cooked foods, *Cancer Res.* **57**:195–198.
6. R. J. Edwards, B. P. Murray, S. Murray, T. Schulz, D. Neubert, et al. (1994). Contribution of CYP1A1 and CYP1A2 to the activation of heterocyclic amines in monkeys and humans, *Carcinogenesis* **15**:829–836.
7. A. R. Boobis, A. M. Lynch, S. Murray, R. de la Torre, A. Solans, et al. (1994). CYP1A2-catalyzed conversion of dietary heterocyclic amines to their proximate carcinogens is their major route of metabolism in humans, *Cancer Res.* **54**:89–94.
8. H. Nagaoka, K. Wakabayashi, S. B. Kim, I. S. Kim, Y. Tanaka, et al. (1992). Adduct formation at C-8 of guanine on in vitro reaction of the ultimate form of 2-amino-1-methyl-6-phenylimidazo[4,5-*b*]pyridine with 2′-deoxyguanosine and its phosphate esters, *Jpn. J. Cancer Res.* **83**:1025–1029.

9. G. J. Dutton (1980). *Glucuronidation of Drugs and Other Compounds.* CRC Press, Boca Raton.
10. R. Meech and P. I. Mackenzie (1997). Structure and function of uridinediphosphate glucuronosyltransferases, *Clin. Exp. Pharmacol. Physiol.* **24**:907–915.
11. A. Radominska-Pandya, P. J. Czernik, J. M. Little, E. Battaglia, and P. I. Mackenzie (1999). Structural and functional studies of UDP-glucuronosyltransferases, *Drug Metab. Rev.* **31**:817–899.
12. J. K. Ritter (2000). Roles of glucuronidation and UDP-glucuronosyltransferases in xenobiotic bioactivation reactions, *Chem.-Biol. Interactions* **129**:171–193.
13. M. B. Fisher, M. F. Paine, T. J. Strelevitz, and S. A. Wrighton (2001). The role of hepatic and extrahepatic UDP-glucuronosyltransferase in human drug metabolism, *Drug Metab. Rev.* **33**:273–297.
14. Q. H. Gong, J. W. Cho, T. Huang, C. Potter, N. Gholami, et al. (2001). Thirteen UDP-glucuronosyltransferase genes are encoded at the human *UGT1* gene complex locus, *Pharmacogenetics* **11**:357–368.
15. I. S. Owens and J. K. Ritter (1995). Gene structure at the human *UGT1* locus creates diversity in isozyme structure, substrate specificity, and regulation, *Prog. Nucleic Acid Res. Mol. Biol.* **51**:305–338.
16. C. Guillemette (2003). Pharmacogenomics of human UDP-glucuronosyltransferase enzymes, *Pharmacogenomics J.* **3**:136–158.
17. P. I. Mackenzie (1990). Expression of chimeric cDNAs in cell culture defines a region of UDP-glucuronosyltransferase involved in substrate selection, *J. Biol. Chem.* **265**: 3432–3435.
18. J. O. Miners, P. A. Smith, M. J. Sorich, R. A. McKinnon, and P. I. Mackenzie (2004). Predicting human drug glucuronidation parameters: Application of in vitro and in silico modeling approaches, *Annu. Rev. Pharmacol. Toxicol.* **44**:1–25.
19. R. H. Tukey and C. P. Strassburg (2000). Human UDP-glucuronosyltransferases: Metabolism, expression, disease, *Annu. Rev. Pharmacol. Toxicol.* **40**:581–616.
20. P. A. Munzel, S. Schmohl, H. Heel, K. Kålberer, B. S. Bock-Hennig, and K. W. Bock (1999). Induction of human UDP-glucuronosyltransferases (UGT1A6, UGT1A9 and UGT2B7) by *t*-butylhydroquinone and 2,3,7,8-tetrachlorodibenzo-*p*-dioxin in Caco-2 cells, *Drug Metab. Dispos.* **27**:569–573.
21. M.-F. Yueh, Y. H. Huang, A. Hiller, S. Chen, N. Nguyen, and R. H. Tukey (2003). Involvement of the xenobiotic response element (XRE) in Ah receptor-mediated induction of human UDP-glucuronosyltransferase 1A1, *J. Biol. Chem.* **278**:15001–15006.
22. K. W. Bock, T. Eckle, M. Ouzzine, and S. Fournel-Gigleux (2000). Coordinate induction by antioxidants of UDP-glucuronosyltransferase UGT1A6 and the apical conjugate export pump MRP2 (multidrug resistance protein 2) in Caco-2 cells, *Biochem. Pharmacol.* **59**:467–470.
23. J. K. Ritter, F. Kessler, M. T. Thompson, A. D. Grove, D. J. Auyeung, and R. A. Fisher (1999). Expression and inducibility of the human bilirubin UDP-glucuronosyltransferase UGT1A1 in liver and cultured primary hepatocytes: Evidence for both genetic and environmental influences, *Hepatology* **30**:476–484.
24. U. Walle and T. Walle (2002). Induction of UDP-glucuronosyltransferase UGT1A1 by flavonoids-structural requirements, *Drug Metab. Dispos.* **30**:564–569.

25. E. M. J. van der Logt, H. M. J. Roelofs, F. M. Nagengast, and W. H. M. Peters (2003). Induction of rat hepatic and intestinal UDP-glucuronosyltransferase by naturally occurring dietary anticarcinogens, *Carcinogenesis* **24**:1651–1656.

26. J. Sugatani, H. Kojima, A. Ueda, S. Kakizaki, K. Yoshinari, et al. (2001). The phenobarbital response enhancer module in the human bilirubin UDP-glucuronosyltransferase *UGT1A1* gene and regulation by the nuclear receptor CAR, *Hepatology* **33**:1232–1238.

27. W. Xie, M. F. Yeuh, A. Radominska-Pandya, S. P. Saini, Y. Negishi, et al. (2003). Control of steroid, heme, and carcinogen metabolism by nuclear pregnane X receptor and constitutive androstane receptor, *Proc. Nat. Acad. Sci. U.S.A.* **100**:4150–4155.

28. P. Duvaldestin, J. L. Mahu, A. M. Preaux, and P. Berthelot (1976). Novobiocin-inhibition and magnesium-interaction of rat liver microsomal bilirubin UDP-glucuronosyltransferase, *Biochem. Pharmacol.* **25**:2587–2592.

29. M. Krcmery and D. Zakim (1993). Effects of oleoyl-CoA on the activity and functional state of UDP-glucuronosyltransferase, *Biochem. Pharmacol.* **46**:897–904.

30. M. Csala, G. Banhegyi, T. Kardon, R. Fulceri, A. Gamberucci, R. A. B. Giunti, and J. Mandl (1996). Inhibition of glucuronidation by an Acyl-CoA-mediated indirect mechanism, *Biochem. Pharmacol.* **52**:1127–1131.

31. C. P. Strassburg, M. P. Manns, and R. H. Tukey (1997). Differential down-regulation of the UDP-glucuronosyltransferase 1A locus is an early event in human liver and bilary cancer, *Cancer Res.* **57**:2979–2985.

32. C. R. Bhasker, W. McKinnon, A. Stone, A. C. Lo, T. Kubota, T. Ishizaki, and J. O. Miners (2000). Genetic polymorphisms of UDP-glucuronosyltransferase 2B7 (UGT2B7) at amino acid 268: Ethnic diversity of alleles and potential clinical significance, *Pharmacogenetics* **10**:670–685.

33. R. H. Tukey and C. P. Strassburg (2001). Genetic multiplicity of the human UDP-glucuronosyltransferases and regulation in the gastrointestinal tract, *Mol. Pharmacol.* **59**:405–414.

34. C. P. Strassburg, S. Kneip, J. Topp, P. Obermayer-Straub, A. Barut, R. H. Tukey, and M. P. Manns (2000). Polymorphic gene regulation and interindividual variation of UDP-glucuronosyltransferase activity in human small intestine, *J. Biol. Chem.* **275**:36164–36171.

35. J. O. Minors, R. A. McKinnon, and P. I. Mackenzie (2002). Genetic polymorphisms of UDP-glucuronosyltransferase and their functional significance, *Toxicology* **181–182**:453–456.

36. D. J. Grant and D. A. Bell (2000). Bilirubin UDP-glucuronosyltransferase 1A1 gene polymorphisms: Susceptibility to oxidative damage and cancer, *Mol. Carcinog.* **29**: 198–204.

37. C. P. Strassburg, A. Vogel, S. Kneip, R. H. Tukey, and M. P. Manns (2002). Polymorphisms of the human UDP-glucuronosyltransferase (UGT) 1A7 gene in colorectal cancer, *Gut* **50**:851–856.

38. J. Ockenga, A. Vogel, N. Teich, V. Keim, M. P. Manns, and C. P. Strassburg (2003). UDP-glucuronosyltransferase (*UGT1A7*) gene polymorphisms increase the risk of chronic pancreatitis and pancreatic cancer, *Gastroenterology* **124**:1802–1808.

39. B. Burchell (2003). Genetic variation of human UDP-glucuronosyltransferase: Implications in disease and drug glucuronidation, *Am. J. Pharmacogenomics* **3**:37–52.

40. O. J. Adegoke, X. O. Shu, Y. T. Gao, Q. Cai, J. Breyer, J. Smith, and W. Zheng (2004). Genetic polymorphisms in uridine diphospho-glucuronosyltransferase 1A1 (UGT1A1) and risk of breast cancer, *Breast Cancer Res. Treat.* **85**:239–245.
41. J. K. Ritter, J. M. Crawford, and I. S. Owens (1991). Cloning of two human liver bilirubin UDP-glucuronosyltransferase cDNAs with expression in COS-1 cells, *J. Biol. Chem.* **266**:1043–1047.
42. P. J. Bosma, J. Seppen, B. Goldhoorn, C. Bakker, R. P. Oude Elferink, J. Roy-Chowdhury, N. Roy-Chowdhury, and P. L. Jansen (1994). Bilirubin UDP-glucuronosyltransferase 1 is the only relevant bilirubin glucuronidating isoform in man, *J. Biol. Chem.* **269**: 17960–17964.
43. P. J. Bosma, J. R. Chowdhury, C. Bakker, S. Gantla, A. deBoer, et al. (1995). The genetic basis of the reduced expression of bilirubin UDP-glucuronosyltransferase 1 in Gilbert's syndrome, *N. Engl. J. Med.* **333**:1171–1175.
44. S. K. Rauchschwalbe, M. T. Zuhlsdorf, U. Schuhly, and J. Kuhlman (2002). Predicting the risk of sporadic elevated bilirubin levels and diagnosing Gilbert's syndrome by genotyping UGT1A1*28 promoter polymorphism, *Int. J. Clin. Pharmacol. Ther.* **40**:233–240.
45. P. I. Mackenzie, J. O. Miners, and R. A. McKinnon (2000). Polymorphisms in UDP-glucuronosyltransferase genes: Functional consequences and clinical relevance, *Clin. Chem. Lab. Med.* **38**:889–892.
46. J.-L. Fang and P. Lazarus (2004). Correlation between the UDP-glucuronosyltransferase (UGT1A1) TATAA box polymorphism and carcinogen detoxification phenotype: Significantly decreased glucuronidating activity against benzo(*a*)pyrene-7,8-dihydrodiol(–) in liver microsomes from subjects with the UGT1A1*28 variant, *Cancer Epidemiol. Biomarkers Prev.* **13**:102–109.
47. A. Orzechowski, D. Schrenk, B. S. Bock-Hennig, and K. W. Bock (1994). Glucuronidation of carcinogenic arylamines and their *N*-hydroxy derivatives by rat and human phenol UDP-glucuronosyltransferases of the *UGT1* gene complex, *Carcinogenesis* **15**:1549–1553.
48. J. K. Ritter, Y. Y. Sheen, and I. S. Owens (1990). Cloning and expression of human liver glucuronosyltransferase in COS-1 cells. 3,4-Catechol estrogens and estriol as primary substrates, *J. Biol. Chem.* **265**:7900–7906.
49. B. L. Coffman, C. D. King, G. R. Rios, and T. R. Tephly (1998). The glucuronidation of opioids, other xenobiotics, and androgens by human UGT2B7Y(268) and UGT2B7H(268), *Drug Metab. Dispos.* **26**:73–77.
50. E. Levesque, M. Beaulieu, M. D. Green, T. R. Tephly, A. Belanger, and D. W. Hum (1997). Isolation and characterization of UGT2B15(Y85): A UDP-glucuronosyltransferase encoded by a polymorphic gene, *Pharmacogenetics* **7**:317–325.
51. Z. Hu and P. G. Wells (1992). In vitro and in vivo biotransformation and covalent binding of benzo(*a*)pyrene in Gunn and RHA rats with a genetic deficiency in bilirubin uridine diphosphate-glucuronosyltransferase, *J. Pharmacol. Exp. Ther.* **263**:334–342.
52. Z. Hu and P. G. Wells (2004). Human interindividual variation in lymphocyte UDP-glucuronosyltransferases as a determinant of in vitro benzo[*a*]pyrene covalent binding and cytotoxicity, *Toxicol. Sci.* **78**:32–40.
53. T. Sugimura and S. Sato (1983). Mutagens-carcinogens in foods, *Cancer. Res. Suppl.* **43**:2415s–2421s.

54. K. Wakabayashi, M. Nagao, H. Esumi, and T. Sugimura (1992). Food-derived mutagens and carcinogens, *Cancer Res. Suppl.* **52**:2092s–2098s.

55. R. H. Adamson, J.-Å. Gustafsson, N. Ito, M. Nagao, T. Sugimura, K. Wakabayashi, and Y. Yamazoe (eds.) (1995). *Heterocyclic Amines in Cooked Foods: Possible Human Carcinogens.* Princeton Scientific Publishing,Princeton, N.J.

56. E. C. Miller and J. Miller (1986). Carcinogens and mutagens that may occur in foods, *Cancer* **58**:1795–1803.

57. T. Sugimura and K. Wakabayashi (1990). Mutagens and carcinogens in foods. In*: Mutagens and Carcinogens in the Diet*, M. W. Pariza, ed. Wiley-Liss, Hoboken, NJ. pp. 1–18.

58. E. M. P. Widmark (1939). Presence of cancer-producing substances in roasted food, *Nature* **143**:984.

59. M. Nagao, M. Honda, Y. Seino, T. Yahagi, and T. Sugimura (1977). Mutagenicities of smoke condensates and the charred surface of fish and meats, *Cancer Lett.* **2**:221–226.

60. B. Commoner, A. Vithayathil, P. Dolara, S. Nair, and P. Madyastha (1978). Formation of mutagens in beef and beef extracts during cooking, *Science* **201**:913–916.

61. J. S. Felton, M. G. Knize, N. H. Shen, B. D. Andresen, L. F. Bjeldanes, and F. T. Hatch (1986). Identification of the mutagens in cooked beef, *Environ. Health Perspect.* **67**:17–24.

62. T. Sugimura (1997). Overview of carcinogenic heterocyclic amines, *Mutat. Res.* **376**:211–219.

63. K. W. Turteltaub, K. H. Dingley, K. D. Curtis, M. A. Malfatti, R. J. Turesky, R. C. Garner, J. S. Felton, and N. P. Lang (1999). Macromolecular adduct formation and metabolism of heterocyclic amines in humans and rodents at low doses, *Cancer Lett.* **143**: 149–156.

64. R. J. Mauthe, K. H. Dingley, S. H. Leveson, S. P. Freeman, R. J. Turesky, R. C. Garner, and K. W. Turteltaub (1999). Comparison of DNA-adduct and tissue-available dose levels of MeIQx in human and rodent colon following adminstration of a very low dose, *Int. J. Cancer Res.* **80**:539–545.

65. R. C. Garner, T. J. Lightfoot, B. C. Cupid, D. Russell, J. M. Coxhead, et al. (1999). Comparative biotransformation studies of MeIQx and PhIP in animal models and humans, *Cancer Lett.* **143**:161–165.

66. R. Sinha, M. Kulldorff, W. H. Chow, J. Denobile, and N. Rothman (2001). Dietary intake of heterocyclic amines, meat-derived mutagenic activity, and risk of colorectal adenomas, *Cancer Epidemiol. Biomarkers Prev.* **10**:559–562.

67. J. S. Felton, M. G. Knize, N. H. Shen, P. R. Lewis, B. D. Anderson, J. Happe, and F. T. Hatch (1986). The isolation and identification of a new mutagen from fried ground beef: 2-amino-1-methyl-6-phenylimidazo[4,5-*b*]pyridine (PhIP), *Carcinogenesis* **7**:1081–1086.

68. H. Esumi, H. Ohgaki, E. Kohzen, S. Takayama, and T. Sugimura (1989). Induction of lymphoma in CDF1 mice by the food mutagen, 2-amino-1-methyl-6-phenylimidazo[4,5-*b*]pyridine, *Jpn. J. Cancer. Res. Gann.* **80**:1176–1178.

69. K. L. Dooley, L. S. Von Tungeln, T. Bucci, P. P. Fu, and F. F. Kadlubar (1992). Comparative carcinogenicity of 4-aminobiphenyl and the food pyrolysates, Glu-P-1, IQ, PhIP, and MeIQx in the neonatal B6C3F$_1$ male mouse, *Cancer Lett.* **62**:205–209.

70. L. H. Thompson, J. D. Tucker, S. A. Stewart, M. L. Christensen, E. P. Salazar, A. V. Carrano, and J. S. Felton (1987). Genotoxicity of compounds from cooked beef in repair-deficient CHO cells versus *Salmonella* mutagenicity, *Mutagenesis* **2**:483–487.

71. M. H. Buonarati, J. D. Tucker, J. L. Minkler, R. W. Wu, L. H. Thompson, and J. S. Felton (1991). Metabolic activation and cytogenetic effects of 2-amino-1-methyl-6-phenylimidazo[4,5-*b*]pyridine (PhIP) in Chinese hamster overy cells expressing murine cytochrome P4501A2, *Mutagenesis* **6**:253–259.

72. M. H. Buonarati, K. W. Turteltaub, N. H. Shen, and J. S. Felton (1990). Role of sulfation and acetylation in the activation of 2-amino-1-methyl-6-phenylimidazo[4,5-*b*]pyridine to intermediates which bind DNA, *Mutat. Res.* **245**:185–190.

73. L. O. Dragsted, H. Frandsen, R. Reistad, J. Alexander, and J. C. Larsen (1995). DNA-binding and disposition of 2-amino-1-methyl-6-phenylimidazo[4,5-*b*]pyridine (PhIP) in the rat, *Carcinogenesis* **16**:2785–2793.

74. L. Fan, H. A. J. Schut, and E. G. Snyderwine (1995). Cytotoxicity, DNA adduct formation and DNA repair induced by 2-hydroxyamino-3-methylimidazo[4,5-*f*]quinoline and 2-hydroxyamino-1-methyl-6-phenylimidazo[4,5-*b*]pyridine in cultered human mammary epithelial cells, *Carcinogenesis* **16**:775–779.

75. K. H. Dingley, K. D. Curtis, S. Nowell, J. S. Felton, N. P. Lang, and K. W. Turteltaub (1999). DNA and protein adduct formation in the colon and blood of humans after exposure to a dietary-relevant dose of 2-amino-1-methyl-6-phenylimidazo[4,5-*b*]pyridine, *Cancer Epidemiol. Biomarkers Prev.* **8**:507–512.

76. W. Zheng, D. R. Gustafson, R. Sinha, J. R. Cerhan, D. Moore, et al. (1998). Well-done meat intake and the risk of breast cancer, *J. Natl. Cancer Inst.* **90**:1724–1729.

77. F. G. Crofts, T. R. Sutter, and P. T. Strickland (1998). Metabolism of 2-amino-1-methyl-6-phenylimidazo[4,5-*b*]pyridine by human cytochrome P4501A1, P4501A2 and P4501B1, *Carcinogenesis* **19**:1969–1973.

78. R. Kato and Y. Yamazoe (1987). Metabolic activation and covalent binding to nucleic acids of carcinogenic heterocyclic amines from cooked foods and amino acid pyrolysates, *Jpn. J. Cancer Res.* **78**:297–311.

79. S. Ozawa, H.-C. Chou, F. F. Kadlubar, K. Nagata, Y. Yamazoe, and R. Kato (1994). Activation of 2-hydroxyamino-1-methyl-6-phenylimidazo[4,5-*b*]pyridine by cDNA-expressed human and rat arylsulfotransferases, *Jpn. J. Cancer Res.* **85**:1220–1228.

80. C. D. Davis, H. A. J. Schut, and E. G. Snyderwine (1993). Enzymatic phase II activation of the *N*-hydroxylamines of IQ, MeIQx and PhIP by various organs of monkeys and rats, *Carcinogenesis* **14**:2091–2096.

81. J. Alexander, H. Wallin, O. J. Rossland, K. Solberg, J. A. Holme, G. Becher, R. Andersson, and S. Grivas (1991). Formation of a glutathione conjugate and a semistable transportable glucuronide conjugate of N2-oxidized species of 2-amino-1-methyl-6-phenylimidazo[4,5-*b*]pyridine (PhIP) in rat liver, *Carcinogenesis* **12**:2239–2245.

82. K. R. Kaderlik, G. J. Mulder, R. J. Turesky, N. P. Lang, C. H. Teitel, M. P. Chiarelli, and F. F. Kadlubar (1994). Glucuronidation of *N*-hydroxy heterocyclic amines by human and rat liver microsomes, *Carcinogenesis* **15**:1695–1701.

83. M. Buonarati, M. Roper, C. Morris, J. Happe, M. Knize, and J. Felton (1992). Metabolism of 2-amino-1-methyl-6-phenylimidazo[4,5-*b*]pyridine (PhIP) in mice, *Carcinogenesis* **13**:621–627.

84. B. E. Watkins, M. Suzuki, H. Wallin, K. Wakabayashi, J. Alexander, M. Vanderlaan, T. Sugimura, and H. Esumi (1991). The effect of dose and enzyme inducers on the metabolism of 2-amino-1-methyl-6-phenylimidazo[4,5-*b*]pyridine (PhIP) in rats, *Carcinogenesis* **12**:2291–2295.

85. B. P. Murray, R. J. Edwards, S. Murray, A. M. Singleton, D. S. Davies, and A. R. Boobis (1993). Human hepatic CYP1A1 and CYP1A2 content, determined with specific anti-peptide antibodies, correlates with the mutagenic activation of PhIP, *Carcinogenesis* **14**:585–592.

86. G. J. Hammons, D. Milton, K. Stepps, F. P. Guengerich, R. H. Tukey, and F. F. Kadlubar (1997). Metabolism of carcinogenic heterocyclic and aromatic amines by recombinant human cytochrome P450 enzymes, *Carcinogenesis* **18**:851–854.

87. P. B. Styczynski, R. C. Blackmon, J. D. Groopman, and T. W. Kensler (1993). The direct glucuronidation of 2-amino-1-methyl-6-phenylimidazo[4,5-*b*]pyridine (PhIP) in human and rabbit liver microsomes, *Chem. Res. Toxicol.* **6**:846–851.

88. M. A. Malfatti, K. S. Kulp, M. G. Knize, C. Davis, J. P. Massengill, et al. (1999). The identification of [2-^{14}C]2-amino-1-methyl-6-phenylimidazo[4,5-*b*]pyridine metabolites in humans, *Carcinogenesis* **20**:705–713.

89. K. S. Kulp, M. G. Knize, M. A. Malfatti, C. P. Salmon, and J. S. Felton (2000). Identification of urine metabolites of 2-amino-1-methyl-6-phenylimidazo[4,5-*b*]pyridine following consumption of a single cooked chicken meal in humans, *Carcinogenesis* **21**:2065–2072.

90. S. A. Nowell, J. S. Massengill, S. Williams, A. Radominska-Pandya, T. R. Tephly, et al. (1999). Glucuronidation of 2-hydroxyamino-1-methyl-6-phenylimidazo[4, 5-*b*]pyridine by human microsomal UDP-glucuronosyltransferases: Identification of specific UGT1A family isoforms involved, *Carcinogenesis* **20**:1107–1114.

91. M.-F. Yueh, N. Nguyen, M. Famourzadeh, C. P. Strassburg, Y. Oda, P. F. Guengerich, and R. H. Tukey (2001). The contribution of UDP-glucuronosyltransferase 1A9 on CYP1A2-mediated genotoxicity by aromatic and heterocyclic amines, *Carcinogenesis* **22**:943–950.

92. M. A. Malfatti and J. S. Felton (2001). *N*-Glucuronidation of 2-amino-1-methyl-6-phenylimidazo[4,5-*b*]pyridine (PhIP) and *N*-hydroxy-PhIP by specific human UDP-glucuronosyltransferases, *Carcinogenesis* **22**:1087–1093.

93. M. A. Malfatti and J. S. Felton (2004). Human UDP-glucuronosyltransferase 1A1 is the primary enzyme responsible for the N-glucuronidation of *N*-hydroxy-PhIP in vitro, *Chem. Res. Toxicol.* **17**:1137–1144.

94. F. F. Kadlubar, M. A. Butler, K. R. Kaderlik, H.-C. Chou, and N. P. Lang (1992). Polymorphisms for aromatic amine metabolism in humans: relevance for human carcinogenesis, *Environ. Health Perspect.* **98**:69–74.

95. R. J. Turesky (2002). Heterocyclic aromatic amine metabolism, DNA adduct formation, mutagenesis, and carcinogenesis, *Drug Metab. Dispos.* **34**:625–650.

96. M. A. Malfatti, E. A. Ubick, and J. S. Felton (2005). The impact of glucuronidation on the bioactivation of the cooked-food carcinogen 2-amino-1-methyl-6-phenylimidazo [4,5-*b*]pyridine *in vivo*, *Carcinogenesis* **26**:2019–2028.

97. C. G. Dietrich, R. Ottenhoff, D. R. de Waart, and R. P. J. Oude Elferink (2001). Lack of UGT1 isoform in Gunn rats changes metabolic ratio and facilitates excretion of the food-derived carcinogen 2-amino-1-methyl-6-phenylimidazo[4,5-*b*]pyridine, *Toxicol. Appl. Pharmacol.* **170**:137–143.

98. M. A. Malfatti, R. W. Wu, and J. S. Felton (2005). The effect of UDP-glucuronosyltransferase 1A1 expression on the mutagenicity and metabolism of the cooked-food carcinogen 2-amino-1-methyl-6-phenylimidazo[4,5-*b*]pyridine in CHO cells, *Mutat. Res.* **570**: 205–214.

16

THE INFORMATICS AND BIOINFORMATICS INFRASTRUCTURE OF A NUTRIGENOMICS BIOBANK

Warren A. Kibbe

Director of Bioinformatics, Robert H. Lurie Comprehensive Cancer Center, Center for Genetic Medicine, and NIH Center for Neurogenomics at Northwestern University, Chicago, Illinois

16.1 INTRODUCTION

Biobanking is one of those intriguing activities that is so simple in concept but can be so difficult in practice. Pharmacogenomics and nutrigenomics biobanks in particular present some difficult issues such as properly gathering and annotating disease and nutrition information for each participant, designing consents and procedures (and obtaining proper approvals) for discovery banks (versus a hypothesis driven bank), and navigating the ethical, regulatory, funding, and intellectual property ramifications of a genetically focused bank. Add to the mix the complexities of mining medical records, maintaining patient confidentiality, providing robust sample handling and tracking, and finally providing robust phenotype mining tools and analytics for genotype/phenotype association studies and you get a sense of the potential complexities underlying the establishment of a nutrigenomics biobank. However, a single nutrigenomics biobank or a coalition of biobanks focused on nutrigenomics has an enormous potential for contributing to our understanding of the complex interactions between our environment, our genetics, and our diet.

Nutritional Genomics: Discovering the Path to Personalized Nutrition
Edited by Jim Kaput and Raymond L. Rodriguez

Although from the title of this chapter you might expect a set of "best practices" for building the core infrastructure necessary for a biobanking initiative, I have tried to highlight open initiatives as well as to describe the design principles and philosophies that we have used in building our informatics infrastructure at Northwestern University. We have tried to anticipate change as much as possible, as computing technologies, the regulatory environment, and our use of, application of, and access to technology will continue to evolve. I have collected our thoughts about building the necessary systems and discuss our implementation strategies behind several tissue repositories and our population-based DNA banking initiative, NUgene. I will also bring in examples from the NCI's caBIG initiative, of which Northwestern is a member in several workspaces, including the clinical trials, architecture, and tissue banking/pathology workspaces. The key driving force behind caBIG is the creation of set of recommendations and standards for interoperability that will enable a national informatics platform for basic science and translational and clinical research for the cancer community. Clinical trials, biobanks, and molecular markers including genetic information are explicitly highlighted as direct targets for caBIG collaborations. Interoperability, enterprise vocabulary, modularity, and defined data elements are all hallmarks of the caBIG architecture and these principles have important implications for any biobank.

Before going further in defining how to build a successful infrastructure for a biobank, defining what I mean by a biobank is important. At Northwestern, a biobank collects samples into a *repository* under a specific consent that allows, in general, patient follow-up, contact for follow-on studies, and the use of specimens in the biobank for ancillary studies. A biobank also has one or more *data collection* instruments, with these data relating to one or more samples and with data collection occurring at one or more sites. In contrast, the repository may contain samples derived from different types of specimens, such as blood or tissues, processed samples such as DNA or tissue microarrays, and the specimens and/or samples may be stored differently, such as paraffin embedded blocks or fresh-frozen material. Typically, only some samples in a repository are available for further analysis or inclusion in other studies. Another way to describe a biobank is that it is a collection of samples in one or more repositories that are consistently annotated and consented, whereas a repository is a physical location of samples that may be coming from differing studies and with differing consents and annotations. In a repository (or a biobank) it is possible that some of the material is anonymous. From a regulatory perspective, it is very important that any identified sample have a signed consent linked to one or more IRB approved protocols that clearly state the allowed uses for the sample. For samples collected after HIPAA went into effect, each participant must also sign a HIPAA authorization form, and at least in the Robert H. Lurie Comprehensive Cancer Center at Northwestern all studies use the same HIPAA authorization form. Tracking regulatory approvals, including consent, retrospective tissue collection on an IRB-approved protocol, and other aspects of tracking levels of consent will be discussed in some detail and strategies presented for maintaining this information. Another important issue for a nutrigenomics biobank is how to provide the users of the biobank with the appropriate de-identified data that enables research while protecting patient confidentiality. From a nutrigenomics perspective, there is additional information regarding diet, diet history, and nutritional information. By necessity, these data are self-reported, and the nature of these data must be taken into account during analysis. For some good guides on designing these data collection forms, see [1–15]. I will discuss how these forms can be incorporated into the data model in a flexible yet informative manner, enabling researchers to make better associations between the self-reported data and trends in diet, nutritional behavior, disease outcomes, and genotype.

As you have probably read, many academic biobanks and repositories have gotten bad reviews in the past few years [1, 8–15], particularly as older tissue-based repositories have been repositioned as genetic repositories. In concept, biobanks are very straightforward—they are repositories of biological specimens coupled with important identifiers and clinical data that enable clinical and translational research. However, it is very important to realize that most biobanks really started off focused on a single hypothesis rather than focused on discovery, and that most biobanks have a very limited set of data describing either the state of the sample or the clinical history of the patient from whom a specimen was taken. For instance, at Northwestern we have an extensive set of prostate tissue and prostate fluid taken from prostate cancer patients and normal controls. We have more than 10,000 specimens dating back to the late 1960s; however, the potential value of these samples is compromised by the lack of clinical and pathological data that was captured at the point of collection of these samples. Both fortunately and unfortunately, the majority of these samples are de-identified and there is no way to further characterize the samples. Likewise, for many of our disease-specific banks, we have extensive characterization of the participants for that particular disease, but relatively little information on disease history (other than for the specific disease under examination), treatments received, treatment response, and adverse events to therapeutic agents. From a genetic investigation standpoint, the ability to correlate genotype with a breadth of clinical and epidemiological data, including diet history and nutritional correlates, will enable entirely new avenues of discovery, prediction, and intervention for health care. The promise of "personalized medicine" to revolutionize human health care cannot be overstated, although it can be overhyped, oversold, and the complexities oversimplified. Like any avenue of scientific discovery (or any other human endeavor!), it has limitations and the results can be misunderstood and misapplied. In part because of the misleading implications of the term "personalized medicine," we prefer the term "genetic medicine," as the results will be based on large genetic association studies and will enable health professionals to choose the appropriate, and just as importantly exclude the inappropriate, therapy and diet based on appropriate genetic and environmental factors. The coming wave of nutrigenomics and pharmacogenomics research that will utilize the next generation of biobanks will unravel complex interactions between multiple haplotypes, long-term health outcomes, diet, and other health variables in large cohort studies. Although single gene studies are still the norm, we are involved in a number of studies involving known pathways, or genes encoding gene products known to interact with a specific pharmaceutical or metabolite. However, for these studies to reach full potential, whole-genome, or at least detailed haplotype, analysis needs to be performed and associated with the clinical and epidemiological data. The downside of this approach is the cost, complexity, and regulatory hurdles that need to be overcome to initiate and complete these studies. For that reason, it is imperative that collections initiated by different organizations be able to combine results. Several biobanking initiatives are now underway with these objectives explicitly in mind. There are several relevant discussions of how best to approach these important issues [16, 17].

16.2 NEXT GENERATION BIOBANKS

We believe that at Northwestern University we have a "next generation" biobank, NUgene. For next generation biobanks, there is a focus on discovery and mechanisms for enriching the sample annotations with time. What differentiates NUgene from other biobanks at Northwestern is that after the point of consent and collection of self-reported data and blood sample, we both retrospectively and prospectively mine patient electronic medical records (eMRs), and as genotyping methodologies become cheaper and endpoints better defined, we will refine the genotype of the participants. Thus, rather than samples and data being "point in time" and relevant only for a particular clinical question, the richness of the clinical phenotype and genotype associated with each participant will continue to grow. We believe that continued annotation is a necessary attribute for any next generation biobank. Other attributes for a next generation genetic biobank include providing a platform for discovery of new genetic associations, the incorporation of finely detailed phenotype information from correlative or ancillary studies, and providing a mechanism for building an arbitrarily complex (of arbitrary granularity) representation of phenotype. For these attributes to contribute to the overall value of the biobank and in particular for either one biobank to span institutions or for multiple biobanks to interoperate, annotations have to be done using a set of well-characterized vocabulary, and ideally each vocabulary element maps onto one or more ontologies. By building on common data elements from a controlled vocabulary that can be mapped onto an ontology, the semantic meaning of the sample annotations can be mapped and overlaps found using standard graph theory. The benefits of this approach are a minimization of the missing data problem and the ability to compare two arbitrarily granular but related facts. There are some excellent examples of using Gene Ontology terms to annotate gene products with the biological process, molecular function, and cellular components based on experimental and computational evidence. These annotations have been very useful in understanding evolutionary and biological relationships in an organism and for comparative analysis across species. The community of Model Organism Databases have used the features of a shared vocabulary in the form of the Gene Ontology to share data and biological knowledge between organisms, and this has enabled entirely new areas of research. For clinical phenotype relationships, researchers at Northwestern have taken a complementary approach in organizing and annotating diseases into an ontology called Disease Ontology, freely available at http://diseaseontology.sourceforge.net/. Disease Ontology was initially conceived of and designed by Rex Chisholm, the PI and Founder of the NUgene Project and also an early member of the Gene Ontology Consortium [18]. Not surprisingly, this ontology has been used to organize data from billing records and electronic medical records to build a "Disease Phenotype" for participants in the NUgene Project.

The *N*orthwestern *U*niversity *gene* bank (NUgene) Project is a large population-based DNA bank at Northwestern University. The design of the biobank and the associated informatics are in alignment with the NIH's goal of translating information from the human genome sequence into clinically relevant disease knowledge to better human health [16, 17]. NUgene combines a centralized genomic DNA sample collection and storage system with a broad-based consent that enables NUgene to update participants' health status with periodic data updates from electronic medical records (eMRs) [19, 20]. The purpose of NUgene is to provide an ethnically and medically diverse population of samples and associated medical information to facilitate the association of specific gene variants with disease and therapeutic outcomes. In concept at least, the associated data could include dietary and nutritional variables.

The NUgene Project is one of the initiatives of the Center for Genetic Medicine at Northwestern University in collaboration with Northwestern's health-care affiliates, Northwestern Memorial Hospital, the Northwestern Medical Faculty Foundation, Children's Memorial Hospital, and Evanston Northwestern Healthcare. Together these institutions provide health-care to over 2 million Chicago residents. An important feature of these health care providers is their commitment to the use of electronic medical records systems. The joint commitment to the electronic data capture of clinical care has enabled the vision of an automated large-scale genetic bank to take shape. The goal of NUgene is to enroll 100,000 volunteers from the Chicago population. We seek to achieve a population similar to that found in the metropolitan area at large to assure representation of a wide range of ethnic and cultural groups.

16.3 INTENDED AUDIENCE FOR THIS CHAPTER

The focus of this chapter is on the infrastructure required to manage a biobank and house the data surrounding the samples housed in the bank. The association and analysis of data acquired and aggregated in the biobank is outside the scope of this chapter. I will also assume that you have reviewed the existing state of the art for open source and commercial biobanking informatics offerings and are interested in building your own biobanking infrastructure, either using some of these systems as building blocks or as examples for your own system. Assuming that your institution/group has the proper expertise and motivation to create or assemble the necessary infrastructure, you will get a better product that conforms more precisely to your needs than you can get with the current generation of commercial products. However, products that are focused on providing flexible workflow management tools that are data and database agnostic are rapidly maturing, and the incorporation of some of these tools for managing the work and data flow for a biobank should be seriously considered. If you have not looked at the existing tools, please see the following web sites or google these companies/products:

Open Source Commercial Companies [21]

Visitrial http://www.visitrial.com/

PhOSCo http://www.phosco.com/

OpenClinica by Akaza Research http://www.openclinica.org/

Closed Source Commercial Companies

Ardais http://www.ardais.com/

PHT http://www.pht.com/

phaseForward http://www.phaseforward.com/

Oracle Clinical http://oracle.com/

Standards

Clinical Data Interchange Standards Consortium http://www.cdisc.org/

Health Level Seven http://www.hl7.org/

Medical Dictionary for Regulatory Activities http://www.fda.gov/medwatch/report/meddra.htm

16.4 ASSUMPTIONS, USE CASES, AND DESIGN CRITERIA

For clarity, we will develop a "prototypical" biobank, with the following attributes, most of which are modeled on the NUgene Project:

- Consent and blood samples will come from a single point in contact, although consent could be a multiple-step process.
- The consent enables automated and ongoing mining of patient electronic medical records, including billing records and lab results.
- Data collection forms may be at the time of collection or from further contacts or ancillary studies involving a subset of the biobank participants.
- Blood specimens will be collected, processed into DNA, and stored.
- DNA samples will be used to generate genetic information. This information will be in the form of DNA sequence, SNP markers, or haplotypes, and these results will become part of the biobank informatics respository.
- Quality of the DNA extraction process can be monitored.
- All sample handling, survey forms, and consent forms are independently barcoded and the identity protected and the workflow trail tracked.
- All tubes including the blood samples, intermediate tubes for DNA extraction, DNA samples, aliquots, and shipping containers will all be barcoded and location tracked for identity, audit trail, and quality purposes.

- Automated acquisition of clinical, dietary, and epidemiological variables from electronic medical records from health-care affiliates will be employed.
- Data collection forms for self-reported information are likely to have both universal and clinic-based variables.
- Automated acquisition and integration of lab data concerning participants will be employed.
- Automated acquisition and integration of billing data concerning participants will be employed.
- All data acquisition will run through a cleaning and de-identification process. The identification process removes Protected Health Information (HIPAA defined identifiers collectively called PHI) but associates clinical, billing, dietary, and epidemiological data with the proper DNA sample. The individual associations are still present but PHI has been removed.
- Association studies and analysis will be performed on de-identified data.
- Samples and annotations can be selected based on clinical, epidemiological, and genetic variables. Annotations will be kept so that participant samples can be chosen for controls based on evidence that a participant does not have or is unlikely to have a condition. Inherent in this selection is an estimation of the likelihood of missing clinical data. That is, if there is only limited clinical data available for a particular patient, it is harder to rule out that the patient does not have a disease. Conversely, if there are billing and eMR data indicating a participant receives primary care at the institution and the data goes back many years, the absence of a diagnosis is more compelling that the patient does not have a disease.
- A given sample may have additional data aggregated that comes from ancillary or follow-on studies, and the origin of these data are all tracked and audited.
- Genotype data that is redeposited in the biobank is reassociated securely and accurately with the proper DNA sample based on the barcode/aliquot ID of "shipped" DNA samples. For the purpose of this reassociation, even in-house genotyping or sequencing is treated as "shipped" from a workflow and sample auditing perspective, since it requires handling the DNA samples.
- Access to any data in the system is audited.
- Access to any data requires secure authentication.
- Access to any data in the system is role based, and access to a given data element requires that an individual have the proper role(s) assigned.
- Variables in the system have detailed metadata; when possible, these variables come from common data elements and standard vocabularies defined in systems such as the NCI Enterprise Vocabulary System, the NCI caDSR, ICD-9, SNOMED, HL7, and other messaging or data exchange formats.
- Disease phenotype is annotated using Disease Ontology, with the evidence behind the assignment tracked, auditable, and able to be further curated.

16.5 REGULATORY AND POLICY ENVIRONMENT

CFR 21 Part 11—Electronic Signatures

The government regulations surrounding CFR 21 Part 11 (electronic signatures and electronic audit trails) are designed to help organizations understand what features are necessary for computerized systems to replace paper audit trails and signatures. The guiding principles behind these regulations are:

1. Individuals signing in to a system are uniquely identified.
2. There are procedures in place for granting and revoking privileges of individuals to the system.
3. There are clear guidelines for training and proper use of the system.
4. All activities that can create or change data and import or export data in the system are logged/tracked by user and date. This entails a complete audit trail, showing who has been in the system and each record accessed, created, changed, approved, or marked as deleted. It is important from an auditing perspective that no data ever be deleted, merely marked as deleted. From the user perspective this results in those data being removed from the interface and removed from standard reports.

In addition to these regulations, there are numerous other regulations covering the details of what needs to be audited and the length of time these records must be kept. At least in the pharmaceutical industry, the procedure is to destroy documents and audit trails as soon as the requirements for retention have expired. For many clinical research databases, these trails may be destroyed after an appropriate length of time, but the underlying documents are kept in perpetuity. For biobanks, these documents correspond to all the data associated with specimens in the biobank, and these would naturally be kept for the life of the bank.

16.6 HIPAA—HEALTH INSURANCE PORTABILITY AND ACCOUNTABILITY ACT OF 1996

There are two major aims of the Health Insurance Portability and Accountability Act of 1996. Title one of HIPAA protects health insurance coverage for workers and their families when they change or lose their jobs. There are numerous rules that cover preexisting conditions and portability of health insurance coverage. Title II of HIPAA, The Administrative Simplification provisions, requires the Department of Health and Human Services to establish national standards for electronic health-care transactions and national identifiers for providers, health plans, and employers. There are components of Title II that also address the security and privacy of health data. One of the hopes of HIPAA was that the adoption of these standards would improve the efficiency and effectiveness of the nation's health-care system by encouraging the widespread use of electronic data interchange in health care. A by-product of the way that most organizations have interpreted HIPAA is that researchers not directly involved in the day-to-day care of patients cannot approach patients for participation in research studies. Some guidance that has come out around HIPAA includes:

- *Ownership*: While the health-care organization owns the health record, the information in that record remains the patient's personal property.
- *Confidentiality*: Institutional policies on confidentiality and release of information must be consistent with state and federal regulations.
- *Access*: Within health-care organizations, personal information contained in medical records is reviewed not only by physicians and nurses, but also by professionals in areas such as social work, case management, rehabilitation, pharmacy, accounting, and quality assurance.
- *Third Party/Business Partners*: The need to access by external parties has escalated dramatically and include attorneys, employers, media representatives, government agencies, and third-party payers. These relationships must be codified and documented.
- *Privacy*: Very sensitive information requires special security, including psychiatric, genetic, HIV status, and substance abuse treatment information.
- Patients should be ensured access to their own records.

HIPAA is a broad topic and has many aspects that affect the design and development of systems behind a biobank. Some of the implications include use of identifiers, coding standards for content and structure, messaging standards for interoperability and federation of physically, organizationally, or conceptually distinct systems, and of course confidentiality, data security, authentication, and authorization. Some additional guidance comes from the DHHS in the form of the National Health Information Infrastructure (NHII) meeting each year. A large focus for this meeting is providing recommendations for developing architecture and standards that comprise the NHII. Some of these recommendations include the adoption of HL7 Version 3 as a messaging standard and the content/vocabularies/coding systems of DICOM, SNOMED, and LOINC as appropriate, and the continuity of health records over a patient's lifetime. The latter recommendation has broad, positive implications for the aggregation of patient data for uses like a nutrigenomics biobank, as well as implications for a more standardized infrastructure for the authorization of health-care and research organizations to have access to an individual's health records. Due to the sensitivity and broad implications of this recommendation, a national health record is unlikely to emerge in the short term. An earlier attempt to create community health information networks (CHINs) by creating Local Health Information Infrastructure Initiatives failed for the following reasons [22]:

- Lack of buy-in because of conflicting missions and poorly conceived objectives by the CHIN members.
- Perceived loss of control and lack of trust in the process.

- Lack of clear ownership over data systems and information.
- Lack of a financial model for supporting the CHIN.
- Lack of adequate technology along with the perceived need for a centralized community-based data repository.

It is very important to realize that although the environment for creating a NHII has changed, some of these barriers to building distributed networks for sharing health information have not changed. However, many of the standards that have been adopted and are being adopted by the community will enable these networks to be built with little additional infrastructure, but will still require an institutional will to share and collaborate.

16.7 GMPs, GLPs, AND GCPs

While incorporating good practices in the setup and the running of a biobank may seem to be overdesign, it will ensure the maximum utility of the bank by the broader community. By not incorporating good practices and engaging members of the clinical trials and pharmaceutical industries, the end uses of the specimens collected by the bank may be of limited value. For instance, for NUgene we found that we needed physically separate spaces for our DNA extraction and handling operation and our genotyping and RT-PCR equipment, as the FDA (and the pharmaceutical industry) is very concerned about cross-contamination of DNA during storage. We had anticipated these concerns with RNA but had not realized the same issues would surface with DNA cross-contamination.

16.8 FUNDING OF BIOBANKS

NIH or other government sources for the funding of biobanks have relatively few programs for the funding of broadly focused biobanks. Recent recommendations by joint NAS and NAE panels [23] have identified the lack of funding for large-scale projects and the importance of large-scale projects in the continued success of life science research. The report also highlighted the need for increased funding for bioinformatics and information dissemination. The recommendations discussed the importance of interinstitutional collaborations and the availability and distribution of information and reagents, a category that certainly includes DNA and genetic biobanks.

16.9 BIOBANKING IN CLINICAL TRIALS

NUgene is an IRB approved population-based DNA bank that enables automated mining of participants' medical records and other institutional information sources including billing records and laboratory results. In our current thinking, NUgene provides an ideal source for controls for specific clinical trials as well as a recruitment tool for identifying patient populations meeting specific target criteria. In order for NUgene to efficiently and effectively develop a detailed picture of participants' clinical histories, we will be mining coded data present in the electronic medical record for each participant. It is very important that this process be secure and that the participants' identifiers be protected (confidentiality maintained) and as automated as possible.

16.10 DATA STANDARDS/SEMANTIC INTEROPERABILITY

The NCI and the FDA are reviewing interoperability as part of the Interagency Oncology Task Force. The goals of this task force are to streamline processing for electronic submissions of Investigational New Drugs (INDs), New Drug Applications (NDAs), Biologics License Applications (BLAs), and amendments including serious adverse events as well as providing capabilities for a clinical investigator repository (based on the NCI investigator repository) and work with the FDA's Critical Path Initiative.

Underlying much of the NCI interoperability movement is the development of Common Data Elements and the technology necessary to make them available between systems and institutions. Much of the technology is available through the caCORE, which has a software developers kit (SDK) and three main components—caDSR, EVS, and caBIO. The caDSR is best thought of as the repository for the metadata surrounding a particular data element or vocabulary structure. caDSR implements ISO/IEC 11179 Parts 1–6, which describe the specification and standardization of data elements, in particular, providing standard methods to convey semantic, syntactic, and lexical meaning. The caDSR provides a metamodel for "data element" metadata. Data elements in the caDSR are both human and machine understandable and can be unambiguously interpreted. Data from caCORE can be accessed using standard formats such as OWL (Ontological Web Language) and RDF, and standard activities such as the Life Science Identifiers (LSIDs) can also be represented in caCORE [4]. The relevant ISO 11179 standard parts are listed below (ISO 2000):

ISO/IEC 11179 Part 1: Framework for the Specification and Standardization of Data Elements

ISO/IEC 11179 Part 2: Classification for Data Elements

ISO/IEC 11179 Part 3: Registry Metamodel and Basic Attributes

ISO/IEC 11179 Part 4: Rules and Guidelines for the Formulation of Data Elements

ISO/IEC 11179 Part 5: Naming and Identification Principles for Data Elements

ISO/IEC 11179 Part 6: Registration of Data Elements

In the design of one aspect of caCORE [6], the caDSR, some of the design principles were:

- Goal: "Semantically unambiguous, interoperability."
- Data Element curators are not necessarily vocabulary experts.
- NCI had a terminology and vocabulary services group: EVS.
- Semantic integration is achieved by tying Standard vocabulary identifier codes to the caDSR metadata.
- The ISO 11179 provides the framework.

The Enterprise Vocabulary Service [7], or EVS, is a terminology server

EVS provides services for synonymy, mapping between vocabularies, hierarchical structures, Subconcepts, Superconcepts, Roles, Semantic type, and so on.

All of caCORE is available as an open source download. One of the key components of the caCORE is the caDSR Toolbox. This toolbox helps developers manage CDEs, metadata, and synchronizing and accessing of CDEs. Some of the caDSR toolbox features are:

Simplifies development and creation of ISO/IEC 11179 compliant metadata by Data.

Simplifies consumption of Data Elements by end users and application developers.

Enhances reuse of Data Elements for all caDSR users.

Enables semantic consistency across research domains.

Supports metadata life cycle and governance processes.

CDE Curation Tool allows nonontology savvy users to Create Data Elements.

Admin Tool allows one to Curate and Administer caDSR—"Power Users."

Provides aframework for storing and deploying any UML Structure.

Sentinel Tool (3.0) generates end user "Alerts" triggered by metadata changes.

Batch Load features the import of items—Excel Loader (MS Excel), UML Loader (XMI), Case Report Form Loader (MS Excel).

The ISO 11179 Wizard allows the construction of ISO compliant Data Elements by building up the pieces. The Wizard builds Names and Definitions from underlying components. The Wizard leverages the ISO structure to enable the sophisticated retrieval related CDEs.

One of the more interesting features of both ISO 11179 and the caDSR is that it provides a framework for storing and deploying any UML structure.

16.11 OTHER STANDARDS BODIES: CDISC

CDISC has several major sections, and a working team devoted to each section. The teams are:

The Operational Data Modeling Team (ODM)
Submissions Data Standards Team (SDS)
Analysis Dataset Model (ADaM) Team
Laboratory Standards Team (LAB)
Joint HL7/CDISC Protocol Representation Group
The CDISC define.xml Team

CDISC is now working closely with HL7 to provide subject matter expertise to HL7 for laboratory and clinical protocol subjects. CDISC has published a number of reference documents for implementing the CDISC standard. The documents are available at the main CDISC web site, http://www.cdisc.org/. These include documents defining the current implementation of the CDISC Operation Data Model (developed by the ODM) and the Study Data Tabulation Model (SDTM) (developed by the SDS). In particular, there is a Submission Data Domain Standards document that contains two sections: the Study Data Tabulation Model (SDTM), which represents the underlying conceptual model behind the SDS standards, and the CDISC SDTM Implementation Guide (SDTM-IG), which includes the detailed domain descriptions, assumptions, and examples. The Analysis Dataset Model (ADaM) Team has also published a document, *Statistical Analysis Dataset Model: General Considerations Version 1.0*, that provides additional guidelines for implementation of the CDISC standards and incorporation of CDISC into analysis datasets, and specifies the general content, structure, and metadata for Analysis Datasets. Each of the existing four CDISC statistical models (Change from Baseline, Survival Analysis, Categorical and Linear Models) are discussed. Other statistical models, including Subject Baseline Characteristics and Adverse Events are under development. Each model has additional documentation that provides additional guidance and examples for the specific statistical methodology.

The CDISC LAB team has released a document for working with the ECG extension of the LAB model. This model complements the HL7 XML ECG Waveform standard by providing for the additional transfer of details on the interpretations and measurements made during the analysis of an ECG Waveform.

HL7/CDISC's Protocol Representation (PR) Group has jointly released the documentation for Standard Protocol Elements for Regulated Clinical Trials. The two element lists, along with descriptive documentation, define key elements of clinical protocols, as identified by the PR Group thus far. This version is based on the ICH guidance for good clinical practice, with special emphasis on ICH E6, E3, and E9. See also the HL7 Clinical Document Architecture model (www.hl7.org). Case Report Tabulation Data Definition Specification (define.xml) includes recommendations from the Health Level 7 (HL7) Regulated Clinical Research Information Management Committee (RCRIM) as well as the CDISC ODM.

16.12 INFORMATICS INFRASTRUCTURE

The minimal underlying IT infrastructure that a biobank requires is a high-speed secure network (if operating on the open Internet, then the servers must be behind a firewall), a file server for storing data coming from lab equipment if sequencing or genotyping are performed in house, barcode scanners and readers, a robust and secure database server, and desktop or laptop computers for data coordinators and research staff. In the case of NUgene, our infrastructure consists of a production data center with an Oracle database server (if your institution does not have Oracle, PostGRES is a very good open source alternative), gigabit Ethernet, a middleware server for the web server/ColdFusion (if you wish to use open source, PHP, Python, or Tomcat/Java are all good choices), a firewall, and a backup server for backing up the database and the middleware server. In addition to the production data center, we also have a second server room located in a different building with similar capabilities and configuration, including a database server, middleware servers, backup devices, and firewall, providing both off-site storage as well as failover capability. All user connections to the database/NUgene informatics system are made through https (128 bit SSL).

Specific NUgene Infrastructure

The database is Oracle 10 g, the database server is a dual processor 2.4 GHz Xeon (2 MB L2 cache) with 12 GB of RAM Dell 6450, and the middleware server is a dual processor 2.2 GHz Xeon with 4 GB RAM Dell 2850. The file and backup servers are Dell 2650s or 2850s, with SCSI attached Arena IDE RAID units with >5 TB of RAID 5 formatted storage. The firewalls are Watchguard Firebox ×1000s and the gigabit switches are unmanaged Dell or Netgear. Backups are performed using Dantz Retrospect to online storage. Archival copies are made either to Quantum SDLT tapes or burned to DVD-R.

16.13 SYSTEM ARCHITECTURE

A system architecture that allows the clean separation of the data model, the business/scientific logic, and the interface is desirable and should be a design goal. The ability to build a highly secure, flexible, HIPAA compliant, CFR 11 Part 21 compliant system requires good design practices, well-documented processes for the users, and integrated auditing. We have approached this using what is now an accepted and standard model for decomposing the design into modular components, building an extensive testing suite for each component that becomes the statement of work for each further design and coding task, and then building the components using a

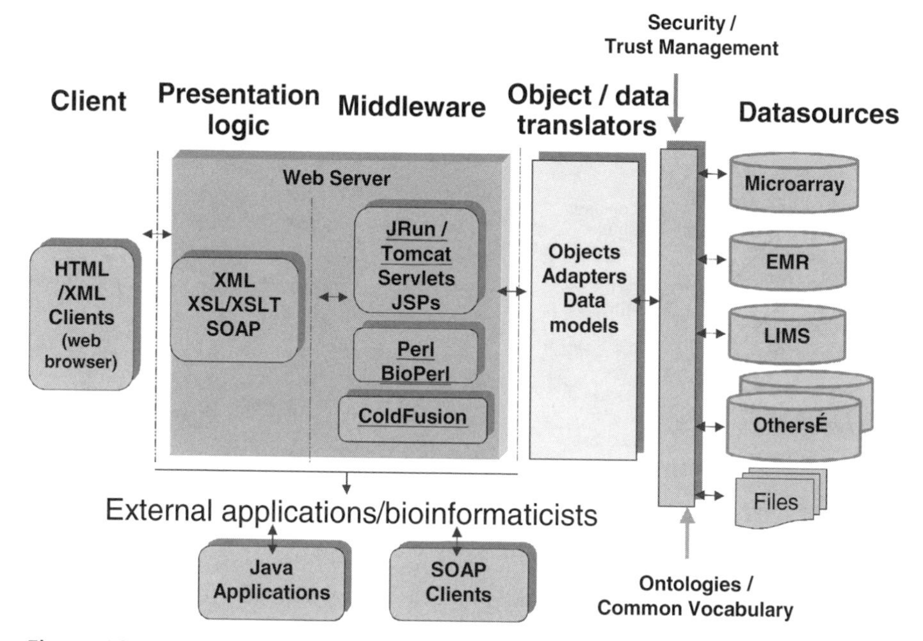

Figure 16.1. Implementation framework showing the connections between existing data sources, the security authentication layer, including interapplication security (enabling single sign-on), common vocabularies, and the middleware components.

standard IDE, code repository (currently CVS, which we are transitioning to Subversion), and build scripts (Ant). A generic structure for this architecture is shown in Figure 16.1.

For programmatic interoperability, the specific layers should use common data standards such as SNOMED and common exchange standards such as HL7. To improve user level interoperability and the user experience, interfaces can either emit data as XML/web services or provide html files that are web browsable or can be saved as Microsoft Excel files. At least in the case of our architecture, these Excel files are actually HTML-formatted files. The latest version of Excel (and many other programs) can now load and be used to view and manipulate XML files.

16.14 SEPARATION OF THE CLINICAL TRIAL/PATIENT IDENTITY MANAGEMENT FROM THE GENOTYPE/PHENOTYPE REPOSITORY

In order to build a de-identified data repository where de-identified phenotype data (the clinical, billing, survey, dietary, nutritional, and epidemiological data) can be associated with genotype data for analysis, we have used an existing clinical trial management system, NOTIS, for clinical trial management and to act as the driver

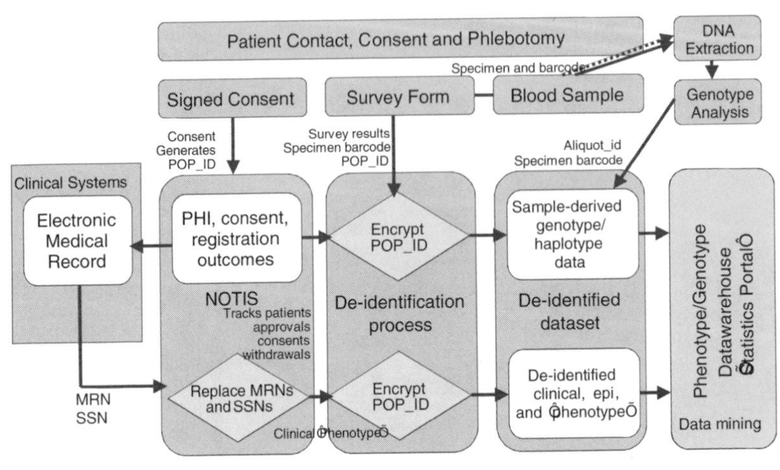

Figure 16.2. NUgene Workflow Model.

for the automated interrogation of the clinical, billing, and lab systems. A diagram of the flow is shown in Figure 16.2.

An extremely important part of the design of the NUgene and NOTIS interactions is that NOTIS houses all the patient health identifiers and NUgene holds the de-identified data mined from medical and billing records from affiliated health-care providers. The resulting warehouse of clinical, self-reported information and genotype is de-identified, expediting IRB approval for research studies using these data. In addition, this warehouse can provide data feeds to external systems. Providing these data feeds via web services with registered common data elements allows a series of similar databases to feed a common repository or support various distributed or federation models. Figure 16.2 shows the workflow process involved in the various aspects of the NUgene registration and data aggregation process, including the de-identification process and data flow.

16.15 DATABASE ARCHITECTURE/DATA MODELING

In addition to the capabilities for mining, de-identification, and housing heterogeneous data sets in NUgene, NOTIS provides a simple, robust, and scalable platform for managing clinical trials. The core data model for NOTIS is very simple—patients are represented once and institutions are represented once, as are research protocols. The intersection of these three entities is a patient on a protocol and represents the action of a patient consenting to enroll (or be screened) in a protocol at an institution. This action is assigned a POP_ID and the value of a POP_ID record is unique throughout NOTIS and identifies uniquely and precisely that consenting action. Around this simple but powerful construct of a POP_ID are the patient and study management features. In addition to NUgene, which uses an encrypted POP_ID to link patient identity for data mining with data in NUgene, numerous other databases are connected to NOTIS with various levels of de-identification. The ancillary repositories house disease and purpose-specific information that can be coupled through NOTIS to provide a complete picture of data available on a study, across studies by patient, or by outcomes or diagnoses either inside a single study or potentially across studies, assuming the protocol definitions, treatment parameters, and other aspects of each study permit the coanalysis of these variables. These data can all be pulled into the NUgene repository, if permitted by the consent for these other studies. Figure 16.3 highlights some of the features of the way we have designed our application architecture.

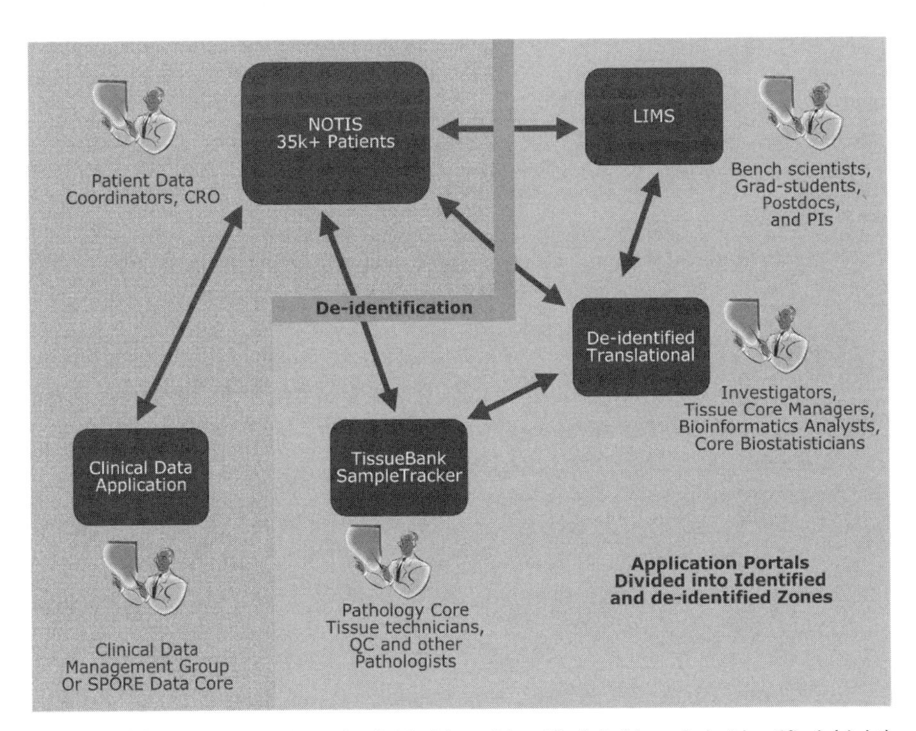

Figure 16.3. Application portals divided into identified (left) and de-identified (right) zones. NOTIS houses the consent and registration data for more than 60,000 cancer patients. The TissueBank application portal houses pathology data on the samples and also tracks the physical location of each sample in a freezer. It is in the green zone because the tissue samples are stored without real-world patient identifiers. The de-identified data reduces the barriers for research access to those data and also allows the access to PHI coupled with clinical and laboratory data to be closely and carefully monitored. The LIMS (top right) facilitates the upload and annotation of laboratory data sets. The Statistics View portal can combine data from any of these sources with PHI removed, providing reports for investigator inspection or statistical analysis. This synthesized view of data is also suitable for producing reports in XML format for distributing to other systems.

16.16 DESIGN PRACTICES

As we have developed better coding practices, we have adopted many agile software methodologies and practices, which we have found very beneficial in developing clinical, translational, and basic science tools and software. Although these practices have worked well for us, software development practices are very dependent on the skill and local environment of the programming team. There is a great deal of literature, primarily in trade journals, on good software design practices and how to build an effective team and an effective environment. I will not attempt to review that body of work; however, good results require good people and effective teams. The purpose of good software practices is to foster teamwork and provide guidelines for interaction between team members on a project. This means, in a scientific venture like a biobank, *all* the people involved in the project, not just the developers. Providing coding standards and guidelines for achieving interoperability and defining project development processes such as versioning and code release all are steps in providing and maintaining a productive coding environment. Specifically, we have adopted the following practices in our group at Northwestern:

- Iterative development process, where the development team reevaluates decisions and directions in conjunction with the oversight and user groups for a project.
- Frequent user meetings during the development, testing, and roll-out phases of a project. These meetings identify requirements for each iteration and are used to introduce prototypes, encourage feedback, and identify problems and changes in the project as early as possible.
- Test driven design (TDD), where the developers write tests before writing code. The test design process is an important way to capture requirements as well as document the expectations for each section of code written.
- Consistent use of metadata. Not an Agile requirement, but by developing consistent data dictionaries the quality and usefulness of the data are maximized.
- Well-documented data and system architecture. We use ERDs or UML diagrams to help document the design and modeling process. However, we do not believe that the software design process should be dictated by modeling.
- We assume that there will be a constant refining of the process model and data model as the project progresses, although we use best practices and design patterns to anticipate likely "flex points" and accommodate those in all phases of development.
- User acceptance testing and quality control. We use "test first" strategies and unit testing during development and have an explicit user acceptance testing phase during each iteration.

We also use a formulation of the development team based on agile practices, breaking the software engineering activities into four distinct roles as illustrated in Figure 16.4. The roles are the project manager, who is in charge of developing milestones, breaking down development activities into tasks, assigning anticipated weights and times to the tasks, and monitoring the development process. The project manager is also responsible for identifying risks to time lines, acquiring resources, managing team activities, and scheduling and running meetings. The QA/User Acceptance Testing (UAT) role is in charge of quality assurance activities, including recommending or building out the testing infrastructure, making sure both automated (unit) testing and user testing are integrated into the iterative development cycle. The Analyst is responsible for working with the users/user community to define the use cases and model the requirements for the application. The Developer is responsible for the actual coding. The developer role may require multiple skills sets, including formal coding methodologies, database design and architecture skills, database tuning, data querying, and vocabulary expertise as well as familiarity with the actual coding languages. For instance, in our group developers use one or more coding languages such as Java, scripting languages such as Perl, ColdFusion, shell scripts, and use development tools such as Eclipse, Ant, CVS, Subversion, unit testing frameworks, and Fitness (http://www.fitnesse.org/) to develop an application. The development "team" can be as simple as one person serving all four roles or can have multiple developers, QA, and analysts on a single team. A "classical" bioinformatics team defines only an analyst and perhaps a developer role.

In general, we use n-tier architecture where $n \geqslant 3$. The architecture is explained in greater detail in Section 16.12. We also focus on web-based user interface design and our tools enable the production of interfaces and publication of data using HTML, XML, and web services as well as the consumption of data from remote sites available using those protocols.

Tightly integrated with our design and development practices is our use of CVS for versioning of code and Ant for managing the deployment of code to our development, staging, and production servers. Development servers are typically used by

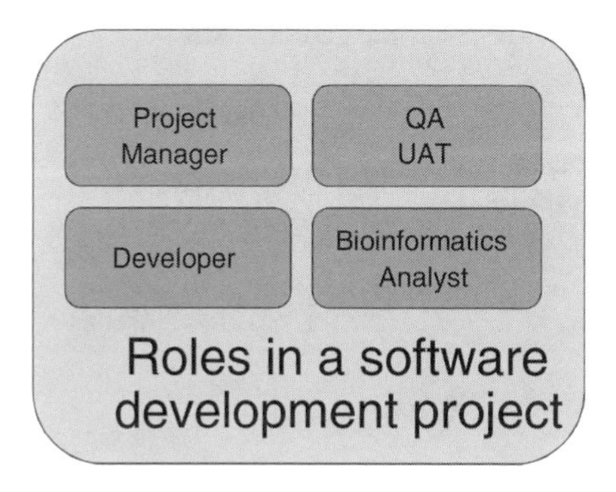

Figure 16.4. Development team roles.

our internal developers and development servers are where automated testing occurs. Iterations are released to staging, where automated testing and UAT occur. The final stage of a release is the movement of code and data model changes to production, where automated and final testing is repeated. This is shown graphically in Figure 16.5. Driving each release cycle is the iteration cycle itself, which is based on an incremental set of tasks prioritized jointly by the users and the developers. The iteration cycle and the components of that cycle are shown in Figure 16.6.

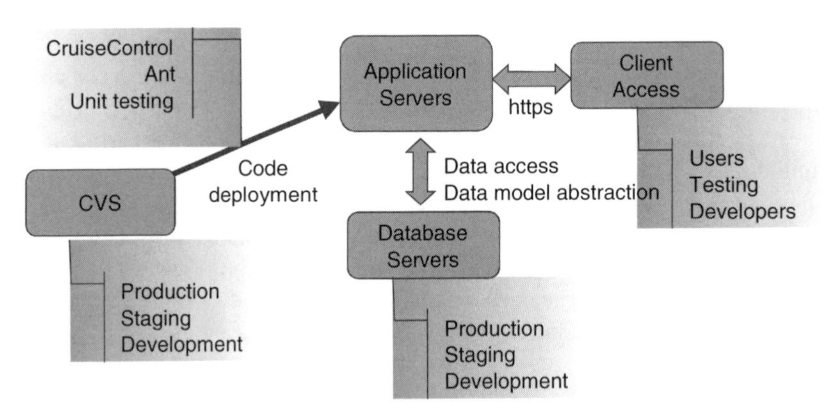

Figure 16.5. Software deployment methodology.

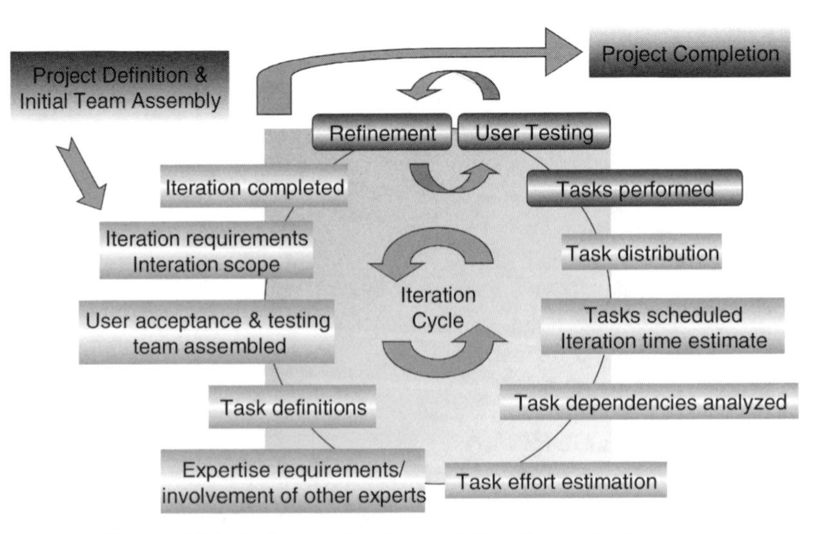

Figure 16.6. Software development iteration cycle diagram.

16.17 CONCLUSION

As I stated at the beginning, I do not consider the principles outlined here as definitive: good design practices will continue to evolve as the software community evolves, databases evolve, and our use of informatics and bioinformatics in the scientific community evolves. I would say, however, that a number of fundamental points are currently true. (1) Have a well-defined project with clear goals, thought leaders, and champions. Scope, use cases, and priorities will all emerge naturally if the project is well defined and the principals involved agree on the project goals. (2) Build a tight working relationship between all project members. The direct, *daily* involvement of an "end user" for each aspect of the project who is willing to *be* part of the development process is critical for success. (3) Build light iterations with frequent releases. This is similar to design principles in the 1980s, where the mantra was prototype, prototype, prototype. However, with today's design tools and platforms, these iterations and releases are not prototypes—they are real production quality releases. Modular design and interoperability are not just buzz words—they enable rapid application development and deployment. (4) Constantly revisit initial decisions and the impact of inevitable changes to the work plan on the data model, the implementation of that model, and the application architecture. Accommodate that change as rapidly as possible, and refactor as necessary. New tools, good object design principles, coding practices, and good database design all help minimize the impact of midcourse corrections; however, responding promptly and crisply to these changes will also help minimize the impact of these changes, put pieces of the application in the users' hands faster, and improve the effectiveness and efficiency of the development team.

REFERENCES

1. M. Anderlik (2003). Commercial biobanks and genetic research: Ethical and legal issues, *Am. J. Pharmacogenomics* **3**(3):203–215.
2. Title 21 Code of Federal Regulations Part 11—electronic signatures. http://www.fda.gov/ora/compliance_ref/part11/.
3. http://www.fda.gov/cder/gmp/index.htm.
4. F. Hartel, D. B. Warzel, and P. Covitz (2004). http://lists.w3.org/Archives/Public/public-swls-ws/2004Sep/att-0039/W3_Position_Paper.htm.
5. ISO11179. http://isotc.iso.ch/livelink/livelink/fetch/2000/2489/Ittf_Home/PubliclyAvailableStandards.htm??Redirect=1.
6. P. A. Covitz, F. Hartel, C. Schaefer, S. De Coronado, G. Fragoso, H. Sahni, S. Gustafson, and K. H. Buetow (2003). caCORE: A common infrastructure for cancer informatics, *Bioinformatics* **19**(18):2404–2412.
7. J. Golbeck, G. Fragoso, F. Hartel, J. Hendler, J. Oberthaler, and B. Parsia (2003). The National Cancer Institute's Thesaurus and Ontology, *J. Web Semantics* **1**(1):75–80.
8. T. Caulfield, R. E. G. Upshur, and A. Daar (2003). *BMC Medi. Ethics* **4**:1.

9. K. Jamrozik, D. P. Weller, and R. F. Heller (2005). Biobank: Who'd bank on it? *Med. J. Aust.* **182**(2):56–57.
10. K. Hoeyer, B. O. Olofsson, T. Mjorndal, and N. Lynoe (2005). The ethics of research using biobanks: Reason to question the importance attributed to informed consent, *Arch. Intern. Med.* **165**(1):97–100.
11. G. Williams and D. Schroeder (2004). Human genetic banking: Altruism, benefit and consent, *New Genet. Soc.* **23**(1):89–103.
12. B. Godard, J. Marshall, C. Laberge, and B. M. Knoppers (2004). Strategies for consulting with the community: The cases of four large-scale genetic databases, *Sci. Eng. Ethics* **10**(3):457–477.
13. S. Thomas, D. Porteous, and P. M. Visscher (2004). Power of direct vs. indirect haplotyping in association studies, *Genet. Epidemiol.* **26**(2):116–124.
14. A. Cambon-Thomsen (2004). The social and ethical issues of post-genomic human biobanks, *Nat. Rev. Genet.* **11**:866–873.
15. I. Hirtzlin, C. Dubreuil, N. Preaubert, J. Duchier, B. Jansen, J. Simon, P. Lobato De Faria, A. Perez-Lezaun, B. Visser, G. D. Williams, A. Cambon-Thomsen, and EUROGENBANK Consortium (2003). An empirical survey on biobanking of human genetic material and data in six EU countries, *Eur. J. Hum. Genet.* **11**(6):475–488.
16. F. S. Collins, E. D. Green, A. E. Guttmacher, and M. S. Guyer (2003). A vision for the future of genomics research: A blueprint for the genomic era, *Nature* **422**:835–847.
17. F. S. Collins (2004). The case for a U.S. prospective cohort study of genes and environment, *Nature* **429**:475–477.
18. The International HapMap Consortium (2003). The International HapMap Project, *Nature* **426**:789–796.
19. F. S. Collins (2004). Nature—What we do and don't know about "race," "ethnicity," genetics and health at the dawn of the genome era, *Genetics* **36**:S13–S15.
20. M. Mitka (2002). Banking (on) genes, *JAMA* **288**:2951–2952.
21. M. D. Uehling (2004). Is there such a thing as free EDC? BioIT World http://www.bio-itworld.com/news/061704_report5380.html.
22. N. Lorenzi (2004). *Strategies for Creating Successful Local Health Information Infrastructure Initiatives.* DHHS NHII Report, January 2004. http://aspe.hhs.gov/sp/nhii/LHII-Lorenzi-12.16.03.pdf.
23. S. J. Nass and B. W. Stillman (eds) (2003). *Large-Scale Biomedical Science: Exploring Strategies for Future Research.* National Academies Press, Washington, DC. http://www.nap.edu/books/0309089123/html/.

17

BIOCOMPUTATION AND THE ANALYSIS OF COMPLEX DATA SETS IN NUTRITIONAL GENOMICS

Kevin Dawson,[1,2,4] Raymond L. Rodriguez,[1,2,4] Wayne Chris Hawkes,[3,4] and Wasyl Malyj[1,2,4]

[1] *Section of Molecular and Cellular Biology, University of California-Davis, Davis, California*
[2] *Bioinformatics Shared Resources Core, University of California-Davis, Davis, California*
[3] *Western Human Nutrition Research Center, University of California-Davis, Davis, California*
[4] *Center of Excellence in Nutritional Genomics, University of California-Davis, Davis, California*

17.1 INTRODUCTION

Nutritional genomics is a high-throughput systems biology science, which focuses on environment and, specifically, diet–gene interactions on a whole-genome scale [1–3]. Results from nutrigenomics will provide the knowledge base for developing innovative solutions to health promotion, maintenance, and disease prevention, diagnosis, and treatment. Diet–gene interactions affecting human health have been well characterized at the protein level and, in some cases, at the gene level: lactose intolerance [4], phenylketonuria [5], galactosemia [6], gluten-sensitive enteropathy

Nutritional Genomics: Discovering the Path to Personalized Nutrition
Edited by Jim Kaput and Raymond L. Rodriguez

[7], and familial hypercholesterolemia [8]. In these classic examples, disease-specific genetic polymorphisms have been identified, the frequency of alleles in various populations analyzed, and clinical dietary guidelines developed for prevention and treatment. However, many common chronic diseases such as obesity [9], diabetes [10, 11], cardiovascular disease [12], breast cancer [13], and prostate cancer [14–16] are the result of many genes interacting with each other and with nutrients and other bioactive compounds in food. Analyzing these polygenic diseases is more complex because of differences in genetic makeup and the complexities of diets available in developed countries. Nevertheless, the mechanistic explanations of the diet–gene interactions underlying complex phenotypes have become possible with recent advances in postgenomic technologies.

High-throughput technologies, such as microarrays and SNP arrays, as well as proteomic and metabolomic profiling, make it possible to characterize aspects of chronic diseases at very high resolution. However, highly parallel measurements of many genes, proteins, or metabolites do not necessarily provide a better understanding of the underlying biological phenomena. The large data sets generated by the "omic" technologies have very high dimensionality (i.e., expression level of each of 20,000+ genes) but lack information density. (What does the expression level of one gene, one set of genes, or all genes reveal about the biology of the system at the time of sampling?) The key challenge in reducing the dimensions from many to a few is to retain the underlying biological information. Linear and nonlinear data reduction methods are used to analyze biological data sets not only to compress gene expression and proteomics data sets but also to determine the dimensionality of the underlying biological process. Instead of graphing 20,000+ individual expression levels, dimensionality reduction methods identify, group, and graph genes or samples with similar properties into one, two, three, or more dimensions. That is, dimensionality reduction algorithms discover low-dimensional structures in high-dimensional data sets. Such low-dimensional structures may identify relatedness of groups of genes, proteins, or metabolites that define disease processes, genetic regulatory networks, or disease subtypes. Disease subtypes (i.e., disease stratification) refer to the grouping of all cases of a chronic disease that may be caused by one of several different molecular mechanisms. Deciphering disease stratification may result in developing individualized therapy for drugs (i.e., pharmacogenomics) and genome-based dietary recommendations (i.e., nutritional genomics).

17.2 NUTRITIONAL GENOMICS: PART OF HIGH-THROUGHPUT BIOLOGY

Reductionism was the principal guiding paradigm in the practice of biomedical sciences for almost a century. Molecular biologists applied this paradigm by focusing on one gene or at most a few genes within a gene family. Attempts were made to know everything about these genes: their paralogues and orthologues, their interactions, and the processes in which they played roles. However, the overarching experimental design was to examine each gene in isolation either molecularly or as a single step in a complex pathway. The other genes were usually beyond the focus of the individual scientist. The molecular biology paradigm acknowledged that, in addition to the known genes, there was a large wealth of *unknown* present in biological systems. The benefit of considering the "unknown" is the chance to discover new biological insights.

Although the reductionist experimental paradigm will always be needed to understand the details of individual genes and the reactions of their products, high-throughput technologies are ushering in a holistic approach to the study of biological systems. The Human Genome Project was the first of these high-throughput technologies that resulted in the identification of almost all gene sequences but not yet all gene functions [17]. In addition, several other mammalian genomes have also been sequenced: interspecies genomic comparisons aid in identifying gene functions and chromosomal loci contributing to phenotypes. Biomedical scientists can now focus attention on complex phenotypes resulting from participation and interaction of many genes and their interactions with environmental factors. The solution to these complex biological problems is often located at the interface of large sets of data, such as the set of all human genes, their expression levels, and polymorphisms. The paradigm of experimental science is becoming increasingly holistic rather than a patchwork of independent but loosely connected building blocks as it was in the pregenomic era. Data sets are becoming progressively large, complex, and noisy. Nutritional genomics is part of this evolution because it is integrating genomic data and knowledge with nutritional data and knowledge using high-throughput technologies.

As the human genome was being sequenced, scientists began to focus attention beyond the genome [18, 19]. The postgenomic era spawned the "omics" technologies: proteomics [20, 21], metabolomics [22], and over a hundred other subdisciplines [23]. Although many consider these technologies as concepts, the goal of nutritional genomics is conceptual in nature: How does the environment influence the expression of genetic information in each individual at the molecular level, and conversely, how do an individual's genes metabolize nutrients and other bioactive compounds? The "omic" technologies provide the tools for addressing the concept that analyzing genes, proteins, or metabolites in the absence of environmental factors is insufficient to understand biological processes. This specialization of tools and technologies is seemingly contrary to the more integrated view in biological sciences described earlier. However, the specialization of the "omics" sciences is fundamentally different from the specialization of reductionist biology, which organized its knowledge by individual macromolecules. For example, gene-based categories were nuclear receptors, protein kinases, or G proteins. In contrast, high-throughput biology organizes its knowledge in functional categories. In the case of nutritional genomics, the functional category is the interaction of genes with nutrients but the unit of focus is not a small cluster of genes and nutrients but all the genes and nutrients.

In addition to the functional differences among the different "omics" technologies, these tool sets share some common characteristics. They extensively identify and classify possible items in a set. For example, the mission of the Human Genome

Project was to identify all genes in the genome by sequencing the entire human genome. Others could then use the resulting information for other technologies and analyses. One example is the microarray, which can place all the human genes and their variants on a chip for expression analysis or genetic variation analysis. "Omics" sciences generate exhaustive lists of all items in a set. Including all items in a set avoids the need to calculate for unknown (not yet discovered) variables in every experiment.

The result of this holistic approach is that data sets in high-throughput biology are of a very high dimensionality. Gene expression arrays, which analyze mRNA abundance of tens of thousands of genes simultaneously, are a case in point. The sum of the results from all the "omics" technologies and their interactions is systems biology, which can be seen as the superset of all biological information. The ultimate goal of systems biology is the creation of a meta-database that will lead to a complete understanding of the biology of the system or organism. One potential result of this complete description of the biological organism and its workings would be the development of in silico, models. Such models can be used for a formal representation of life, which could test ideas, drug candidates, diagnostic assays, and treatment options in silico, reducing the risks and increasing the benefits for humans and other organisms.

Common Characteristics of 'Omics"Sciences

- "Omics" sciences extensively identify and classify components in a set.
- "Omics" sciences generate complete lists of items in a set.
- "Omics" sciences develop ontology systems describing belongingness of and interactions between list items.
- Data sets are of very high dimensionality.
- Systems biology, a superset of all "omics" sciences, generates the information for understanding biological processes with the promise of in silico modeling of those processes.

17.3 GENE EXPRESSION ARRAYS

The gene expression microarray is an assay that measures expression levels of tens of thousands of genes in parallel on a single chip. Microarray assays can be performed from a very small amount of a biological sample, thus allowing for an experimental design involving many sample groups, repeats, dense time series, and samples collected at high granularity (i.e., highly detailed) from various anatomic locations. Today, the cost of microarrays is the principal factor limiting the number of samples that can be examined in a particular experiment. In spite of the high cost of microarrays, two-thirds of researchers surveyed by GenomeWeb had performed more than 200 microarrays and 57% spent more than $100,000 on microarrays in 2003 [24]. Sixty-eight percent were oligonucleotide arrays, mostly Affymetrix chips.

With the widespread use of microarrays in basic research and their increasing use in medical diagnostics, biomedical researchers can anticipate lower costs that will lead to more studies utilizing hundreds, if not thousands, of samples. This expansion in sample size will provide researchers with higher resolution insights into biological processes reflected in temporal, spatial, and functional patterns in microarray data sets. To reveal these patterns, several clustering techniques have been developed and applied to microarray data.

A common task in the analysis of large microarray data sets is sample classification based on gene expression patterns. This process can be divided into two steps: class prediction and class discovery. During class prediction, samples are assigned to predefined sample classes; whereas class discovery is the process of establishing new sample classes. While supervised methods are used for class prediction, unsupervised methods are needed for class discovery. For example, when gene expression arrays are used for cancer classification, class prediction assigns tumor samples into preexisting groups of malignancies, while class discovery reveals previously unknown cancer subtypes [25]. The newly discovered tumor subtypes may have different clinical patterns, respond differently to certain drugs, and require more aggressive or less aggressive surgical and radiological treatments. Class discovery may also reveal previously unknown processes in cancer biology and define more specific indications for certain drugs or nutrients. Specific drugs or dietary interventions may be used to target newly discovered tumor subtypes, thus facilitating pharmacogenomic drug design and development. Such goals will soon become readily achievable from microarray studies using large samples. Class prediction and class discovery using large data sets will require the evaluation, adaptation, and development of robust mathematical, statistical, and computational tools.

17.4 PROTEOMICS AND METABOLOMICS DATA

Recent advances in applying mass spectroscopic methods to biological molecules and metabolites represent a paradigm shift for the life sciences. Mass spectroscopic methods are being used in conjunction with proteomic, genomic, and informatics techniques to identify and measure multiple analytes in a highly parallel fashion. These methods will ultimately provide insights into how cells regulate gene expression in response to their environment.

The four main challenges facing proteomics and metabolomics research today are (1) the differing chemistries of the molecules to be analyzed, (2) instrumentation, (3) data acquisition and storage, and (4) data analysis, visualization, and interpretation. To detect and quantify RNA and DNA levels is relatively simple compared to proteins and metabolites. For instance, the chemical structure of a nucleic acid encoding hemoglobin (a soluble protein) is essentially identical to the chemistry of a nucleic acid encoding collagen (a fibrous, nonsoluble protein). The differences in chemistries between different protein classes are a distinct disadvantage for high-throughput proteomic analyses. Similar concerns apply to certain metabolites, although in many cases the biological fluids analyzed for metabolomics, such as serum or urine, contain mainly soluble metabolites. In terms of data analysis and interpretation, the specific challenge is to transform large data sets into information and information into knowledge. This necessitates the development of novel informatics tools and theories that can reveal the nonlinear principal dimensions embedded in complex, high-dimensional data sets.

17.5 SOURCES OF COMPLEXITY IN NUTRITIONAL GENOMICS

In nutritional genomics, as in many other "omics" sciences, the data sets are large, complex, and noisy. First, we analyze the sources of this complexity and search for solutions to manage it. Human diets consist of a large variety of foods containing a large variety of nutrients and bioactive compounds. Farmers and ranchers produce many naturally occurring, selectively bred, and genetically modified varieties. The food industry contributes to this comlexity by extracting and combining macro- and micronutrients from plants and animals in ways not found in nature. Food additives, conservation methods, and preparation techniques add complexity and variety to modern diets. Nutrient contents of foods may vary with seasons [26] and environments. Price, custom, growing season, and other factors drive the consumption of certain seasonal foods. Consumers are also influenced by the media, advertisements, and fads. The impact of well-publicized clinical studies linking food to health outcomes is also considerable. Consumers want to eat food that is affordable, full of flavor, healthy, and not fattening. Since such "one-fits-all" food does not exist, consumers are uncertain about their choices and look for guidance. In the United States, the effect of some fads can be large enough to measurably affect public health or send food companies into bankruptcy. Food preparation varies based on ethnicity, cultural background, the available free time for food preparation, and cooking skills. The size and frequency of meals may also vary from individual to individual.

Where we live has a tremendous influence on what we eat. The country, the region, the stores available in our neighborhood, the family situation, the time available for shopping, cooking, and consuming the meal, and the resources that can be spent on food choices all affect what we eat. The country of origin influences the cuisine that we eat and many religions have restrictions on certain foods. Some holidays are associated with the consumption or overconsumption of certain foods, which may have large regional variations. In Christian cultures [27] the abstinence from some foods is encouraged (e.g., during Lent). In Muslim and Judaic cultures, fasting may be daily for a month (e.g., Ramadan [28]) or for certain High Holy Days (e.g., Yom Kippur [29]). Children and the elderly have different nutrient requirements. Food requirements also vary dependent on life style and exercise. Many athletes consume functional foods and dietary supplements. One of the goals of nutritional genomics is to determine the food components that prevent or mitigate disease and promote health in each individual. "Healthy food" may be interpreted differently by individuals. For example, milk cannot be considered a healthy food for someone with lactose intolerance and appropriate amounts of red, nonfatty meat may be a very healthy food for someone with hypochromic anemia and normal serum cholesterol level.

In addition to the complexities of food and food preparation, genetic makeup constitutes a significant variable in the study of diet–genotype interactions. In a few isolated cases, variations in single genes cause variation in response to a particular food ingredient. For example, lactose tolerance results from a SNP variation that arose in the European population [30]. Southeast Asians and other populations geographically removed from the founding mutation have the normal mammalian response of lactose intolerance in adulthood. However, in most of the other cases, the genetic variability is a complex factor itself. Multigenic diseases, such as type 2 diabetes mellitus, obesity, cardiovascular disease, and stroke, are all influenced by diet. Some diseases have their own complexities. All chronic diseases (e.g., type 2 diabetes mellitus and obesity) and many types of cancer are groups of several individual diseases probably caused by different but overlapping groups of genes. Obesity, for example, consists of about 30 Mendelian phenotypes for which there are over 300 genetic markers known [31]. Access to health care is another important variable affecting the health status of individuals, which can be a factor that needs to be included in clinical studies on the health effects of foods.

Sources of Complexity for Nutritional Genomics Data Sets

1. Variety of foods.
2. Seasonal variations in selection and content of food.
3. Advertisement, fads, public response to news and studies on foods and health.
4. Food preparation and cooking.
5. Genetic background.
6. Complexity of disease.
7. Socioeconomic status, income, geographic environment.
8. Cultural and religious background.
9. Age and health status.
10. Exercise and life style.
11. Access to health care.

These sources of variance stand not in isolation but are highly interrelated. For example, income influences which food we can buy, which health insurance plan we can afford, and whether we can allocate time on physical exercise. The genetic and cultural backgrounds commonly overlap with each other and affect the choice of foods and the cuisine. There are multiple interactions between these variables [32]. One of the goals of nutritional genomics is to model this network of interactions.

17.6 DATA SETS IN NUTRITIONAL GENOMICS

The long-range goal of nutritional genomics is to develop a network of a few highly interconnected databases as outlined above. To start building this network, it is necessary to focus on the analysis expression of genetic differences with diet as a variable among individuals. These data sets commonly include gene expression data, SNP variations, and proteomics and metabolomics data.

The flow of genetic information can be modeled with the central dogma, which states that the genetic information is stored in DNA, which is transcribed into pre-mRNA. This is then spliced into mature mRNA. mRNA is translated into proteins and the proteins may be processed or post-translationally modified. This adds to the complexity of the biological information.

High-throughput biology has developed techniques of detecting the different macromolecules in a highly parallel fashion. The DNA of an individual or a tumor can be analyzed with sequencing or the individual genotypes can be determined with several SNP platforms. mRNA expression levels in donor blood or tissues are typically determined using quantitative real-time PCR or gene expression arrays. High-throughput methods can also be used to analyze alternate splicing. For example, although the commonly used Affymetrix chips (e.g., the HGU133plus2.0 array) are not specifically designed to detect splice variants, in some instances the probes query an exon that may be missing in a splice variant (Figure 17.1). Affymetrix is developing microarrays that are specifically designed to detect open reading frames, which helps to detect many more splice variants of genes. Protein expression levels may be studied with a variety of mass spectrometry assays (e.g., SELDI-TOF and MALDI-TOF). Small molecule metabolite concentrations in body fluids or dietary nutrient concentrations are measured by various analytical instruments. Individual expressed phenotypes can be described using anatomical, physiological, and clinical characterization; and this information can be linked to the SNP genotype data in clinical or population genetics studies [33].

DNA, RNA, protein, and metabolites and the processes that connect them are influenced by environmental factors; nutrients are the most important of these environmental factors. In addition to these areas, nutritional genomics also uses informatics applications for the study of information in textual format, such as the scientific literature, the Internet, patents, clinical study documents, and regulatory guidelines.

Data Types in Nutritional Genomics

- Genotype (SNP, tumor genotype)
- Gene expression (tissues, blood)
- Proteomics, SELDI-TOF, MALDI-TOF spectral elements (serum)
- Metabolite concentrations in body fluids
- Dietary nutrient concentrations
- Text information (scientific literature, Internet, patents, patient records, clinical studies, regulatory guidelines)

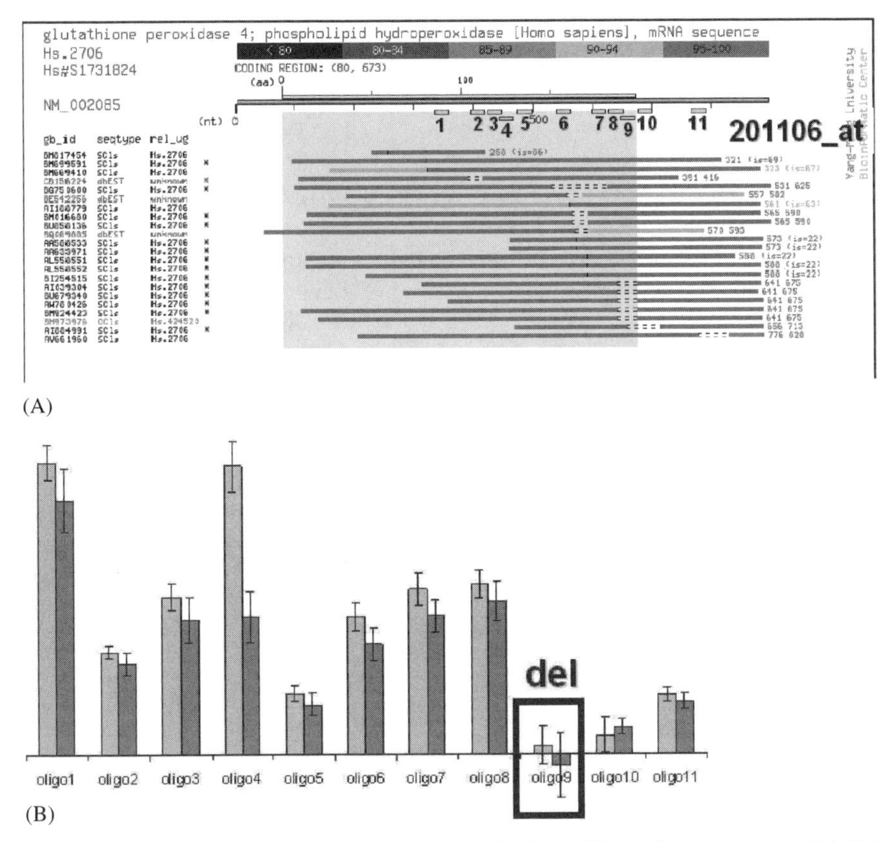

Figure 17.1. Alternate splicing can be detected using Affymetrix microarrays. (A) This example shows the known splice variants of the human glutathione peroxidase 4 (GPX4) gene from the PALSdb database aligned with the Affymetrix 201106_at oligo set. (B) Gene expression signals of the 11 PM oligonucleotides are shown in an experiment. Oligo 9 was missing in all the samples. (Values are expressed as RMA (log-2 intensity) values ± S.D.).

17.7 LEVEL OF COMPLEXITY IN GENE EXPRESSION EXPERIMENTS

To introduce the concept of dimensionality reduction, we use the example of a gene expression experiment. However, our conclusions can also be applied to other types of data. In a well-designed gene expression experiment, a small set of variables is allowed or made to change. Formulating a well-defined question and experimental design is an obvious requirement for producing reliable and interpretable data. However, high-throughput experiments produce data of very high complexity (the results of mRNA abundance as analyzed by the microarray) in the experiment. A typical gene expression study could easily mean tens of thousands of genes and tens or hundreds of microarrays. Dimensionality reduction methods are required to extract information from these large high-dimensional data sets without losing information. The goal is to identify the biological signal while filtering out the biological and measurement noise.

A typical gene expression data analysis starts with data acquisition and preprocessing (normalization, background correction, and gene calling) followed by data reduction and information extraction, such as clustering, principal component analysis (PCA), and multidimensional scaling (MDS). Most commonly, the analysis is designed to identify genes that are differentially expressed: upregulated or downregulated relative to a control or between two physiological conditions. Changes in gene expression contribute to changes in phenotype (health to disease, for example) but may be either a cause or an effect. Many interpret gene expression changes as causal for a phenotypic process. If truly causal, these genes may be targets for drugs or their expression levels can be used as diagnostic markers (e.g., tumor signatures).

The challenge in gene expression experiments is also its greatest strength: thousands or tens of thousands of genes are measured simultaneously. However, since there is always sample-to-sample variation and measurement error present in the studies, the commonly used p-value of 0.05 results in including about 50 false positive genes out of 1000 gene expression data points. Bonferroni correction and its variants are used for increasing the stringency of selection of differentially expressed genes [34]. Alternatively, increasing the sample number would improve the strength of the analysis. However, this is usually not possible because of the high price of the assay.

One of the solutions to this challenge is the application of "filters" that limit the number of genes taken into consideration. Affymetrix arrays contain 11 or 12 perfect match (PM) probes in every probe set. These are 25 mer oligonucleotides specifically hybridizing to a 25 nt. piece of the mRNA being probed. For each of these PM oligonucleotides, the array also contains a mismatch (MM), negative control oligonucleotide. Every MM oligonucleotide was altered in its center to serve as a negative control for the respective PM oligonucleotide. The difference between the PM and MM intensities can be used for statistical calculation to determine the presence or absence of a certain gene in the biological sample (gene calling). Another way of improving statistical power by restricting the number of analyzed genes is the application of tissue-specific and pathway-specific filters (gene enrichment methods) [35]. However, these techniques require the application of prior knowledge to the data analysis, which has the potential to take the analysis to predetermined paths and introduce the subjective preconceived bias of the investigator. This is a classical issue in data analysis and is not specific for microarray studies. The classical problem is the choice between idea-driven and data-driven approaches or, in other words, the application of supervised or unsupervised methods. Idea-driven methods use prior knowledge (e.g., tissue- and pathway-specific filters or enrichment) to improve the sensitivity of the data analysis. The disadvantage of this approach is that it may introduce subjective bias. It may also filter out the noise of the apparently nonimportant variables at the same time that it misses some biologically valid results. The data-driven approach, on the other hand, is descriptive, avoids subjective bias, and allows class discovery. The disadvantage of an unsupervised method is that it has a high false discovery rate. Furthermore, the results from this latter approach cannot always be explained using prior knowledge.

17.8 DIMENSIONALITY REDUCTION METHODS

One of the advantages of using dimensionality reduction methods is that we can combine the benefits of the idea-driven and data-driven approaches. In a well-designed gene expression experiment, the study is designed in a way that aims to test a well-defined hypothesis. With this in mind, the experiment can be planned as straightforward and as simple as possible. Then the investigator can execute the steps of data collection, preprocessing, and gene expression value calculation in an unsupervised fashion. In both the supervised and unsupervised methods, a high level of complexity and a high dimensionality result from analyzing thousands of genes. For instance, with the Affymetrix HGU133plus2.0 arrays, every sample creates a 54,675-dimensional data point. To manage and analyze this large data set, dimensionality reduction algorithms can be extremely useful. These methods can translate the 54,675-dimensional data set into a lower-dimensional data space.

To demonstrate dimensionality reduction methods, the "Swiss Roll" data set (with a small amount of added "noise" to simulate experimental error) is often used [36, 37]. Swiss Rolls are convenient examples because the three-dimensional shape is formed by a thin sheet rolled into a spiral cylindrical form. Figure 17.2 shows a three-dimensional plot of 2400 data points from the Swiss Roll. By changing the angle of reference (Figure 17.3), the data points define a curved two-dimensional surface. The goal of dimensionality reduction is to deduce a simpler underlying structure in this three-dimensional problem using the computer directly, rather than relying on manipulation of the data by the researcher.

For the Swiss Roll example, we first apply principal component analysis (PCA), mentioned earlier, a powerful traditional method validated by many decades of research and development. PCA discovers three principal components; the first explains 88% of the variance, the first and second together explain 90% of the variance, and the three principal components together explain 99.6% of the variance (Figure 17.4). In summary, principal components analysis concludes that the data set is three dimensional.

Other nonlinear dimensionality reduction methods could also be used to analyze these data. Our bioinformatics core has been applying a method called Isomap to complex biological data sets. The Isomap algorithm first calculates the distances (usually Euclidean) between each data point and its immediate neighbors (typically 3–12 neighbors). The algorithm then calculates the "geodesic" distance between any two nonneighboring points by summing together the appropriate nearest-neighbor distances. By using this approach, Isomap deduces that the Swiss Roll is really a two-dimensional problem imbedded in three dimensions, a result that differs from that obtained with PCA. Figure 17.5 shows how Isomap has "unrolled" the underlying structure of Figure 17.2. The residual variance (i.e., the mean squared deviation

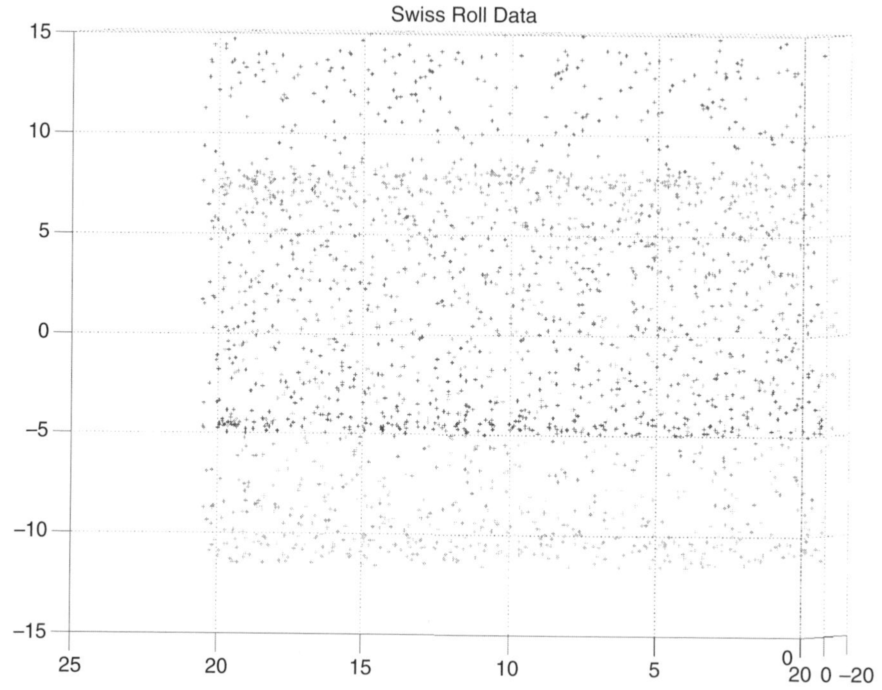

Figure 17.2. The Swiss Roll data set. The figure shows 2400 points from the Swiss Roll data set plotted in three dimensions.

about a model) versus dimensionality using Isomap is indicated by the diamonds in Figure 17.4. The primary or first nonlinear principal component is the length of the unrolled sheet of data and the second component is its width (Figure 17.6).

In the intervening few years since the invention of Isomap, several improvements have been proposed. Figure 17.6 compares Isomap [36] and a method published simultaneously called locally linear embedding (LLE) [37], to one of these improved variants: Hessian LLE (HLLE) [38]. To illustrate the utility of HLLE, a modified version of the Swiss Roll, where a rectangular patch of data has been removed, is employed. The analysis and visualization by Isomap shows that Isomap has detected the anomaly and has exaggerated its extent but is unable to replicate its shape, denoting a circle instead of a square. Similarly, LLE detects the anomaly but distorts it. The new method, HLLE, both detects the anomaly and replicates its shape. Although this predesigned structure (the rectangular patch) is a simulation, it reveals that different algorithms must be used for analyzing different types of high-dimensional data sets. However, HLLE is significantly more computationally intensive than Isomap, so for the remainder of this chapter we use Isomap as our nonlinear dimensionality reduction method. As biocomputational hardware infrastructures become faster and more powerful, we expect new generations of nonlinear dimensionality reduction algorithms to supersede our current tools.

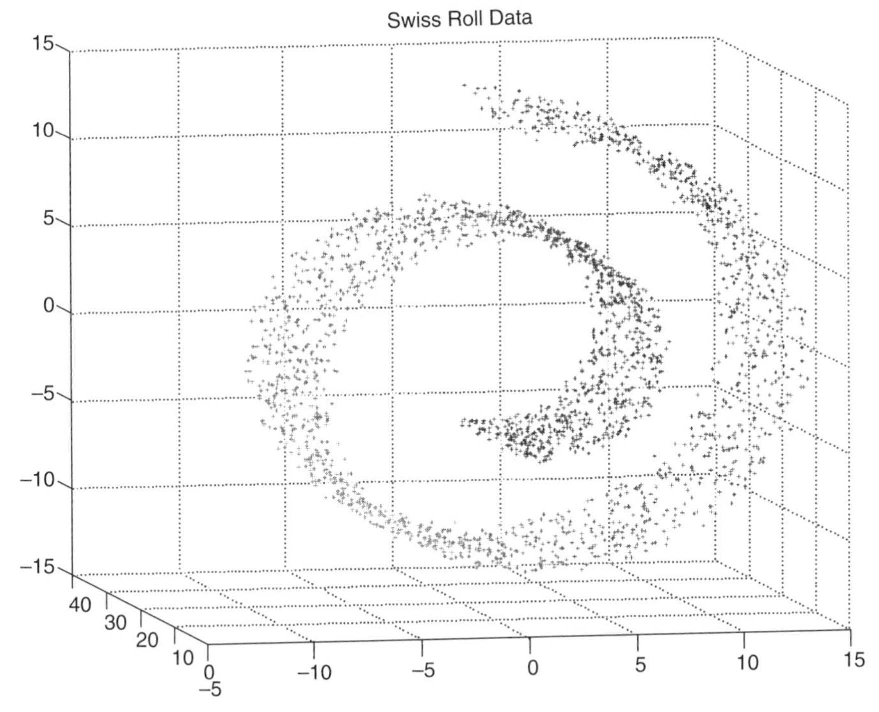

Figure 17.3. A different perspective of the Swiss Roll data set. The viewpoint of Figure 17.2 has been changed and we can see that the 2400 points lie along a spiral sheet in the three-dimensional space.

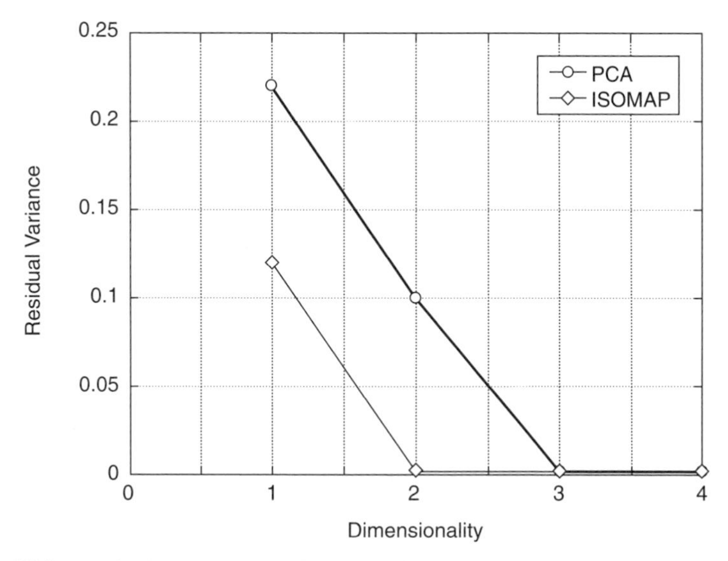

Figure 17.4. Residual variance as a function of dimensionality. The upper curve shows the residual variance using a principal component model; the lower curve displays the residual variance of an Isomap model.

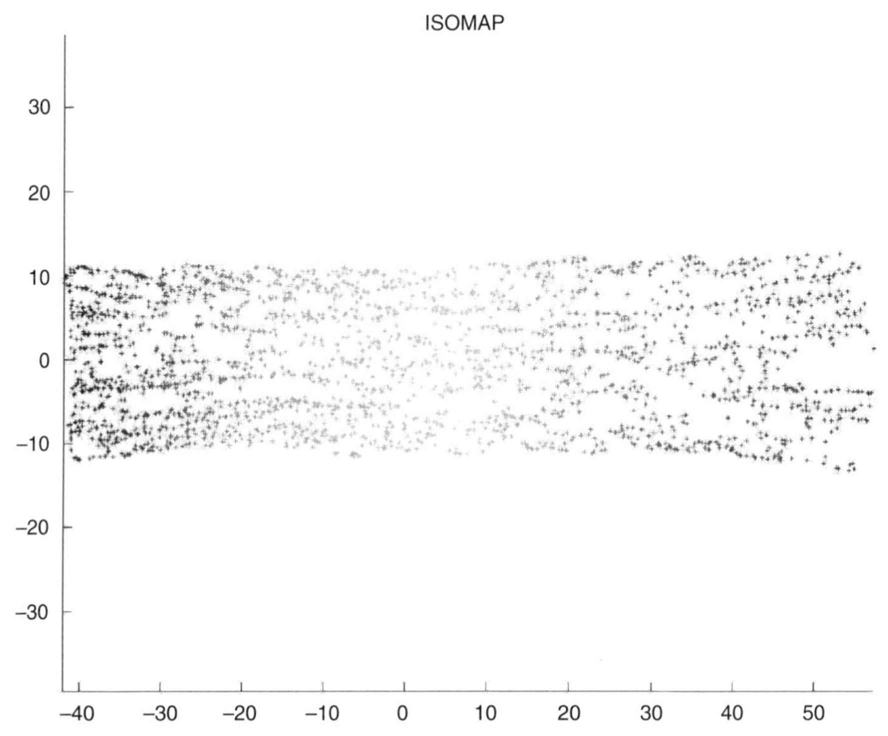

Figure 17.5. The Swiss Roll data set "unrolled" by Isomap. The Isomap method has deduced that the Swiss Roll appears to be a two-dimensional sheet embedded in a three-dimensional space.

The Swiss Roll example has demonstrated that high-dimensional data sets may contain simpler, lower-dimensional structures. Similarly, high-dimensional gene expression data sets typically contain lower-dimensional structures. The goal of a dimensionality reduction method is to determine the dimensionality of the underlying biological. For instance, with the help of a dimensionality reduction algorithm (e.g., Isomap) on a gene expression data set, the residual variance can be calculated after the application of low-dimensional models with one, two, and three . . . dimensions. In an example with a kidney tissue data set, we were able to explain more than 96% of the variance with a two-dimensional Isomap model (Figure 17.7). Higher-dimensional models, on the other hand, could not explain much more variance of the same gene expression data set. In other words, Isomap discovered that the true underlying dimensionality of the data set is only two. In general, we can successfully transform a high-dimensional data set into a two- or three-dimensional space, and we can then visualize the samples in this low-dimensional space. Such reduction methods are useful for human interpretation of the data. For instance, a three-dimensional visualization of the data in the previous example can reveal that one of the discovered principal components describes a known phenotypic parame-

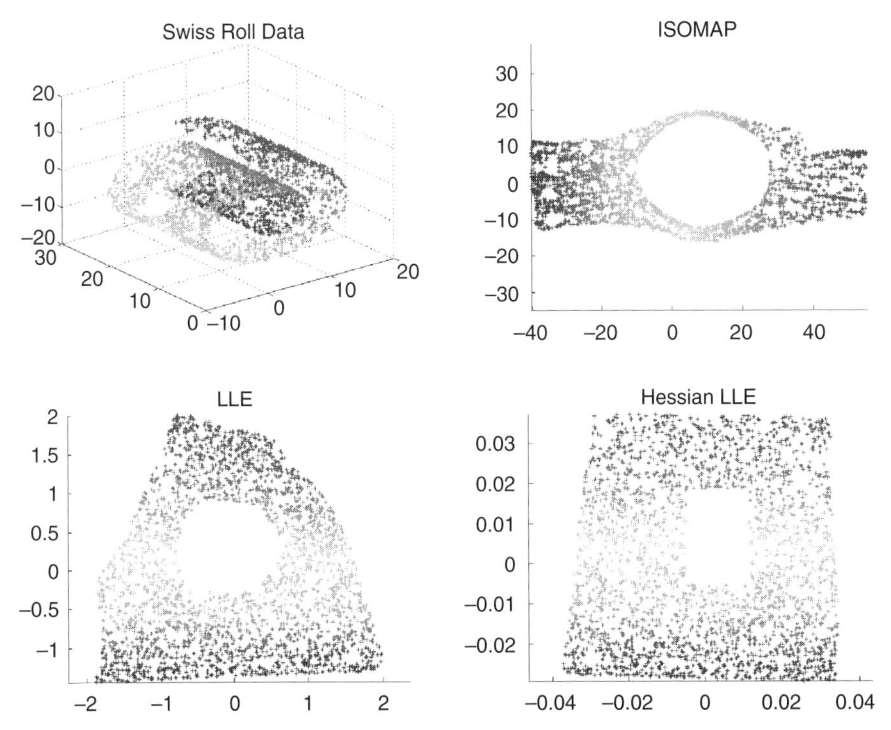

Figure 17.6. A more challenging Swiss Roll data set. The top left panel illustrates the Swiss Roll data set analyzed previously, but with a rectangular strip of data removed from the center. The top right panel shows that Isomap detects the anomaly but fails to replicate its shape. The bottom left panel shows that the LLE method, like Isomap, fails to replicate the shape of the anomaly. The bottom right panel illustrates the performance of the Hessian LLE method, which detects the anomaly and correctly replicates its shape.

ter. In this example, the second principal component refers to the difference between the two major phenotypes (normal kidney tissue versus clear cell carcinoma of the kidney) (Figure 17.8). The Isomap algorithm discovered the principal component and the scientist could explain the underlying biological cause (the difference between the pathological and normal tissue samples).

Dimensionality reduction methods can be applied not only to gene expression data but also to other high-throughput "omics" data sets [59]. In another example, we analyzed single nucleotide polymorphism (SNP) data collected by the SNP Consortium [39]. Three-dimensional principal component analysis (PCA) discovered three distinct clusters in this data set (Figure 17.9). PCA classified individuals into three clusters that turned out to be representative of their continental origin. As expected, Yoruban Nigerian and African American individuals mapped at close locations in the three-dimensional PCA model. This particular example does not describe genetic distances between groups of individuals because the distribution of individuals enrolled in this study is not representative of any region.

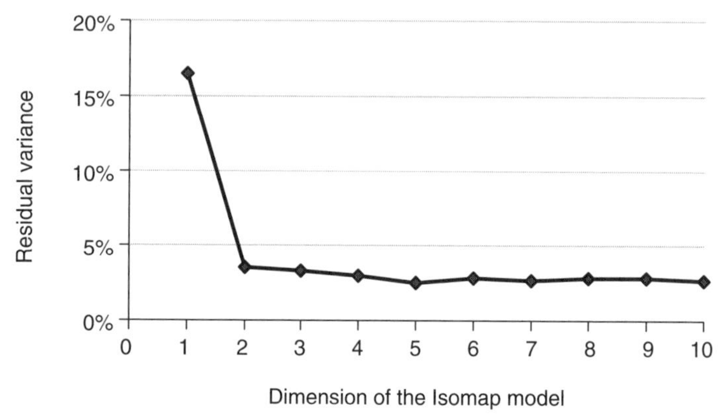

Figure 17.7. Residual variance of an Isomap model of different dimensions. The NCBI/GEO GSE4 gene expression data set was analyzed using Isomap at dimensions 1 through 10. A two-dimensional Isomap model was able to explain more than 96% of the variance. However, higher-dimensional models could not explain much more additional variance of the gene expression data set.

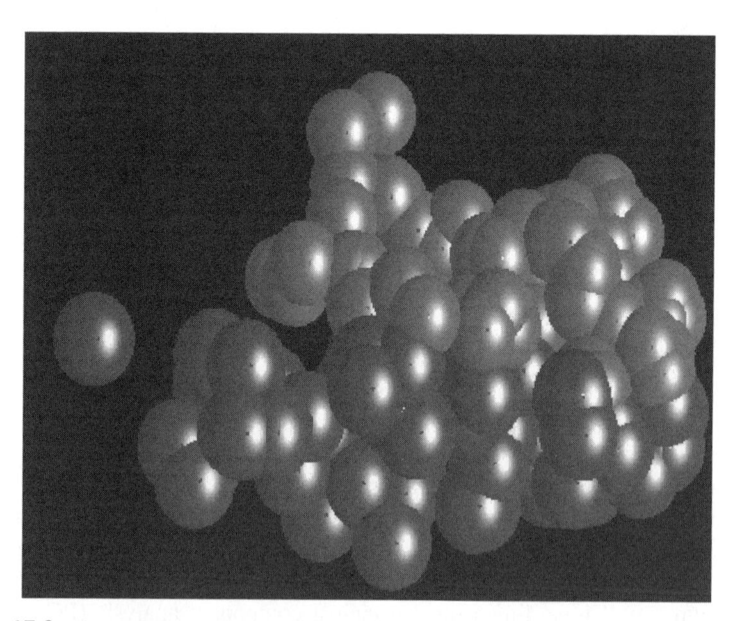

Figure 17.8. Three-dimensional visualization of the Isomap model. The two shades show samples with two different phenotypes (grey, normal kidney tissue; black, clear cell carcinoma of the kidney). The principal components on the *x*- and *z*-axes describe unknown attributes, while the *y*-axis shows the difference between the two phenotypes. (See insert for color representation.)

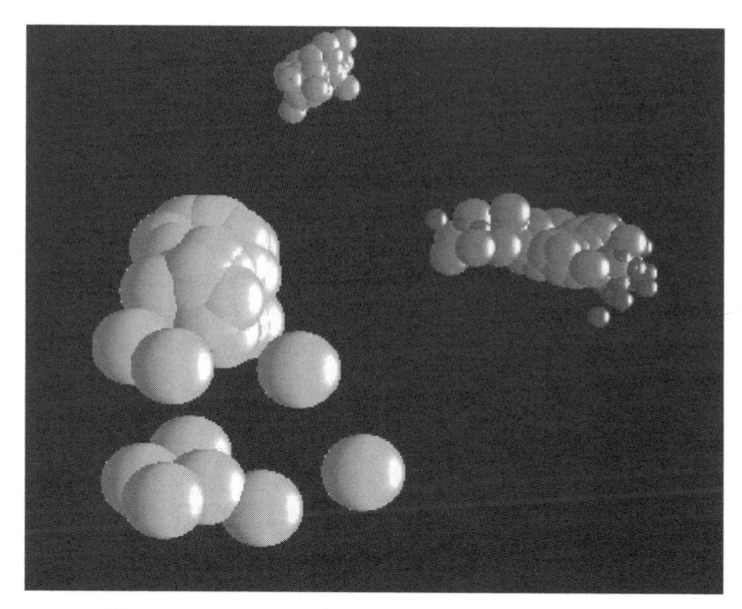

Figure 17.9. Principal component analysis (PCA) identifies three clusters in a single nucleotide polymorphism (SNP) data set. Individual SNP patterns of Asian, Caucasian, African, and African-American individuals were analyzed using PCA. A three-dimensional PCA model discovers three clusters present in the data set. (See insert for color representation.)

These two examples illustrate the application of dimensionality reduction algorithms to genomic data. In general, dimensionality reduction methods can be used to summarize large data sets, identify structures and trends in the data, identify redundancy and correlation in data, and produce "insightful" graphical displays of the results. These techniques are commonly applied to problems of pattern recognition [37], handwriting recognition [36], speech recognition [40], signal processing [41], cognitive science [36], marketing intelligence [42], and high-throughput biology [43].

Methods of Dimensionality Reduction

- Principal component analysis (PCA) [44] and singular value decomposition (SVD) [45] find the n dimensions that explain most of the variance.
- Multidimensional scaling (MDS) [46] states that the interpoint distance in the low-dimensional space is as close as possible to the interpoint distance in the high-dimensional space.
- Linear discriminant analysis (LDA) [47] takes a weighted sum of values of the variables that determine a classification. The value of the weighted sum is then used to determine the classification. Usually, it is discovered by training samples with known classes.

- Self-organizing maps (SOMs) [48]—a data visualization technique—is based on cluster analysis.
- Charting finds local neighborhoods in the low-dimensional space and mixes them while trying to minimize distortion.
- Locally linear embedding (LLE) [37] is similar to Isomap but differs in focus on analyzing distances to nearest neighbors and mapping them to a smooth nonlinear manifold of lower dimensionality.
- Hessian eigenmaps (HLLE) [38] can handle a significantly wider range of situations than the original Isomap algorithm.
- Laplacian eigenmaps [40] are a computationally effective algorithm that has a natural connection to clustering.
- Isomap employs nonlinear embedding; uses geodesic distances [36].

17.9 CASE STUDY (MICROARRAY EXPERIMENT OF A DIETARY INTERVENTION)

To demonstrate the complexity of a high-throughput experiment in nutritional genomics, we are using the example of a dietary intervention study. In this experiment, 16 patients received one of two different diets, high or low in a particular nutrient in an otherwise balanced diet. The particular findings of this preliminary study will not be presented here. Patients' blood samples were collected after overnight fasting and analyzed before and after the dietary intervention. In this study, HGU133A microarrays were analyzed with 22,283 genes on each array.

The first source of variability that must be controlled is the *array-to-array variance*. Such variance may originate from differences in RNA extraction, sample processing, labeling, hybridization, washing, technicians, and scanning. Although Affymetrix microarray technology is considered to be one of the most reproducible gene expression platforms, signal intensity distributions may still be very different from one array to another. Figure 17.10 shows the distribution of intensities in the 32 microarrays analyzed in this experiment. Each transcript was probed with 11 or 12 pairs of oligonucleotides (the probe set) of the HGU133A microarray. The intensity distributions of the mismatched (MM) probes are shown in Figure 17.10A in a log-2 scale. The intensity distributions of the perfectly matched (PM) probes are shown in Figure 17.10B in a log-2 scale. Note the large variations from array to array. Several methods were developed to control this variance and calculate the gene expression values from the measured probe intensities [49–53]. Most laboratories today use the MAS 5.0 method originally available from Affymetrix. The RMA method, on the other hand, offers significant improvements compared to MAS 5.0. RMA normalizes probe-level intensities across all microarrays. Therefore, RMA controls for the sample-to-sample variance caused by differences in sample preparation, labeling, and scanning. These different normalization methods (MAS 5.0, RMA, etc.) were compared by several groups of investigators [54–56]. In our example, the result of these probe-level calculations is a much smaller data set with only 22,283 gene expression values (versus 506,944 values at the probe level), each referring to a probeset that is referring to a transcript. In our example, the distribution of the gene expression values is shown on Figure 17.10C,D after the application of the MAS 5.0 and robust multichip average (RMA) algorithms, respectively. MAS 5.0 values are shown on a log-2 scale; while the RMA values are already log-2 transformed. Note that the difference among the 32 microarray intensity distributions at the expression level is much less (after the application of the MAS 5.0 or RMA algorithms) than at the probe level.

In gene expression studies, a second source of variance originates from the *nonhomogeneity of the source tissue*. Each tissue, including blood, is composed of different types of cells. Dependent on the count of certain leukocyte subclasses in the subject's blood, the gene expression values may differ dramatically (Figure 17.11). In this example, gene expression in CD19+/CD45+ B lymphocytes cluster together with the B lymphocyte-specific genes (Fc fragment of IgE, CD79A, CD79B, CD19, CD22, CD72, immunoglobulin heavy chain γ, B cell scaffold protein, B cell linker, etc.) (Figure 17.11A). Similarly, the CD16+/CD56+ natural killer cells cluster together with the natural killer-specific genes (Figure 17.11B). This example demonstrates that by looking at the gene expression results alone and seeing that a certain set of genes is overexpressed does not necessarily mean a change of the expression levels since different cell types may express different levels of a gene irrespective of the dietary intervention. An equally valid alternative explanation may be that the cell count of a certain leukocyte subtype increased. Similar challenges develop in cell culture experiments when a treatment may affect the cell cycle. However, a change of expression levels of cell cycle-related genes does not necessarily mean that our intervention directly affected these genes. For instance, if we find that genes specific to mitosis are less expressed, we may speculate that the intervention inhibited mitosis. However, the same result could be found if the cells grew somewhat slower and were in a different phase of the cell cycle at the time of harvesting. Similarly, clinical samples may also contain a large array of different cells. Tumor samples may contain surrounding tissues and different parts of the tumor. Purified cell samples may be more homogeneous and will contain less variance of cell type. However, the purification method itself may change the expression levels of many genes because of the washing procedures or simply because of the time between isolation and final analysis. For instance, fluorescence-activated cell sorting (FACS) may provide a very well-defined set of cells. However, the immunocytochemistry labeling and cell sorting will change the expression levels of many genes [57]. However, the benefits of the separation may outweigh the disadvantage of the altered gene expression values caused by the separation method. Similarly, gene expression studies on tumor samples are also affected by the heterogeneity of the tumor and the contaminating other cell types [58].

Nutritional genomics is specifically facing the challenge that dietary interventions usually cause relatively small changes in gene expression levels compared to

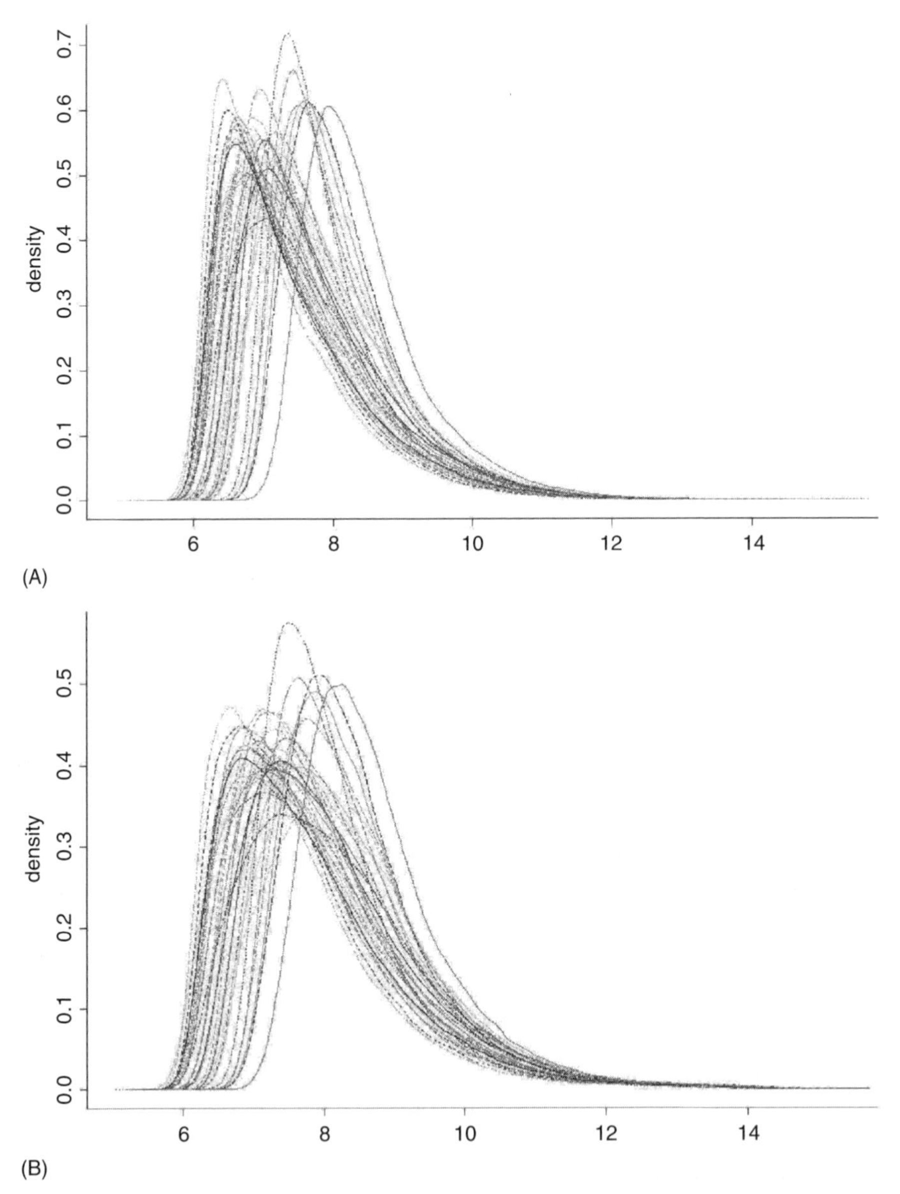

Figure 17.10. Intensity distribution of signals from 32 HGU133A arrays. (A) The intensity distribution of the 247,965 MM (mismatch) oligonucleotide signals is showed on a log-2 scale. (B) The intensity distribution of the 247,965 PM (perfect match) oligonucleotide signals is shown on a log-2 scale. (C) Distribution of the 22,283 expression values calculated by MAS 5.0 is shown on a log-2 scale. (D) Distribution of the 22,283 expression values calculated by RMA.

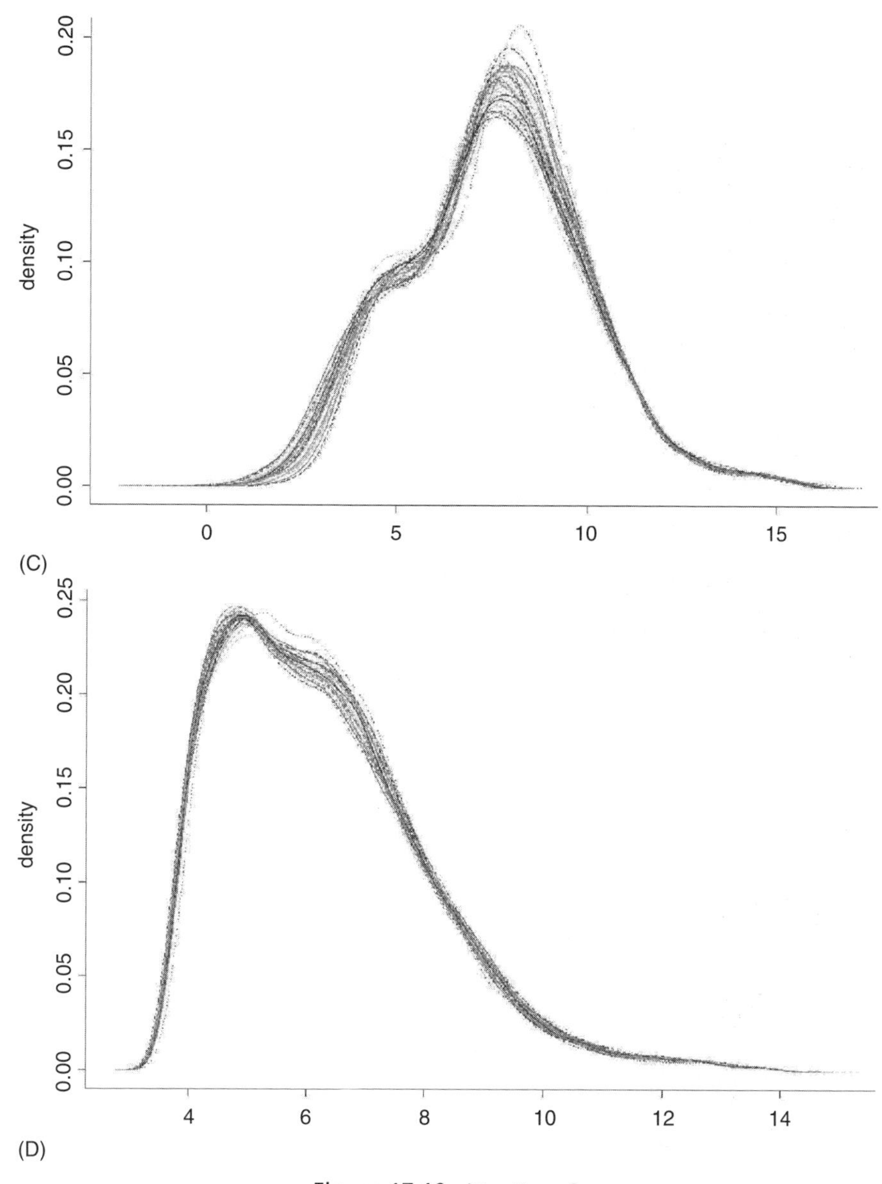

Figure 17.10. *(Continued)*

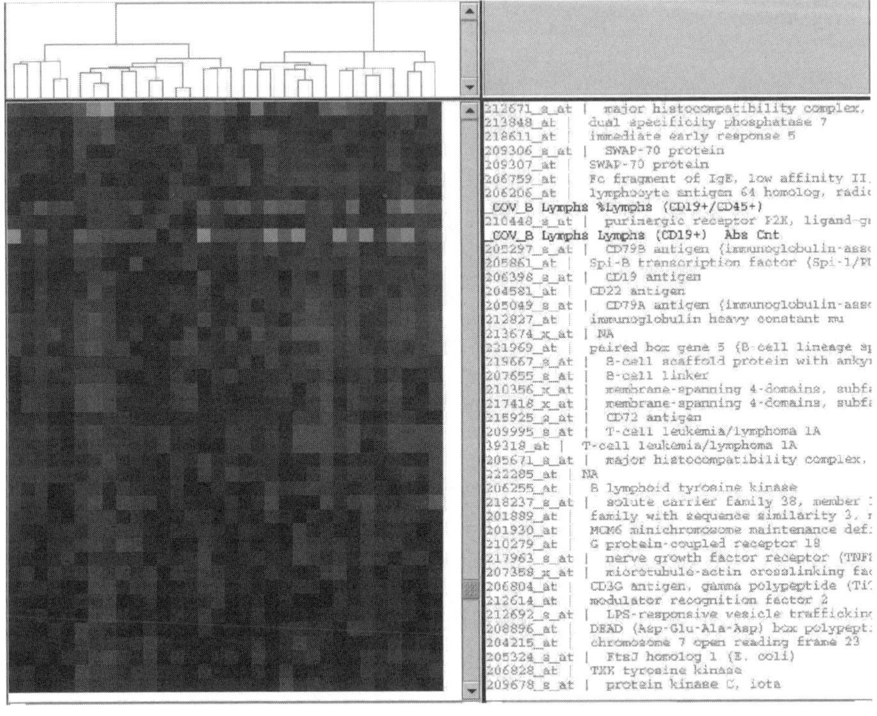

212671_s_at | major histocompatibility complex,
213848_at | dual specificity phosphatase 7
218611_at | immediate early response 5
209306_s_at | SWAP-70 protein
209307_at | SWAP-70 protein
206759_at | Fc fragment of IgE, low affinity II,
206206_at | lymphocyte antigen 64 homolog, radi
_COV_B Lymphs %Lymphs (CD19+/CD45+)
210448_s_at | purinergic receptor P2X, ligand-g
_COV_B Lymphs Lymphs (CD19+) Abs Cnt
205297_s_at | CD79B antigen (immunoglobulin-ass
205861_at | Spi-B transcription factor (Spi-1/PU
206398_s_at | CD19 antigen
204581_at | CD22 antigen
205049_s_at | CD79A antigen (immunoglobulin-ass
212827_at | immunoglobulin heavy constant mu
213674_x_at | NA
221969_at | paired box gene 5 (B-cell lineage s
219667_s_at | B-cell scaffold protein with anky
207655_s_at | B-cell linker
210356_x_at | membrane-spanning 4-domains, subf
217418_x_at | membrane-spanning 4-domains, subf
215925_s_at | CD72 antigen
209995_s_at | T-cell leukemia/lymphoma 1A
39318_at | T-cell leukemia/lymphoma 1A
205671_s_at | major histocompatibility complex,
222285_at | NA
206255_at | B lymphoid tyrosine kinase
218237_s_at | solute carrier family 38, member
201889_at | family with sequence similarity 3,
201930_at | MCM6 minichromosome maintenance def
210279_at | G protein-coupled receptor 18
217963_s_at | nerve growth factor receptor (TNF
207388_x_at | microtubule-actin crosslinking fa
206804_at | CD3G antigen, gamma polypeptide (Ti
212614_at | modulator recognition factor 2
212692_s_at | LPS-responsive vesicle traffickin
208896_at | DEAD (Asp-Glu-Ala-Asp) box polypept
204215_at | chromosome 7 open reading frame 23
205324_s_at | FtsJ homolog 1 (E. coli)
206828_at | TXK tyrosine kinase
209678_s_at | protein kinase C, iota

(A)

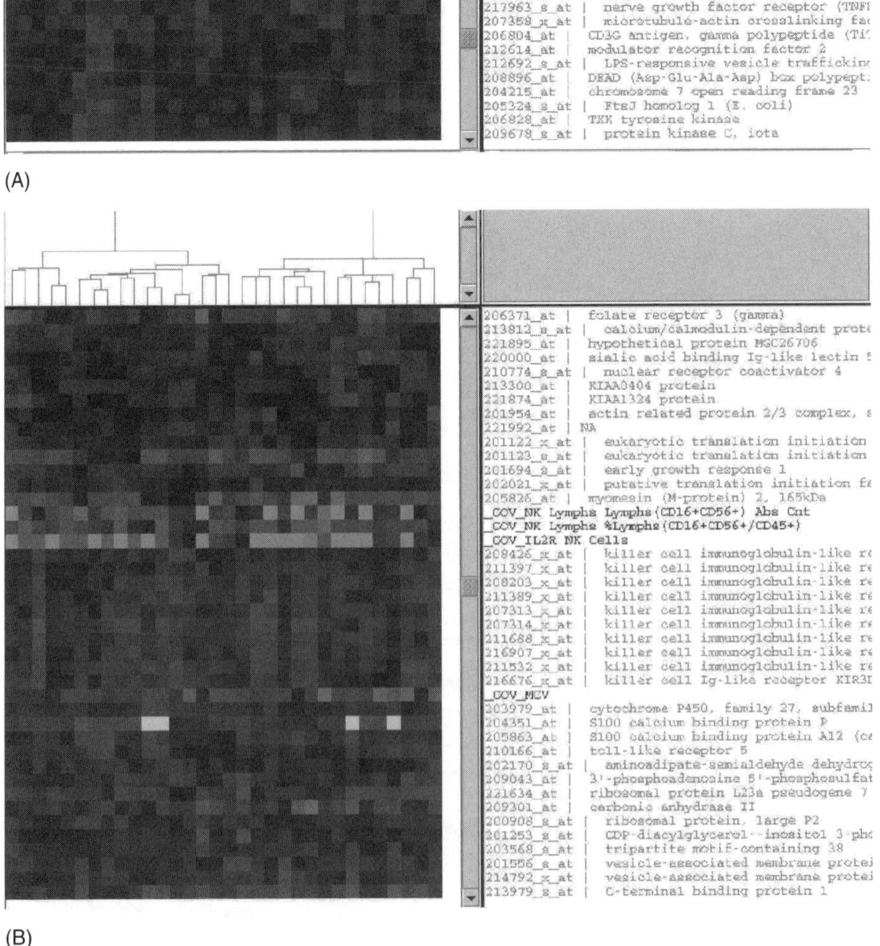

206371_at | folate receptor 3 (gamma)
213812_s_at | calcium/calmodulin-dependent prot
221895_at | hypothetical protein MGC26706
220000_at | sialic acid binding Ig-like lectin
210774_s_at | nuclear receptor coactivator 4
213300_at | KIAA0404 protein
221874_at | KIAA1324 protein
201954_at | actin related protein 2/3 complex,
221992_at | NA
201122_x_at | eukaryotic translation initiation
201123_s_at | eukaryotic translation initiation
201694_s_at | early growth response 1
202021_x_at | putative translation initiation f
205826_at | myomesin (M-protein) 2, 165kDa
_COV_NK Lymphs Lymphs(CD16+CD56+) Abs Cnt
_COV_NK Lymphs %Lymphs(CD16+CD56+/CD45+)
_COV_IL2R NK Cells
208426_x_at | killer cell immunoglobulin-like r
211397_x_at | killer cell immunoglobulin-like r
208203_x_at | killer cell immunoglobulin-like r
211389_x_at | killer cell immunoglobulin-like r
207313_x_at | killer cell immunoglobulin-like r
207314_x_at | killer cell immunoglobulin-like r
211688_x_at | killer cell immunoglobulin-like r
216907_x_at | killer cell immunoglobulin-like r
211532_x_at | killer cell immunoglobulin-like r
216676_x_at | killer cell Ig-like receptor KIR3
_COV_MCV
203979_at | cytochrome P450, family 27, subfami
204351_at | S100 calcium binding protein P
205863_at | S100 calcium binding protein A12 (c
210166_at | toll-like receptor 5
202170_s_at | aminoadipate-semialdehyde dehydro
209043_at | 3'-phosphoadenosine 5'-phosphosulfa
221634_at | ribosomal protein L23a pseudogene 7
209301_at | carbonic anhydrase II
200908_s_at | ribosomal protein, large P2
201253_s_at | CDP-diacylglycerol--inositol 3-ph
203568_s_at | tripartite motif-containing 38
201556_s_at | vesicle-associated membrane prote
214792_x_at | vesicle-associated membrane prote
213979_s_at | C-terminal binding protein 1

(B)

the naturally high *subject-to-subject variance* caused by differences in genetic makeup. Gene expression patterns of different people are usually more diverse than the change of expression patterns caused by a dietary intervention (Figure 17.12A). To manage this situation, we can compare changes of diet-induced gene expression values. An alternative solution is to use analysis of variance (ANOVA) including a patient-specific and a treatment-specific term in the calculations (Figure 17.12B). ANOVA also allows for the control of a batch bias that is a common problem with microarrays. Affymetrix arrays are expensive; therefore, many experiments are done in batches contingent upon the availability of funding. When the separate batches of samples are analyzed with microarrays at different times, the difference between the batches may introduce a significant but unwanted source of variance. Batch-specific bias can be controlled by ANOVA only if the samples were randomly analyzed using arrays from different batches. Designing experiments to minimize systematic errors is an important component of high-throughput analyses.

17.10 CONCLUSION

We have seen that the data sets in nutritional genomics are large, complex, and noisy. Consequently, it is important to simplify the data in a way that the maximum amount of information is extracted from the data without losing information or introducing the investigator's preconceived bias. Several methods of dimensionality reduction and information extraction are available. Not only do many of these methods extract information but some of them can also explain interactions between variables in the data set. Collaborations among biomedical scientists, bioinformaticians, and biostatisticians will help avoid pitfalls that are not easily corrected after the samples are collected.

ACKNOWLEDGMENTS

The project was supported in part by a grant from the National Institutes of Health, P60MD00222. This chapter is dedicated to Professor Bruce N. Ames for his pioneering scientific achievements and leadership in promoting human health through a better understanding of gene–environment interactions.

Figure 17.11. Expression values of blood cell type-specific gene clusters follow the pattern of particular cell types. (A) CD19+/CD45+ B lymphocytes cluster together with the B lymphocyte-specific genes (Fc fragment of IgE, CD79A, CD79B, CD19, CD22, CD72, immunoglobulin heavy chain γ, B cell scaffold protein, B cell linker, etc). (B) Similarly, the CD16+/CD56+ natural killer cells cluster together with the natural killer-specific genes. (See insert for color representation.)

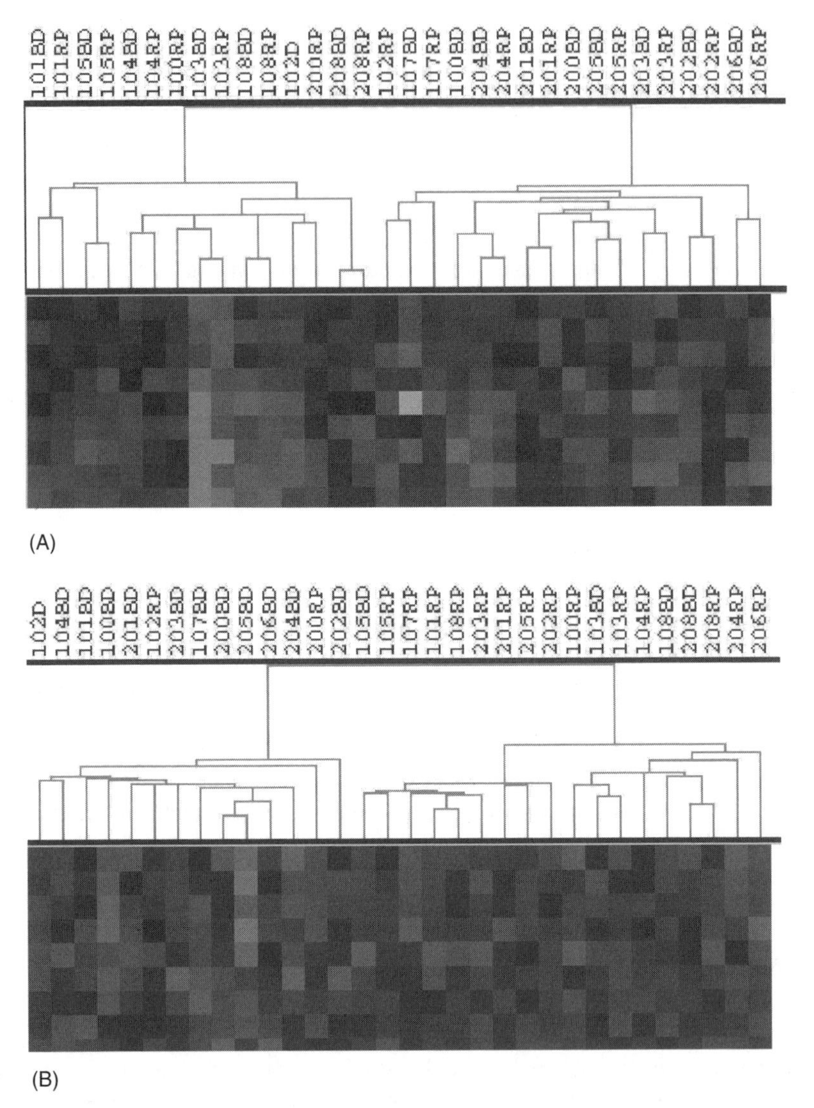

Figure 17.12. Interpatient variance of gene expression is usually higher than the effect of a dietary intervention. (A) Hierarchical clustering of gene expression values of 16 subjects before and after a dietary intervention reveals that the patient-to-patient variance is higher than the treatment-specific variance. The numbers (101, 102, etc.) label different study subjects. Please note that samples from the same patient cluster together. (B) Using analysis of variance (ANOVA), patient-specific variance may be separated from the treatment-specific variance. (See insert for color representation.)

REFERENCES

1. J. Kaput and R. L. Rodriguez (2004). Nutritional genomics: The next frontier in the postgenomic era, *Physiol. Genomics* **16**:166–177.
2. R. M. DeBusk, C. P. Fogarty, J. M. Ordovas, and K. S. Kornman (2005). Nutritional genomics in practice: Where do we begin? *J. Am. Diet. Assoc.* **105**:589–598.
3. J. M. Ordovas and V. Mooser (2004). Nutrigenomics and nutrigenetics, *Curr. Opin. Lipidol.* **15**:101–108.
4. C. Buning, H. Schmidt, H. Lochs, and J. Ockenga (2004). Genetic components of lactose intolerance and community frequency, *J. Bone Miner. Res.* **19**:1746; author reply 1747.
5. L. J. Spaapen and M. E. Rubio-Gozalbo (2003). Tetrahydrobiopterin-responsive phenylalanine hydroxylase deficiency, state of the art, *Mol. Genet. Metab.* **78**:93–99.
6. M. Suzuki, C. West, and E. Beutler (2001). Large-scale molecular screening for galactosemia alleles in a pan-ethnic population, *Hum. Genet.* **109**:210–215.
7. M. Haapalahti, P. Kulmala, T. J. Karttunen, et al. (2005). Nutritional status in adolescents and young adults with screen-detected celiac disease, *J. Pediatr. Gastroenterol. Nutr.* **40**:566–570.
8. G. Miltiadous, M. Cariolou, and M. Elisaf (2004). Genetic polymorphisms affecting the phenotypic expression of patients with molecularly defined familial hypercholesterolaemia, *Atherosclerosis* **177**:217.
9. C. G. Bell, A. J. Walley, and P. Froguel (2005). The genetics of human obesity, *Nat. Rev. Genet.* **6**:221–234.
10. I. Barroso (2005). Genetics of Type 2 diabetes, *Diabet. Med.* **22**:517–535.
11. C. C. Wang, M. L. Goalstone, and B. Draznin (2004). Molecular mechanisms of insulin resistance that impact cardiovascular biology, *Diabetes* **53**:2735–2740.
12. F. Cambien (2005). Coronary heart disease and polymorphisms in genes affecting lipid metabolism and inflammation, *Curr. Atheroscler. Rep.* **7**:188–195.
13. C. B. Ambrosone, S. E. McCann, J. L. Freudenheim, et al. (2004). Breast cancer risk in premenopausal women is inversely associated with consumption of broccoli, a source of isothiocyanates, but is not modified by GST genotype, *J. Nutr.* **134**:1134–1138.
14. J. L. Chin and R. E. Reiter (2004). Molecular markers and prostate cancer prognosis, *Clin. Prostate Cancer* **3**:157–164.
15. K. Huppi and G. V. Chandramouli (2004). Molecular profiling of prostate cancer, *Curr. Urol. Rep.* **5**:45–51.
16. V. M. Adhami, N. Ahmad, and H. Mukhtar (2003). Molecular targets for green tea in prostate cancer prevention, *J. Nutr.* **133**:2417S–2424S.
17. J. L. Ashurst and J. E. Collins (2003). Gene annotation: Prediction and testing, *Annu. Rev. Genomics Hum. Genet.* **4**:69–88.
18. F. S. Collins, M. S. Guyer, and A. Charkravarti (1997). Variations on a theme: cataloging human DNA sequence variation, *Science* **278**:1580–1581.
19. G. D. Schuler, M. S. Boguski, E. A. Stewart, et al. (1996). A gene map of the human genome, *Science* **274**:540–546.
20. A. Persidis (1998). Proteomics, *Nat. Biotechnol.* **16**:393–394.
21. M. J. Geisow (1998). Proteomics: One small step for a digital computer, one giant leap for humankind, *Nat. Biotechnol.* **16**:206.

22. L. M. Raamsdonk, B. Teusink, D. Broadhurst, et al. (2001). A functional genomics strategy that uses metabolome data to reveal the phenotype of silent mutations, *Nat. Biotechnol.* **19**:45–50.

23. E. C. Butcher, E. L. Berg, and E. J. Kunkel (2004). Systems biology in drug discovery, *Nat. Biotechnol.* **22**:1253–1259.

24. Annonymous (2004). Arrays come to age, *Genome Technol.* **42**:38–39.

25. T. R. Golub, D. K. Slonim, P. Tamayo, et al. (1999). Molecular classification of cancer: class discovery and class prediction by gene expression monitoring, *Science* **286**:531–537.

26. U. Palaniappan, R. I. Cue, H. Payette, and K. Gray-Donald (2003). Implications of day-to-day variability on measurements of usual food and nutrient intakes, *J. Nutr.* **133**:232–235.

27. K. O. Sarri, M. K. Linardakis, F. N. Bervanaki, N. E. Tzanakis, and A. G. Kafatos (2004). Greek Orthodox fasting rituals: A hidden characteristic of the Mediterranean diet of Crete, *Br. J. Nutr.* **92**:277–284.

28. R. Roky, I. Houti, S. Moussamih, S. Qotbi, and N. Aadil (2004). Physiological and chronobiological changes during Ramadan intermittent fasting, *Ann. Nutr. Metab.* **48**:296–303.

29. A. Wiser, E. Maymon, M. Mazor, et al. (1997). Effect of the Yom Kippur fast on parturition, *Harefuah* **132**:745–824.

30. M. Kuokkanen, N. S. Enattah, A. Oksanen, et al. (2003). Transcriptional regulation of the lactase-phlorizin hydrolase gene by polymorphisms associated with adult-type hypolactasia, *Gut* **52**:647–652.

31. L. Perusse, T. Rankinen, A. Zuberi, et al. (2005). The human obesity gene map: The 2004 update, *Obesity Res.* **13**:381–490.

32. A. Marti, M. J. Moreno-Aliaga, J. Hebebrand, and J. A. Martinez (2004). Genes, lifestyles and obesity, *Int. J. Obesity-Relat. Metab. Disorders* **28**(Suppl 3):S29–S36.

33. R. Nielsen (2004). Population genetic analysis of ascertained SNP data, *Hum. Genomics* **1**:218–224.

34. S. H. Jung, H. Bang, and S. Young (2005). Sample size calculation for multiple testing in microarray data analysis, *Biostatistics* **6**:157–169.

35. F. Gao, B. C. Foat, and H. J. Bussemaker (2004). Defining transcriptional networks through integrative modeling of mRNA expression and transcription factor binding data, *BMC Bioinformatics* **5**:31.

36. J. B. Tenenbaum, V. de Silva, and J. C. Langford (2000). A global geometric framework for nonlinear dimensionality reduction, *Science* **290**:2319–2323.

37. S. T. Roweis and L. K. Saul (2000). Nonlinear dimensionality reduction by locally linear embedding, *Science* **290**:2323–2326.

38. D. L. Donoho and K. Grimes (2003). *Hessian Eigenmaps: New Locally Linear Embedding Techniques for High-Dimensional Data.* Department of Statistics, Stanford University, Stanford, CA, pp. 1–15.

39. G. A. Thorisson and L. D. Stein (2003). The SNP Consortium website: Past, present and future, *Nucleic Acids Res.* **31**:124–127.

40. M. Belkin and P. Niyogi (2003). Laplacian eigenmaps for dimensionality reduction and data representation, *Neural Comput.* **15**:1373–1396.

41. L. M. Bruce, C. H. Koger, and J. Li (2002). Dimensionality reduction of hyperspectral data using discrete wavelet transform feature extraction, *IEEE Trans. Geosci. Remote Sensing* **40**:2331–2337.

42. W. S. DeSarbo, A. M. Degeratu, M. Wedel, and M. K. Saxton (2001). The spatial representation of market information, *Market. Sci.* **20**:426–441.
43. K. Z. Mao (2005). Identifying critical variables of principal components for unsupervised feature selection, *IEEE Trans. Syst. Man Cybern. Part B Cybern.* **35**:339–344.
44. Y. Tan, L. Shi, W. Tong, and C. Wang (2005). Multi-class cancer classification by total principal component regression (TPCR) using microarray gene expression data, *Nucleic Acids Res.* **33**:56–65.
45. L. Liu, D. M. Hawkins, S. Ghosh, and S. S. Young (2003). Robust singular value decomposition analysis of microarray data, *Proc. Natl. Acad. Sci. U.S.A.* **100**:13167–13172.
46. Y. H. Taguchi and Y. Oono (2005). Relational patterns of gene expression via non-metric multidimensional scaling analysis, *Bioinformatics* **21**:730–740.
47. M. A. Mendez, C. Hodar, C. Vulpe, M. Gonzalez, and V. Cambiazo (2002). Discriminant analysis to evaluate clustering of gene expression data, *FEBS Lett.* **522**:24–28.
48. J. Wang, J. Delabie, H. Aasheim, E. Smeland, and O. Myklebost (2002). Clustering of the SOM easily reveals distinct gene expression patterns: Results of a reanalysis of lymphoma study, *BMC Bioinformatics* **3**:36.
49. E. Hubbell, W. M. Liu, and R. Mei (2002). Robust estimators for expression analysis, *Bioinformatics* **18**:1585–1592.
50. C. Li and W. Hung Wong (2001). Model-based analysis of oligonucleotide arrays: Model validation, design issues and standard error application, *Genome Biol.* **2**: RESEARCH0032.
51. C. Li and W. H. Wong (2001). Model-based analysis of oligonucleotide arrays: expression index computation and outlier detection, *Proc. Natl. Acad. Sci. U.S.A.* **98**:31–36.
52. R. A. Irizarry, B. Hobbs, F. Collin, et al. (2003). Exploration, normalization, and summaries of high density oligonucleotide array probe level data, *Biostatistics* **4**:249–264.
53. W. M. Liu, R. Mei, X. Di, et al. (2002). Analysis of high density expression microarrays with signed-rank call algorithms, *Bioinformatics* **18**:1593–1599.
54. B. M. Bolstad, R. A. Irizarry, M. Astrand, and T. P. Speed (2003). A comparison of normalization methods for high density oligonucleotide array data based on variance and bias, *Bioinformatics* **19**:185–193.
55. R. S. Parrish and H. J. Spencer, 3rd (2004). Effect of normalization on significance testing for oligonucleotide microarrays, *J. Biopharm. Stat.* **14**:575–589.
56. B. Rosati, F. Grau, A. Kuehler, S. Rodriguez, and D. McKinnon (2004). Comparison of different probe-level analysis techniques for oligonucleotide microarrays, *Biotechniques* **36**:316–322.
57. P. Szaniszlo, N. Wang, M. Sinha, et al. (2004). Getting the right cells to the array: gene expression microarray analysis of cell mixtures and sorted cells, *Cytometry Part A* **59**:191–202.
58. D. de Ridder, C. E. van der Linden, T. Schonewille, et al. (2005). Purity for clarity: the need for purification of tumor cells in DNA microarray studies, *Leukemia* **19**:618–627.
59. K. Dawson, R. L. Rodriguez, and W. Malyj (2005). Sample phenotype clusters in high-density oligonucleotide microarray data sets are revealed using isomap, a nonlinear algorithm, *BMC Bioinformatics* **6**:195.

18

CULTURAL HUMILITY: A CONTRIBUTION TO HEALTH PROFESSIONAL EDUCATION IN NUTRIGENOMICS

Melanie Tervalon[1] and Erik Fernandez[2]

[1]*Children's Hospital and Research Center at Oakland, Oakland California*
[2]*University of California–Davis, School of Medicine, Davis, California*

18.1 INTRODUCTION

Nutrigenomics enters the national work of eliminating racial and ethnic health disparities at an exciting juncture in the history of science. Rapid advances in genomic and nutritional sciences are joining breakthroughs in biological and social realms in ways that will improve the health and well-being of groups and individuals, from all racial, ethnic, cultural, and social backgrounds. The emerging conceptual models and data from the biological and social sciences that explain genetic variations are noteworthy. Each arena calls for an interpretation in the context of historical and contemporary definitions of race, ethnicity, and culture. The new biological data can be at odds with existing conceptual social explanations of race, ethnicity, and culture. This milieu provides an excellent environment for a health professionals education program, which explores the strengths and weaknesses of these emerging social and biological models and their applicability to health care and health outcomes [1–3].

This social science–biology interface is demonstrated well by BiDil, a drug that appears to be of particular benefit to individuals who sustain heart failure and identify themselves as having ancestry that can be traced predominantly to Africa. BiDil

Nutritional Genomics: Discovering the Path to Personalized Nutrition
Edited by Jim Kaput and Raymond L. Rodriguez

is an example of the possibilities that exist in the search for solutions for particular population groups, in this case individuals of African descent, who, in the United States, often have poorer health outcomes than individuals in the country without this genetic composition [4]. At the same time, this scientific advance uncovers other unanswered questions that are inevitably linked to the discovery of race- or ethnicity-based medicines: How do we identify those in the country who may have ancestry traced to Africa? Will we rely on self-report for genetic history? And, if self-report is used, will many individuals go untreated who could benefit from this therapy, since individuals may not know, or may not claim the African portion of their heritage? Or can we expect some genetic testing methods to assist us? Will these kinds of discoveries, in pill form, cause researchers, community members, and individuals alike to neglect attending to the social and environmental factors that contribute to the unfortunate matrix of poor health outcomes for distinct racial and ethnic groups and subgroups in the United States [5]?

At the same time, the challenge of informing all stakeholders—bench scientists, health professionals, and consumers—of the possibilities and problems with novel genetic, medical, and nutritional developments can be daunting. Of practical importance is the timely dissemination of new clinical tests, drugs, and food intake recommendations that are expected products in this field. Now that culture, race, and ethnicity are an acknowledged and prominent part of the research agenda, how do we consciously incorporate respectful interaction with individual and group cultural beliefs and practices, which inform this new field and are a prominent part of the clinician–patient, researcher–subject relationship [6]?

It is in the context of emerging knowledge about the links of genetics, society, and environment to health outcomes that we offer the principles and practice of *cultural humility* for use in health and research professional education [7]. This approach helps health professionals to interact with care and reason in clinical practice and scientific research with the elements of culture, race, and ethnicity in the presence of the rapid advances in the field of nutrigenomics. These principles offer a significant contribution to health professional education focused on eliminating racial and ethnic health disparities.

This chapter covers two areas. First, a description of the origin and principles of cultural humility as a teaching approach for the health professions. Second, the practical application of this approach in the Pilot Curriculum of the Center of Excellence in Nutritional Genomics, colocated at Children's Hospital and Research Center at Oakland (CHRCO) and the University of California at Davis [6].

18.2 CULTURAL HUMILITY

The Multicultural Curriculum Program (MCCP) at Children's Hospital at Oakland (1994–1997) created an innovative medical education project with the goal of teaching health professionals—pediatric residents, attendings, community physicians, social workers, occupational therapists, psychologists, and other members of the hospital staff—to skillfully and respectfully embrace issues of multiculturalism in their practices and services. The goals and content of the program are described in Tables 18.1 and 18.2 [8].

The term cultural humility captures the central principles that emerged through the implementation of the educational process. Reference [7] explains these principles in greater detail. Briefly, cultural humility encourages health professionals and staff of all job descriptions to work with the issues of culture and difference in health and research using several principles: self-reflection, lifelong learning, patient-focused interviewing and care, community expertise, redressing power imbalances in patient–provider relationships, and calls for a parallel process within health institutions that models and supports these principles in policies and practices. The descriptions that follow highlight three of these principles: patient-focused interviewing, self-reflection, and lifelong learning [7, 8].

Patient-Focused Interviewing and Care: The Community Teaches

From the earliest phases of development, the MCCP shifted the expertise in issues of culture, language, and difference from the academic institution to the community. Community members participated as program planners and presenters, using their first-person voice and their physical presence to expose hospital personnel to information that is immediately relevant to patients' current cultural realities and life experiences within our society. The community demonstrated its unique position in teaching health professionals and institutions by describing their lived experiences at the complex interfaces of race, racism, ethnic identity, gender, culture, health beliefs, religion, and the demands of day-to-day living in relationship to health, well-being, and health disparities.

For many participants, this was their first opportunity to learn from instructors who explicitly identified themselves with membership in any of several historically

TABLE 18.1. Overall Goals: MCCP 1994–1997

- Provide a safe, intellectual, and practical environment for supporting respect, understanding, and appreciation of difference.
- Establish and sustain a positive sense of the value of difference in our work in health care.
- Provide didactic and experiential information on cultural, racial, and ethnic differences.
- Utilize this information to transform our behaviors and skills in our health-care practices.

TABLE 18.2. Uniform Content of the Learning Modules: MCCP 1994–1997

- Social, cultural, political, and economic history
- Health beliefs/concepts using patient-based explanatory models of illness/wellness
- Cultural perspectives on specific biomedical illness/wellness categories
- Health promotion/advocacy that is community informed
- Communication skills for health practitioners and workers

disenfranchised groups: these included not only racial and ethnic minorities, but also lesbians, gays, the working poor, and very poor people [8–12]. For participants, making community members teachers, and a great source of expertise, communicates and reinforces that people who are traditionally characterized as disenfranchised are capable, necessary, and essential partners in their own health care. The cultural humility model indicates that such an orientation to the patient–provider interaction, that is, patient as teacher and expert, remains an unspoken requirement in the process of understanding the relationship of culture to health, especially given that culture is a dynamic and ever-changing phenomenon. At the same time, this orientation of patient as teacher matches well with the burgeoning literature on quality care in health [13, 14].

The strategy of highlighting, in more than a token sense, the scholarship and expertise in the community is a seldom-used, though often touted, approach in health professions education [13, 15, 16]. Personal narratives from community members, which described gripping stories of illness that range from excellent to suboptimal experiences within our health-care institution, had a distinct impact on participants. These presentations provided some of the most impressionable teaching and learning moments of the curriculum. This is also consistent with the experience of other programs [17–21].

Self-Reflection

The program utilized the concepts and content from the disciplines of history, sociology, psychology, political science, education, and especially medical anthropology to teach health professionals about generally held culture-based health beliefs and practices in large group session. Strategies to avoid the inherent pitfall of stereotyping individuals in teaching generalities of cultural groups were carefully crafted and included (1) up-front, explicit statements regarding the reality of intragroup diversity; (2) case presentations of intergenerational conflict and varying dimensions of socioeconomic class; and (3) speakers, cultural presenters, and readings that reflected the great variation of backgrounds, traditions, and concerns [7, 8, 22]. For example, the 1996 Asian American Families and Issues module included presenters who were of Asian Indian, Chinese, Filipino, Japanese, Korean, Laotian, and Vietnamese descent [8].

In small group sessions, participants were encouraged to reach beyond an intellectually detached description of "The Other" and engage in self-reflection, self-knowledge, and self-critique. Rarely are health professionals given the opportunity to publicly examine their own experiences of culture, race, racism, ethnicity, class, or gender identification in the context of health or in the context of the broader society. Instead, the notion of "color blind" health professionals was rejected and trainees were encouraged to acknowledge and identify the ways in which health professionals in this country have incorporated the insidious social messages of inferiority and lack of ability so often attributed to the disenfranchised. Presenters and program planners persistently and explicitly cited the expanding medical literature and case studies from practice that describe differential treatment of patients by race, ethnicity, linguistic capability, sexual orientation, and class, which demand an urgent and honest appraisal from members of the "helping" professions [14, 22–25].

Facilitators led trainees through the exploration of their own unintentional and usually unconscious processes of racism, classism, and homophobia. Of note, the trainees' self-reflection revealed how these processes get magnified in the context of the culture of our own medical profession, a culture fraught with power imbalances, and how these processes are shaped by time pressures and other challenges common to the practice of medicine in this country in this present era [7, 12, 26–33].

Lifelong Learning: A Process, Not an Endpoint

This principle emphasizes to participants that the development of cultural competence does not begin and end with attendance at one, two, or several diversity seminars. In contrast to the notion of static mastery and expertise, which is typically the gold standard in health professions training and implied by the term "competence," cultural humility reinforces a lifelong process of personal and professional development in which one engages again and again, with patients, colleagues, community members, and oneself with respect to issues of culture and difference. It was made clear that it is unrealistic to teach (or for trainees to master) every nuance and detail of the culture-based health belief practices and life experiences of the many peoples in the United States today because these practices are constantly evolving.

Some of the continuing call to "mechanize" this material reflects a certain resistance to the notion of simply listening to and validating the perspectives and expertise of individual patients and communities. At the same time, it also reflects a resistance to the humility and courage required to think and respond on one's own and anew in each clinical encounter and to say that one does not know, thus becoming the patient's student [7, 8, 29–34].

This process-oriented, lifelong learner model of adult diversity education frees people from the fear and pain of making mistakes publicly, and from the fear and complacency of doing what it takes to learn from those mistakes, as active participants in the learning process [7].

18.3 SUMMARY

In brief, cultural humility is an approach that deemphasizes the notion of static mastery and elevates the dynamic and ever-changing interplay of culture, health, community, and each unique individual [7, 8, 30]. This approach also encourages respect, humility, and courage in every moment of the health-care providers practice so that individual patients and families feel comfortable, validated, and empowered enough to teach us about issues related to culture and difference in every clinical encounter [29–31]. It reminds health professionals to pay close attention to their own multidimensional cultural identities, beliefs, and biases as they operate in the patient–professional encounter. Optimally, these principles encourage the establishment of a complex and mutually beneficial partnership between patient and professional, in the best interest of the patient's health and well-being [7, 21, 35].

There are many students, seasoned clinicians, researchers, and other service providers waiting for constructive intellectual and practical leadership from which they can learn how to talk to one another and with their patients more respectfully, more compassionately, and more effectively about issues of culture, race, racism, ethnicity, sexual orientation, literacy, and economic and environmental conditions to name but a few. The cultural humility model offers this significant contribution to the urgent national mandates for health professionals and educators to address the elimination of racial and ethnic health disparities in this 21st century.

The cultural humility model was not created with the field of nutrigenomics in mind. However, the implications of nutritional genomics advances to health-care practice and research in diverse populations, both locally and internationally, make incorporation of the cultural humility model into health-care practitioner and researcher education more important than ever before. Specifically, gene–environment interactions, the heart of nutrigenomics, may differ among individuals with different genetic ancestries and cultures. Environment consists of not only nutrient intakes but also of social context that includes cultural norms and practices. Therefore, developing solutions to nutritional genomics health-care problems like overweight, type 2 diabetes mellitus, cardiovascular disease, and undernutrition is best approached with a framework in which an expanded view of environment is a central idea. The cultural humility model provides this kind of framework and can assist researchers and clinicians alike who seek to understand the molecular basis of nutrient-gene interactions and how this information can translate into improved health and well-being for patients and families.

Pilot Curriculum: Center of Excellence in Nutritional Genomics 2004–2007

In this section, we describe the construction and implementation of the Pilot Curriculum of the Center for Nutritional Genomics colocated at Children's Hospital and Research Center at Oakland (CHRCO) and University of California at Davis as a present-day example of utilizing the cultural humility model in health professional education.

CHRCO is a private, nonprofit, exclusively pediatric medical center whose patient population reflects the significant racial and ethnic diversity of the surrounding community (Table 18.3). Conversely, like many other medical training institutions in this country, the members of the CHRCO hospital-based teaching faculty are relatively homogenous in racial and ethnic composition [9–11]. While the approximately 70 pediatric residents represent more diverse racial, cultural, and ethnic groups than in many institutions, the combined physician racial, cultural, and ethnic representation is still far from reflective of the patient population (Table 18.4). The anticipated research outcomes in the field of nutrigenomics have wide-ranging social and health implications for racially and ethnically diverse patient populations.

TABLE 18.3. Percentage of Total Population by Race of Alameda County, California,[a] in 2000

Race	White	Asian	Hispanic/ Latino	Black/ African-American	Native Hawaiian/ Pacific Islander	American Indian/ Alaskan Native	Other	Two or More
Percentage of total	40.9	20.3	19	14.6	0.6	0.4	0.3	3.9

[a] Reported in Alameda County Health Status Report 2003 by the Alameda County Public Health Department.

TABLE 18.4. Population of Physicians (Attending/Resident/Intern Levels) Employed by Children's Hospital and Research Center Oakland, as of December 2003, and Divided by Racial Identification[a]

Race	White	Asian	Black	Hispanic	Indian
Percentage of total (Actual number)	57.7 (86)	23.5 (35)	10 (15)	6.7 (10)	2 (3)

[a] Data reported by the CHRCO Human Resources Department.

Therefore, it is important for pediatricians at CHRCO and its collaborative institutions, entrusted with leadership in the field of nutrigenomics research, to be exquisitely educated in cultural diversity and how it intersects with health.

18.4 GOALS AND OBJECTIVES: CURRICULUM CONTENT

There are two main goals for the Pilot Curriculum. The first goal is to design an augmented cultural competence educational experience directed at pediatricians working in CHRCO. The principles of the cultural humility model and the current accumulated knowledge in the field of cultural competence are being taught within the context of the expanding field of nutrigenomics [22, 27, 32, 36–38].

The objectives in achieving this first goal are threefold. The first is to increase participants' baseline knowledge about local and national health disparities, cultural humility, cultural competence, the human genome project, and the concepts of nutrigenomics. The second is to provide participants with frameworks for incorporating this new information into their clinical practice. The last objective is to document the design, implementation strategies, content, and evaluation methods used in this educational curriculum for use and adaptation in other teaching settings.

The second goal is to design a curriculum and evaluation method that will demonstrate the impact of this physician-training program in recognizable, clinically significant terms. The purpose is to demonstrate that participation in the curriculum can lead to improved culture-based awareness, knowledge, and skills for pediatricians and, thus, improve health outcomes and diminish health disparities for diverse patient populations that the physicians serve [39].

There are two patient-centered objectives and one trainee-centered objective under this second goal. The patient-centered objectives are (1) to demonstrate improved patient-based knowledge about current strategies to prevent and reduce obesity and (2) to demonstrate improved cooperation with prevention and intervention strategies once negotiated and agreed upon with the pediatrician. The trainee-centered objective is to demonstrate improved cross-cultural communication skills, in genomic research, nutrition, diet, and prevention of obesity.

18.5 GOALS AND OBJECTIVES: CURRICULUM DESIGN

There are four curriculum design objectives. The first objective is to ensure that the curriculum is reproducible and customized. To that end, the curriculum is presented in modules, or segments, each with its own discrete learning objectives and each able to stand alone as a single educational session. The overall modular design, however, is a progressive one with the content of the modules building on one another. The sequential, modular nature of this design allows individuals and institutions to customize their learning experience by undertaking the entire sequence or only those modules that are useful given their unique finances, time constraints, and institutional considerations.

The second design objective is to incorporate varied and interwoven modalities in the delivery of information. This is based on the understanding that multilevel and integrated teaching methods work best for maximum knowledge retention and skill acquisition for adults [40]. Therefore, the modalities used in the curriculum will include, but are not limited to, lectures, videotaped model patient interviews, self-guided online lessons, film clips, and small group participation.

Another objective is to introduce effective community input and participation in both the design and implementation of curriculum content [7, 8, 12, 20, 21]. This occurs via community cultural leader recommendations, construction, and oversight into the content of online materials as well as community participation in modeling physician–patient interactions in scripted model cases. Participating pediatricians are required to attend community activities organized by the Center for Nutrigenomics Community Outreach and Education Core. These examples are concrete ways in which community inclusion in the design of health professions education can occur. The lessons learned and documented in this process will serve as a template for institutions using this curriculum to maximize community inclusion in the curriculum's implementation.

The final objective is to present culture and culturally influenced health beliefs in the context of power differentials that lead to health disparities [41]. In this paradigm, the organizing principle does not list "accepted" cultural health beliefs for particular groups but instead is focused on showing how health beliefs work with social dynamics and cultural constructs and contribute to health disparities [7, 8, 12, 23, 27, 35, 41].

18.6 CURRICULUM STRUCTURE AND CONTENT: DIDACTICS, SMALL GROUPS, AND VIDEOTAPING

Didactics

Didactic materials are presented online. The participants will access the material at their convenience at home or work, through the CHRCO media resource. The didactic sessions cover the basic information of interfaces of culture, race, ethnicity, and the wide variety of elements of personal makeup, which contribute to differential experiences for patients and providers in health care; the origin, principles, and application of the cultural humility model and the tenets of cultural and linguistic competence; and the relationship between this knowledge and practice, with a focus on obesity. This online portion of the curriculum includes written material in a slide show format, required readings, and selected film clips or originally produced videotaped segments illustrating case examples. This varied presentation delivers the required information in an engaging and clinically applicable way. Each online segment starts with a pretest and ends with a separate post-test, with immediate feedback for the participant.

Past experience with multicultural training programs at CHRCO has highlighted that job and life-style time constraints of physicians may be a barrier to participation in what may appear as "extra" work. We acknowledge that different physicians have different levels of prior training in cultural competence. Self-directed review of didactic modules allows for pediatrician participants to review the material at the pace most conducive to each person's unique experience, life style, and learning style. While all participants are required to review all modules, the self-directed modular format allows each participant to spend more or less time with those modules with which he/she feels more or less comfortable, respectively.

Small Group Sessions

The nature of the discussion of culture and power imbalances, and how the two interact to create health disparities is potentially explosive. This fact is most poignant in health professionals training, where physicians are faced with the realization that their well-intentioned actions may be based on incorrect assumptions and, therefore, complicit in the perpetuation of health disparities [22]. Therefore, discussions among health professionals of our possible role in the creation and fostering of health disparities can best take place in an emotionally safe and confidential arena [29, 42, 43]. Small groups provide an excellent opportunity to practice cross-cultural dialogue, to experience one's own and others' sense of anger, pain, loss, and subsequent relief and freedom when one acknowledges the power imbalance in the patient–health professional interaction [29, 33, 44]. The small group experience keeps the patient–physician interaction from being the first time a physician is challenged to engage with the often highly charged and often subtle presentations of racial, cultural, or class issues that emerge in a patient-centered, patient-focused communication in the clinical encounter.

The small group experience gives the Nutrigenomic Curriculum Pilot plan organizers a strong foundation from which to create the small group session process and content [7, 9, 30]. First, the moderators for the small group sessions are recruited from CHRCO senior pediatricians who have participated in previous small group experiences in multicultural education at CHRCO. Therefore, pediatrician participants benefit from being led in this learning process by trained and experienced colleagues with whom they already have a working relationship.

Second, the sessions are organized around a detailed review of sample cases, centered on nutrigenomics. Case review allows the incorporation of the didactic online learning points with immediate clinical material in a setting designated for observing, modeling, and practicing new communication skills. Third, these sessions are interspersed between online self-directed modules. The timing of small group sessions allows the participants to immediately see the clinical relevance of the didactic material. The small group sessions also mirror the progressive nature of the modular online format.

Last, as in the online format, a pretest and a post-test are given before and after each small group experience, respectively. The results of these evaluations allow small group facilitators to modify or adjust their teaching/facilitation styles for future sessions, to help participants focus on those areas of communication in which they need more work, and to indicate what materials the participants learn, retain, and use over the course of the curriculum.

Each small group session begins with a review of the content and teaching points of the preceding online module topics. Facilitators can then ensure that participants satisfied the curriculum requirements and can clarify any questions that learners may have about the material presented, so that participants start each session with a similar knowledge base.

The facilitators then present a detailed case constructed to illustrate the online module content. Participants discuss the case in the context of the new material with the facilitators guiding the discussion toward an integration and exploration of the implications of the new materials for clinical practice and health-care delivery.

Participants assume the role of physician or patient in an exercise of modeling cross-cultural interviewing skills and recovery skills. The facilitators and fellow participants give constructive criticism and the facilitators guide the discussion that follows.

Each small group session ends with written self-reflection exercises. This gives participants the opportunity to strengthen their ability to identify and address their strengths as well as their own biases, stereotypes, and assumptions that may interfere with the delivery of quality health care. These exercises also provide the participants with the opportunity to identify their strengths and to focus improvement in deficient areas in preparation for subsequent sessions.

Videotaping

Videotaping participants interviewing patients and the patients' families is an important facet in the educational experience. This portion of the curriculum allows trainees to translate the didactic information into attitudes, behaviors, and skills. Trainees witness, critique, and evaluate their own performance, in conjunction with a facilitator-educator. A review of the actual videotaped interview gives the facilitator and learner an opportunity to critically discuss and address specific interviewing and recovery techniques in a real-time clinical setting.

Similar training methods termed OSCEs currently exist for the teaching and evaluation of physical exam skills. Current trials of cultural OSCEs, in which students' interviewing skills are critiqued using standardized actor-patients, are occurring at several medical centers such as the Maimonides Medical Center in Brooklyn, New York, and the University of Arizona Medical School at Tucson. Patients with the primary diagnosis of obesity, followed regularly in the physician participant's continuity clinic are recruited for the videotaping session.

The trainee physician discusses the learning objectives with a facilitator one-on-one prior to the session and the participant's reactions and impressions are discussed in a private meeting after the session. These individual sessions build on the principle of learning about one's own strengths and weaknesses in a safe environment, in the interest of improving patient outcomes. Pre-interviewing and post-interviewing objective lists are created to aid facilitators and participants in assuring that curriculum requirements are being met.

18.7 THE TEACHING STAFF

The online content was constructed by the author of this chapter with the cooperation of colleagues in the Center for Nutritional Genomics, the experience of leaders in the field of cultural competence, and the expertise of community members already engaged in local efforts to reduce racial and ethnic health disparities.

Small group facilitators are attending physicians at CHRCO who have formal training in the small group process. This formal training may have occurred as part of CHRCO's Multicultural Curriculum Program or through an outside agency. These same facilitators will serve as videotape interview preceptors.

Community members and trained actors are hired to act as patients for both the online video segments and practice patient interviews with participants.

18.8 EVALUATION

Two data sources are collected as information for the evaluation. The first source is the various pre- and post-tests described throughout this chapter. These tools provide the ability to evaluate whether participants remember what they were taught. The videotaped patient interviews are the second source of data. This data provides insight into whether or not participants use the information they have been taught in a clinical setting [45].

The most important measure of this curriculum's efficacy in application of the cultural humility model to nutrigenomics education will be evaluation of real patient health outcomes. At the beginning of the two-year curriculum, a patient cohort will be recruited for each of the participants from their continuity clinic. These patients will carry the diagnosis of obesity. It is our hope that we will see changes in patient outcomes, albeit small ones, which can be documented and connected to the materials, learned by the trainees and shared with patients, through the course of the curriculum.

The authors have chosen to focus on the pediatric residents and attending physicians at CHRCO for the sake of having a ready-made and sustainable test cohort to follow over the years of the pilot project and evaluation. This does not mean that the curriculum has been constructed solely for use in physician training. On the contrary, the authors have expressly designed a curriculum whose major subjects and topics (e.g., cultural humility, lifelong learning, and institutional racism) can be generalized to all fields. The specific presentation of the material (e.g., case scenarios, real-life examples of the general principles in action, and speakers used to present the material) can be customized to and resources drawn from each institution and/or field of study using the curriculum. This format allows researchers, health-care providers, administrators, social workers, and participants from all job descriptions within a given institution to participate and benefit from this curriculum once it has been tested with our initial physician cohort. The authors in the Education Core have further ensured exposure of the curriculum to participants other than health-care providers by presenting it to students from many different fields of study as a seminar-based learning experience at the University of California–Davis and by offering one-day training seminars on specific aspects of the curriculum at CHORI.

18.9 CONCLUSION

Today, inside and outside of the university setting, use of the words genomics, race, ethnicity, community, culture, cultural competence, health disparities, health inequities, social class, and many others are all around us. We must, therefore, ask ourselves what, in truth, we are conveying to each other in our discourse and dialogue related to these terms. Are we communicating with vocabulary that has shared meaning and nuance? Or are we approaching these areas through our own professional disciplines, personal interpretation, or lived experience? And how do we arrive at common language and communicate patiently with each other through our varied intellectual and practical meanings?

It is the goal of this innovative curriculum project to address these questions by exploring these terms, their complex layers of meaning, and their contemporary applications in the community *and* in nutritional genomics-based health-care services and research. It is our hope that dialogue and practical work of this kind, across disciplines and meaning, encourages solutions that do indeed eliminate racial and ethnic health disparities in our lifetime.

ACKNOWLEDGMENTS

We wish to thank the California Wellness Foundation for funding the MCCP 1994–1997 and Jann Murray Garcia, MD, MPH for her contributions to the manuscript. This work was supported in part by a grant from the National Center for Minority Health and Health Disparities (P60MD00222).

REFERENCES

1. N. Kreiger (2003). Does racism harm health? Did child abuse exist before 1962? On explicit questions, critical science, and current controversies: An ecosocial perspective, *AJPH* **93**(2):94–199.
2. K. Fuller (2003). Health disparities: Reframing the problem, *Med. Sci. Monit.* **9**(3): SR9–SR15.
3. L. Kolonel, D. Altshuler, and B. Henderson (2004). The multiethnic cohort study: Exploring genes, lifestyle and cancer risk, *Nat. Rev.* **4**:1–9.
4. C. Johnson (2004). Debate erupts over a drug that works better in African-Americans, *Boston Globe*, August 24.
5. M. J. Bamshad, S. Wooding, W. Watkins, et al. (2003). Human population genetic structure and inference of group membership, *Am. J. Hum. Genet.* **72**:578–589.
6. K. Jackson (2003). Nutrigenomics: It's in the genes, *Today's Dietician* **Oct**:27–29.
7. M. Tervalon and J. Murray-García (1998). Cultural humility versus cultural competence: a critical distinction in defining physician training outcomes in medical education, *J. Health Care Poor Underserved* **9**:117–125.
8. M. Tervalon, K. Epstein, and J. Murray-García (2002). Children's Hospital Oakland Multicultural Curriculum Program Portable Curriculum, Oakland, CA.
9. J. Cohen (1998). Time to shatter the glass ceiling for minority faculty, *JAMA* **280**:821–822.
10. P. Gonzalez and B. Stoll (2002). The Color of Medicine: Strategies for Increasing Diversity in the U.S. Physician Workforce. Community Catalyst, supported by a grant from the W.K. Kellogg Foundation. www.communitycatalyst.org. Accessed 4/19/02.
11. A. Palepu, P. Carr, R. Friedman, et al. (1998). Minority faculty and academic rank in medicine, *JAMA* **280**:767–771.
12. W. Ventres and P. Gordon (1990). Communication strategies in caring for the underserved, *J. Health Care Poor Underserved* **1**:305–314.
13. P. Cleary and S. Edgman-Levitan (1997). Health care quality: Incorporating consumer perspectives, *JAMA* **278**:1608–1612.
14. K. Fiscella, P. Franks, M. Gold, et al. (2000). Inequality in quality: Addressing socioeconomic, racial and ethnic disparities in health care, *JAMA* **283**:2579–2584.
15. S. Rankin and M. Kappy (1993). Developing therapeutic relationships in multicultural settings, *Acad. Med.* **68**:826–827.

16. D. Levine, D. Becker, and L. Bone (1992). Narrowing the gap in health status of minority populations: A community–academic medical center partnership, *Am. J. Prev. Med.* **8**:319–323.

17. B. Li, D. Camiano, and R. Comer (1998). A cultural diversity curriculum: Combining didactic, problem-solving, and simulated experiences, *JAMWA* **53**:127–129.

18. J. Morse and P. Field (1995). *Qualitative Research Methods for Health Professionals*. Sage Publications, Thousand Oaks, CA.

19. L. Robbins, J. Fantone, J. Hermann, et al. (1998). Improving cultural awareness and sensitivity training in medical school, *Acad. Med.* **73**:S31–S34.

20. E. A. Jacobs, C. Kohrman, and M. Lemon (2003). Teaching physicians-in-training to address racial disparities in health: A hospital–community partnership, *Public Health Rep.* **118**:349–356.

21. M. Tervalon (2003). Community–physician education partnerships: One strategy to eliminate racial/ethnic health disparities. Commentary, *Public Health Rep.* **118**:357.

22. B. Smedley, A. Stith, and A. Nelson (eds.) (2002). *Unequal Treatment: Confronting Racial and Ethnic Disparities in Health Care*. Institute of Medicine National Academies Press, Washington, DC.

23. J. Carrillo, A. Green, and J. R. Betancourt (1999). Cross-cultural primary care: A patient-based approach, *Ann. Intern. Med.* **130**:829–834.

24. D. B. Smith (1998). Addressing racial inequities in health care: Civil rights monitoring and report cards, *J. Health Politic Policy Law* **23**:75–105.

25. D. Williams (2000). Understanding and addressing racial disparities in health care, *Minority Health Today* **2**:30–39.

26. D. Sue (1991). A model for cultural diversity training, *J. Counseling Dev.* **70**:99–105.

27. J. Kai, J. Spencer, M. Wilkes, et al. (1999). Learning to value ethnic diversity: What, why, how? *Med. Educ.* **33**:616–623.

28. J. Dreachslin and P. Hunt (1996). *Diversity Leadership*. Health Administration Press, Chicago.

29. E. Pinderhughes (1989). *Race, Ethnicity, and Power: Keys to Efficacy in the Clinical Encounter.* The Free Press, New York.

30. J. Murray-García and S. Harrell (2002). *The Small Group Experience in Multicultural Medical Education.* Multicultural Curriculum Program at Children's Hospital Oakland, Oakland, CA, Desktop publication.

31. C. Ridley (1995). *Overcoming Unintentional Racism in Counseling and Therapy: A Practitioner's Guide to Intentional Intervention.* Sage Publications, Thousand Oaks, CA.

32. A. Green, J. Betancourt, and J. Carrillo (2002). Integrating social factors into cross-cultural medical education, *Acad. Med.* **77**:193–197.

33. K. Kavanaugh and P. Kennedy (1992). *Promoting Cultural Diversity: Strategies for Health Care Professionals.* Sage Publications, Thousand Oaks, CA.

34. K. Culhane-Pera, C. Reif, E. Egil, et al. (1997). A curriculum for multicultural education family medicine, *Fam. Med.* **29**:719–723.

35. M. Tervalon (2003). Components of culture in health for medical students' education. Special theme article, *Acad. Med.* **78**(6):June.

36. T. L. Cross, B. J. Bazron, K. W. Dennis, et al. (1989) *Towards a Culturally Competent System of Care: A Monograph on Effective Services for Minority Children Who Are Severely Emotionally Disturbed.* Georgetown University Child Development Center, Washington, DC.
37. G. Flores (2000). Culture and the patient–physician relationship: Achieving cultural competency in health care, *J. Pediatr.* **136**:14–23.
38. C. Brach and I. Fraser (2000). Can cultural competency reduce racial and ethnic health disparities? A review and conceptual model, *Med. Res. Rev.* **57**:181–217.
39. R. H. Loudon (1999). Educating medical students for work in culturally diverse societies. *JAMA* **282**(9): 875–880.
40. F. Paas (2003). Cognitive load measurement as a means to advance cognitive load theory. *Educ. Psychologist* **38**(1): 63–71.
41. D. Wear (2003). Insurgent multiculturalism: Rethinking how and why we teach culture in medical education, *Acad. Med.* **78**:549–554.
42. J. Ring (2000). The long and winding road: Personal reflections of an antiracism trainer, *Am. J. Orthopsychiatry* **70**:73–81.
43. B. D. Tatum (1992). Talking about race, learning about racism: The application of racial identity development theory in the classroom, *Harvard Educ. Rev.* **62**:1–24.
44. A. Abernathy (1995). Managing racial anger: A critical skill in cultural competence. *J. Multicultural Counseling Dev.* **23**:96–102.
45. J. R. Betancourt (2003). Cross-cultural medical education: Conceptual approaches and frameworks for evaluation, *Acad. Med.* **78**:560–569.

19

NUTRIENTS AND NORMS: ETHICAL ISSUES IN NUTRITIONAL GENOMICS

David Castle,[1] Cheryl Cline,[2] Abdallah S. Daar,[2,3]
Charoula Tsamis,[2] and Peter A. Singer[2]

[1] *Department of Philosophy, University of Guelph, Guelph, Ontario, Canada*
[2] *Joint Centre for Bioethics, University of Toronto, Toronto, Ontario, Canada*
[3] *Program in Applied Ethics and Biotechnology, University of Toronto, Toronto, Ontario, Canada*

19.1 PROACTIVE ETHICS AND NUTRITIONAL GENOMICS

In May 2001, a small start-up company in the United Kingdom launched the world's first nutrigenetic testing service. Capitalizing on new advances in nutrigenetic research, Sciona Ltd [1] began offering customized dietary advice to customers on the basis of a life-style questionnaire and genetic test. Sciona offered its tests for sale directly to the public through retail outlets. Within weeks of the products' appearance on store shelves, Sciona suddenly found themselves at the center of national media attention [2]. GeneWatch and the U.K. Consumers' Association had begun a campaign to raise public awareness about the service and how it was being marketed. They raised fears that tests might mislead customers about healthy life-style choices, customers might learn things about their health which they may not want to know about, insurers and employers might gain access to this information

Nutritional Genomics: Discovering the Path to Personalized Nutrition
Edited by Jim Kaput and Raymond L. Rodriguez

in the future, and customer's genetic information might be patented or used for research without their knowledge [3].

Sciona had proactively engaged a bioethics consultant [3]. Because their direct-to-consumer service was offered in a regulatory vacuum, however, there was no place they could turn for certification of good ethical practice in order to reassure the public [3]. Extant British regulations were outdated and ill-equipped to handle the new testing technology. Oversight mechanisms for the emerging direct-to-public genetic test market were also lacking at the time. Consequently, Sciona was left on its own to weather a storm of public controversy. In the ensuing months, the Body Shop contract was cancelled and a number of major drugstore chains publicly pledged not to carry their products. The company eventually abandoned its direct-to-consumer business strategy and began to offer their services through the more conventional route of qualified medical practitioners [4]. Meanwhile, the U.K. Human Genetics Commission conducted a study and subsequently issued a report containing suggestions about how to best regulate genetic tests made available directly to consumers [5].

Sciona is a telling illustration of the setbacks and uncertainties that can arise when scientific and technological innovation falls even slightly out of step with the social factors that influence its acceptance. Despite the best intentions to maximize the benefits and minimize the risks of new science and technology, not all innovation will prove to be socially acceptable. In other cases, innovations will be taken up but will prompt legal and ethical change. Ideally, science, ethics, and law would work synchronously to ensure that these desirable social outcomes are achieved without difficulty, yet this is not always the case. In fact, it appears to be the rare instances when "science" and "society" conjoin in a coordinated effort. More typically, science advances, ethics and law then react.

The fast pace of innovation in genomics and biotechnology need not outstrip their ethical and legal evaluation if the social effects of innovation are fully addressed. For example, the information gained from mapping and sequencing the human genome would have profound and often unexpected implications for individuals, families, and society. This point was recognized in the 1990s by the founders of the Human Genome Project, who set aside funding for the Ethical, Legal and Social Implications (ELSI) program. The mandate of the ELSI program is to identify and address the ethical, legal, and social implications of human genome research at the same time that the basic scientific issues are being studied. In this way, it is hoped that problem areas will be identified and solutions developed before adverse effects occur [6].

As part of the Genome Canada [4] supported GE3LS (Genomics, Ethics, Environment, Economics, Law and Society) initiative, we in the Canadian Program on Genomics and Global Health [5] have embarked on a large-scale program of research guided by a proactive approach [6]. Our project on the ethics of nutritional genomics reflects this programmatic outlook [6]. An international advisory panel has been established to identify anticipated risks and expected benefits of nutrigenomics (Figure 19.1), and focus groups and web-based public consultations have elicited feedback from a number of constituencies. The project recently completed a report

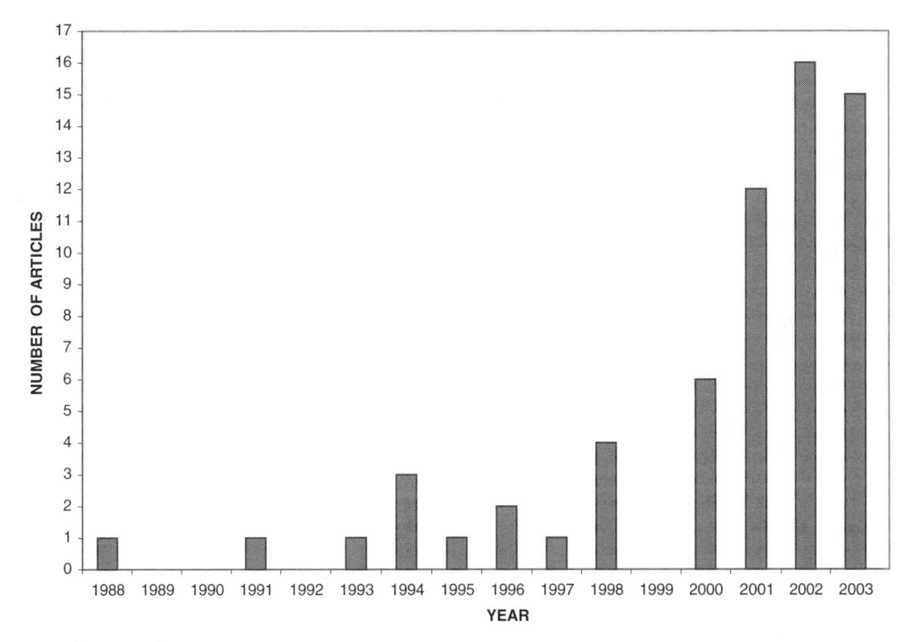

Figure 19.1. Number of articles published reflecting gene–nutrient interactions.

titled *Nutrition and Genes: Science, Society and the Supermarket*, which synthesizes the findings of these different activities. This report, which will be available in 2005, contains an overview and set of summary recommendations and observations on the ethics and regulation of nutrigenomics.

In this chapter, we describe the main ethical issues arising in nutrigenomics, and particularly in nutrigenetic applications. The purpose of airing these issues is to help prevent in the future the kinds of uncertainties faced by Sciona. The five issues that we discuss in this chapter arise in the context of both basic nutrigenomics research and its clinical and commercial uses: (1) claims of health benefits arising from nutrigenomics, (2) managing nutrigenomics information, (3) methods for delivering nutrigenomics services, (4) nutrigenomics products, and (5) access to nutrigenomics.

A final word about the perspective of this chapter is in order. Within the field of bioethics there is much discussion about the correct approach to analyzing ethical issues and what method for providing ethical resolutions should be advocated. Our guiding considerations are the interests of individuals and the public, the safeguarding of individuals from harm, and the promotion of a stable regulatory climate for research and commercialization. We hold the view that since individuals can benefit or be harmed by nutrigenomics, the autonomy of individuals is the appropriate initial "object" of analysis for nutrigenomics [6]. We are also concerned with the ethical principle of justice, particularly in the final section on access to nutrigenomics. We likewise recognize that there are competing methodologies seeking to replace a principles-based approach. We make some concession to pragmatist ethics, particularly where we contend with the public face of nutrigenomics in the science policy that enables the research and development, and the regulations that structure and guide the commercialization of nutritional genomics research.

19.2 CLAIMS OF HEALTH BENEFITS ARISING FROM NUTRIGENOMICS

Nutrigenomics is the study of nutrient–gene interactions. Like toxicogenomics and pharmacogenomics, it aims to understand the impact of exogenous controls on gene expression, as well as the reciprocal effects of genetic variation on nutrient uptake and sensitivity. It is one thing to have discoveries in science and quite another to generate and validate applications of the science being made available to the public. Since the latter goal represents the greatest potential for benefit and harm to individuals, we must be sure that nutrigenomics is a science strong enough to support the claims of health benefits on which these applications rely.

Nutrigenomics only recently became a field of intense, international scientific inquiry, as a PubMed search for human "nutrient–gene interactions" reveals (Fig. 19.1). Nevertheless, the science has powerful tools arising from microarray technology, single nucleotide polymorphism (SNP) analysis, gene expression profiling, proteomics, metabolomics, bioinformatics, and biocomputation [7, 8]. The systems biology approach widely adopted in nutrigenomics is generating new biomarkers that are crucial in the measurement of nutrient–gene interactions [7] and the implementation of dietary changes [8]. Some have argued that nutrigenomics will overhaul nutritional sciences to so great a degree that it should be likened to a new frontier [8] or crossing the Rubicon of nutritional science [9].

The value of nutrigenomics will be clearly demonstrated if it is able to address well-defined clinical problems with nutritional interventions such as inflammatory diseases [9], diabetes [10], obesity [9], osteoporosis, and cardiovascular disease, as well as a number of cancers [9]. Despite the complex nature of nutrient–gene interactions, some remain confident that by abiding by the basic tenets of nutrigenomics—that gene expression is measurably influenced by nutrients and can have long-term effects on disease progression—complex nutrient–gene interactions will be resolvable [9]. Along with other nutritional scientists, however, they also recognize that problems as basic as variation in laboratory techniques and training, small study sample sizes, and a general lack of knowledge about people's dietary habits will have to be addressed first [9].

There is also a general problem about the relationship between nutri*genomics* research and nutri*genetic* research. Some work, for example, the discovery science approach conducted by the Institute for Systems Biology, takes a systems biology approach to studying perturbations of genomes in the way defined by Ideker and co-workers [9]. Most research, however, analyzes the association of SNPs or haplotypes with disease or response to diet. A genetic focus on nutrient–gene interaction takes place against a background of total genetic variation within a genome, which makes it somewhat more difficult to draw a neat line between nutri*genomics* and nutri*genetics*. In large nutrient–gene studies, SNPs can be shown to modulate nutrient uptake and sensitivity, but because they usually have low penetrance, long-term predictions for individuals with the SNP are often unreliable, in some cases because of gene–gene interactions such as epistatic interactions. Individual genes are also associated with multiple phenotypic effects, as in the case of APOE4, which is implicated in Alzheimer disease progression and cardiovascular disease risk susceptibility. Additionally, the presence of a SNP and association with heterogeneous nutrient response neither rules out the influence of other endogenous and exogenous influences on a metabolic pathway, nor resolves the fact that subsets of groups with a disease-associated SNP will manifest the disease phenotype [9].

Searching for key polymorphisms, and then intervening with preventative dietary measures remains a good idea, but what is missing are the ". . . large prospective population studies with carefully collected behavioral, clinical, biochemical, and clinical data" [9], which would validate candidate gene studies and support nutrient interventions. Such data would further integrate nutrient intervention studies, association studies, and estimates for people about their individual disease susceptibility. Because these studies are not available, some commentators have suggested that generalized nutrigenetic knowledge that could be used in clinical contexts remains in the future [9], and others have suggested that even if the clinical validity of nutrigenetics can be demonstrated, its general clinical utility remains unknown [9].

This cursory examination of the strength of nutrigenomics science and the reliability of nutrigenetic applications suggests that nutrigenomics is not yet at the point where the science is strong enough to support a large number of claims of health benefits to individuals. The hope is that, in years to come, genetic susceptibilities to disease will be detected and nutrigenomics will link nutrient interventions with disease susceptibilities. Presumably, nutrigenomic science will be built one brick at a time, rather than the whole edifice going up at once. Although a few of these bricks might be available now, individuals hoping to use nutrigenetic information for health benefits will have to carefully assess the science for each candidate nutrient–gene association and the recommended dietary interventions as the science evolves.

Evaluating nutrigenomic claims of health benefits on a case-by-case basis is a sober way to proceed and reflects the cautious, fallibilist approach endorsed in the science itself. Not everyone wants to wait until there is irrefutable scientific evidence of benefit before instituting a low-risk nutritional measure. It will be important to strike the right balance between the quest for scientific certainty, and the availability of nutrigenomics-related tests, services, and products for informed consumers who want to act on the basis of imperfect information.

19.3 MANAGING NUTRIGENOMICS INFORMATION

Like other genome-based science, nutrigenomics requires the collection of personal genetic information. This is true for both large-scale studies and for personalized nutritional advice. Because of the personal nature of genetic information, it is crucial that a person's right to autonomy regarding decisions having to do with this information be respected. Respect for an individual's autonomy means that information provided for research or testing purposes must be freely given without coercion or manipulation, and this applies equally to the administration of the genetic test as it does to the uses of the resulting information [9]. Consent to genetic testing is also commonly associated with expectations of privacy, which involve freedom from unwanted tests, and the ability to withhold information from third parties [9]. This may include provisions for consent to secondary uses of genetic samples or information, particularly in the case of samples stored in so-called biobanks—electronic repositories of typically anonymized personal information. Genetic information is also expected to remain confidential between health-care providers and patients as directed by existing professional obligations. Finally, consent to genetic testing is considered free if assurances can be provided that test results will not be used to unfairly discriminate against individuals [9].

Concerns about consent, privacy, and confidentiality in the medical setting take on new urgency in the context of genetics. All genetic information is a "family affair" [9]. Individuals who learn about their own genetics also learn something of the rest of their family, which raises issues for health-care providers who may feel compelled to violate confidentiality to warn family members of potential genetic disease. Normally, individual autonomy-based confidentiality precludes sharing genetic information, but this principle is in conflict with a duty to warn and care for others especially if the condition is severe [9].

Because nutrigenetic testing is prevention oriented, it is theoretically better to administer tests early in the life cycle. This has led some members of the nutrigenomics community to recommend genetic profiling for children and infants, possibly as early as birth [9]. A child's consent to medical testing is normally obtained by proxy, and parents generally have wide-ranging discretion about what choices they can make on their behalf. However, particularly when dealing with serious matters of health, this discretion is not unlimited. Whereas screens for some diseases are commonplace, direct asymptomatic tests for minors are often discouraged to prevent the distress of having a medical test, to prevent stigmatization, or to protect against wrongful use of the information. Tests are also strongly prohibited where asymptomatic children would be tested for late-onset disorders for which there is no known intervention or cure [9]. In cases where there is overwhelming indirect evidence that the child has the disease and medical intervention based on a test would have major benefits, genetic tests may be ordered [9].

With children one of the greatest concerns is that they cannot fully consent on their own behalf to the collection and storage of data that might one day have undesirable implications for them. A similar concern about misuse of information faces everyone who takes a genetic test if the sample or information is used by third parties. Use of genetic information by life and disability insurers can deter people from taking genetic tests [9, 10]. Without assurances that third party use can be prevented, individuals might forego the benefits of genetic technologies. At the same time, insurers and employers must discriminate among people to risk rate them or choose appropriate candidates for jobs. The question is whether it is inherently inappropriate to discriminate on genetic grounds, or whether some kinds of genetic discrimination might be appropriate in insurance and employment contexts. One approach is to treat one's genetic inheritance as something over which one has no control, like one's sex or race. Unlike life-style preferences or educational background, people do not choose their genes and so they should not be held responsible for their effects. But another position suggests that the long-term future of insurance requires that insurers collect important genetic information to risk rate individuals in order to avoid the detrimental effects of adverse selection—a situation where people purchase extra insurance using information pertinent to their risk rating but keep it from their insurer. Since no corollary for employment is known, this potentially unique concern of insurers has led to calls for banning genetics-based risk rating until the problem is studied further—even though risk rating on the basis of genetic tests does not currently have strong actuarial support [9, 10].

The ethical considerations outlined above apply equally to other kinds of genomic science as they do to nutrigenomics. However, nutrigenomics gives some unique twists to these familiar themes. To begin with, nutrigenomics addresses susceptibilities for low-penetrance genes in contrast to serious monogenic disorders like Tay Sachs. The lower predictive value has ramifications for individuals' views of the seriousness of how nutrigenetic information is managed and the potential for harm attributable to third parties. That said, as nutrigenomics progresses, more powerful assays of existing data may become possible, and more powerful predictive tools may arise as better biomarkers and bioinformatics techniques become developed. These developments within nutrigenomics must be monitored because they can easily shift the ethical landscape for individuals' assessment of the need for strict controls on the flow of nutrigenetic information.

It appears that consent is easily obtained for nutrigenetic testing because most testing is done at the request of consumers or patients. One criterion of genuine consent is that individuals are fully informed about the risks and benefits of the procedure, but this may not always be possible because of the probabilistic nature of nutrigenetic information. The best physicians can do is distinguish between monogenic, deterministic disorders and disease susceptibilities. Whether individuals can make rationally consistent use of this information would also require careful study. Perhaps given that the tests are noninvasive and the results trade in susceptibilities, only implied consent is necessary. According to Henry Greely, "risks to a patient from genetic tests are not the kinds of direct medical risks, such as death or paralysis, that informed consent usually covers" and we may therefore want to relax consent requirements [9]. But again, this could all change if nutrigenetics gains more potency as the science matures.

If nutrigenetic tests are taken up as wellness tests and not medical diagnostic tests, test takers may not take the possible risks associated with the service seriously enough. Even if the test result does not have high predictive value now, a genetic sample is still collected and stored and the information from it can likewise be biobanked and resampled. DNA has been called a "future diary," holding much currently unknown but potentially highly sensitive information about our medical and social prospects [9]. For this reason, the temptation to treat nutrigenetic tests as mere life-style tests should be resisted. Nutrigenetic tests should be treated with the same respect for conditions of confidentiality any other genetic test would invoke.

Given the relatively low probative value of nutrigenetic testing, it is unlikely that physicians will have a conflict in their duties to patient confidence and to warn others of harm. Such cases usually arise where failure to warn in order to avoid breached confidences results in severe harm or death to third parties. Since the informational value of nutrigenetic tests has lower impact, physicians will not face this burden. However, it may be the case that individuals finding themselves testing positive for gene-based disease susceptibility may wish to communicate this information to family members, particularly if simple dietary modifications would ablate the risk factor. Given the relatively low impact of nutrigenetic information, and since it is generally not of a particularly sensitive or potentially stigmatizing nature, sharing confidences with other family members would not seem to pose significant risks to individuals' well-being. This is particularly likely so long as nutrigenetic testing, as it tends to be commercialized now, focuses on single independent SNPs, rather than identifying combinations of SNPs that may be more strongly predictive.

Special conditions apply in the case of children. The greatest concern is that children's knowledge of their susceptibility or of something being "wrong with them" could have long-term consequences for their emotional well-being and self-esteem. If the benefits of nutrigenetic tests are not immediate and do not require immediate nutritional intervention, there may be no compelling reason to subject children to emotionally compromising tests. Furthermore, parents may fail to understand the probabilistic nature of nutrigenetics and thereby may misinterpret the meaning of their children's susceptibility to disease.

A broader concern is that information arising in nutrigenetic tests from family members may become associated with children with future repercussions for their insurability or employability. When in the future these children are consenting adults, it is quite reasonable to suppose that they might object to how their genetic information was handled in the past. Since nutrigenetics involves discretionary testing for susceptibilities, the magnitude of future harm for children could outweigh the benefits accrued by family members here and now. However, if we find that some of the SNPs are linked to diet, and diet could alter the outcome of disease if the diets are started when a person is young, then there is compelling reason to test early in life.

The specter of the abuse of genetic information by third parties could forestall or prevent people from getting valuable nutrigenetic tests. Equally, one's willingness to contribute to nutrigenomics research programs could be undermined if concerns about third party access are not allayed by assurances that personal data will be anonymized. It is also conceivable that third party access to genetic information might tip public confidence in favor of either private or public testing.

These are, however, worst case scenarios, which do not appear to be imminent for many monogenic disorders, much less nutrigenetic outcomes. If the concerns for high-penetrance genetic diseases are often overstated [9–11], it seems even less likely that nutrigenomics will have much actuarial value or meaning to potential employers in the short term—particularly early adopters would have to be sure that the benefits to them would outweigh potentially serious public backlash. In the long term, this may change if regulations permit genetic discrimination and actuarial value can be attributed to more precise genetic tests. Equally, employers may one day be able to make use of nutrigenetic data to forecast and reduce employee absenteeism, predict early retirement, or control employer health premium contributions [9].

Because nutrigenomics information has risks as well as benefits for the individual and his/her family, such information needs to be collected with consent and protected with confidentiality. At the moment, both the risks and benefits of nutrigenomics information seem modest. As the science evolves, this could change.

19.4 METHODS FOR DELIVERING NUTRIGENOMICS SERVICES

One of the most important issues in nutrigenomics is deciding on the best method for delivering nutrigenetic services. Who is permitted to offer the service determines how the test is administered and the results interpreted, what kind of counseling supports are required, how the genetic information is managed, and whether outside parties can have access to it. Two systems for offering nutrigenetic services have appeared in the marketplace thus far. Initially, a number of start-up firms in the United Kingdom and North America offered their products and services directly to consumers, primarily over the Internet. However, most of these companies have subsequently migrated to having a health-care practitioner administer the test, in what now seems to have become standard industry practice. Other yet-to-be-tried models have their attractions as well and below we consider two of these, a team-based approach to service delivery and nutrigenetic tests offered as part of a public health screening program.

Selling direct to the public via the Internet or mail-order empowers consumers to seek the products and services they want without the assistance of a health intermediary. Respecting consumers' autonomy in this way recognizes that they alone determine what they do to their bodies [9] on the assumption that they are best positioned to judge their own interests [10]. As motivated consumers, people will also adopt an attitude of self-care [10] and will select preventative behavior [11]. The concern, however, is that tests of low predictive value will be offered broadly, and such tests also might be the springboard for tied selling of nutrigenetic products that offer few if any health benefits [11].

In a health practitioner model, regulatory safeguards are introduced by placing nutrigenetic services under the purview of a qualified health-care provider. As mentioned above, many direct-to-consumer nutrigenetic businesses have migrated to this model. Genetic counselors are especially well-equipped to offer a wide range of counseling and interpretive services and are increasingly trained to explain probabilistic statements about disease susceptibility [11]. The problem, however, is that there is a lack of trained specialists who have knowledge about nutritional sciences and the resource base to offer nutrigenetic services. Primary care deliverers may be called on to respond to the demand for nutrigenetic services, which has the benefit of fostering long-term oversight of nutrigenetic interventions, but which puts pressure on medical schools to include nutritional sciences and genetics in already heavily loaded curricula [11]. Not all physicians may be willing to expand their range of services, in large part because they did not have training in basic genetics, probability risk, and nutritional sciences [11–13]. The reverse problem is cause for concern if nutrition specialists are asked to provide nutrigenetic services. They are not generally conversant with genetic issues and would require additional training to better understand genetic risks and gene–environment interactions, to become familiar with the ethical, legal, and social implications involved in genetic testing, and to learn how to communicate all of this to patients [11, 12].

Since it may be unreasonable to expect any one group of health practitioners to deliver nutrigenetic services in light of the diverse range of competencies involved, a team of differently skilled practitioners may need to be coordinated. A referral system or jointly staffed clinic may meet individuals' needs to find credible information quickly and efficiently from well-trained and reliable health-care providers. Such a system would likely achieve optimal results as long as individuals received the complete service within a tightly coordinated framework, but this is not assured unless direct-to-consumer services wither away. A blended service delivery model mirrors the complexities of nutrigenetic science in the branches of the services it would have to provide. Individuals seeking nutrigenetic services because they are interested in gathering wellness or life-style advice might be alienated by the complexity of this service delivery model and its location in a medical setting.

If the science bears out, nutrigenetic testing could eventually have its biggest impact less in the clinical setting and more so in the area of public health [11]. Public health measures encourage health-promoting behavior [11]. Nutrigenomics screening could be offered as part of a population-based prevention program. Theoretically, nutrigenetic tests could play a role in the prevention of disease by offering entire populations or targeted subgroups assessments of their susceptibilities to certain common diseases, combined with dietary advice. Universal public screening programs could have the effect of broadly distributing the benefits of nutrigenomics information to large segments of the population [11]. However, there is always the risk of untoward segmentation of populations into ethnic groups, further stigmatized by socioeconomic class in a public health model.

When choosing between different methods of service delivery, a complex array of factors come into play. The strength of the science will likely be the strongest determinant in deciding who should deliver nutrigenetic tests. Tests for common polymorphisms that have high predictive value and for which there are nutritional interventions with demonstrated clinical validity will generate the greatest benefits for consumers. Health practitioners would likely be interested in delivering tests with this level of usefulness in the clinic. There are concerns, however, that common, low-penetrance SNPs will dominate nutrigenetic services [11], particularly in jurisdictions lacking regulations that might otherwise only permit highly predictive tests. Because low-penetrance genes are prevalent in the general population, testing for them will be particularly attractive to commercial service providers. The kind of regulatory environment will likewise determine the extent to which health-care practitioners are involved in nutrigenetic service provision, as well as the requirement for monitoring and surveillance of nutrigenomics services. Another important determinant of service quality lies in the availability of skilled personnel who have the professional competence to deliver tests and test results, offer dietary advice, and explain the probabilistic nature of nutrigenetic claims. To a great extent, having highly qualified personnel to deliver nutrigenetic services will depend on whether long-term health-care money—private or public—is there to support nutrigenomics. Finally, the dynamics of professional culture jockeying to "own" or avoid the nutrigenetic territory will have an enormous impact on how end users experience nutrigenetics [11]. Some professions may doubt the utility of nutrigenetics and may have no further interest in becoming better educated about it, whereas other professionals may welcome the opportunity to expand their portfolio of services [11].

Bearing these considerations in mind, can a service delivery model that best optimizes delivery of nutrigenetic services be chosen? We have suggested that purely direct-to-consumer service has the advantage of empowering consumers, but the social climate does not yet appear ready for direct-to-consumer provision of nutrigenetic tests in all countries. Health practitioners may be the most adaptable and responsive group to offer services if they see value in them. Their interest in doing so will be predictably correlated with their training and the validity and utility of nutrigenetic tests. Blended services may prove to be unwieldy and could displace the virtues of nutrigenetics as a highly personalized and motivating method of self-care. Compliance with public health recommendations for dietary patterns is rarely above 60% [11]. Regardless of which model is judged most appropriate for a particular jurisdiction, the foregoing makes it clear that the provision of nutrigenetic services requires greater familiarity with nutritional sciences, genomics, and genetics on the part of regulators and health-care providers than currently exists to ensure that the interests of the public are well-protected.

19.5 NUTRIGENOMICS PRODUCTS

So far we have concentrated on ethical issues associated with offering genetic tests for gene-based disease susceptibilities for which nutrient interventions might mitigate or prevent the disease. In this section we consider what nutrient interventions might look like, and whether the method by which we intervene has ethical ramifications. Nutrient intervention can take many different forms and with varying degrees of efficacy, not all of which is always under regulatory control. Often what is at issue is the gray area lying between the regulation of some products as foods or medicines. Nutrigenomics also encourages a different perspective on foods, which may have social effects individuals do not anticipate when they embrace nutrigenetic testing.

The market for health products offers an impressive array of products that are intended to have effects very much like drugs but without being classified as drugs [11]. For example, nutraceuticals are isolates of bioactive ingredients like fish oils in capsule form. When foods are believed to have an effect on a disease comparable to a pharmaceutical, such as the HeartBar [11], they are described as medicinal foods. Dietary supplements are not foods but are treated as dietary adjuncts and usually include vitamins and minerals.

Functional foods are a very important group of foods, lauded for their having some bioactive ingredient associated with disease prevention, growth, or development [12]. Health Canada proposed the definition of a functional food as: "similar in appearance to a conventional food, to be consumed as a part of the usual diet, to demonstrate physiological benefits, and/or to reduce the risk of chronic disease beyond basic nutritional functions" [12]. Functional foods are produced either by maximizing beneficial food components or by minimizing nonbeneficial components [12]. Probiotics, prebiotics, and phytochemicals are currently being studied for beneficial health effects [12].

Regulators in many jurisdictions have struggled to impose some order on the many products available by categorizing the health claims for foods. These include structure/function claims, health claims that relate food components to disease, medicinal food claims for foods associated with particular diseases, and disease reduction claims. Despite the known bioactivity and association with particular diseases, foods and their derivatives are distinguished from drugs. Foods are presumed to be safe, are not tested like drugs, and are consumed unsupervised. Overconsumption of certain foods is not thought of as a drug overdose and food–food or food–drug interactions are generally overlooked [12]. And yet people seek out these products precisely because of their drug-like efficacy [13].

There is an obvious need to protect people from unsubstantiated or false health claims made about foods, food components, or food derivatives. It would be intolerable if people put stock in ineffective products, possibly excluding in the process therapy with conventional drugs prescribed and monitored in a conventional medical context. Controversy about the alleged antidepressive properties of various herbal remedies is a case in point [13]. As more is learned about the bioactive properties of foods, however, the problem might be reversed: the actual risk to individuals is that they consume products as efficacious and targeted as some drugs.

Nutrigenomics will accelerate the identification of a broader array of bioactive compounds, and nutrigenetic services will expose individuals to selected compounds known to have a high level of interactivity with their specific genotype. In this respect, nutrigenomics increases the chance that individuals might be harmed by foods or their subcomponents. These risk can be mitigated with oversight of nutrigenetic services and by vigilant regulation of food products claimed to have health benefits.

The blurring of the food–drug distinction has also raised concerns that nutrigenomics could change how people enjoy and interact with food. Cultural associations with food are an important part of day-to-day living. How we consume food requires many decisions about what to eat, when and with whom to eat, and where we purchase the food [13]. Nutrigenomics, some have argued, may disrupt these social interactions in two possible ways. First, the trend toward individualized diets could disrupt the idea of shared meals in families if no one consumes the same thing at the same time [13]. Second, if food is "medicalized" it may lose some of the cultural associations it would otherwise have [13]. It should be emphasized that the socially disruptive effects of individualizing and medicalizing food would be worse in cases where nutrigenetic advice recommended drastic dietary modifications involving functional foods. In this scenario, complete meals would be individualized. A more likely scenario is that conventional cultural ties with local diets and family food choices will continue, but nutrigenetic advice encourages the use of like nutraceuticals or other supplements.

19.6 ACCESS TO NUTRIGENOMICS

Nutrigenomics promises significant health benefits, but the question, "For whom?" remains. Theoretically, improved dietary guidelines based on nutrigenomics could be broadly distributed through public health programs. Nutrigenomics could also reach a smaller segment of the public through the provision of personalized nutritional services. Ideally, all would benefit from the development of nutrigenomics, but there are a number of indicators which suggest that nutrigenomics will benefit only select groups of people. Inequities in this new field are foreseeable given the way that other health technologies are often isolated in industrialized countries, or find their way only to global elites with access to the newest and best health technologies wherever they reside. A significant part of the explanation for each source of inequity lies in the fact that research and development for new health biotechnologies are associated with controls on access to knowledge and technologies, often in the form of intellectual property.

Nutrigenomics may have the capacity to lessen global health inequities, but a number of obstacles stand in the way. Currently, 90% of the world's health research resources are focused on 10% of the world's population (the so-called 10/90 gap); the United States alone spends more than half of the total world expenditure on health, and two billion people still have no access to low-cost essential medicines [13]. A study conducted by the University of Toronto's Joint Centre for Bioethics used a number of criteria to identify the top ten biotechnologies likely to contribute to improvements in global health equity [13]. These criteria can be similarly applied to assess nutrigenomics. The first critical step is to evaluate the potential health impact of nutrigenomics. Given that nutrigenomics is a broadly generalizable science, lessons learned can be applied to all people. Second, nutrigenomics may not be an appropriate technology where nutritional interventions and improved dietary practices might be a luxury. This will be true wherever subsistence or rudimentary diets with little choice of staple foods are the norm, where health improvements are systematically undermined by infrastructure or political strife, or where provision of conventional nutrition information is already a challenge [13]. Third, it is unlikely that most developing countries will be able to adopt nutrigenomics in a relatively short time frame because of a lack of resources and preexisting healthcare system deficiencies.

Finally, it is worth considering nutrigenomics' long-term potential for health and economic benefits for developing countries since the focus is on prevention using foods costing less than pharmaceuticals. In many developing countries, noncommunicable diseases such as cardiovascular disease and diabetes are becoming more significant causes of mortality and morbidity than infectious diseases [13]. As this trend progresses, population-level dietary changes will have significant impacts on health-care costs, and there is some preliminary evidence that knowledge about genetic susceptibility will contribute to lowering costs, especially if the science focuses on conditions involving genetic variants that are commonly found in disadvantaged subpopulations [13].

In industrialized countries, significant disparities in access to health-care resources and in health indicators exist between different socioeconomic classes, with the poor and often minorities being the most disadvantaged. Dietary habits and access to health care are key determinants of health, but genetic differences are also thought to play a role. Nutrigenomics research geared toward the health needs of the poor and of minority populations has the potential to help reduce these health inequities if it focuses on genetic variations that increase susceptibility to disease [13]. Of course, it will only make this contribution so long as changes in diet can be encouraged by public health officials. For the same reasons that the preventative benefits of nutrigenomics might be encouraged in developing countries, industrialized countries may be persuaded to ameliorate skyrocketing health-care costs with strategic investments in nutrigenomics.

In industrialized countries, the group identified as most likely to embrace nutrigenomics first are typically female, university-educated women between 50 and 64, who are willing and able to pay for nutrigenetic tests and have a history of specialty and health food purchasing preferences [13]. If this trend continues and the market segment does not broaden quickly, nutrigenetic products will be commercialized as wellness or life-style products in a boutique context. Tied selling of health products and programs with nutrigenetic tests is already available. Even if regulations intended to protect consumers by limiting direct-to-consumer and tied selling are put in place, this same market segment is already predisposed to seek out nutrigenetics as a form of personalized nutrition from physicians and nutritionists. How the market for nutrigenetic products and services will grow and whether it will become further segmented are currently unclear.

An endemic feature of health biotechnology innovation, which may limit access to nutrigenomics research and its clinical applications, is intellectual property. While there is much dispute at the theoretical level about whether patents in particular really do create an anti-commons for biotechnology innovation [13], examples like the BRCA test for breast cancer patented by Myriad Genetics raise concerns that gene patents are hindering access to the fruits of biomedical research, interfering with patient care, contributing to the global genomics divide, and creating tensions among international trading partners [13]. While there is some evidence that applications for new gene patents are declining [13], the effect of the incentive to patent, and hence to limit access to knowledge and technology through a temporary monopoly, is to drive up costs of health biotechnology, including nutrigenomics [13].

Other factors complicate the access to nutrigenetic services and products. The United States currently holds 80% of all gene sequence patents, which means that reward for innovation is concentrated on the priorities of one industrialized country [13]. Another problem is that while gene patents have dominated health biotechnology, the food industry interests in nutrigenomics will more likely be tied up in their traditional forms of intellectual property protection: trade secret and industrial processing design. Since neither of the latter forms of intellectual property protection require public access to the information, valuable knowledge about nutrient–gene interactions may remain proprietary.

It is currently too early to judge whether nutrigenomics will improve or aggravate global health inequities that exist within and between countries. Admittedly the science appears to offer a boon to health equity since to a greater or lesser extent most people consume food everyday. Whether people can access the foods that would make a difference to their genetic constitution, and whether they can find information that would lead them to modify their food intake to take advantage of nutrigenomics is another issue. What is clear is that the role of nutrigenomics on global health equity will need explicit attention, otherwise the technology will likely increase the "genomics divide." The role of nutrigenomics in developing countries should be a funded research priority as the field develops.

19.7 CONCLUSION

When new fields of science or technology are considered by ethicists and regulators, the presumption is often that hard matters of fact have been resolved in the science, and these now await the consideration of the soft decision making process located in the social sciences and humanities. The reverse is actually true, as Ravetz and Funtowicz have pointed out, "all too often, we must make hard policy decisions where our only scientific inputs are irremediably soft" [13]. This is the situation with nutrigenomics. While the rapidly evolving science of nutrigenomics holds great promise for the detection of susceptibilities to disease and for health improvements that link genes to nutrition, it is still an emerging science that lacks much of the clinical evidence necessary to validate claims of health benefits and disease prevention. This assessment of the science suggests that nutrigenomics could become better than "irremediably soft" with the right evidence, but will likely always remain a science of probabilities since it trades in disease susceptibility arising from low-penetrance polymorphisms.

The point to be drawn from the above is not that nutrigenomics is "soft" science that one should be wary of, but that what one says about the ethical and regulatory issues associated with nutrigenomics is somewhat contingent on the nature and strength of the science. This is an inevitable consequence of a methodology of proactive bioethics, which takes developments within the science seriously. It implies that what might be an ethically defensible position about nutrigenomics now might change as the science develops. As the science unfolds, so too should the ethical and regulatory appraisal of that science. After all, some of the ethical issues considered relevant at one time may influence "hard" regulatory decisions that will no longer apply to a maturing science later on. Ongoing interdisciplinary dialogue about the latest opportunities and limitations of nutrigenomics and its clinical and commercial applications is required in order to foster greater public awareness and understanding of the potential risks and benefits of this emerging field.

REFERENCES

1. www.sciona.com.
2. James Meek (2002). Public misled by gene test hype: Scientists cast doubt on "irresponsible" claims for checks offered by Body Shop, *The Guardian*, March 12.
3. BBC (2003). New controls on genetic tests, Wednesday February 5.
4. http://www.genomecanada.ca/.
5. See http://www.utoronto.ca/jcb/genomics/index.html.
6. R. Chadwick (2004). Nutrigenomics, individualism and public health, *Proc. Nutr. Soc.* **63**:161–166. For the relationship between individuals and communities, see M. Korthals, J. Keulartz, and T. Swierstra (2003). You only live twice: ethical deficiencies in dealing with nutrigenomics, *Eursafe* 152–155.
7. B. van Ommen (2004). Nutrigenomics: Exploiting systems biology in the nutrition and health arenas, *Nutrition* **20**:4–8.
8. T. Peregrin (2001). The new frontier of nutrition science: Nutrigenomics, *J. Am. Diet. Assoc.* **101**:1306.
9. M. Stevenson (1999). Commercializing safety and efficacy of home genetic tests, *J. Biolaw Business* **3**(1):29–39.
10. C. Powell (2003). Piloting home tests, *The Beacon Journal*. Ohio. P. Webb, E. Anionwu, et al. (2002). *The Supply of Genetic Tests Direct to the Public*. Human Genetics Commission, United Kingdom.
11. Lois. Skelly (1999). Functional foods: Over-the-counter medicine in a meal, *Healthform* **5**(7):1.
12. C. L. Kruger and S. W. Mann (2003). Safety evaluation of functional ingredients, *Food Chem. Toxicol.* **41**:795.
13. See http://www.nusap.net/sections.php?op=viewarticle&artid=13.

GLOSSARY

Adipogenesis The formation of adipose tissue (fat cells). May also refer to the production of fat, either fatty degeneration or fatty infiltration.

Adipose tissue Specialized tissue that stores fat.

Agonist-induced activation For nuclear receptors, a small molecule binds to an inactive nuclear receptor causing conformational changes that result in protein–DNA interaction, recruitment of cofactors, and transcription factors ultimately leading to gene transcription. After dissociation of the agonist, the nuclear receptor may return to its inactive state.

Alleles Alternate forms of the same gene. For organisms with two sets of chromosomes like humans, an individual can have two copies of the same allele (i.e., homozygous for the allele) or two different alleles (i.e., heterozygous for the alleles).

Alpha-linolenic acid (18:3) A polyunsaturated fatty acid with 18 carbon atoms; the only omega-3 fatty acid found in vegetable products; it is most abundant in canola oil; a fatty acid essential for nutrition. The term "omega-3" (also referred to as "n-3") refers to the third carbon atom from the beginning of the fatty acid chain.

American Society for Testing and Materials (ASTM) A section of the American National Standards Institute (ANSI) that has a subcommittee (E31) for general health-care informatics. This E31 Subcommittee on Healthcare Informatics develops standards related to the architecture, content, storage, security, confidentiality, functionality, and communication of information used within health care and health-care decision making, including patient-specific information and knowledge. (www.astm.ong)

Aneuploidy An abnormal number of chromosomes. In the case of humans, any chromosome number more or less than 46 is referred to as an aneuploid condition.

Antioxidant Chemical that inhibits oxidation and reacts with free radicals to form a harmless product. An antioxidant will have one or more unpaired electrons.

Apolipoprotein The protein component that combines with a lipid to form a lipoprotein.

Apoptosis Programmed cell death. A normal cellular sequence of reactions that destroys the cell without releasing harmful substances into the surrounding area.

Array analysis A solid support on which a collection of gene-specific nucleic acids are placed at defined locations, either by spotting or direct chemical synthesis. In array analysis, a nucleic acid in the sample is labeled and then hybridized with the gene-specific targets on the array. Based on the amount of probe hybridized to each target spot, information is gained about the specific identify and quantity of the nucleic acid in the sample.

Nutritional Genomics: Discovering the Path to Personalized Nutrition
Edited by Jim Kaput and Raymond L. Rodriguez

The advantage of arrays is that they allow target sequences to be interrogated by the thousands instead of individually.

Atherosclerosis The progressive narrowing and hardening of the arteries over time; often used to describe a condition where lipids (fats) collect under the inner lining of damaged artery walls.

Bifidobacteria Lactic acid producing microorganisms that inhabit the gastrointestinal tract.

Bilirubin A chemical found in bile that is the normal degradation product of hemoglobin and other heme-containing proteins.

Bioactivation The conversion of an inactive compound into an active one within a living organism.

Bioactive A chemical (often found in various foods) that interacts with the molecular components of a living organism.

Biobank A collection of biological specimens; often used to describe a collection of DNA samples.

Biocomputation Analyses of biological data usually with algorithms or programs for reducing complexity of the data.

Bioinformatics The collection, storage, manipulation, management, and retrieval of biological data.

Biological system An organism and its design properties, which specify phenotypic outputs given perturbation inputs (see Chapter 5).

Biomarker A substance sometimes found in the blood, other body fluids, or tissues that can be used to measure the presence or progress of disease or the effects of treatment.

BMI (body mass index) BMI is calculated by dividing a person's weight (in kilograms) by height in meters squared (BMI = kilogram/meter^2).

Bowman–Birk inhibitor A protease inhibitor of trypsin and chymotrypsin found in soybeans.

Buffering The process by which a robust system absorbs changing inputs from the environment, while maintaining stable outputs (see Chapter 5); based on the concept of a chemical buffer, which is a solution (usually) consisting of a weak acid and its salt or a weak base and its salt that maintain changes in pH.

Buffering capacity The amount of stability/robustness that a gene or genetic module imparts on the system in response to a particular perturbation input; reflects the strength of genetic interaction (see Chapter 5).

Buffering specificity The combined selectivity and capacity of genetic interactions across a series of different perturbations (see Chapter 5).

Calorie (food) The amount of heat (i.e., energy) needed to raise the temperature of 1 gram or water from 15 to 16°C or 4.184 absolute joules. One food calories is equivalent to 1000 calories or 1 kilocalorie. Calories are used to describe the energy content of various foods. For nutrition, a food calorie can be defined practically as the amount of energy consumed by a 150 lb individual during 1 minute of sleep.

Carcinogen A substance or chemical agent that perturbs normal cellular processes leading to unscheduled cell division and cancer, or the increased risk of cancer.

Carcinogenesis The molecular processes that result in cancer.

Catechins Natural plant compounds belonging to the class of polyphenols present in high concentrations in green tea. Catechins like EGCG are potent antioxidants and have been shown to block signaling pathways that lead to cell proliferation in human and animal cells.

Cellular context The perturbation state of the cell, which changes as a function of genetic and environmental alterations. Genetic interaction, and thus genetic buffering, is always measured with respect to cellular context (see Chapter 5).

Chemopreventive agent A chemical constituent, drug, or food supplement that prevents disease by interrupting deleterious biological reactions or processes.

Chimera An organism, organ, or part consisting of two or more tissues of different genetic composition, produced as a result of organ transplant, grafting, or genetic engineering.

Current Procedural Terminology (CPT) A uniform set of codes that identify each service, procedure, or supply for diagnosis, symptom, condition, or problem. Current Procedural Terminology was developed by the American Medical Association in 1966. http://www.amaassn.org/ama/pub/category/3113.html. See also *International Classification of Diseases*.

Cytochrome P450s A family of liver enzymes involved in metabolism of exogenous and endogenous chemicals.

Data model A conceptual framework for the development of a new or enhanced software application. The purpose of data modeling is to develop an accurate model, which may be shown in a graphical representation such as UML, of the information needs and business processes addressed by a particular application or connected set of operations (see Chapter 16).

Dauer Metabolically dormant larva stage.

Detoxification Any process that removes a toxin; in biological systems, a process that usually requires enzymes for modifying reactive chemicals. For example, the liver enzyme NAT2 can detoxify aromatic amines in tobacco smoke.

Diet The sum total of all the nourishing materials (food, drink, and supplements) consumed by an organism. In humans (and most animals), a proper diet requires certain essential vitamins, minerals, proteins, and fats. The balance between starvation and obesity depends on the amount of nourishing materials consumed as fuel and the amount of energy expended.

Dietary phytosterols Plant-derived chemicals with a typical ring structure similar to sterols found in animals.

Digital Imaging and Communications in Medicine (DICOM) The Digital Imaging and Communications in Medicine (DICOM) Standard was developed for the transmission of images and is used internationally for Picture Archiving and Communication Systems (PACS). This standard was developed by the joint committee of the ACR (the American College of Radiology) and NEMA (the National Electrical Manufacturers Association) to meet the needs of manufacturers and users of medical imaging equipment for interconnection of devices on standard networks.

Dosage compensation Any system that equalizes the amount of product produced by genes present in different numbers. In mammals, it describes the X–inactivation mechanism that ensures equal amounts of X–specific gene activity in XY male and XX female cells.

Dyslipidemia Abnormal lipid profiles usually characterized by high triglyceride concentrations, low HDL cholesterol, and increased concentrations of small, dense LDL; associated with metabolic disorder such as insulin resistance, obesity, and type 2 diabetes.

Ectopic gene expression Expression of a gene in a cell where it is typically not expressed.

Endothelial cells Cells that line the interior surface of heart, blood vessels, and serous cavities of the body.

Enhancers A set of short sequence elements that stimulate transcription of a gene and whose function is not critically dependent on their precise position or orientation.

Epidemiology Classically, the study of the occurrence of a disease in a population, especially the factors that influence incidence, severity, and distribution.

Epigenetics The study of heritable changes in gene function that occur without a change in the sequence of nuclear DNA.

Epistasis In its strictest classical genetic definition, it is the interaction of one gene (or locus) with another.

Etiology The causes or origins of disease.

Exon DNA sequences that occur in mRNAs, which contain ribosomal binding sites, protein coding sequences, and information for mRNA stability and perhaps cellular location.

Folic acid A yellowish-orange compound, $C_{19}H_{19}N_7O_6$, of the vitamin B complex group, occurring in green plants, fresh fruit, liver, and yeast; also called folacin, folate, and vitamin B_c.

Gene activity Transcription, translation, stability, physical association, or enzymatic function of molecules ultimately attributable to the same segment of DNA.

Genetic association studies Statistical analyses that link chromosomal regions with disease subphenotypes or incidence.

Genetic buffering A property of biological systems whereby stability (robustness) of phenotypic outputs is conferred by gene activities that interact to absorb system perturbations (see Chapter 5).

Genetic buffering capacity The strength of interaction between a gene, or genetic module, and a particular perturbation. Strong interactions indicate genes with high buffering capacity, meaning that the gene activity is required for phenotypic stability in response to even weak intensity of perturbation (see Chapter 5).

Genetic buffering selectivity The qualitative pattern of interactions for a gene or genetic module, where gene interaction is measured in the context of different perturbation types (see Chapter 5).

Genetic buffering protocol The organization of genetic buffering modules and their molecular activities that confer cellular robustness in response to a perturbations. Connectivity between genetic interaction modules together with molecular genetic knowledge of the functions of genes in each module facilitates hypotheses about genetic buffering protocols (see Chapter 5).

Genetic buffering specificity The combined buffering capacity and selectivity of a gene or genetic module. Genetic interaction modules are defined on the basis of shared genetic buffering specificity between individual genes. Hierarchical clustering of quantitative interaction data is presented as one strategy for depicting genetic buffering specificity and identifying genetic interaction modules (see Chapter 5).

Genetic epidemiology The study of genetic components in a complex biological system.

Genetic interaction The nonadditive effect that variation (gene deletion) at one genetic locus has on the phenotypic response of a biological system to a defined perturbation. Interactions can be synergistic (enhancing effect of perturbation) or antagonistic (suppressing effect of perturbation).

Genetic interaction module A set of genes sharing the same genetic buffering specificity. Gene interaction modules are experimentally defined using knockout strains in a co-isogenic genetic background, are dynamic with respect to the perturbations and gene deletions tested, and are also dependent on the method used for classifying genes according to their shared buffering specificity (e.g., hierarchical clustering) (see Chapter 5).

Genetic interaction network An experimentally determined set of genetic interactions derived by quantitative phenotypic analysis of single gene knockout strains in comparison with an isogenic reference (i.e., no deletion) strain. Genetic interaction networks are dynamic, depending on the panel of knockout strains used and the genetic and/or environmental cellular contexts tested (see Chapter 5).

Genetic polymorphism The difference in DNA sequence from a reference sequence.

Genetic system The genetic underpinnings of a biological system, that is, the attributes of gene activities and the organization of their interactions that confer properties on biological systems (see Chapter 5).

Genistein A chemical of the isoflavone class found in plants that has a structure similar to estrogen. Hence, it is a phytoestrogen.

Genomics The high-throughput, highly parallel study of all the genes (and gene products—RNA and proteins) as a dynamic system, over time, determining how they interact and influence biological pathways, networks, and physiology, in a global sense.

Gut microflora All of the microbes in the gastrointestinal tract.

Haplosufficient When one functional allele suffices as well as two to produce the normal phenotype.

Haplotype A contraction of the phrase "haploid genotype." A specific collection of linked polymorphisms (e.g., SNPs, simple tandem repeats, or insertions and deletions) within a cluster of related genes or region of a chromosome.

Hardy–Weinberg Equilibrium The stable frequency distribution of genotypes, AA, Aa, and aa, in the proportions p2, 2pq, and q2, respectively (where p and q are the frequencies of the alleles, A and a). These distributions are a consequence of random mating in the absence of mutation, migration, natural selection, or random drift.

Health Information Standards Board (HISB) A subgroup of the American National Standards Institute (ANSI). The American National Standards Institute's Healthcare Informatics Standards Board (ANSI HISB) provides an open, public forum for the voluntary coordination of health-care informatics standards among all U.S. standard developing organizations. www.ansi.org/.

Health Insurance Portability and Accountability Act (HIPAA) The administrative simplification provisions of the Health Insurance Portability and Accountability Act of 1996 are intended to reduce the costs and administrative burdens of health care by making possible the standardized, electronic transmission of many administrative and financial transactions that are currently carried out manually on paper. HIPAA also has some significant implications for the solicitation of participants to become research subjects in research settings.

Health Level 7 (HL7) (www.hl7.org) A standards development organization formed in 1987 to produce a standard for hospital information systems. HL7 received ANSI accreditation as an Accredited Standards Development Organization in 1994. The HL7 standard is an American National Standard for electronic data exchange in health care that enables disparate computer applications to exchange key sets of clinical and administrative information. HL7 is primarily concerned with movement within institutions of orders; clinical observations and data, including test results, admission, transfer and discharge records, and charge and billing information (coordinating here with X12). HL7 is the selected standard for the interfacing of clinical data for most health-care institutions.

Hepatocarcinogenesis The formation of cancer in the liver.

Hepatocyte A cell of the liver.

Heritability The degree to which the variance in the distribution of a phenotype is attributable to genetics. For example, height and weight in humans are from 40% to 70% heritable or due to genetics.

Heterocyclic amines A class of chemicals with a ring structure and amine groups; can be formed by cooking food.

Heterologous genes Usually refers to introduction of a foreign gene in another organism.

HL7 Reference Information Model (www.hl7.org) A conceptual model that defines all the information from which the data content of HL7 messages is drawn. The HL7 Version 3 RIM has many features necessary for representing and modeling nearly any process in a health-care setting.

Homologous genes Two or more genes whose sequences are significantly related because of a close evolutionary relationship, either between species or within a species.

Hyperforin A chemical found in St. John's wort (and other plants) that is bioactive. It can bind and activate expression of genes involved in drug metabolism.

Hypertension Arterial disease in which chronic high blood pressure is the primary symptom.

Imprinting Inheritance of parental *germline* DNA methylation patterns by offspring; determination of the expression of a gene by its parental origin.

Insulin resistance Reduced sensitivity to insulin by the body's insulin-dependent processes (as glucose uptake, lipolysis, and inhibition of glucose production by the liver) that results in lowered activity of these processes or an increase in insulin production or both. Insulin resistance is typical of type 2 diabetes but often occurs in the absence of diabetes.

International Classification of Diseases (ICD-9-CM) The International Classification of Diseases, Ninth Revision, Clinical Modification (ICD-9-CM) was developed in the United States to provide a way to classify morbidity data for indexing of medical records, medical case reviews, and ambulatory and other medical care programs, as well as for basic health statistics. It is based on the World Health Organization (WHO) international ICD-9. A version based on ICD-10 (ICD-10-CM) is in preparation. http://www.who.int/whosis/icd10/othercla.htm. See *Current Procedural Terminology*.

IRB protocol Institutional review board (IRB) is a committee of physicians, statisticians, researchers, community advocates, and others that ensures that a clinical trial is ethical and that the rights of study participants are protected, Procedures for studies must be approved by the IRB before being carried out. IRB approval is required for clinical trials and any research involving humans.

Isoelectric point The pH at which an amphoteric molecule has a net charge equal to zero.

Isoflavones A class of plant metabolites similar in structure to estrogen. The isoflavones, diadzin, and genistein are found in soybeans and other plants and are believed to provide health benefits.

Knockout (KO) The targeted inactivation of a gene typically in a cell culture, plant, or experimental animal.

Ligand A molecule that binds to a specific site on a protein such as a nuclear hormone receptor, membrane receptor, enzyme, or antibody. A ligand can be an activator or inhibitor.

Linkage disequilibrium (LD) Describes a condition where alleles occur together more frequently than can be accounted for by chance. When alleles (or genetic markers) are in strong LD, this indicates that the two alleles are physically close to each other on the chromosome. Genetic markers in strong LD with an inherited disease can be used to may and characterize candidate genes involved in that disease.

Lipogenesis The synthesis of lipids from nonlipid precursors.

Logical Observations—Identifiers, Names, Codes (LOINC) Coding system for the electronic exchange of laboratory test results and other observations. LOINC development involved a public–private partnership comprised of several federal agencies, academia, and the vendor community. This model can be applied to other standards setting domains. www.loinc.org.

Long terminal repeats (LTRs) Long repeating sequences of DNA usually at either end of a DNA sequence. They encode sites of recombinase binding and action or transcriptional activity.

Lumen The central cavity of a cell.

Mass spectroscopy (MS) A technique for separating ions based on their mass-to-charge ratios.

Maternal effect The contribution that the mother makes to embryo survival.

Metabolic fingerprinting or profiling Classifying a sample by the types and amounts of metabolites relative to a reference sample(s).

Metabolite Any substance produced by metabolism or by a metabolic process.

Metabolome The sum total of all metabolites in a cell, tissue, organ, or organism.

Metabolomics Defined as the global analysis of metabolites—small molecules generated in the process of metabolism—that represent the sum total of all the metabolic pathways in an organism, with a focus on the identification of each pathway and its role in an organism's function.

Micronutrient A substance, such as a vitamin or mineral, that is essential in minute amounts for the proper growth and metabolism of a living organism.

Molecular buffering The biochemical basis for genetic buffering. Genetic buffering is defined experimentally by genetic interaction networks; however, hypotheses about the molecular basis for connectivity between genetic interaction modules can be generated based on information about the molecular functions of individual genes and their positions on the interaction network (see Chapter 5).

Molecular epidemiology A science that focuses on the contribution of potential genetic and environmental risk factors, identified at the molecular level, to the etiology, distribution, and prevention of disease within families and across populations.

Molecular markers Any DNA, protein, RNA, or metabolite that is used as a surrogate for a phenotype or more complex biological process.

NAD Nicotinamide adenine dinucleotide (oxidized form); coenzyme that acts as electron and hydrogen carrier in some oxidation–reduction reactions.

NADH Nicotinamide adenine dinucleotide (reduced form).

National Council for Prescription Drug Programs (NCPDP) Founded in 1978, the NCPDP focuses on prescription drug messages and works to create and promote data interchange and processing standards for the pharmacy services sector of the health-care industry. This is the standard for billing retail drug sales. The NCPDP is currently working on a standard for physicians to submit prescriptions electronically. www.ncpdp.org.

Natural selection The process in nature by which only the organisms best adapted to their environment tend to survive and transmit their genetic information in increasing numbers to succeeding generations, while those less adapted tend to be eliminated.

Nicotinamide A form of B complex vitamin niacin; a component of NAD that is involved in a wide range of biological processes such as energy production and production of fatty acids, steroids, and cholesterol.

Nuclear magnetic resonance spectroscopy (NMR) A form of spectroscopy which depends on the absorption and emission of energy from changes in spin states of the nucleus of an atom. Absorption and emission are affected by the local chemical environment.

Nuclear receptors Ligand-inducible transcription factors. Their natural ligands are in various chemical classes such as steroid hormones, dietary lipids and their derivatives, oxysterols, bile acids, eicosanoids, and lipid-soluble vitamins.

Nucleotide excision repair pathway A sequence of enzymatic reactions involved in the repair of DNA.

Nutrigenetics A subdiscipline of nutritional genomics usually referring to the association of a gene variant (see SNP—single nucleotide polymorphism) with an intermediate risk factor (e.g., cholesterol level or glucose response) that is influenced by a particular nutrients (e.g., saturated fat). Because of genes–genes interactions (see epistasis), nutrigenetics is most informative when viewed in the context of the entire genome.

Nutrigenomics The use of genomics to investigate diet and gene interactions involved in health or idsease.

Nutrition A three-step process by which nourishing materials (food, drink, and supplements) are ingested, broken down, and utilized in metabolism, to achieve health by sustaining normal cellular activity. As a discipline, nutrition is the study of these processes in the context of health and disease.

Nutritional genomics or nutrigenomics The study of how foods affect the expression of genetic information in an individual and how an individual's genetic makeup metabolizes and responds to nutrients and bioactives.

Oligonucleotides Short single stranded DNA sequences.

Oligosaccbarides Compounds containing 2–10 monosaccharides linked in a linear or branched chain.

Omega-3 fatty acid Any of various polyunsaturated fatty acids that are found primarily in fish, fish oils, vegetable oils, and leafy green vegetables. See *Alpha-linolenic acid.*

"Omics" technologies High-throughput technologies to analyze simultaneously various kinds of macromolecules. For example, transcriptomics measures many transcripts, proteomics measures many proteins, and metabolomics measures many metabolites.

Oncogene A gene that can cause cancer or the transformation of normal cells into cancer cells. Normally, an oncogene is an altered version of a normal gene.

Ontology A controlled vocabulary that describes objects and the relations between them in a formal way. Ontologies have a grammar for using the vocabulary terms to express something meaningful within a specified domain of interest.

Osteoporosis A condition characterized by a decrease in bone mass and density.

Oxidative damage Damage to cells caused by oxidants.

Pathophysiology The physiology of abnormal states; specifically; the functional changes that accompany a particular syndrome or disease.

Penetrance The frequency with which a genotype manifests itself in a given phenotype.

Perturbation A changed input into a biological system; typically an environmental or genetic alteration in the cell (see Chapter 5).

Pharmacogenetics and pharmacogenomics The convergence of pharmacology and genetics dealing with genetically determined responses to drugs.

Phenotype The observable characteristics of an organism produced by the interaction of the organism's genotype and its environment. Height, eye color, enzyme activity, or disease state are all examples of phenotype. In the context of genetic buffering, phenotypes reflect the reproductive fitness (fitness traits) such as cell proliferation.

Phytochemicals A nonnutritive bioactive plant substance, such as a flavonoid or carotenoid, considered to be beneficial to human health.

Phytonutrients Nonessential bioactive dietary chemicals derived originally from the plant kingdom.

Pluripotent cells Cells capable of differentiating into numerous cell types.

Polymorphisms Differences between otherwise identical macromolecules; usually refers to changes in DNA. See *Single nucleotide polymorphisms.*

Polyphenol A group of chemicals usually from plants characterized by the presence of more than one phenolic group.

Positional isomers Chemicals having the same atomic composition but differing in one position.

Postprandial state The physiological and metabolic state following food intake.

Power Statistical power is the capability of a test to detect a significant effect. The components of a power calculation are sample size, effect size (magnitude of trend), alpha level (odds of concluding the presence of an effect due to chance only), and power (odds of finding the presence of an effect when there is one).

Prebiotics A dietary constituent or food supplement that nourishes and promotes the growth of beneficial bacteria already in the digestive system.

Probiotics Live, active cultures of necessary bacteria that are actually ingested and promote health. Live culture yogurt is an example of a probiotic.

Promoter A region in DNA, (usually) 5′ to the coding sequence of the gene, that encodes regulatory sequences needed for gene transcription.

Proteome The total collection of proteins in a cell or cellular substructure.

Proteomics The study of all proteins in a cell or organism.

Proteosome A collection of proteins in a defined structure involved in degradation of proteins.

Pseudogenes A nonfunctional gene derived from an ancestral active gene.

Recommended Daily Allowance (RDA) The amount of essential nutrients that the Food and Drug Administration considers adequate to meet the nutrient requirement of nearly all healthy individuals in particular life stage and gender group.

Recommended Daily Intake (RDI) Estimates of daily minimal dietary intake of established nutrients provided by the Food and Nutrition Board of the National Research Council. Optimal levels have not been formally established.

Repository (biobank) The physical storage of biological samples.

Response elements Response elements are sequence-specific recognition sites of transcription factors. Many response elements are located within 1 kb from the transcriptional start site.

Retinol A 20 carbon primary alcohol; a fat-soluble vitamin or a mixture of vitamins, especially vitamin A_1 or a mixture of vitamins A_1 and A_2, occurring principally in fish-liver oils, milk, and some yellow and dark green vegetables, and functioning in normal cell growth and development; also called vitamin A.

Retrotransposon A transposable DNA element that transposes by means of an RNA intermediate. Retrotransposons encode a reverse transcriptase that acts on the RNA transcript to make a cDNA copy, which then integrates into chromosomal DNA at a different location.

Robustness Constancy or resilience to change. Cellular robustness is the maintenance of stable phenotypic outputs in response to perturbation, which is presumed to result from natural selection for buffering interactions between genes, other genes, and the environment (see Chapter 5).

RT-PCR Real-time polymerase chain reaction; a method for monitoring the amplification of RNA molecules from a particular sample into many copies of DNA molecules.

Signal transduction A network of cellular enzymatic reactions stimulated by an extracellular signal interacting at the cell surface. The reactions alter signaling molecules that cause changes in the level of other proteins, enzymes, or metabolites and ultimately effects a change in the cell's function.

Single nucleotide polymorphism (SNP) Any variation of a single nucleotide in an otherwise identical DNA sequence.

Sir2 An NAD^+-dependent histone deacetylase and member of the sirtuin gene family.

Systematized Nomenclature of Medicine (SNOMED) SNOMED-CT (Clinical Terminology) has been created from the combination of SNOMED-RT (Reference Terminology) and Read codes. NLM and others are working to bring coding systems such as this SNOMED-CT (Clinical Terminology) into the public domain. snomed.org/.

Systems biology An approach of studying biological systems that analyzes multiple macromolecular species (DNA polymorphisms, RNA, protein, metabolites, etc.) in one experiment; a holistic approach to studying biological systems.

Transactivation Activation of gene transcription, which is induced by binding of a transcription factor or nuclear receptor to a DNA regulatory sequence.

Transcription coactivator Can potentiate transcriptional properties, mediated by nuclear receptors. Many coactivators are associated in large multiprotein complexes.

Transcription corepressor Certain nuclear receptors, when bound in absence of their ligand to their target genes, can inhibit gene transcription initiation.

Transcription factor Generally, a protein that functions to initiate, enhance, or inhibit the transcription of a gene. Transcription factors bind to DNA regulatory sequences of target genes to modify the rate of gene transcription initiation.

Trans fatty acid A fatty acid that has been produced by hydrogenating an unsaturated fatty acid (and so changing its shape); found in processed foods such as margarine and fried foods and puddings and commercially baked goods and partially hydrogenated vegetable oils.

Transfection Introduction of (usually) foreign DNA into a eukaryotic cell.

Transgenerational epigenetic inheritance Inheritance of somatic DNA methylation patterns from parents to offspring.

Transrepression Processes through which nuclear receptors can antagonize the transcriptional activity of other transcription factors without DNA binding or altering the DNA activity.

Ubiquitin pathway A series of enzymatic reactions that ultimately degrade proteins. Ubiquitin is added to proteins targeting them for degradation.

Unified Modeling Language (UML) A graphical language for visualizing, specifying, constructing, and documenting the artifacts of a software system.

Unified Modeling Language System (UMLS) Developed by the National Library of Medicine as a standard health vocabulary that enables cross-referencing to other terminology and classification systems. Includes a metathesaurus, a semantic network, and an information sources map. The purpose of UMLS is to help health professionals and researchers retrieve and integrate electronic biomedical information from a variety of sources, irrespective of the variations in the way similar concepts are expressed in different sources and classification systems. Has incorporated many source vocabularies including SNOMED, ICD-9, and CPT. www.nlm.nih.gov/researeh/umls.

Xenobiotic chemicals A chemical that is not a natural component of the organism or its diet.

X12N Dominant standard for electronic commerce. The American National Standards Institute Accredited Standards Committee X12 (ASC X12) selected X12N as the standard for electronic data interchange (EDI) used in administrative and financial health-care transactions (excluding retail pharmacy transactions) in compliance with the Health Insurance Portability and Accountability Act of 1996. Used for external financial transactions, financial coverage verification, and insurance transactions and claims. www.x12.org/x12org/index.cfm.

Zoonutrients Nonessential bioactive dietary chemicals derived solely from animal tissue.

INDEX

Note: OMIM (Online Mendelian Inheritance of Man) is supported by the NIH and the National Library of Medicine. It provides information about the genetic basis of disease, provides links to gene and protein sequences, and can accessed at: *http://www.ncbi.nlm.nih.gov/entrez/query.fcgi?db=OMIM*

Nutritional Genomics: Discovering the Path to Personalized Nutrition
Edited by Jim Kaput and Raymond L. Rodriguez